Student Solutions Manual

for
Gustafson and Frisk's

Beginning and Intermediate Algebra
An Integrated Approach

Fourth Edition

Michael G. Welden
Rock Valley College

THOMSON

BROOKS/COLE

Australia • Canada • Mexico • Singapore • Spain • United Kingdom • United States

For more information about our products,
contact us at:
Thomson Learning Academic Resource Center
1-800-423-0563

For permission to use material from this text or
product, submit a request online at
http://www.thomsonrights.com.
Any additional questions about permissions can be
submitted by email to **thomsonrights@thomson.com.**

Thomson Brooks/Cole
10 Davis Drive
Belmont, CA 94002-3098
USA

Asia
Thomson Learning
5 Shenton Way #01-01
UIC Building
Singapore 068808

Australia/New Zealand
Thomson Learning
102 Dodds Street
Southbank, Victoria 3006
Australia

Canada
Nelson
1120 Birchmount Road
Toronto, Ontario M1K 5G4
Canada

Europe/Middle East/South Africa
Thomson Learning
High Holborn House
50/51 Bedford Row
London WC1R 4LR
United Kingdom

Latin America
Thomson Learning
Seneca, 53
Colonia Polanco
11560 Mexico D.F.
Mexico

Spain/Portugal
Paraninfo
Calle/Magallanes, 25
28015 Madrid, Spain

Preface

This manual contains detailed solutions to all of the odd exercises of the text *Beginning and Intermediate Algebra*, fourth edition, by R. David Gustafson and Peter D. Frisk. It also contains solutions to all chapter summary, chapter test, and cumulative review exercises found in the text.

Many of the exercises in the text may be solved using more than one method, but it is not feasible to list all possible solutions in this manual. Also, some of the exercises may have been solved in this manual using a method that differs slightly from that presented in the text. There are a few exercises in the text whose solutions may vary from person to person. Some of these solutions may not have been included in this manual. For the solution to an exercise like this, the notation "answers may vary" has been included.

Please remember that only reading a solution does not teach you how to solve a problem. To repeat a commonly used phrase, mathematics is not a spectator sport. You MUST make an honest attempt to solve each exercise in the text without using this manual first. This manual should be viewed more or less as a last resort. Above all, DO NOT simply copy the solution from this manual onto your own paper. Doing so will not help you learn how to do the exercise, nor will it help you to do better on quizzes or tests.

I would like to thank the members of the mathematics faculty at Rock Valley College and Rebecca Subity of Brooks/Cole Publishing Company for their help and support. This solutions manual was prepared using EXP 5.1.

This book is dedicated to Peggy, Murray, and Marsha Welden, as well as to my niece Hannah Welden.

May your study of this material be successful and rewarding.

Michael G. Welden

Contents

Exercise 1.1 (page 10)

1. set   **3.** whole   **5.** subset   **7.** prime

9. is not equal to   **11.** is greater than or equal to   **13.** number   **15.** natural: $1, 2, 6, 9$

17. positive integers: $1, 2, 6, 9$ **19.** integers: $-3, -1, 0, 1, 2, 6, 9$

21. real: $-3, -\frac{1}{2}, -1, 0, 1, 2, \frac{5}{3}, \sqrt{7}, 3.25, 6, 9$ **23.** odd integers: $-3, -1, 1, 9$

25. composite: $6, 9$ **27.** $4 + 5 = 9$; natural, odd, composite, whole

29. $15 - 15 = 0$
even, whole

31. $3 \cdot 8 = 24$
natural, even, composite, whole

33. $24 \div 8 = 3$
natural, odd, prime, whole

35. $5 \boxed{\phantom{<}} 3 + 2$
$5 \boxed{=} 5$

37. $25 \boxed{<} 32$

39. $5 + 7 \boxed{\phantom{<}} 10$
$12 \boxed{>} 10$

41. $3 + 9 \boxed{\phantom{<}} 20 - 8$
$12 \boxed{=} 12$

43. $4 \cdot 2 \boxed{\phantom{<}} 2 \cdot 4$
$8 \boxed{=} 8$

45. $8 \div 2 \boxed{\phantom{<}} 4 + 2$
$4 \boxed{<} 6$

47. $3 + 2 + 5 \boxed{\phantom{<}} 5 + 2 + 3$
$10 \boxed{=} 10$

49. $7 > 3$

51. $8 \leq 8$ **53.** $3 + 4 = 7$ **55.** $\sqrt{2} \approx 1.41$

57. $3 \leq 7 \Rightarrow \boxed{7 \geq 3}$ **59.** $6 > 0 \Rightarrow \boxed{0 < 6}$ **61.** $3 + 8 > 8 \Rightarrow \boxed{8 < 3 + 8}$

63. $6 - 2 < 10 - 4 \Rightarrow \boxed{10 - 4 > 6 - 2}$ **65.** $2 \cdot 3 < 3 \cdot 4 \Rightarrow \boxed{3 \cdot 4 > 2 \cdot 3}$

67. $\dfrac{12}{4} < \dfrac{24}{6} \Rightarrow \boxed{\dfrac{24}{6} > \dfrac{12}{4}}$

69.

6 is greater than 3.   6 is to the right of 3.

71.

11 is greater than 6.   11 is to the right of 6.

73.

2 is greater than 0.   2 is to the right of 0.

75.

8 is greater than 0.   8 is to the right of 0.

77.

79.

81.

83.

85. ←———(———)———→
1   5

87. ←———]———[———→
-3   3

89. $|36| = 36$

91. $|0| = 0$

93. $|-230| = 230$

95. $|12 - 4| = |8| = 8$

97. Answers may vary.

99. Answers may vary.

101. Answers may vary.

Exercise 1.2 (page 24)

1. true

3. false; 21 has factors of 3 and 7.

5. true; 8 is to the right of -2.

7. true; $|-9| = 9$, so $9 \leq |-9|$.

9. $3 + 7 \boxed{=} 10$

11. $|-2| = 2$, so $|-2| \boxed{=} 2$

13. numerator

15. simplify

17. proper

19. 1

21. multiply

23. numerators, denominator

25. plus (or $+$)

27. repeating

29. $\dfrac{6}{12} = \dfrac{1 \cdot \overset{1}{\cancel{6}}}{2 \cdot \cancel{6}_{1}} = \dfrac{1}{2}$

31. $\dfrac{15}{20} = \dfrac{3 \cdot \overset{1}{\cancel{5}}}{4 \cdot \cancel{5}_{1}} = \dfrac{3}{4}$

33. $\dfrac{24}{18} = \dfrac{4 \cdot \overset{1}{\cancel{6}}}{3 \cdot \cancel{6}_{1}} = \dfrac{4}{3}$

35. $\dfrac{72}{64} = \dfrac{9 \cdot \overset{1}{\cancel{8}}}{8 \cdot \cancel{8}_{1}} = \dfrac{9}{8}$

37. $\dfrac{1}{2} \cdot \dfrac{3}{5} = \dfrac{1 \cdot 3}{2 \cdot 5} = \dfrac{3}{10}$

39. $\dfrac{4}{3} \cdot \dfrac{6}{5} = \dfrac{4 \cdot 6}{3 \cdot 5} = \dfrac{4 \cdot 2 \cdot \overset{1}{\cancel{3}}}{\cancel{3} \cdot 5}_{1} = \dfrac{8}{5}$

41. $\dfrac{5}{12} \cdot \dfrac{18}{5} = \dfrac{5 \cdot 18}{12 \cdot 5} = \dfrac{\overset{1}{\cancel{5}} \cdot 3 \cdot \overset{1}{\cancel{6}}}{2 \cdot \cancel{6}_{1} \cdot \cancel{5}_{1}} = \dfrac{3}{2}$

43. $\dfrac{17}{34} \cdot \dfrac{3}{6} = \dfrac{17 \cdot 3}{34 \cdot 6} = \dfrac{\overset{1}{\cancel{17}} \cdot \overset{1}{\cancel{3}}}{2 \cdot \cancel{17} \cdot 2 \cdot \cancel{3}}_{1} = \dfrac{1}{4}$

45. $12 \cdot \dfrac{5}{6} = \dfrac{12}{1} \cdot \dfrac{5}{6} = \dfrac{12 \cdot 5}{1 \cdot 6} = \dfrac{2 \cdot \overset{1}{\cancel{6}} \cdot 5}{1 \cdot \underset{1}{\cancel{6}}}$

$\qquad\qquad = \dfrac{10}{1} = 10$

47. $\dfrac{10}{21} \cdot 14 = \dfrac{10}{21} \cdot \dfrac{14}{1} = \dfrac{10 \cdot 14}{21 \cdot 1} = \dfrac{10 \cdot 2 \cdot \overset{1}{\cancel{7}}}{3 \cdot \underset{1}{\cancel{7}}}$

$\qquad\qquad = \dfrac{20}{3}$

49. $\dfrac{3}{5} \div \dfrac{2}{3} = \dfrac{3}{5} \cdot \dfrac{3}{2} = \dfrac{3 \cdot 3}{5 \cdot 2} = \dfrac{9}{10}$

51. $\dfrac{3}{4} \div \dfrac{6}{5} = \dfrac{3}{4} \cdot \dfrac{5}{6} = \dfrac{3 \cdot 5}{4 \cdot 6} = \dfrac{\overset{1}{\cancel{3}} \cdot 5}{4 \cdot 2 \cdot \underset{1}{\cancel{3}}} = \dfrac{5}{8}$

53. $\dfrac{2}{13} \div \dfrac{8}{13} = \dfrac{2}{13} \cdot \dfrac{13}{8} = \dfrac{2 \cdot 13}{13 \cdot 8} = \dfrac{\overset{1}{\cancel{2}} \cdot \overset{1}{\cancel{13}}}{\underset{1}{\cancel{13}} \cdot 4 \cdot \underset{1}{\cancel{2}}}$

$\qquad\qquad = \dfrac{1}{4}$

55. $\dfrac{21}{35} \div \dfrac{3}{14} = \dfrac{21}{35} \cdot \dfrac{14}{3} = \dfrac{21 \cdot 14}{35 \cdot 3}$

$\qquad\qquad = \dfrac{\overset{1}{\cancel{7}} \cdot \overset{1}{\cancel{3}} \cdot 14}{5 \cdot \underset{1}{\cancel{7}} \cdot \underset{1}{\cancel{3}}} = \dfrac{14}{5}$

57. $6 \div \dfrac{3}{14} = \dfrac{6}{1} \div \dfrac{3}{14} = \dfrac{6}{1} \cdot \dfrac{14}{3} = \dfrac{6 \cdot 14}{1 \cdot 3} = \dfrac{2 \cdot \overset{1}{\cancel{3}} \cdot 14}{1 \cdot \underset{1}{\cancel{3}}} = \dfrac{28}{1} = 28$

59. $\dfrac{42}{30} \div 7 = \dfrac{42}{30} \div \dfrac{7}{1} = \dfrac{42}{30} \cdot \dfrac{1}{7} = \dfrac{42 \cdot 1}{30 \cdot 7} = \dfrac{\overset{1}{\cancel{6}} \cdot \overset{1}{\cancel{7}} \cdot 1}{5 \cdot \underset{1}{\cancel{6}} \cdot \underset{1}{\cancel{7}}} = \dfrac{1}{5}$

61. $\dfrac{3}{5} + \dfrac{3}{5} = \dfrac{3+3}{5} = \dfrac{6}{5}$

63. $\dfrac{4}{13} - \dfrac{3}{13} = \dfrac{4-3}{13} = \dfrac{1}{13}$

65. $\dfrac{1}{6} + \dfrac{1}{24} = \dfrac{1 \cdot \mathbf{4}}{6 \cdot \mathbf{4}} + \dfrac{1}{24} = \dfrac{4}{24} + \dfrac{1}{24}$

$\qquad\qquad = \dfrac{4+1}{24} = \dfrac{5}{24}$

67. $\dfrac{3}{5} + \dfrac{2}{3} = \dfrac{3 \cdot \mathbf{3}}{5 \cdot \mathbf{3}} + \dfrac{2 \cdot \mathbf{5}}{3 \cdot \mathbf{5}} = \dfrac{9}{15} + \dfrac{10}{15}$

$\qquad\qquad = \dfrac{9+10}{15} = \dfrac{19}{15}$

69. $\dfrac{9}{4} - \dfrac{5}{6} = \dfrac{9 \cdot \mathbf{3}}{4 \cdot \mathbf{3}} - \dfrac{5 \cdot \mathbf{2}}{6 \cdot \mathbf{2}} = \dfrac{27}{12} - \dfrac{10}{12} = \dfrac{27-10}{12} = \dfrac{17}{12}$

71. $\dfrac{7}{10} - \dfrac{1}{14} = \dfrac{7 \cdot \mathbf{7}}{10 \cdot \mathbf{7}} - \dfrac{1 \cdot \mathbf{5}}{14 \cdot \mathbf{5}} = \dfrac{49}{70} - \dfrac{5}{70} = \dfrac{49-5}{70} = \dfrac{44}{70} = \dfrac{22 \cdot \overset{1}{\cancel{2}}}{35 \cdot \underset{1}{\cancel{2}}} = \dfrac{22}{35}$

73. $3 - \dfrac{3}{4} = \dfrac{3}{1} - \dfrac{3}{4} = \dfrac{3 \cdot \mathbf{4}}{1 \cdot \mathbf{4}} - \dfrac{3}{4} = \dfrac{12}{4} - \dfrac{3}{4} = \dfrac{12-3}{4} = \dfrac{9}{4}$

75. $\dfrac{17}{3} + 4 = \dfrac{17}{3} + \dfrac{4}{1} = \dfrac{17}{3} + \dfrac{4 \cdot \mathbf{3}}{1 \cdot \mathbf{3}} = \dfrac{17}{3} + \dfrac{12}{3} = \dfrac{17+12}{3} = \dfrac{29}{3}$

77. $4\dfrac{3}{5} + \dfrac{3}{5} = \left(4 + \dfrac{3}{5}\right) + \dfrac{3}{5} = \left(\dfrac{20}{5} + \dfrac{3}{5}\right) + \dfrac{3}{5} = \dfrac{23}{5} + \dfrac{3}{5} = \dfrac{26}{5} = 5\dfrac{1}{5}$

79. $3\frac{1}{3} - 1\frac{2}{3} = \left(3 + \frac{1}{3}\right) - \left(1 + \frac{2}{3}\right) = \left(\frac{9}{3} + \frac{1}{3}\right) - \left(\frac{3}{3} + \frac{2}{3}\right) = \frac{10}{3} - \frac{5}{3} = \frac{5}{3} = 1\frac{2}{3}$

81. $3\frac{3}{4} - 2\frac{1}{2} = \left(3 + \frac{3}{4}\right) - \left(2 + \frac{1}{2}\right) = \left(\frac{12}{4} + \frac{3}{4}\right) - \left(\frac{8}{4} + \frac{2}{4}\right) = \frac{15}{4} - \frac{10}{4} = \frac{5}{4} = 1\frac{1}{4}$

83. $8\frac{2}{9} - 7\frac{2}{3} = \left(8 + \frac{2}{9}\right) - \left(7 + \frac{2}{3}\right) = \left(\frac{72}{9} + \frac{2}{9}\right) - \left(\frac{63}{9} + \frac{6}{9}\right) = \frac{74}{9} - \frac{69}{9} = \frac{5}{9}$

85.
```
      2  3  .  4  5
  +   1  3  5  .  2
  ─────────────────
      1  5  8  .  6  5
```

87.
```
                 11
           6     7   13
      6    7  .  2   3   5
  -   2    2  .  4   5
  ────────────────────────
      4    4  .  7   8   5
```

89.
```
                 3.  4
      ×        1 3.  2
      ───────────────
                 6  8
            1  0  2
            3  4
      ───────────────
            4  4  8  8
```
Put two digits to the right of the decimal point. Answer: 44.88

91. $0.\,2\,3\,\overline{\big)\,1.\,0\,4\,6\,5}$

Move decimal points 2 places right.
```
                  4.  5  5
      2  3. ) 1  0  4.  6  5
              9  2
      ────────────────
              1  2  6
              1  1  5
      ────────────────
                 1  1  5
                 1  1  5
      ────────────────
                       0
```

Problems 93-99 are to be solved using a calculator. The keystrokes needed to solve each problem using a TI-83 graphing calculator appear in each solution. There may be other solutions. Keystrokes for other calculators may be slightly different.

93. $\boxed{3}\boxed{2}\boxed{3}\boxed{.}\boxed{2}\boxed{4}\boxed{+}\boxed{2}\boxed{7}\boxed{.}\boxed{2}\boxed{5}\boxed{4}\boxed{3}\boxed{\text{ENTER}}$ {350.4943} $\Rightarrow$ 350.49

95. $\boxed{5}\boxed{5}\boxed{.}\boxed{7}\boxed{7}\boxed{4}\boxed{4}\boxed{3}\boxed{-}\boxed{.}\boxed{5}\boxed{6}\boxed{8}\boxed{2}\boxed{4}\boxed{5}\boxed{\text{ENTER}}$
{55.206185} $\Rightarrow$ 55.21

97. $\boxed{2}\boxed{5}\boxed{.}\boxed{2}\boxed{5}\boxed{\times}\boxed{1}\boxed{3}\boxed{2}\boxed{.}\boxed{1}\boxed{7}\boxed{9}\boxed{\text{ENTER}}$ {3337.51975} $\Rightarrow$ 3,337.52

99. $\boxed{4}\boxed{.}\boxed{5}\boxed{6}\boxed{9}\boxed{4}\boxed{3}\boxed{2}\boxed{3}\boxed{\div}\boxed{.}\boxed{4}\boxed{5}\boxed{6}\boxed{\text{ENTER}}$ {10.02068487} $\Rightarrow$ 10.02

101. $43\frac{1}{2} - 12\frac{1}{3} = 43 + \frac{1}{2} - 12 - \frac{1}{3} = \frac{258}{6} + \frac{3}{6} - \frac{72}{6} - \frac{2}{6} = \frac{187}{6} = 31\frac{1}{6}$ acres

103. $14 \cdot 3\frac{1}{4} = \frac{14}{1} \cdot \frac{13}{4} = \frac{14 \cdot 13}{1 \cdot 4} = \frac{7 \cdot \overset{1}{\cancel{2}} \cdot 13}{1 \cdot 2 \cdot \underset{1}{\cancel{2}}} = \frac{91}{2} = 45\frac{1}{2}$ yd

105. 24% of $42,500 = 0.24(42,500) = \$10,200$

107. $0.23(17,500) = 4,025$ defective $\Rightarrow 17,500 - 4025 = 13,475$ acceptable units

109. $0.12(18,700,000) = 2,244,000$ increase $\Rightarrow$ sales $= 18,700,000 + 2,244,000 = \$20,944,000$

111. # gallons $= 15,675.2 \div 25.5 = 614.7137255 \Rightarrow$ cost $= 614.7137255(1.67) \approx \$1,026.57$

113. Area $=$ length $\cdot$ width $= (253.5 \text{ ft})(178.5 \text{ ft}) = 45,249.75 \text{ ft}^2$
Drums of sealer $= 45,249.75 \div 4,000 \approx 11.3 \Rightarrow$ needs 12 drums; Cost $= 12(97.50) = \$1,170$

115. Standard $= 37.50(2,530) = \$94,875$; High-capacity $= 57.35(1,670) = \$95,774.50$
The high-capacity order will produce the greater profit.

117. Silage per cow $= 0.57(12,000) = 6840$ pounds; $30(6,840) = 205,200$ lb of silage

119. Regular $= 1,730 + 36(107.75) = 1,730 + 3,879 = 5,609$
High $= 4,170 + 36(57.50) = 4,170 + 2,070 = 6,240$
The high-efficiency furnace will be more expensive after 3 years.

121. Answers may vary.   **123. Answers may vary.**   **125. Answers may vary.**

127. No. Each proper fraction is less than 1. When a number is multiplied by a number less than 1, the result is smaller than the original number.

Exercise 1.3 (page 35)

1.

3. 17 is a prime number.

5. exponent   **7.** multiplication   **9.** $P = 4s$ **11.** $P = 2l + 2w$

13. $P = a + b + c$ **15.** $P = a + b + c + d$ **17.** $C = \pi D$, or $C = 2\pi r$ **19.** $V = lwh$

21. $V = \frac{1}{3}Bh$ **23.** $V = \frac{4}{3}\pi r^3$ **25.** $4^2 = 4 \cdot 4 = 16$ **27.** $6^2 = 6 \cdot 6 = 36$

29. $\left(\dfrac{1}{10}\right)^4 = \dfrac{1}{10} \cdot \dfrac{1}{10} \cdot \dfrac{1}{10} \cdot \dfrac{1}{10} = \dfrac{1}{10,000}$

Problems 31-33 are to be solved using a calculator. The keystrokes needed to solve each problem using a TI-83 graphing calculator appear in each solution. There may be other solutions. Keystrokes for other calculators may be slightly different.

31. $\boxed{7}\;\boxed{.}\;\boxed{9}\;\boxed{\wedge}\;\boxed{3}\;\boxed{\text{ENTER}}$
$\{493.039\}$

33. $\boxed{2}\;\boxed{5}\;\boxed{.}\;\boxed{3}\;\boxed{\wedge}\;\boxed{2}\;\boxed{\text{ENTER}}$
$\{640.09\}$

35. $x^2 = x \cdot x$ **37.** $3z^4 = 3 \cdot z \cdot z \cdot z \cdot z$ **39.** $(5t)^2 = 5t \cdot 5t$

SECTION 1.3

41. $5(2x)^3 = 5 \cdot 2x \cdot 2x \cdot 2x$ **43.** $4(3)^2 = 4 \cdot 9 = 36$ **45.** $(5 \cdot 2)^3 = 10^3 = 1{,}000$

47. $2(3^2) = 2 \cdot 9 = 18$ **49.** $(3 \cdot 2)^3 = 6^3 = 216$ **51.** $3 \cdot 5 - 4 = 15 - 4 = 11$

53. $3(5 - 4) = 3(1) = 3$ **55.** $3 + 5^2 = 3 + 25 = 28$ **57.** $(3 + 5)^2 = (8)^2 = 64$

59. $2 + 3 \cdot 5 - 4 = 2 + 15 - 4 = 17 - 4 = 13$ **61.** $64 \div (3 + 1) = 64 \div 4 = 16$

63. $(7 + 9) \div (2 \cdot 4) = 16 \div 8 = 2$ **65.** $(5 + 7) \div 3 \cdot 4 = 12 \div 3 \cdot 4 = 4 \cdot 4 = 16$

67. $24 \div 4 \cdot 3 + 3 = 6 \cdot 3 + 3 = 18 + 3 = 21$

69. $3^2 + 2(1 + 4) - 2 = 9 + 2(5) - 2 = 9 + 10 - 2 = 19 - 2 = 17$

71. $5^2 - (7 - 3)^2 = 5^2 - 4^2 = 25 - 16 = 9$ **73.** $(2 \cdot 3 - 4)^3 = (6 - 4)^3 = 2^3 = 8$

75. $\dfrac{3}{5} \cdot \dfrac{10}{3} + \dfrac{1}{2} \cdot 12 = \dfrac{3}{5} \cdot \dfrac{10}{3} + \dfrac{1}{2} \cdot \dfrac{12}{1} = \dfrac{\overset{1}{\cancel{3}}}{\underset{1}{\cancel{5}}} \cdot \dfrac{2 \cdot \overset{1}{\cancel{5}}}{\underset{1}{\cancel{3}}} + \dfrac{1}{\underset{1}{\cancel{2}}} \cdot \dfrac{6 \cdot \overset{1}{\cancel{2}}}{1} = \dfrac{2}{1} + \dfrac{6}{1} = 2 + 6 = 8$

77. $\left[\dfrac{1}{3} - \left(\dfrac{1}{2}\right)^2\right]^2 = \left[\dfrac{1}{3} - \dfrac{1}{4}\right]^2 = \left[\dfrac{1 \cdot 4}{3 \cdot 4} - \dfrac{1 \cdot 3}{4 \cdot 3}\right]^2 = \left[\dfrac{4}{12} - \dfrac{3}{12}\right]^2 = \left[\dfrac{1}{12}\right]^2 = \dfrac{1}{144}$

Problems 79-85 are to be solved using a calculator. The keystrokes needed to solve each problem using a TI-83 graphing calculator appear in each solution. There may be other solutions. Keystrokes for other calculators may be slightly different.

79. numerator: $\boxed{(}\ \boxed{3}\ \boxed{+}\ \boxed{5}\ \boxed{)}\ \boxed{x^2}\ \boxed{+}\ \boxed{2}\ \boxed{\text{ENTER}}\ \{66\}$ fraction $= \dfrac{66}{6} = 11$
denominator: $\boxed{2}\ \boxed{\times}\ \boxed{(}\ \boxed{8}\ \boxed{-}\ \boxed{5}\ \boxed{)}\ \boxed{\text{ENTER}}\ \{6\}$

81. numerator: $\boxed{(}\ \boxed{5}\ \boxed{-}\ \boxed{3}\ \boxed{)}\ \boxed{x^2}\ \boxed{+}\ \boxed{2}\ \boxed{\text{ENTER}}\ \{6\}$ fraction $= \dfrac{6}{6} = 1$
denominator: $\boxed{4}\ \boxed{x^2}\ \boxed{-}\ \boxed{(}\ \boxed{8}\ \boxed{+}\ \boxed{2}\ \boxed{)}\ \boxed{\text{ENTER}}\ \{6\}$

83. numerator: $\boxed{2}\ \boxed{\times}\ \boxed{(}\ \boxed{4}\ \boxed{+}\ \boxed{2}\ \boxed{\times}\ \boxed{(}\ \boxed{3}\ \boxed{-}\ \boxed{1}\ \boxed{)}\ \boxed{)}\ \boxed{\text{ENTER}}\ \{16\}$
denominator: $\boxed{3}\ \boxed{\times}\ \boxed{(}\ \boxed{3}\ \boxed{\times}\ \boxed{(}\ \boxed{2}\ \boxed{\times}\ \boxed{3}\ \boxed{-}\ \boxed{4}\ \boxed{)}\ \boxed{)}\ \boxed{\text{ENTER}}\ \{18\}$
fraction $= \dfrac{16}{18} = \dfrac{8}{9}$

85. numerator: $\boxed{3}\ \boxed{\times}\ \boxed{7}\ \boxed{-}\ \boxed{5}\ \boxed{\times}\ \boxed{(}\ \boxed{3}\ \boxed{\times}\ \boxed{4}\ \boxed{-}\ \boxed{1}\ \boxed{1}\ \boxed{)}\ \boxed{\text{ENTER}}\ \{16\}$
denominator: $\boxed{4}\ \boxed{\times}\ \boxed{(}\ \boxed{3}\ \boxed{+}\ \boxed{2}\ \boxed{)}\ \boxed{-}\ \boxed{3}\ \boxed{x^2}\ \boxed{+}\ \boxed{5}\ \boxed{\text{ENTER}}\ \{16\}$
fraction $= \dfrac{16}{16} = 1$

87. $39 = (3 \cdot 8) + (5 \cdot 3)$

89. $87 = (3 \cdot 8 + 5) \cdot 3$

91. $P = 4s = 4(4 \text{ in.}) = 16 \text{ in.}$

93. $P = a + b + c = 3 \text{ m} + 5 \text{ m} + 7 \text{ m}$
$= 15 \text{ m}$

95. $A = s^2 = (5 \text{ m})^2 = 25 \text{ m}^2$

97. $A = bh = (6 \text{ ft})(10 \text{ ft}) = 60 \text{ ft}^2$

99. $C = \pi D \approx \dfrac{22}{7}(28 \text{ m}) = \dfrac{22}{7} \cdot \dfrac{28}{1} \text{ m} = \dfrac{22}{\underset{1}{7}} \cdot \dfrac{4 \cdot \overset{1}{7}}{1} \text{ m} = 88 \text{ m}$

101. $A = \pi r^2 \approx \dfrac{22}{7}(21 \text{ ft})^2 = \dfrac{22}{7}\left(441 \text{ ft}^2\right) = \dfrac{22}{7} \cdot \dfrac{441}{1} \text{ ft}^2 = \dfrac{22}{\underset{1}{7}} \cdot \dfrac{63 \cdot \overset{1}{7}}{1} \text{ ft}^2 = 1{,}386 \text{ ft}^2$

103. $V = \frac{1}{3}Bh = \frac{1}{3}(3 \text{ cm})^2(2 \text{ cm}) = \frac{1}{3}(9 \text{ cm}^2)(2 \text{ cm}) = (3 \text{ cm}^2)(2 \text{ cm}) = 6 \text{ cm}^3$

105. $V = \dfrac{4}{3}\pi r^3 \approx \dfrac{4}{3} \cdot \dfrac{22}{7}(6 \text{ m})^3 = \dfrac{88}{21}\left(216 \text{ m}^3\right) = \dfrac{88}{21} \cdot \dfrac{216}{1} \text{ m}^3 \approx 905 \text{ m}^3$

107. Cylinder: $V = Bh = \pi(4 \text{ cm})^2(14 \text{ cm}) \approx \dfrac{22}{7} \cdot \dfrac{16}{1} \text{ cm}^2 \cdot \dfrac{14}{1} \text{ cm} = \dfrac{22 \cdot 16 \cdot 2 \cdot \overset{1}{7}}{\underset{1}{7} \cdot 1 \cdot 1} \text{ cm}^3 = 704 \text{ cm}^3$

Cone: $V = \frac{1}{3}Bh = \frac{1}{3}\pi(4 \text{ cm})^2(21 \text{ cm}) \approx \dfrac{1}{3} \cdot \dfrac{22}{7} \cdot \dfrac{16}{1} \text{ cm}^2 \cdot \dfrac{21}{1} \text{ cm} = \dfrac{22 \cdot 16 \cdot \overset{1}{\underset{1}{3}} \cdot \overset{1}{7}}{\underset{1}{3} \cdot \underset{1}{7} \cdot 1 \cdot 1} \text{ cm}^3 = 352 \text{ cm}^3$

Total $= 704 \text{ cm}^3 + 352 \text{ cm}^3 = 1{,}056 \text{ cm}^3$

109. $P = 4s = 4\left(30\frac{2}{5} \text{ m}\right) = 4\left(30 + \frac{2}{5} \text{ m}\right) = 4\left(\dfrac{150}{5} + \dfrac{2}{5} \text{ m}\right) = \dfrac{4}{1}\left(\dfrac{152}{5} \text{ m}\right) = \dfrac{608}{5} \text{ m} = 121\frac{3}{5} \text{ m}$

111. $V = \frac{4}{3}\pi r^3 = \frac{4}{3}\pi(21.35 \text{ ft})^3 \approx 40{,}764.51 \text{ ft}^3$

113. $V = lwh = (40 \text{ ft})(40 \text{ ft})(9 \text{ ft}) = 14{,}400 \text{ ft}^3$; Per student $= 14{,}400 \text{ ft}^3 \div 30 = 480 \text{ ft}^3$ per student

115. $f = \dfrac{rs}{(r+s)(n-1)} = \dfrac{(8)(12)}{(8+12)(1.6-1)} = \dfrac{96}{(20)(0.6)} = \dfrac{96}{12} = 8$

117. Answers may vary.

119. Increasing powers produce larger numbers.

Exercise 1.4 (page 45)

1. $5 + 3(7-2) = 5 + 3(5) = 5 + 15 = 20$

3. $5 + 3(7) - 2 = 5 + 21 - 2 = 26 - 2 = 24$

5. arrows

7. subtract, greater

9. add, opposite

11. $4 + 8 = +(4 + 8) = +12$

13. $(-3) + (-7) = -(3 + 7) = -10$

15. $6 + (-4) = +(6 - 4) = +2$

17. $9 + (-11) = -(11 - 9) = -2$

19. $(-0.4) + 0.9 = +(0.9 - 0.4) = +0.5$

21. $\dfrac{1}{5} + \left(+\dfrac{1}{7}\right) = \dfrac{7}{35} + \left(+\dfrac{5}{35}\right) = +\left(\dfrac{7}{35} + \dfrac{5}{35}\right) = +\dfrac{12}{35}$

23. $5 + (-4) = +(5 - 4) = +1$

25. $-1.3 + 3.5 = +(3.5 - 1.3) = +2.2$

27. $5 + [4 + (-2)] = 5 + [2] = 7$

29. $-2 + (-4 + 5) = -2 + 1 = -1$

31. $(-3 + 5) + 2 = 2 + 2 = 4$

33. $-15 + (-4 + 12) = -15 + 8 = -7$

35. $[-4 + (-3)] + [2 + (-2)] = [-7] + [0]$
$= -7$

37. $-4 + (-3 + 2) + (-3) = -4 + (-1) + (-3)$
$= -5 + (-3) = -8$

39. $-|8 + (-4)| + 7 = -|4| + 7 = -4 + 7 = 3$

41. $-5.2 + |-2.5 + (-4)| = -5.2 + |-6.5|$
$= -5.2 + 6.5 = 1.3$

43. $8 - 4 = 8 + (-4) = 4$

45. $8 - (-4) = 8 + (+4) = 12$

47. $0 - (-5) = 0 + (+5) = 5$

49. $\dfrac{5}{3} - \dfrac{7}{6} = \dfrac{10}{6} - \dfrac{7}{6} = \dfrac{10}{6} + \left(-\dfrac{7}{6}\right) = \dfrac{3}{6} = \dfrac{1}{2}$

51. $-3\dfrac{1}{2} - 5\dfrac{1}{4} = -\dfrac{7}{2} - \dfrac{21}{4} = -\dfrac{14}{4} - \dfrac{21}{4} = -\dfrac{14}{4} + \left(-\dfrac{21}{4}\right) = -\dfrac{35}{4} = -8\dfrac{3}{4}$

53. $-6.7 - (-2.5) = -6.7 + (+2.5) = -4.2$

55. $8 - 4 = 8 + (-4) = 4$

57. $-10 - (-3) = -10 + (+3) = -7$

59. $+3 - [(-4) - 3] = +3 - [(-4) + (-3)] = +3 - [-7] = +3 + [+7] = 10$

61. $(5 - 3) + (3 - 5) = [5 + (-3)] + [3 + (-5)] = [2] + [-2] = 0$

63. $5 - [4 + (-2) - 5] = 5 - [4 + (-2) + (-5)] = 5 - [-3] = 5 + [+3] = 8$

65. $\dfrac{5 - (-4)}{3 - (-6)} = \dfrac{5 + (+4)}{3 + (+6)} = \dfrac{9}{9} = 1$

67. $\dfrac{-4 - 2}{-[2 + (-3)]} = \dfrac{-4 + (-2)}{-[-1]} = \dfrac{-6}{+1} = -6$

69. $\left(\dfrac{5}{2} - 3\right) - \left(\dfrac{3}{2} - 5\right) = \left(\dfrac{5}{2} - \dfrac{6}{2}\right) - \left(\dfrac{3}{2} - \dfrac{10}{2}\right) = \left(-\dfrac{1}{2}\right) - \left(-\dfrac{7}{2}\right) = \left(-\dfrac{1}{2}\right) + \left(+\dfrac{7}{2}\right)$
$= \dfrac{6}{2} = 3$

71. $(5.2 - 2.5) - (5.25 - 5) = [5.2 + (-2.5)] - [5.25 + (-5)] = 2.7 - [0.25] = 2.7 + (-0.25)$
$$= 2.45$$

Problems 73-75 are to be solved using a calculator. The keystrokes needed to solve each problem using a TI-83 graphing calculator appear in each solution. There may be other solutions. Keystrokes for other calculators may be slightly different.

73. | 2 | . | 3 | 4 | ^ | 3 | − | 3 | . | 4 | 7 | + | . | 7 | 2 | x^2 | ENTER |
$\{9.861304\} \Rightarrow 9.9$

75. | 2 | . | 3 | 4 | x^2 | − | 3 | . | 4 | 7 | x^2 | − | . | 7 | 2 | x^2 | ENTER |
$\{-7.0837\} \Rightarrow -7.1$

77. $(-575) + (+400) = -175$
She still owes $175.

79. $(+13) + (-4) = +9$

81. $(-14) + 10 = -4°$

83. $1700 - (-300) = 2000$ years

85. $(-2,300) + (1,750) + (1,875) = +1,325$ m

87. $32,000 - 28,000 = 4,000$ ft

89. $+32 - (+27) = 5°$

91. $12,153 - 23 + 57 = 12,187$

93. $500 \cdot 2 - 300 = 1000 - 300 = 700$ shares

95. $437.45 + 25.17 + 37.93 + 45.26 - 17.13 - 83.44 - 22.58 = \$422.66$

97. $115,000 - 78 - 446 - 216 - 7,612.32 - 23,445.11 + 223 = \$83,425.57$

99. **Answers may vary.**

101. The answers agree if the two numbers have the same sign. The answers do not agree if the numbers have opposite signs.

Exercise 1.5 (page 53)

1. $30 \cdot 37\frac{1}{2} = \frac{30}{1} \cdot \frac{75}{2} = \frac{15 \cdot \overset{1}{\cancel{2}} \cdot 75}{1 \cdot \underset{1}{\cancel{2}}} = 1,125$ lb

3. $5^3 - 8(3)^2 = 125 - 8(9) = 125 - 72 = 53$

5. positive

7. positive

9. positive

11. a

13. 0

15. $(+6)(+8) = 48$

17. $(-8)(-7) = 56$

19. $(+12)(-12) = -144$

SECTION 1.5

21. $\left(\dfrac{1}{2}\right)(-32) = -\dfrac{1}{2} \cdot \dfrac{32}{1} = -\dfrac{32}{2} = -16$

23. $\left(-\dfrac{3}{4}\right)\left(-\dfrac{8}{3}\right) = +\dfrac{3}{4} \cdot \dfrac{8}{3} = \dfrac{24}{12} = 2$

25. $(-3)\left(-\dfrac{1}{3}\right) = +\dfrac{3}{1} \cdot \dfrac{1}{3} = \dfrac{3}{3} = 1$

27. $(3)(-4)(-6) = (-12)(-6) = 72$

29. $(-2)(3)(4) = (-6)(4) = -24$

31. $(2)(-5)(-6)(-7) = (-10)(-6)(-7)$
$= (+60)(-7) = -420$

33. $(-2)(-2)(-2)(-3)(-4) = (+4)(-2)(-3)(-4) = (-8)(-3)(-4) = (+24)(-4) = -96$

35. $(-2)^2 = (-2)(-2) = 4$

37. $(-4)^3 = (-4)(-4)(-4) = (+16)(-4)$
$= -64$

39. $(-7)^2 = (-7)(-7) = 49$

41. $-(-3)^2 = -(-3)(-3) = -(+9) = -9$

43. $(-1)(2^3) = (-1)(8) = -8$

45. $2 + (-1)(-3) = 2 + 3 = 5$

47. $(-1+2)(-3) = 1(-3) = -3$

49. $[-1-(-3)][-1+(-3)] = [-1+3][-4]$
$= [2][-4] = -8$

51. $-1(2) + (2)(-3) = -2 + (-6) = -8$

53. $(-1)(2)(-3) + 6 = (-2)(-3) + 6$
$= 6 + 6 = 12$

55. $(-1)^2[2-(-3)] = (-1)(-1)[2+(+3)] = 1[5] = 5$

57. $(-1)^3(-2)^2 + (-3)^2 = (-1)(-1)(-1)(-2)(-2) + (-3)(-3) = -4 + 9 = 5$

59. $\dfrac{80}{-20} = -4$

61. $\dfrac{-110}{-55} = 2$

63. $\dfrac{-160}{40} = -4$

65. $\dfrac{320}{-16} = -20$

67. $\dfrac{8-12}{-2} = \dfrac{-4}{-2} = 2$

69. $\dfrac{20-25}{7-12} = \dfrac{-5}{-5} = 1$

71. $\dfrac{3(4)}{-2} = \dfrac{12}{-2} = -6$

73. $\dfrac{5(-18)}{3} = \dfrac{-90}{3} = -30$

75. $\dfrac{4+(-18)}{-2} = \dfrac{-14}{-2} = 7$

77. $\dfrac{(-2)(5)(4)}{-3+1} = \dfrac{-40}{-2} = 20$

79. $\dfrac{1}{2} - \dfrac{2}{3} = \dfrac{3}{6} + \left(-\dfrac{4}{6}\right) = -\dfrac{1}{6}$

81. $\dfrac{1}{2} - \dfrac{2}{3} - \dfrac{3}{4} = \dfrac{6}{12} + \left(-\dfrac{8}{12}\right) + \left(-\dfrac{9}{12}\right)$
$= -\dfrac{11}{12}$

83. $\left(\dfrac{1}{2} - \dfrac{2}{3}\right)\left(\dfrac{1}{2} + \dfrac{2}{3}\right) = \left(\dfrac{3}{6} - \dfrac{4}{6}\right)\left(\dfrac{3}{6} + \dfrac{4}{6}\right) = \left(-\dfrac{1}{6}\right)\left(\dfrac{7}{6}\right) = -\dfrac{7}{36}$

Problems 85-87 are to be solved using a calculator. The keystrokes needed to solve each problem using a TI-83 graphing calculator appear in each solution. There may be other solutions. Keystrokes for other calculators may be slightly different.

85. $\boxed{(}\ \boxed{(}\ \boxed{(-)}\ \boxed{6}\ \boxed{)}\ \boxed{\wedge}\ \boxed{3}\ \boxed{+}\ \boxed{4}\ \boxed{\times}\ \boxed{(}\ \boxed{(-)}\ \boxed{3}\ \boxed{)}\ \boxed{\wedge}\ \boxed{3}\ \boxed{)}\ \boxed{\div}\ \boxed{(}\ \boxed{4}\ \boxed{-}\ \boxed{6}$
$\boxed{)}\ \boxed{x^2}\ \boxed{\text{ENTER}}\ \{-81\}$

87. $\boxed{(}\ \boxed{4}\ \boxed{\times}\ \boxed{(}\ \boxed{(-)}\ \boxed{6}\ \boxed{)}\ \boxed{x^2}\ \boxed{\times}\ \boxed{(-)}\ \boxed{3}\ \boxed{+}\ \boxed{4}\ \boxed{x^2}\ \boxed{\times}\ \boxed{(-)}\ \boxed{6}\ \boxed{)}\ \boxed{\div}\ \boxed{(}\ \boxed{2}$
$\boxed{\times}\ \boxed{(-)}\ \boxed{6}\ \boxed{-}\ \boxed{2}\ \boxed{\times}\ \boxed{(-)}\ \boxed{3}\ \boxed{)}\ \boxed{\text{ENTER}}\ \{88\}$

89. $(+2)(+3) = +6$

91. $(-30)(+15) = -450$

93. $(+23)(-120) = -2,760$

95. $\dfrac{-18}{-3} = +6$

97. $\dfrac{(+26) + (+35) + (+17) + (-25) + (-31) + (-12) + (-24)}{7} = \dfrac{-14}{7} = -2 \text{ per day}$

99. $613.50(18) = \$11,043 \Rightarrow$ enough $\$$

101. Answers may vary.

103. If the quotient is undefined, then the denominator must equal 0, and the product of the two numbers is 0.

105. If x^5 is negative, then x must be negative.

Exercise 1.6 (page 61)

1. $0.14 \cdot 3,800 = 532$

3. $\dfrac{-4 + (7 - 9)}{(-9 - 7) + 4} = \dfrac{-4 + (-2)}{-16 + 4} = \dfrac{-6}{-12} = \dfrac{1}{2}$

5. sum

7. multiplication

9. algebraic

11. variables

13. $x + y$

15. $x(2y)$

17. $y - x$

19. $\dfrac{y}{x}$

21. $z + \dfrac{x}{y}$

23. $z - xy$

25. $3xy$

27. $\dfrac{x + y}{y + z}$

29. $xy + \dfrac{y}{z}$

31. the sum of x and 3

33. the quotient obtained when x is divided by y

35. the product of 2, x and y (or twice the product of x and y)

37. the quotient obtained when 5 is divided by the sum of x and y

39. the quotient obtained when the sum of 3 and x is divided by y

41. the product of x, y and the sum of x and y

43. $x + z = 8 + 2 = 10$

45. $y - z = 4 - 2 = 2$

47. $yz - 3 = (4)(2) - 3 = 5$

49. $\dfrac{xy}{z} = \dfrac{(8)(4)}{2} = 16$

51. $x + y = (-2) + 5 = 3$

53. $xyz = (-2)(5)(-3) = (-10)(-3) = 30$

55. $\dfrac{yz}{x} = \dfrac{(5)(-3)}{-2} = \dfrac{-15}{-2} = \dfrac{15}{2}$

57. $\dfrac{3(x+z)}{y} = \dfrac{3[(-2)+(-3)]}{5} = \dfrac{3[-5]}{5}$
$$= \dfrac{-15}{5} = -3$$

59. $\dfrac{3(x+z^2)+4}{y(x-z)} = \dfrac{3\left[(-2)+(-3)^2\right]+4}{5[(-2)-(-3)]} = \dfrac{3(-2+9)+4}{5(1)} = \dfrac{3(7)+4}{5} = \dfrac{21+4}{5} = \dfrac{25}{5} = 5$

61. $6d$: 1 term; coef. $= 6$

63. $-xy - 4t + 35$: 3 terms; coef. $= -1$

65. $3ab + bc - cd - ef$: 4 terms; coef. $= 3$

67. $-4xyz + 7xy - z$: 3 terms; coef. $= -4$

69. $3x + 4y + 2z + 2$: 4 terms; coef. $= 3$

71. 3rd term: $19x$; factors: $19, x$

73. 1st term: $29xyz$; factors: $29, x, y, z$

75. 1st term: $3xyz$; factors: $3, x, y, z$

77. 3rd term: $17xz$; factors: $17, x, z$

79. coefficients: 5, 1 and 8

81. x and y are common to the 1st and 3rd terms.

83. coefficients: 3, 1 and 25; $3 \cdot 1 \cdot 25 = 75$

85. x and y are common to the 1st and 3rd terms.

87. $c + 4$

89. $9,987t$

91. $\dfrac{x}{5}$

93. $3d + 5$

95. $\dfrac{N(N-1)}{2} = \dfrac{10,000(10,000-1)}{2} = \dfrac{10,000(9,999)}{2} = \dfrac{99,990,000}{2} = 49,995,000$ comparisons

97. Answers may vary.

99. Answers may vary.

101. $37x \Rightarrow 37(2x)$
$37(2x) = 2(37x)$
$37x$ is doubled.

Exercise 1.7 (page 68)

1. $x + y^2 \geq z$

3. $|x| \geq \boxed{0}$

5. positive

7. real

9. $a + b = b + \underline{a}$

11. $(a + b) + c = a + \underline{(b + c)}$

13. $a(b + c) = ab + \underline{ac}$

15. $a \cdot 1 = \underline{a}$

17. element, multiplication

SECTION 1.7

19. $a, \dfrac{1}{a}$

21. $x + y = 12 + (-2) = 10$

23. $xy = 12(-2) = -24$

25. $x^2 = 12^2 = 144$

27. $\dfrac{x}{y^2} = \dfrac{12}{(-2)^2} = \dfrac{12}{4} = 3$

29. $x + y = 5 + 7 = 12$
$\ y + x = 7 + 5 = 12$

31. $3x + 2y = 3(5) + 2(7) = 15 + 14 = 29$
$\ 2y + 3x = 2(7) + 3(5) = 14 + 15 = 29$

33. $x(x + y) = 5(5 + 7) = 5(12) = 60$
$\ (x + y)x = (5 + 7)5 = (12)5 = 60$

35. $(x + y) + z = [2 + (-3)] + 1 = -1 + 1 = \boxed{0}$; $\ x + (y + z) = 2 + (-3 + 1) = 2 + (-2) = \boxed{0}$

37. $(xz)y = [2(1)](-3) = [2](-3) = \boxed{-6}$; $\ x(yz) = 2[-3(1)] = 2[-3] = \boxed{-6}$

39. $x^2(yz^2) = 2^2\left[-3(1)^2\right] = 4[-3(1)] = 4[-3] = \boxed{-12}$
$\ (x^2 y)z^2 = \left[2^2(-3)\right](1)^2 = [4(-3)](1) = [-12](1) = \boxed{-12}$

41. $3(x + y) = 3x + 3y$

43. $x(x + 3) = x \cdot x + x \cdot 3 = x^2 + 3x$

45. $-x(a + b) = (-x)a + (-x)b = -ax - bx$

47. $4(x^2 + x) = 4x^2 + 4x$

49. $-5(t + 2) = -5t + (-5)(2) = -5t - 10$

51. $-2a(x + a) = -2ax + (-2a)a$
$\qquad\qquad\ = -2ax - 2a^2$

53. additive inverse: -2
multiplicative inverse: $\dfrac{1}{2}$

55. additive inverse: $-\dfrac{1}{3}$
multiplicative inverse: 3

57. additive inverse: 0
multiplicative inverse: none

59. additive inverse: $\dfrac{5}{2}$
multiplicative inverse: $-\dfrac{2}{5}$

61. additive inverse: 0.2
multiplicative inverse: -5

63. additive inverse: $-\dfrac{4}{3}$
multiplicative inverse: $\dfrac{3}{4}$

65. commutative property of addition

67. commutative property of multiplication

69. distributive property

71. commutative property of addition

73. identity for multiplication

75. additive inverse

77. $3(x + 2) = 3x + 3(2) = 3x + 6$

79. $y^2 x = xy^2$

81. $(x + y)z = (y + x)z$

83. $(xy)z = x(yz)$

85. $0 + x = x$

87. **Answers may vary.**

13

89. Closure for addition would not be true (odd number plus odd number equals even number).
Closure for multiplication would be true (odd number times odd number equals odd number).
There would be no additive identity (0 is an even number).
There would be a multiplicative identity, 1 (1 is an odd number).

Chapter 1 Summary (page 70)

1. natural: $1, 2, 3, 4, 5$

2. prime: $2, 3, 5$

3. odd, natural: $1, 3, 5$

4. composite: 4

5. integers: $-6, 0, 5$

6. rational: $-6, -\dfrac{2}{3}, 0, 2.6, 5$

7. prime: 5

8. real: $-6, -\dfrac{2}{3}, 0, \sqrt{2}, 2.6, \pi, 5$

9. even integers: $-6, 0$

10. odd integers: 5

11. not rational: $\sqrt{2}, \pi$

12. $-5 \boxed{\phantom{<}} 12 - 12$
$-5 \boxed{<} 0$

13. $\dfrac{24}{6} \boxed{\phantom{<}} 5$
$4 \boxed{<} 5$

14. $13 - 13 \boxed{\phantom{<}} 5 - \dfrac{25}{5}$
$0 \boxed{} 5 - 5$
$0 \boxed{=} 0$

15. $\dfrac{21}{7} \boxed{} -33$
$3 \boxed{>} -33$

16. $-(-8) = +8$

17. $-(12 - 4) = -(8) = -8$

18.

$$14 \quad 16 \quad 18 \quad 20$$

19.

$$19 \quad 21 \quad 23 \quad 25$$

20.

$$-3 \qquad 2$$

21.

$$-4 \qquad 3$$

22. $|53 - 42| = |11| = 11$

23. $|-31| = 31$

24. $\dfrac{45}{27} = \dfrac{5 \cdot \overset{1}{\cancel{9}}}{3 \cdot \underset{1}{\cancel{9}}} = \dfrac{5}{3}$

25. $\dfrac{121}{11} = \dfrac{11 \cdot \overset{1}{\cancel{11}}}{1 \cdot \underset{1}{\cancel{11}}} = \dfrac{11}{1} = 11$

26. $\dfrac{31}{15} \cdot \dfrac{10}{62} = \dfrac{\overset{1}{\cancel{31}} \cdot \overset{1}{\cancel{2}} \cdot \overset{1}{\cancel{5}}}{3 \cdot \underset{1}{\cancel{5}} \cdot \underset{1}{\cancel{2}} \cdot \underset{1}{\cancel{31}}} = \dfrac{1}{3}$

27. $\dfrac{25}{36} \cdot \dfrac{12}{15} \cdot \dfrac{3}{5} = \dfrac{\overset{1}{\cancel{5}} \cdot \overset{1}{\cancel{5}} \cdot \overset{1}{\cancel{12}} \cdot \overset{1}{\cancel{3}}}{3 \cdot \underset{1}{\cancel{12}} \cdot 3 \cdot \underset{1}{\cancel{5}} \cdot \underset{1}{\cancel{5}}} = \dfrac{1}{3}$

28. $\dfrac{18}{21} \div \dfrac{6}{7} = \dfrac{18}{21} \cdot \dfrac{7}{6} = \dfrac{\overset{1}{\cancel{3}} \cdot \overset{1}{\cancel{6}} \cdot \overset{1}{\cancel{7}}}{\underset{1}{\cancel{3}} \cdot \underset{1}{\cancel{7}} \cdot \underset{1}{\cancel{6}}} = \dfrac{1}{1} = 1$

29. $\dfrac{14}{24} \div \dfrac{7}{12} \div \dfrac{2}{5} = \dfrac{14}{24} \cdot \dfrac{12}{7} \cdot \dfrac{5}{2}$

$= \dfrac{\overset{1}{\cancel{2}} \cdot \overset{1}{\cancel{7}} \cdot \overset{1}{\cancel{12}} \cdot 5}{\underset{1}{\cancel{2}} \cdot \underset{1}{\cancel{12}} \cdot \underset{1}{\cancel{7}} \cdot 2} = \dfrac{5}{2}$

30. $\dfrac{7}{12} + \dfrac{9}{12} = \dfrac{7+9}{12} = \dfrac{16}{12} = \dfrac{4 \cdot \overset{1}{\cancel{4}}}{3 \cdot \underset{1}{\cancel{4}}} = \dfrac{4}{3}$

31. $\dfrac{13}{24} - \dfrac{5}{24} = \dfrac{13-5}{24} = \dfrac{8}{24} = \dfrac{\overset{1}{\cancel{8}}}{3 \cdot \underset{1}{\cancel{8}}} = \dfrac{1}{3}$

32. $\dfrac{1}{3} + \dfrac{1}{7} = \dfrac{1 \cdot \mathbf{7}}{3 \cdot \mathbf{7}} + \dfrac{1 \cdot \mathbf{3}}{7 \cdot \mathbf{3}} = \dfrac{7}{21} + \dfrac{3}{21}$
$= \dfrac{7+3}{21} = \dfrac{10}{21}$

33. $\dfrac{5}{7} + \dfrac{4}{9} = \dfrac{5 \cdot \mathbf{9}}{7 \cdot \mathbf{9}} + \dfrac{4 \cdot \mathbf{7}}{9 \cdot \mathbf{7}} = \dfrac{45}{63} + \dfrac{28}{63}$
$= \dfrac{45+28}{63} = \dfrac{73}{63}$

34. $\dfrac{2}{3} - \dfrac{1}{7} = \dfrac{2 \cdot \mathbf{7}}{3 \cdot \mathbf{7}} - \dfrac{1 \cdot \mathbf{3}}{7 \cdot \mathbf{3}} = \dfrac{14}{21} - \dfrac{3}{21}$
$= \dfrac{14-3}{21} = \dfrac{11}{21}$

35. $\dfrac{4}{5} - \dfrac{2}{3} = \dfrac{4 \cdot \mathbf{3}}{5 \cdot \mathbf{3}} - \dfrac{2 \cdot \mathbf{5}}{3 \cdot \mathbf{5}} = \dfrac{12}{15} - \dfrac{10}{15}$
$= \dfrac{12-10}{15} = \dfrac{2}{15}$

36. $3\dfrac{2}{3} + 5\dfrac{1}{4} = \dfrac{11}{3} + \dfrac{21}{4} = \dfrac{11 \cdot \mathbf{4}}{3 \cdot \mathbf{4}} + \dfrac{21 \cdot \mathbf{3}}{4 \cdot \mathbf{3}} = \dfrac{44}{12} + \dfrac{63}{12} = \dfrac{44+63}{12} = \dfrac{107}{12} = 8\dfrac{11}{12}$

37. $7\dfrac{5}{12} - 4\dfrac{1}{2} = \dfrac{89}{12} - \dfrac{9}{2} = \dfrac{89}{12} - \dfrac{9 \cdot \mathbf{6}}{2 \cdot \mathbf{6}} = \dfrac{89}{12} - \dfrac{54}{12} = \dfrac{89-54}{12} = \dfrac{35}{12} = 2\dfrac{11}{12}$

38. $32.71 + 15.9 = 48.61$

39. $27.92 - 14.93 = 12.99$

40. $5.3 \cdot 3.5 = 18.55$

41. $21.83 \div 5.9 = 3.7$

42. $2.7(4.92 - 3.18) = 2.7(1.74) \approx 4.70$

43. $\dfrac{3.3 + 2.5}{0.22} = \dfrac{5.8}{0.22} \approx 26.36$

44. $\dfrac{12.5}{14.7 - 11.2} = \dfrac{12.5}{3.5} \approx 3.57$

45. $(3 - 0.7)(3.63 - 2) = (2.3)(1.63) \approx 3.75$

46. $\text{avg.} = \dfrac{5.2 + 4.7 + 9.5 + 8}{4} = \dfrac{27.4}{4}$
$= 6.85 \text{ hours}$

47. $0.15(380) = 57$

48. Front/Back: $2(2.7 + 2.7 + 4.2) = 2(9.6) = 19.2$ ft.   TOTAL $= 19.2 + 13.2 + 7.8 = 40.2$ ft.
Top/Bottom: $2(1.2 + 1.2 + 4.2) = 2(6.6) = 13.2$ ft.
Sides: $2(1.2 + 2.7) = 2(3.9) = 7.8$ ft.

49. $3^4 = 3 \cdot 3 \cdot 3 \cdot 3 = 81$

50. $\left(\dfrac{2}{3}\right)^2 = \dfrac{2}{3} \cdot \dfrac{2}{3} = \dfrac{4}{9}$

51. $(0.5)^2 = (0.5)(0.5) = 0.25$

52. $5^2 + 2^3 = 5 \cdot 5 + 2 \cdot 2 \cdot 2 = 25 + 8 = 33$

53. $3^2 + 4^2 = 9 + 16 = 25$

54. $(3 + 4)^2 = 7^2 = 49$

55. $V = Bh = \pi r^2 h = \pi \left(\dfrac{32.1}{2} \text{ ft} \right)^2 (18.7 \text{ ft}) = \pi \left(257.6025 \text{ ft}^2 \right)(18.7 \text{ ft}) \approx 15{,}133.6 \text{ ft}^3$

56. $5 + 3^3 = 5 + 27 = 32$

57. $7 \cdot 2 - 7 = 14 - 7 = 7$

58. $4 + (8 \div 4) = 4 + 2 = 6$

59. $(4 + 8) \div 4 = 12 \div 4 = 3$

60. $5^3 - \dfrac{81}{3} = 125 - 27 = 98$

61. $(5 - 2)^2 + 5^2 + 2^2 = 3^2 + 5^2 + 2^2$
$$= 9 + 25 + 4 = 38$$

62. $\dfrac{4 \cdot 3 + 3^4}{31} = \dfrac{12 + 81}{31} = \dfrac{93}{31} = 3$

63. $\dfrac{4}{3} \cdot \dfrac{9}{2} + \dfrac{1}{2} \cdot 18 = \dfrac{2 \cdot \overset{1}{\cancel{2}} \cdot 3 \cdot \overset{1}{\cancel{3}}}{\underset{1}{\cancel{3}} \cdot \underset{1}{\cancel{2}}} + \dfrac{1 \cdot 9 \cdot \overset{1}{\cancel{2}}}{\underset{1}{\cancel{2}} \cdot 1}$
$$= \dfrac{6}{1} + \dfrac{9}{1} = 15$$

64. $8^2 - 6 = 64 - 6 = 58$

65. $(8 - 6)^2 = 2^2 = 4$

66. $\dfrac{6 + 8}{6 - 4} = \dfrac{14}{2} = 7$

67. $\dfrac{6(8) - 12}{4 + 8} = \dfrac{48 - 12}{12} = \dfrac{36}{12} = 3$

68. $2^2 + 2(3)^2 = 4 + 2(9) = 4 + 18 = 22$

69. $\dfrac{2^2 + 3}{2^3 - 1} = \dfrac{4 + 3}{8 - 1} = \dfrac{7}{7} = 1$

70. $(+7) + (+8) = +(7 + 8) = 15$

71. $(-25) + (-32) = -(25 + 32) = -57$

72. $(-2.7) + (-3.8) = -(2.7 + 3.8) = -6.5$

73. $\dfrac{1}{3} + \dfrac{1}{6} = \dfrac{2}{6} + \dfrac{1}{6} = \dfrac{3}{6} = \dfrac{1}{2}$

74. $(+12) + (-24) = -(24 - 12) = -12$

75. $(-44) + (+60) = +(60 - 44) = 16$

76. $3.7 + (-2.5) = +(3.7 - 2.5) = 1.2$

77. $-5.6 + (+2.06) = -(5.6 - 2.06)$
$$= -3.54$$

78. $15 - (-4) = 15 + (+4) = 19$

79. $-12 - (-13) = -12 + (+13) = 1$

80. $[-5 + (-5)] - (-5) = [-10] + (+5) = -5$

81. $1 - [5 - (-3)] = 1 - [5 + (+3)]$
$$= 1 - [8] = -7$$

82. $\dfrac{5}{6} - \left(-\dfrac{2}{3}\right) = \dfrac{5}{6} + \dfrac{2}{3} = \dfrac{5}{6} + \dfrac{4}{6} = \dfrac{9}{6} = \dfrac{3}{2}$

83. $\dfrac{2}{3} - \left(\dfrac{1}{3} - \dfrac{2}{3}\right) = \dfrac{2}{3} - \left(-\dfrac{1}{3}\right)$

$$= \dfrac{2}{3} + \dfrac{1}{3} = \dfrac{3}{3} = 1$$

84. $\left|\dfrac{3}{7} - \left(-\dfrac{4}{7}\right)\right| = \left|\dfrac{3}{7} + \dfrac{4}{7}\right| = \left|\dfrac{7}{7}\right| = |1| = 1$

85. $\dfrac{3}{7} - \left|-\dfrac{4}{7}\right| = \dfrac{3}{7} - \left(+\dfrac{4}{7}\right)$

$$= \dfrac{3}{7} + \left(-\dfrac{4}{7}\right) = -\dfrac{1}{7}$$

86. $(+3)(+4) = +12$

87. $(-5)(-12) = +60$

88. $\left(-\dfrac{3}{14}\right)\left(-\dfrac{7}{6}\right) = +\dfrac{3}{14} \cdot \dfrac{7}{6}$

$$= \dfrac{\overset{1}{\cancel{3}} \cdot \overset{1}{\cancel{7}}}{2 \cdot 7 \cdot 2 \cdot \underset{1}{\cancel{3}}} = +\dfrac{1}{4}$$

89. $(3.75)(0.37) = +1.3875$

90. $5(-7) = -35$

91. $(-15)(7) = -105$

92. $\left(-\dfrac{1}{2}\right)\left(\dfrac{4}{3}\right) = -\dfrac{1}{2} \cdot \dfrac{4}{3} = -\dfrac{1 \cdot 2 \cdot \overset{1}{\cancel{2}}}{\underset{1}{\cancel{2}} \cdot 3} = -\dfrac{2}{3}$

93. $(-12.2)(3.7) = -45.14$

94. $\dfrac{+25}{+5} = +5$

95. $\dfrac{-14}{-2} = +7$

96. $\dfrac{(-2)(-7)}{4} = \dfrac{+14}{4} = +\dfrac{7 \cdot \overset{1}{\cancel{2}}}{2 \cdot \underset{1}{\cancel{2}}} = +\dfrac{7}{2}$

97. $\dfrac{-22.5}{-3.75} = +6$

98. $\dfrac{-25}{5} = -5$

99. $\dfrac{(-3)(-4)}{(-6)} = \dfrac{+12}{-6} = -2$

100. $\left(\dfrac{-10}{2}\right)^2 - (-1)^3 = (-5)^2 - (-1)^3$

$$= 25 - (-1)$$
$$= 25 + 1 = 26$$

101. $\dfrac{[-3 + (-4)]^2}{10 + (-3)} = \dfrac{[-7]^2}{7} = \dfrac{49}{7} = 7$

102. $\left(\dfrac{-3 + (-3)}{3}\right)\left(\dfrac{-15}{5}\right) = \left(\dfrac{-6}{3}\right)\left(\dfrac{-15}{5}\right)$

$$= (-2)(-3) = 6$$

103. $\dfrac{-2 - (-8)}{5 + (-1)} = \dfrac{-2 + (+8)}{4} = \dfrac{6}{4} = \dfrac{3}{2}$

104. xz

105. $x + 2y$

106. $2(x + y)$

107. $x - yz$

108. the product of 3, x and y

109. 5 decreased by the product of y and z

110. 5 less than the product of y and z

111. the quotient obtained when the sum of x, y and z is divided by twice their product

112. $y + z = -3 + (-1) = -4$

113. $x + y = 2 + (-3) = -1$

114. $x + (y + z) = 2 + [-3 + (-1)]$
$\qquad = 2 + [-4] = -2$

115. $x - y = 2 - (-3) = 2 + (+3) = 5$

116. $x - (y - z) = 2 - [-3 - (-1)]$
$\qquad = 2 - [-3 + (+1)]$
$\qquad = 2 - [-2] = 2 + (+2) = 4$

117. $(x - y) - z = [2 - (-3)] - (-1)$
$\qquad = [2 + (+3)] + (+1)$
$\qquad = 5 + 1 = 6$

118. $xy = (2)(-3) = -6$

119. $yz = (-3)(-1) = 3$

120. $x(x + z) = 2[2 + (-1)] = 2[1] = 2$

121. $xyz = (2)(-3)(-1) = (-6)(-1) = 6$

122. $y^2 z + x = (-3)^2(-1) + 2$
$\qquad = 9(-1) + 2 = -9 + 2 = -7$

123. $yz^3 + (xy)^2 = (-3)(-1)^3 + [2(-3)]^2$
$\qquad = (-3)(-1) + [-6]^2$
$\qquad = 3 + 36 = 39$

124. $\dfrac{xy}{z} = \dfrac{(2)(-3)}{(-1)} = \dfrac{-6}{-1} = 6$

125. $\dfrac{|xy|}{3z} = \dfrac{|2(-3)|}{3(-1)} = \dfrac{|-6|}{-3} = \dfrac{6}{-3} = -2$

126. three terms

127. 7

128. 1

129. $2 + 4 + 3 = 9$

130. closure property

131. commutative property of multiplication

132. associative property of addition

133. distributive property

134. commutative property of addition

135. associative property of multiplication

136. commutative property of addition

137. identity for multiplication

138. additive inverse

139. identity for addition

Chapter 1 Test (page 76)

1. $31, 37, 41, 43, 47$

2. 2

3.

4.

5. $-|23| = -(+23) = -23$

6. $-|7| + |-7| = -(+7) + (+7) = -7 + 7 = 0$

7. $3(4-2) \boxed{} -2(2-5)$
  $3(2) \boxed{} -2(-3)$
  $6 \boxed{=} 6$

8. $1 + 4 \cdot 3 \boxed{} -2(-7)$
  $1 + 12 \boxed{} +14$
  $13 \boxed{<} 14$

9. 25% of 136 $\boxed{}$ $\dfrac{1}{2}$ of 66
  $0.25(136) \boxed{} \dfrac{1}{2}(66)$
  $34 \boxed{>} 33$

10. $-13.7 \boxed{} -|-13.7|$
  $-13.7 \boxed{} -(+13.7)$
  $-13.7 \boxed{=} -13.7$

11. $\dfrac{26}{40} = \dfrac{13 \cdot \overset{1}{\cancel{2}}}{20 \cdot \underset{1}{\cancel{2}}} = \dfrac{13}{20}$

12. $\dfrac{7}{8} \cdot \dfrac{24}{21} = \dfrac{\overset{1}{\cancel{7}} \cdot \overset{1}{\cancel{3}} \cdot \overset{1}{\cancel{8}}}{\underset{1}{\cancel{8}} \cdot \underset{1}{\cancel{3}} \cdot \underset{1}{\cancel{7}}} = \dfrac{1}{1} = 1$

13. $\dfrac{18}{35} \div \dfrac{9}{14} = \dfrac{18}{35} \cdot \dfrac{14}{9} = \dfrac{2 \cdot \overset{1}{\cancel{9}} \cdot 2 \cdot \overset{1}{\cancel{7}}}{5 \cdot \underset{1}{\cancel{7}} \cdot \underset{1}{\cancel{9}}} = \dfrac{4}{5}$

14. $\dfrac{24}{16} + 3 = \dfrac{3 \cdot \overset{1}{\cancel{8}}}{2 \cdot \underset{1}{\cancel{8}}} + \dfrac{3}{1} = \dfrac{3}{2} + \dfrac{3 \cdot \mathbf{2}}{1 \cdot \mathbf{2}} = \dfrac{3}{2} + \dfrac{6}{2} = \dfrac{3+6}{2} = \dfrac{9}{2} \left(\text{or } 4\dfrac{1}{2} \right)$

15. $\dfrac{17-5}{36} - \dfrac{2(13-5)}{12} = \dfrac{12}{36} - \dfrac{2(8)}{12} = \dfrac{12}{36} - \dfrac{16}{12} = \dfrac{\overset{1}{\cancel{12}}}{3 \cdot \underset{1}{\cancel{12}}} - \dfrac{4 \cdot \overset{1}{\cancel{4}}}{3 \cdot \underset{1}{\cancel{4}}} = \dfrac{1}{3} - \dfrac{4}{3} = \dfrac{1-4}{3} = \dfrac{-3}{3} = -1$

16. $\dfrac{|-7-(-6)|}{-7-|-6|} = \dfrac{|-7+(+6)|}{-7-(+6)} = \dfrac{|-1|}{-7+(-6)} = \dfrac{1}{-13} = -\dfrac{1}{13}$

17. $0.17(457) = 77.69 \approx 77.7$

18. $A = lw = (23.56 \text{ ft})(12.8 \text{ ft}) = 301.568 \text{ ft}^2$
  $\approx 301.57 \text{ ft}^2$

19. $A = \dfrac{1}{2}bh = \dfrac{1}{2}(16 \text{ cm})(8 \text{ cm}) = \dfrac{1}{2}(128 \text{ cm}^2)$
  $= 64 \text{ cm}^2$

20. $V = Bh = \pi r^2 h = \pi(7 \text{ in.})^2(10 \text{ in.})$
  $= \pi(49 \text{ in.}^2)(10 \text{ in.})$
  $= \pi(490 \text{ in.}^3) \approx 1{,}539 \text{ in.}^3$

21. $xy + z = (-2)(3) + 4 = -6 + 4 = -2$

22. $x(y + z) = -2(3 + 4) = -2(7) = -14$

23. $\dfrac{z+4y}{2x} = \dfrac{4+4(3)}{2(-2)} = \dfrac{4+12}{-4} = \dfrac{16}{-4} = -4$

24. $|x^3 - z| = |(-2)^3 - 4| = |-8 - 4| = |-12|$
  $= 12$

25. $x^3 + y^2 + z = (-2)^3 + (3)^2 + 4$
  $= -8 + 9 + 4 = 5$

26. $|x| - 3|y| - 4|z| = |-2| - 3|3| - 4|4|$
  $= 2 - 3(3) - 4(4)$
  $= 2 - 9 - 16 = -23$

27. $\dfrac{xy}{x+y}$

28. $5y - (x+y)$

29. $x(12+12) + y(7+7) = 24x + 14y$

30. $12a + 8b$

31. 3 **32.** 4 terms **33.** 0 **34.** 5

35. commutative property of multiplication **36.** distributive property

37. commutative property of addition **38.** multiplicative inverse property

Exercise 2.1 (page 89)

1. $\dfrac{4}{5} + \dfrac{2}{3} = \dfrac{4\cdot 3}{5\cdot 3} + \dfrac{2\cdot 5}{3\cdot 5} = \dfrac{12}{15} + \dfrac{10}{15} = \dfrac{22}{15}$

3. $\dfrac{5}{9} \div \dfrac{3}{5} = \dfrac{5}{9} \cdot \dfrac{5}{3} = \dfrac{25}{27}$

5. $2 + 3\cdot 4 = 2 + 12 = 14$

7. $3 + 4^3(-5) = 3 + 64(-5) = 3 + (-320)$
$$= -317$$

9. equation **11.** equivalent **13.** equal **15.** equal

17. regular price **19.** 100 **21.** $x = 2$; equation

23. $7x < 8$
not an equation

25. $x + 7 = 0$
equation

27. $1 + 1 = 3$
equation

29. $x + 2 = 3$
$1 + 2 \overset{?}{=} 3$
$3 = 3$
1 is a solution.

31. $a - 7 = 0$
$-7 - 7 \overset{?}{=} 0$
$-14 \neq 0$
-7 is not a solution.

33. $\dfrac{y}{7} = 4$
$\dfrac{28}{7} \overset{?}{=} 4$
$4 = 4$
28 is a solution.

35. $\dfrac{x}{5} = x$
$\dfrac{0}{5} \overset{?}{=} 0$
$0 = 0$
0 is a solution.

37. $3k + 5 = 5k - 1$
$3(3) + 5 \overset{?}{=} 5(3) - 1$
$9 + 5 \overset{?}{=} 15 - 1$
$14 = 14$
3 is a solution.

39. $\dfrac{5+x}{10} - x = \dfrac{1}{2}$
$\dfrac{5+0}{10} - 0 \overset{?}{=} \dfrac{1}{2}$
$\dfrac{5}{10} - 0 \overset{?}{=} \dfrac{1}{2}$
$\dfrac{1}{2} = \dfrac{1}{2}$
0 is a solution.

41. $x + 7 = 13$ $\quad x + 7 = 13$
$x + 7 - 7 = 13 - 7$ $\quad 6 + 7 \overset{?}{=} 13$
$x = 6$ $\quad 13 = 13$

43. $y - 7 = 12$ $\quad y - 7 = 12$
$y - 7 + 7 = 12 + 7$ $\quad 19 - 7 \overset{?}{=} 12$
$y = 19$ $\quad 12 = 12$

45.
$$1 = y - 5 \qquad 1 = y - 5$$
$$1 + \mathbf{5} = y - 5 + \mathbf{5} \qquad 1 \overset{?}{=} 6 - 5$$
$$6 = y \qquad 1 = 1$$

47.
$$p - 404 = 115$$
$$p - 404 + \mathbf{404} = 115 + \mathbf{404}$$
$$p = 519$$
Check: $\quad p - 404 = 115$
$$519 - 404 \overset{?}{=} 115$$
$$115 = 115$$

49.
$$-37 + z = 37$$
$$-37 + \mathbf{37} + z = 37 + \mathbf{37}$$
$$z = 74$$
Check: $\quad -37 + z = 37$
$$-37 + 74 \overset{?}{=} 37$$
$$37 = 37$$

51.
$$-57 = b - 29$$
$$-57 + \mathbf{29} = b - 29 + \mathbf{29}$$
$$-28 = b$$
Check: $\quad -57 = b - 29$
$$-57 \overset{?}{=} -28 - 29$$
$$-57 = -57$$

53.
$$\frac{4}{3} = -\frac{2}{3} + x$$
$$\frac{4}{3} + \frac{\mathbf{2}}{\mathbf{3}} = -\frac{2}{3} + \frac{\mathbf{2}}{\mathbf{3}} + x$$
$$\frac{6}{3} = x, \text{ or } x = 2$$
Check: $\quad \dfrac{4}{3} = -\dfrac{2}{3} + x$
$$\frac{4}{3} \overset{?}{=} -\frac{2}{3} + 2$$
$$\frac{4}{3} \overset{?}{=} -\frac{2}{3} + \frac{6}{3}$$
$$\frac{4}{3} = \frac{4}{3}$$

55.
$$d + \frac{2}{3} = \frac{3}{2}$$
$$d + \frac{2}{3} - \frac{\mathbf{2}}{\mathbf{3}} = \frac{3}{2} - \frac{\mathbf{2}}{\mathbf{3}}$$
$$d = \frac{9}{6} - \frac{4}{6} = \frac{5}{6}$$
Check: $\quad d + \dfrac{2}{3} = \dfrac{3}{2}$
$$\frac{5}{6} + \frac{2}{3} \overset{?}{=} \frac{3}{2}$$
$$\frac{5}{6} + \frac{4}{6} \overset{?}{=} \frac{3}{2}$$
$$\frac{9}{6} \overset{?}{=} \frac{3}{2}$$
$$\frac{3}{2} = \frac{3}{2}$$

57.
$$-\frac{3}{5} = x - \frac{2}{5}$$
$$-\frac{3}{5} + \frac{\mathbf{2}}{\mathbf{5}} = x - \frac{2}{5} + \frac{\mathbf{2}}{\mathbf{5}}$$
$$-\frac{1}{5} = x$$
Check: $\quad -\dfrac{3}{5} = x - \dfrac{2}{5}$
$$-\frac{3}{5} \overset{?}{=} -\frac{1}{5} - \frac{2}{5}$$
$$-\frac{3}{5} = -\frac{3}{5}$$

59.
$$r - \frac{1}{5} = \frac{3}{10}$$
$$r - \frac{1}{5} + \frac{\mathbf{1}}{\mathbf{5}} = \frac{3}{10} + \frac{\mathbf{1}}{\mathbf{5}}$$
$$r = \frac{3}{10} + \frac{2}{10} = \frac{5}{10} = \frac{1}{2}$$
Check: $\quad r - \dfrac{1}{5} = \dfrac{3}{10}$
$$\frac{1}{2} - \frac{1}{5} \overset{?}{=} \frac{3}{10}$$
$$\frac{5}{10} - \frac{2}{10} \overset{?}{=} \frac{3}{10}$$
$$\frac{3}{10} = \frac{3}{10}$$

61.
$$\frac{x}{5} = 5 \qquad \frac{x}{5} = 5$$
$$5 \cdot \frac{x}{5} = \mathbf{5} \cdot 5 \qquad \frac{25}{5} \stackrel{?}{=} 5$$
$$x = 25 \qquad 5 = 5$$

63.
$$\frac{x}{32} = -2 \qquad \frac{x}{32} = -2$$
$$32 \cdot \frac{x}{32} = \mathbf{32} \cdot (-2) \qquad \frac{-64}{32} \stackrel{?}{=} -2$$
$$x = -64 \qquad -2 = -2$$

65.
$$\frac{b}{3} = 5 \qquad \frac{b}{3} = 5$$
$$3 \cdot \frac{b}{3} = \mathbf{3} \cdot 5 \qquad \frac{15}{3} \stackrel{?}{=} 5$$
$$b = 15 \qquad 5 = 5$$

67.
$$-3 = \frac{s}{11} \qquad -3 = \frac{s}{11}$$
$$11 \cdot (-3) = \mathbf{11} \cdot \frac{s}{11} \qquad -3 \stackrel{?}{=} \frac{-33}{11}$$
$$-33 = s \qquad -3 = -3$$

69.
$$6x = 18 \qquad 6x = 18$$
$$\frac{6x}{6} = \frac{18}{6} \qquad 6(3) \stackrel{?}{=} 18$$
$$x = 3 \qquad 18 = 18$$

71.
$$-4x = 36 \qquad -4x = 36$$
$$\frac{-4x}{-4} = \frac{36}{-4} \qquad -4(-9) \stackrel{?}{=} 36$$
$$x = -9 \qquad 36 = 36$$

73.
$$4t = 108 \qquad 4t = 108$$
$$\frac{4t}{4} = \frac{108}{4} \qquad 4(27) \stackrel{?}{=} 108$$
$$t = 27 \qquad 108 = 108$$

75.
$$11x = -121 \qquad 11x = -121$$
$$\frac{11x}{11} = \frac{-121}{11} \qquad 11(-11) \stackrel{?}{=} -121$$
$$x = -11 \qquad -121 = -121$$

77.
$$2x = \frac{1}{7} \qquad 2x = \frac{1}{7}$$
$$\frac{1}{2} \cdot 2x = \frac{1}{2} \cdot \frac{1}{7} \qquad 2 \cdot \frac{1}{14} \stackrel{?}{=} \frac{1}{7}$$
$$x = \frac{1}{14} \qquad \frac{2}{14} \stackrel{?}{=} \frac{1}{7}$$
$$\frac{1}{7} = \frac{1}{7}$$

79.
$$5x = \frac{5}{8} \qquad 5x = \frac{5}{8}$$
$$\frac{1}{5} \cdot 5x = \frac{1}{5} \cdot \frac{5}{8} \qquad 5 \cdot \frac{1}{8} \stackrel{?}{=} \frac{5}{8}$$
$$x = \frac{1}{8} \qquad \frac{5}{8} = \frac{5}{8}$$

81.
$$\frac{z}{7} = 14 \qquad \frac{z}{7} = 14$$
$$7 \cdot \frac{z}{7} = \mathbf{7} \cdot 14 \qquad \frac{98}{7} \stackrel{?}{=} 14$$
$$z = 98 \qquad 14 = 14$$

83.
$$\frac{w}{7} = \frac{5}{7} \qquad \frac{w}{7} = \frac{5}{7}$$
$$7 \cdot \frac{w}{7} = \mathbf{7} \cdot \frac{5}{7} \qquad \frac{5}{7} = \frac{5}{7}$$
$$w = 5$$

85.
$$\frac{s}{-3} = -\frac{5}{6} \qquad \frac{s}{-3} = -\frac{5}{6}$$
$$-6 \cdot \frac{s}{-3} = \mathbf{-6} \cdot \left(-\frac{5}{6}\right) \qquad \frac{\frac{5}{2}}{-3} \stackrel{?}{=} -\frac{5}{6}$$
$$2s = 5 \qquad \frac{\frac{5}{2} \cdot 2}{-3 \cdot 2} \stackrel{?}{=} -\frac{5}{6}$$
$$\frac{2s}{2} = \frac{5}{2} \qquad \frac{5}{-6} = -\frac{5}{6}$$
$$s = \frac{5}{2}$$

87.
$$-32z = 64 \qquad -32z = 64$$
$$\frac{-32z}{-32} = \frac{64}{-32} \qquad -32(-2) \stackrel{?}{=} 64$$
$$z = -2 \qquad 64 = 64$$

89.
$$18z = -9$$
$$\frac{18z}{18} = \frac{-9}{18} \qquad 18\left(-\frac{1}{2}\right) \stackrel{?}{=} -9$$
$$z = -\frac{1}{2} \qquad -9 = -9$$

91.
$$0.25x = 1228$$
$$\frac{0.25x}{0.25} = \frac{1228}{0.25} \qquad 0.25(4912) \stackrel{?}{=} 1228$$
$$x = 4912 \qquad 4912 = 4912$$

93.
$$\frac{b}{3} = \frac{1}{3} \qquad \frac{b}{3} = \frac{1}{3}$$
$$3 \cdot \frac{b}{3} = 3 \cdot \frac{1}{3} \qquad \frac{1}{3} = \frac{1}{3}$$
$$b = 1$$

95.
$$-1.2w = -102$$
$$\frac{-1.2w}{-1.2} = \frac{-102}{-1.2} \qquad -1.2(85) \stackrel{?}{=} -102$$
$$w = 85 \qquad -102 = -102$$

97.
$$\frac{u}{5} = -\frac{3}{10} \qquad \frac{u}{5} = -\frac{3}{10}$$
$$10 \cdot \frac{u}{5} = 10 \cdot \left(-\frac{3}{10}\right) \qquad \frac{-\frac{3}{2}}{5} \stackrel{?}{=} -\frac{3}{10}$$
$$2u = -3 \qquad \frac{-\frac{3}{2} \cdot 2}{5 \cdot 2} \stackrel{?}{=} -\frac{3}{10}$$
$$\frac{2u}{2} = \frac{-3}{2} \qquad \frac{-3}{10} = -\frac{3}{10}$$
$$u = -\frac{3}{2}$$

99.
$$\frac{p}{0.2} = 12 \qquad \frac{p}{0.2} = 12$$
$$0.2 \cdot \frac{p}{0.2} = 0.2 \cdot 12 \qquad \frac{2.4}{0.2} \stackrel{?}{=} 12$$
$$p = 2.4 \qquad 12 = 12$$

101.
$$rb = a$$
$$40\% \cdot 200 = a$$
$$0.40(200) = a$$
$$80 = a$$

103.
$$rb = a$$
$$50\% \cdot 38 = a$$
$$0.50(38) = a$$
$$19 = a$$

105.
$$rb = a$$
$$15\% \cdot b = 48$$
$$0.15b = 48$$
$$\frac{0.15b}{0.15} = \frac{48}{0.15}$$
$$b = 320$$

107.
$$rb = a$$
$$35\% \cdot b = 133$$
$$0.35b = 133$$
$$\frac{0.35b}{0.35} = \frac{133}{0.35}$$
$$b = 380$$

109.
$$rb = a$$
$$28\% \cdot b = 42$$
$$0.28b = 42$$
$$\frac{0.28b}{0.28} = \frac{42}{0.28}$$
$$b = 150$$

111.
$$rb = a$$
$$r(357.5) = 71.5$$
$$\frac{r(357.5)}{357.5} = \frac{71.5}{357.5}$$
$$r = 0.20$$
$$r = 20\%$$

113.
$$rb = a$$
$$r(4) = 0.32$$
$$\frac{r(4)}{4} = \frac{0.32}{4}$$
$$r = 0.08$$
$$r = 8\%$$

115.
$$rb = a$$
$$r(17) = 34$$
$$\frac{r(17)}{17} = \frac{34}{17}$$
$$r = 2.00$$
$$r = 200\%$$

117. Let r = the regular price. Then

$$\boxed{\text{Sale price}} = \boxed{\text{Regular price}} - \boxed{\text{Markdown}}$$
$$7995 = r - 1350$$
$$7995 + \mathbf{1350} = r - 1350 + \mathbf{1350}$$
$$9345 = r$$

The regular price is \$9,345.

119. Let w = the wholesale price. Then

$$\boxed{\text{Retail price}} = \boxed{\text{Wholesale price}} + \boxed{\text{Markup}}$$
$$175 = w + 85$$
$$175 - \mathbf{85} = w + 85 - \mathbf{85}$$
$$90 = w$$

The wholesale price is \$90.

121.
$$A = p + i$$
$$5010 = 4750 + i$$
$$5010 - \mathbf{4750} = 4750 - \mathbf{4750} + i$$
$$260 = i$$
The deposit earned \$260 in interest.

123.
$$v = p + a$$
$$150000 = p + 57000$$
$$150000 - \mathbf{57000} = p + 57000 - \mathbf{57000}$$
$$93000 = p$$
The original price was \$93,000.

125.
$$c = p + t$$
$$512 = 317 + t$$
$$512 - \mathbf{317} = 317 - \mathbf{317} + t$$
$$195 = t$$
The carpet cost \$195 to install.

127. Let $c =$ the condominium price. Then
$$\boxed{\begin{array}{c}\text{Condo}\\\text{price}\end{array}} = \boxed{\begin{array}{c}\text{House}\\\text{price}\end{array}} - 57595$$
$$c = 202744 - 57595$$
$$c = 145149$$
The price of the condominium is \$145,149.

129. Let $x =$ the original number in the audience.
$$\frac{1}{3} \cdot \boxed{\begin{array}{c}\text{Original}\\\text{audience}\end{array}} = \boxed{\begin{array}{c}\text{Number}\\\text{who left}\end{array}}$$
$$\frac{1}{3}x = 78$$
$$3 \cdot \frac{1}{3}x = 3 \cdot 78$$
$$x = 234$$
There were originally 234 in the audience.

131. Let $x =$ the number in the senior class.
$$\frac{1}{7} \cdot \boxed{\begin{array}{c}\text{Number}\\\text{in class}\end{array}} = \boxed{\begin{array}{c}\text{Number}\\\text{off-campus}\end{array}}$$
$$\frac{1}{7}x = 217$$
$$\mathbf{7} \cdot \frac{1}{7}x = \mathbf{7} \cdot 217$$
$$x = 1519$$
There are 1,519 in the senior class.

133. Let $r =$ the percentage not pleased.
$$a = 9200 - 4140 = 5060$$
$$rb = a$$
$$r(9200) = 5060$$
$$\frac{r(9200)}{9200} = \frac{5060}{9200}$$
$$r = 0.55$$
$$r = 55\%$$
55% of those surveyed were not pleased.

135. Let $x =$ the selling price.
$$rb = a$$
$$0.05(x) = 13.50$$
$$\frac{0.05x}{0.05} = \frac{13.50}{0.05}$$
$$x = 270$$
The selling price was \$270.

137. Let $x =$ the total number of patients.
$$rb = a$$
$$0.18(x) = 1008$$
$$\frac{0.18x}{0.18} = \frac{1008}{0.18}$$
$$x = 5600$$
The hospital treated 5,600 patients.

139. Answers may vary.

141. $A_{\text{circle}} = A_{\text{square}}$
$$\pi r^2 = s^2$$
$$\pi(4.5)^2 = 8^2$$
$$\pi(20.25) = 64$$
$$\frac{\pi(20.25)}{20.25} = \frac{64}{20.25}$$
$$\pi \approx 3.16$$

Exercise 2.2 (page 98)

1. $P = 2l + 2w = 2(8.5\,\text{cm}) + 2(16.5\,\text{cm}) = 17\,\text{cm} + 33\,\text{cm} = 50\,\text{cm}$

3. $A = \frac{1}{2}h(b + d) = \frac{1}{2}(8.5\,\text{in.})(6.7\,\text{in.} + 12.2\,\text{in.}) = \frac{1}{2}(8.5\,\text{in.})(18.9\,\text{in.}) = \frac{1}{2}(160.65\,\text{in.}^2) = 80.325\,\text{in.}^2$

5. cost

7. percent

9.
$$5x - 1 = 4$$
$$5x - 1 + 1 = 4 + 1$$
$$5x = 5$$
$$\frac{5x}{5} = \frac{5}{5}$$
$$x = 1$$

$$5x - 1 = 4$$
$$5(1) - 1 \overset{?}{=} 4$$
$$5 - 1 \overset{?}{=} 4$$
$$4 = 4$$

11.
$$6x + 2 = -4$$
$$6x + 2 - 2 = -4 - 2$$
$$6x = -6$$
$$\frac{6x}{6} = \frac{-6}{6}$$
$$x = -1$$

$$6x + 2 = -4$$
$$6(-1) + 2 \overset{?}{=} -4$$
$$-6 + 2 \overset{?}{=} -4$$
$$-4 = -4$$

13.
$$3x - 8 = 1$$
$$3x - 8 + 8 = 1 + 8$$
$$3x = 9$$
$$\frac{3x}{3} = \frac{9}{3}$$
$$x = 3$$

$$3x - 8 = 1$$
$$3(3) - 8 \overset{?}{=} 1$$
$$9 - 8 \overset{?}{=} 1$$
$$1 = 1$$

15.
$$11x + 17 = -5$$
$$11x + 17 - 17 = -5 - 17$$
$$11x = -22$$
$$\frac{11x}{11} = \frac{-22}{11}$$
$$x = -2$$

$$11x + 17 = -5$$
$$11(-2) + 17 \overset{?}{=} -5$$
$$-22 + 17 \overset{?}{=} -5$$
$$-5 = -5$$

17.
$$43t + 72 = 158$$
$$43t + 72 - 72 = 158 - 72$$
$$43t = 86$$
$$\frac{43t}{43} = \frac{86}{43}$$
$$t = 2$$

$$43t + 72 = 158$$
$$43(2) + 72 \overset{?}{=} 158$$
$$86 + 72 \overset{?}{=} 158$$
$$158 = 158$$

19.

$$-47 - 21s = 58$$
$$-47 + 47 - 21s = 58 + 47$$
$$-21s = 105$$
$$\frac{-21s}{-21} = \frac{105}{-21}$$
$$s = -5$$

$$-47 - 21s = 58$$
$$-47 - 21(-5) \overset{?}{=} 58$$
$$-47 + 105 \overset{?}{=} 58$$
$$58 = 58$$

21.

$$2y - \frac{5}{3} = \frac{4}{3}$$
$$2y - \frac{5}{3} + \frac{5}{3} = \frac{4}{3} + \frac{5}{3}$$
$$2y = \frac{9}{3}$$
$$2y = 3$$
$$\frac{2y}{2} = \frac{3}{2}$$
$$y = \frac{3}{2}$$

$$2y - \frac{5}{3} = \frac{4}{3}$$
$$2\left(\frac{3}{2}\right) - \frac{5}{3} \overset{?}{=} \frac{4}{3}$$
$$3 - \frac{5}{3} \overset{?}{=} \frac{4}{3}$$
$$\frac{9}{3} - \frac{5}{3} \overset{?}{=} \frac{4}{3}$$
$$\frac{4}{3} = \frac{4}{3}$$

23.

$$-0.4y - 12 = -20$$
$$-0.4y - 12 + 12 = -20 + 12$$
$$-0.4y = -8$$
$$\frac{-0.4y}{-0.4} = \frac{-8}{-0.4}$$
$$y = 20$$

$$-0.4y - 12 = -20$$
$$-0.4(20) - 12 \overset{?}{=} -20$$
$$-8 - 12 \overset{?}{=} -20$$
$$-20 = -20$$

25.

$$\frac{x}{3} - 3 = -2$$
$$\frac{x}{3} - 3 + 3 = -2 + 3$$
$$\frac{x}{3} = 1$$
$$3 \cdot \frac{x}{3} = 3 \cdot 1$$
$$x = 3$$

$$\frac{x}{3} - 3 = -2$$
$$\frac{3}{3} - 3 \overset{?}{=} -2$$
$$1 - 3 \overset{?}{=} -2$$
$$-2 = -2$$

27.

$$\frac{z}{9} + 5 = -1$$
$$\frac{z}{9} + 5 - 5 = -1 - 5$$
$$\frac{z}{9} = -6$$
$$9 \cdot \frac{z}{9} = 9(-6)$$
$$z = -54$$

$$\frac{z}{9} + 5 = -1$$
$$\frac{-54}{9} + 5 \overset{?}{=} -1$$
$$-6 + 5 \overset{?}{=} -1$$
$$-1 = -1$$

29.

$$\frac{b}{3} + 5 = 2$$
$$\frac{b}{3} + 5 - 5 = 2 - 5$$
$$\frac{b}{3} = -3$$
$$3 \cdot \frac{b}{3} = 3(-3)$$
$$b = -9$$

$$\frac{b}{3} + 5 = 2$$
$$\frac{-9}{3} + 5 \overset{?}{=} 2$$
$$-3 + 5 \overset{?}{=} 2$$
$$2 = 2$$

31.

$$\frac{s}{11} + 9 = 6$$
$$\frac{s}{11} + 9 - 9 = 6 - 9$$
$$\frac{s}{11} = -3$$
$$11 \cdot \frac{s}{11} = 11(-3)$$
$$s = -33$$

$$\frac{s}{11} + 9 = 6$$
$$\frac{-33}{11} + 9 \overset{?}{=} 6$$
$$-3 + 9 \overset{?}{=} 6$$
$$6 = 6$$

33.
$$\frac{k}{5} - \frac{1}{2} = \frac{3}{2}$$
$$\frac{k}{5} - \frac{1}{2} + \frac{1}{2} = \frac{3}{2} + \frac{1}{2}$$
$$\frac{k}{5} = \frac{4}{2}$$
$$\frac{k}{5} = 2$$
$$5 \cdot \frac{k}{5} = 5 \cdot 2$$
$$k = 10$$

$$\frac{k}{5} - \frac{1}{2} = \frac{3}{2}$$
$$\frac{10}{5} - \frac{1}{2} \overset{?}{=} \frac{3}{2}$$
$$2 - \frac{1}{2} \overset{?}{=} \frac{3}{2}$$
$$\frac{4}{2} - \frac{1}{2} \overset{?}{=} \frac{3}{2}$$
$$\frac{3}{2} = \frac{3}{2}$$

35.
$$\frac{w}{16} + \frac{5}{4} = 1$$
$$\frac{w}{16} + \frac{5}{4} - \frac{5}{4} = 1 - \frac{5}{4}$$
$$\frac{w}{16} = \frac{4}{4} - \frac{5}{4}$$
$$\frac{w}{16} = -\frac{1}{4}$$
$$16 \cdot \frac{w}{16} = 16\left(-\frac{1}{4}\right)$$
$$w = -4$$

$$\frac{w}{16} + \frac{5}{4} = 1$$
$$\frac{-4}{16} + \frac{5}{4} \overset{?}{=} 1$$
$$-\frac{1}{4} + \frac{5}{4} \overset{?}{=} 1$$
$$\frac{4}{4} \overset{?}{=} 1$$
$$1 = 1$$

37.
$$\frac{b+5}{3} = 11$$
$$3 \cdot \frac{b+5}{3} = 3 \cdot 11$$
$$b + 5 = 33$$
$$b + 5 - 5 = 33 - 5$$
$$b = 28$$

$$\frac{b+5}{3} = 11$$
$$\frac{28+5}{3} \overset{?}{=} 11$$
$$\frac{33}{3} \overset{?}{=} 11$$
$$11 = 11$$

39.
$$\frac{r+7}{3} = 4$$
$$3 \cdot \frac{r+7}{3} = 3 \cdot 4$$
$$r + 7 = 12$$
$$r + 7 - 7 = 12 - 7$$
$$r = 5$$

$$\frac{r+7}{3} = 4$$
$$\frac{5+7}{3} \overset{?}{=} 4$$
$$\frac{12}{3} \overset{?}{=} 4$$
$$4 = 4$$

41.
$$\frac{u-2}{5} = 1$$
$$5 \cdot \frac{u-2}{5} = 5 \cdot 1$$
$$u - 2 = 5$$
$$u - 2 + 2 = 5 + 2$$
$$u = 7$$

$$\frac{u-2}{5} = 1$$
$$\frac{7-2}{5} \overset{?}{=} 1$$
$$\frac{5}{5} \overset{?}{=} 1$$
$$1 = 1$$

43.
$$\frac{x-4}{4} = -3$$
$$4 \cdot \frac{x-4}{4} = 4(-3)$$
$$x - 4 = -12$$
$$x - 4 + 4 = -12 + 4$$
$$x = -8$$

$$\frac{x-4}{4} = -3$$
$$\frac{-8-4}{4} \overset{?}{=} -3$$
$$\frac{-12}{4} \overset{?}{=} -3$$
$$-3 = -3$$

45.
$$\frac{3x}{2} - 6 = 9$$
$$\frac{3x}{2} - 6 + 6 = 9 + 6$$
$$\frac{3x}{2} = 15$$
$$2 \cdot \frac{3x}{2} = 2 \cdot 15$$
$$3x = 30$$
$$\frac{3x}{3} = \frac{30}{3}$$
$$x = 10$$

$$\frac{3x}{2} - 6 = 9$$
$$\frac{3(10)}{2} - 6 \overset{?}{=} 9$$
$$\frac{30}{2} - 6 \overset{?}{=} 9$$
$$15 - 6 \overset{?}{=} 9$$
$$9 = 9$$

47.
$$\frac{3y}{2} + 5 = 11$$
$$\frac{3y}{2} + 5 - 5 = 11 - 5$$
$$\frac{3y}{2} = 6$$
$$2 \cdot \frac{3y}{2} = 2 \cdot 6$$
$$3y = 12$$
$$\frac{3y}{3} = \frac{12}{3}$$
$$y = 4$$

$$\frac{3y}{2} + 5 = 11$$
$$\frac{3(4)}{2} + 5 \overset{?}{=} 11$$
$$\frac{12}{2} + 5 \overset{?}{=} 11$$
$$6 + 5 \overset{?}{=} 11$$
$$11 = 11$$

49.

$$\frac{2x}{3} + \frac{1}{2} = 3$$

$$\frac{2x}{3} + \frac{1}{2} - \frac{1}{2} = 3 - \frac{1}{2}$$

$$\frac{2x}{3} = \frac{5}{2}$$

$$3 \cdot \frac{2x}{3} = 3 \cdot \frac{5}{2}$$

$$2x = \frac{15}{2}$$

$$\frac{1}{2} \cdot 2x = \frac{1}{2} \cdot \frac{15}{2}$$

$$x = \frac{15}{4}$$

$$\frac{2x}{3} + \frac{1}{2} = 3$$

$$\frac{2\left(\frac{15}{4}\right)}{3} + \frac{1}{2} \overset{?}{=} 3$$

$$\frac{\frac{15}{2}}{3} + \frac{1}{2} \overset{?}{=} 3$$

$$\frac{15}{6} + \frac{1}{2} \overset{?}{=} 3$$

$$\frac{5}{2} + \frac{1}{2} \overset{?}{=} 3$$

$$\frac{6}{2} \overset{?}{=} 3$$

$$3 = 3$$

51.

$$\frac{3x}{4} - \frac{2}{5} = 2$$

$$\frac{3x}{4} - \frac{2}{5} + \frac{2}{5} = 2 + \frac{2}{5}$$

$$\frac{3x}{4} = \frac{12}{5}$$

$$4 \cdot \frac{3x}{4} = 4 \cdot \frac{12}{5}$$

$$3x = \frac{48}{5}$$

$$\frac{1}{3} \cdot 3x = \frac{1}{3} \cdot \frac{48}{5}$$

$$x = \frac{16}{5}$$

$$\frac{3x}{4} - \frac{2}{5} = 2$$

$$\frac{3\left(\frac{16}{5}\right)}{4} - \frac{2}{5} \overset{?}{=} 2$$

$$\frac{\frac{48}{5}}{4} - \frac{2}{5} \overset{?}{=} 2$$

$$\frac{48}{20} - \frac{2}{5} \overset{?}{=} 2$$

$$\frac{12}{5} - \frac{2}{5} \overset{?}{=} 2$$

$$\frac{10}{5} \overset{?}{=} 2$$

$$2 = 2$$

53.

$$\frac{3x - 12}{2} = 9$$

$$2 \cdot \frac{3x - 12}{2} = 2 \cdot 9$$

$$3x - 12 = 18$$

$$3x - 12 + 12 = 18 + 12$$

$$3x = 30$$

$$\frac{3x}{3} = \frac{30}{3}$$

$$x = 10$$

$$\frac{3x - 12}{2} = 9$$

$$\frac{3(10) - 12}{2} \overset{?}{=} 9$$

$$\frac{30 - 12}{2} \overset{?}{=} 9$$

$$\frac{18}{2} \overset{?}{=} 9$$

$$9 = 9$$

55.

$$\frac{5k - 8}{9} = 1$$

$$9 \cdot \frac{5k - 8}{9} = 9 \cdot 1$$

$$5k - 8 = 9$$

$$5k - 8 + 8 = 9 + 8$$

$$5k = 17$$

$$\frac{5k}{5} = \frac{17}{5}$$

$$k = \frac{17}{5}$$

$$\frac{5k - 8}{9} = 1$$

$$\frac{5 \cdot \frac{17}{5} - 8}{9} \overset{?}{=} 1$$

$$\frac{17 - 8}{9} \overset{?}{=} 1$$

$$\frac{9}{9} \overset{?}{=} 1$$

$$1 = 1$$

57.

$$\frac{3z + 2}{17} = 0$$

$$17 \cdot \frac{3z + 2}{17} = 17 \cdot 0$$

$$3z + 2 = 0$$

$$3z + 2 - 2 = 0 - 2$$

$$3z = -2$$

$$\frac{3z}{3} = \frac{-2}{3}$$

$$z = -\frac{2}{3}$$

$$\frac{3z + 2}{17} = 0$$

$$\frac{3\left(-\frac{2}{3}\right) + 2}{17} \overset{?}{=} 0$$

$$\frac{-2 + 2}{17} \overset{?}{=} 0$$

$$\frac{0}{17} \overset{?}{=} 0$$

$$0 = 0$$

59.

$$\frac{17k - 28}{21} + \frac{4}{3} = 0$$

$$\frac{17k - 28}{21} + \frac{4}{3} - \frac{4}{3} = 0 - \frac{4}{3}$$

$$\frac{17k - 28}{21} = -\frac{4}{3}$$

$$21 \cdot \frac{17k - 28}{21} = 21\left(-\frac{4}{3}\right)$$

$$17k - 28 = -28$$

$$17k - 28 + 28 = -28 + 28$$

$$17k = 0$$

$$\frac{17k}{17} = \frac{0}{17}$$

$$k = 0$$

$$\frac{17k - 28}{21} + \frac{4}{3} = 0$$

$$\frac{17 \cdot 0 - 28}{21} + \frac{4}{3} \stackrel{?}{=} 0$$

$$\frac{0 - 28}{21} + \frac{4}{3} \stackrel{?}{=} 0$$

$$-\frac{4}{3} + \frac{4}{3} \stackrel{?}{=} 0$$

$$0 = 0$$

61.

$$-\frac{x}{3} - \frac{1}{2} = -\frac{5}{2}$$

$$-\frac{x}{3} - \frac{1}{2} + \frac{1}{2} = -\frac{5}{2} + \frac{1}{2}$$

$$-\frac{x}{3} = -\frac{4}{2}$$

$$-\frac{x}{3} = -2$$

$$-3\left(-\frac{x}{3}\right) = -3(-2)$$

$$x = 6$$

$$-\frac{x}{3} - \frac{1}{2} = -\frac{5}{2}$$

$$-\frac{6}{3} - \frac{1}{2} \stackrel{?}{=} -\frac{5}{2}$$

$$-2 - \frac{1}{2} \stackrel{?}{=} -\frac{5}{2}$$

$$-\frac{4}{2} - \frac{1}{2} \stackrel{?}{=} -\frac{5}{2}$$

$$-\frac{5}{2} = -\frac{5}{2}$$

63.

$$\frac{9 - 5w}{15} = \frac{2}{5}$$

$$15 \cdot \frac{9 - 5w}{15} = 15 \cdot \frac{2}{5}$$

$$9 - 5w = 6$$

$$9 - 9 - 5w = 6 - 9$$

$$-5w = -3$$

$$\frac{-5w}{-5} = \frac{-3}{-5}$$

$$w = \frac{3}{5}$$

$$\frac{9 - 5w}{15} = \frac{2}{5}$$

$$\frac{9 - 5\left(\frac{3}{5}\right)}{15} \stackrel{?}{=} \frac{2}{5}$$

$$\frac{9 - 3}{15} \stackrel{?}{=} \frac{2}{5}$$

$$\frac{6}{15} \stackrel{?}{=} \frac{2}{5}$$

$$\frac{2}{5} = \frac{2}{5}$$

65. Let x = the original number.

Then $3x - 6$ = the other number.

$$\boxed{\text{The other number}} = 9$$

$$3x - 6 = 9$$

$$3x - 6 + 6 = 9 + 6$$

$$3x = 15$$

$$\frac{3x}{3} = \frac{15}{3}$$

$$x = 5$$

The original number is 5.

67. Let x = her former rent.

Then $2x - 100$ = the new rent.

$$\boxed{\text{The new rent}} = 400$$

$$2x - 100 = 400$$

$$2x - 100 + 100 = 400 + 100$$

$$2x = 500$$

$$\frac{2x}{2} = \frac{500}{2}$$

$$x = 250$$

Her former rent was $250.

69. Let $x = $ the number of days.

Then $16 + 12x = $ the total cost.

$$\boxed{\text{The total cost}} = 100$$
$$16 + 12x = 100$$
$$16 - 16 + 12x = 100 - 16$$
$$12x = 84$$
$$\frac{12x}{12} = \frac{84}{12}$$
$$x = 7$$

The owner was gone for 7 days.

71. Let $x = $ the # of minutes (after the 1st).

Then $0.85 + 0.27x = $ the total cost.

$$\boxed{\text{The total cost}} = 8.50$$
$$0.85 + 0.27x = 8.50$$
$$0.85 - 0.85 + 0.27x = 8.50 - 0.85$$
$$0.27x = 7.65$$
$$\frac{0.27x}{0.27} = \frac{7.65}{0.27}$$
$$x \approx 28.3$$

She can talk for 28 minutes **after the first minute**, for a total of 29 minutes.

73. Let $x = $ the money from ticket sales.

Then $1500 + 0.20x = $ the total income.

$$\boxed{\text{The total income}} = 2980$$
$$1500 + 0.20x = 2980$$
$$1500 - 1500 + 0.20x = 2980 - 1500$$
$$0.20x = 1480$$
$$\frac{0.20x}{0.20} = \frac{1480}{0.20}$$
$$x = 7400$$

The total ticket sales were $7,400.

75. Let $x = $ the score on the fifth exam.

$$\boxed{\text{Average score}} = 90$$
$$\frac{85 + 80 + 95 + 78 + x}{5} = 90$$
$$\frac{338 + x}{5} = 90$$
$$5 \cdot \frac{338 + x}{5} = 5 \cdot 90$$
$$338 + x = 450$$
$$338 - 338 + x = 450 - 338$$
$$x = 112$$

It is impossible to receive an A.

77. Let $x = $ the regular price.

Then $0.80x = $ the sale price.

$$\boxed{\begin{array}{c}\text{Final} \\ \text{price}\end{array}} = 0.90 \cdot \boxed{\begin{array}{c}\text{Sale} \\ \text{price}\end{array}}$$
$$36 = 0.90(0.80x)$$
$$36 = 0.72x$$
$$\frac{36}{0.72} = \frac{0.72x}{0.72}$$
$$50 = x$$

The original price was $50.

79. For a purchase of $100:

$$\boxed{\text{Markdown}} = \boxed{\begin{array}{c}\text{Percent} \\ \text{markdown}\end{array}} \cdot \boxed{\begin{array}{c}\text{Regular} \\ \text{price}\end{array}}$$
$$15 = r \cdot 100$$
$$\frac{15}{100} = \frac{r \cdot 100}{100}$$
$$0.15 = r$$
$$15\% = r$$

For a purchase of $250:

$$\boxed{\text{Markdown}} = \boxed{\begin{array}{c}\text{Percent} \\ \text{markdown}\end{array}} \cdot \boxed{\begin{array}{c}\text{Regular} \\ \text{price}\end{array}}$$
$$15 = r \cdot 250$$
$$\frac{15}{250} = \frac{r \cdot 250}{250}$$
$$0.06 = r$$
$$6\% = r$$

The range of the percent discount is from 6% to 15%.

81. Answers may vary.

83.
$$\frac{7x + \#}{22} = \frac{1}{2}$$
$$\frac{7(1) + \#}{22} = \frac{1}{2}$$
$$\frac{7 + \#}{22} = \frac{1}{2}$$
$$22 \cdot \frac{7 + \#}{22} = 22 \cdot \frac{1}{2}$$
$$7 + \# = 11$$
$$7 - 7 + \# = 11 - 7$$
$$\# = 4$$

The original equation was $\dfrac{7x + 4}{22} = \dfrac{1}{2}$.

Exercise 2.3 (page 105)

1. $x^2 z(y^3 - z) = (-3)^2(0)\left[(-5)^3 - 0\right] = 0$

3. $\dfrac{x - y^2}{2y - 1 + x} = \dfrac{-3 - (-5)^2}{2(-5) - 1 + (-3)} = \dfrac{-3 - (+25)}{-10 - 1 + (-3)} = \dfrac{-28}{-14} = 2$

5. $\dfrac{6}{7} - \dfrac{5}{8} = \dfrac{6 \cdot 8}{7 \cdot 8} - \dfrac{5 \cdot 7}{8 \cdot 7} = \dfrac{48}{56} - \dfrac{35}{56}$
$$= \dfrac{48 - 35}{56} = \dfrac{13}{56}$$

7. $\dfrac{6}{7} \div \dfrac{5}{8} = \dfrac{6}{7} \cdot \dfrac{8}{5} = \dfrac{48}{35}$

9. variables, like

11. identity

13. $3x + 17x = (3 + 17)x = 20x$

15. $8x^2 - 5x^2 = (8 - 5)x^2 = 3x^2$

17. $9x + 3y \Rightarrow$ not like terms

19. $3(x + 2) + 4x = 3 \cdot x + 3 \cdot 2 + 4x$
$$= 3x + 6 + 4x = 7x + 6$$

21. $5(z - 3) + 2z = 5 \cdot z - 5 \cdot 3 + 2z$
$$= 5z - 15 + 2z = 7z - 15$$

23. $12(x + 11) - 11 = 12 \cdot x + 12 \cdot 11 - 11$
$$= 12x + 132 - 11$$
$$= 12x + 121$$

25. $8(y + 7) - 2(y - 3) = 8 \cdot y + 8 \cdot 7 + (-2) \cdot y + (-2)(-3) = 8y + 56 - 2y + 6 = 6y + 62$

27. $2x + 4(y - x) + 3y = 2x + 4 \cdot y + 4(-x) + 3y = 2x + 4y - 4x + 3y = -2x + 7y$

29. $(x + 2) - (x - y) = 1(x + 2) - 1(x - y) = 1 \cdot x + 1 \cdot 2 + (-1) \cdot x + (-1)(-y) = x + 2 - x + y$
$$= y + 2$$

31. $2\left(4x + \dfrac{9}{2}\right) - 3\left(x + \dfrac{2}{3}\right) = 2 \cdot 4x + 2 \cdot \dfrac{9}{2} + (-3) \cdot x + (-3) \cdot \dfrac{2}{3} = 8x + 9 - 3x - 2 = 5x + 7$

33. $8x(x + 3) - 3x^2 = 8x \cdot x + 8x \cdot 3 - 3x^2 = 8x^2 + 24x - 3x^2 = 5x^2 + 24x$

35.
$$3x + 2 = 2x$$
$$3x - 3x + 2 = 2x - 3x$$
$$2 = -x$$
$$-1(2) = -1(-x)$$
$$-2 = x$$

$$3x + 2 = 2x$$
$$3(-2) + 2 \overset{?}{=} 2(-2)$$
$$-6 + 2 \overset{?}{=} -4$$
$$-4 = -4$$

37.
$$5x - 3 = 4x$$
$$5x - 5x - 3 = 4x - 5x$$
$$-3 = -x$$
$$-1(-3) = -1(-x)$$
$$3 = x$$

$$5x - 3 = 4x$$
$$5(3) - 3 \overset{?}{=} 4(3)$$
$$15 - 3 \overset{?}{=} 12$$
$$12 = 12$$

39.
$$9y - 3 = 6y$$
$$9y - 9y - 3 = 6y - 9y$$
$$-3 = -3y$$
$$\dfrac{-3}{-3} = \dfrac{-3y}{-3}$$
$$1 = y$$

$$9y - 3 = 6y$$
$$9(1) - 3 \overset{?}{=} 6(1)$$
$$9 - 3 \overset{?}{=} 6$$
$$6 = 6$$

41.
$$8y - 7 = y$$
$$8y - 8y - 7 = y - 8y$$
$$-7 = -7y$$
$$\dfrac{-7}{-7} = \dfrac{-7y}{-7}$$
$$1 = y$$

$$8y - 7 = y$$
$$8(1) - 7 \overset{?}{=} 1$$
$$8 - 7 \overset{?}{=} 1$$
$$1 = 1$$

43.
$$9 - 23w = 4w$$
$$9 - 23w + 23w = 4w + 23w$$
$$9 = 27w$$
$$\dfrac{9}{27} = \dfrac{27w}{27}$$
$$\dfrac{1}{3} = w$$

$$9 - 23w = 4w$$
$$9 - 23\left(\dfrac{1}{3}\right) \overset{?}{=} 4\left(\dfrac{1}{3}\right)$$
$$\dfrac{27}{3} - \dfrac{23}{3} \overset{?}{=} \dfrac{4}{3}$$
$$\dfrac{4}{3} = \dfrac{4}{3}$$

45.
$$22 - 3r = 8r$$
$$22 - 3r + 3r = 8r + 3r$$
$$22 = 11r$$
$$\dfrac{22}{11} = \dfrac{11r}{11}$$
$$2 = r$$

$$22 - 3r = 8r$$
$$22 - 3(2) \overset{?}{=} 8(2)$$
$$22 - 6 \overset{?}{=} 16$$
$$16 = 16$$

47.
$$3(a + 2) = 4a$$
$$3a + 6 = 4a$$
$$3a - 3a + 6 = 4a - 3a$$
$$6 = a$$

$$3(a + 2) = 4a$$
$$3(6 + 2) \overset{?}{=} 4(6)$$
$$3(8) \overset{?}{=} 24$$
$$24 = 24$$

49.
$$5(b + 7) = 6b$$
$$5b + 35 = 6b$$
$$5b - 5b + 35 = 6b - 5b$$
$$35 = b$$

$$5(b + 7) = 6b$$
$$5(35 + 7) \overset{?}{=} 6(35)$$
$$5(42) \overset{?}{=} 210$$
$$210 = 210$$

51.
$$2 + 3(x - 5) = 4(x - 1)$$
$$2 + 3x - 15 = 4x - 4$$
$$3x - 13 = 4x - 4$$
$$3x - 3x - 13 = 4x - 3x - 4$$
$$-13 = x - 4$$
$$-13 + 4 = x - 4 + 4$$
$$-9 = x$$

$$2 + 3(x - 5) = 4(x - 1)$$
$$2 + 3(-9 - 5) \stackrel{?}{=} 4(-9 - 1)$$
$$2 + 3(-14) \stackrel{?}{=} 4(-10)$$
$$2 + (-42) \stackrel{?}{=} -40$$
$$-40 = -40$$

53.
$$10x + 3(2 - x) = 5(x + 2) - 4$$
$$10x + 6 - 3x = 5x + 10 - 4$$
$$7x + 6 = 5x + 6$$
$$7x - 5x + 6 = 5x - 5x + 6$$
$$2x + 6 = 6$$
$$2x + 6 - 6 = 6 - 6$$
$$2x = 0$$
$$\frac{2x}{2} = \frac{0}{2}$$
$$x = 0$$

$$10x + 3(2 - x) = 5(x + 2) - 4$$
$$10(0) + 3(2 - 0) \stackrel{?}{=} 5(0 + 2) - 4$$
$$0 + 3(2) \stackrel{?}{=} 5(2) - 4$$
$$0 + 6 \stackrel{?}{=} 10 - 4$$
$$6 = 6$$

55.
$$3(a + 2) = 2(a - 7)$$
$$3a + 6 = 2a - 14$$
$$3a - 2a + 6 = 2a - 2a - 14$$
$$a + 6 = -14$$
$$a + 6 - 6 = -14 - 6$$
$$a = -20$$

$$3(a + 2) = 2(a - 7)$$
$$3(-20 + 2) \stackrel{?}{=} 2(-20 - 7)$$
$$3(-18) \stackrel{?}{=} 2(-27)$$
$$-54 = -54$$

57.
$$9(x + 11) + 5(13 - x) = 0$$
$$9x + 99 + 65 - 5x = 0$$
$$4x + 164 = 0$$
$$4x + 164 - 164 = 0 - 164$$
$$4x = -164$$
$$\frac{4x}{4} = \frac{-164}{4}$$
$$x = -41$$

$$9(x + 11) + 5(13 - x) = 0$$
$$9(-41 + 11) + 5[13 - (-41)] \stackrel{?}{=} 0$$
$$9(-30) + 5[13 + 41] \stackrel{?}{=} 0$$
$$-270 + 5(54) \stackrel{?}{=} 0$$
$$-270 + 270 \stackrel{?}{=} 0$$
$$0 = 0$$

59.
$$\frac{3(t - 7)}{2} = t - 6$$
$$2 \cdot \frac{3(t - 7)}{2} = 2(t - 6)$$
$$3(t - 7) = 2(t - 6)$$
$$3t - 21 = 2t - 12$$
$$3t - 2t - 21 = 2t - 2t - 12$$
$$t - 21 = -12$$
$$t - 21 + 21 = -12 + 21$$
$$t = 9$$

$$\frac{3(t - 7)}{2} = t - 6$$
$$\frac{3(9 - 7)}{2} \stackrel{?}{=} 9 - 6$$
$$\frac{3(2)}{2} \stackrel{?}{=} 9 - 6$$
$$\frac{6}{2} \stackrel{?}{=} 3$$
$$3 = 3$$

61.

$$\frac{5(2-s)}{3} = s+6$$

$$3 \cdot \frac{5(2-s)}{3} = 3(s+6)$$

$$5(2-s) = 3(s+6)$$

$$10 - 5s = 3s + 18$$

$$10 - 5s + 5s = 3s + 5s + 18$$

$$10 = 8s + 18$$

$$10 - 18 = 8s + 18 - 18$$

$$-8 = 8s$$

$$\frac{-8}{8} = \frac{8s}{8}$$

$$-1 = s$$

$$\frac{5(2-s)}{3} = s+6$$

$$\frac{5[2-(-1)]}{3} \stackrel{?}{=} -1+6$$

$$\frac{5(2+1)}{3} \stackrel{?}{=} 5$$

$$\frac{5(3)}{3} \stackrel{?}{=} 5$$

$$\frac{15}{3} \stackrel{?}{=} 5$$

$$5 = 5$$

63.

$$\frac{4(2x-10)}{3} = 2(x-4)$$

$$3 \cdot \frac{4(2x-10)}{3} = 3 \cdot 2(x-4)$$

$$4(2x-10) = 6(x-4)$$

$$8x - 40 = 6x - 24$$

$$8x - 6x - 40 = 6x - 6x - 24$$

$$2x - 40 = -24$$

$$2x - 40 + 40 = -24 + 40$$

$$2x = 16$$

$$\frac{2x}{2} = \frac{16}{2}$$

$$x = 8$$

$$\frac{4(2x-10)}{3} = 2(x-4)$$

$$\frac{4[2(8)-10]}{3} \stackrel{?}{=} 2(8-4)$$

$$\frac{4(16-10)}{3} \stackrel{?}{=} 2(4)$$

$$\frac{4(6)}{3} \stackrel{?}{=} 8$$

$$\frac{24}{3} \stackrel{?}{=} 8$$

$$8 = 8$$

65.

$$3.1(x-2) = 1.3x + 2.8$$

$$3.1x - 6.2 = 1.3x + 2.8$$

$$3.1x - 1.3x - 6.2 = 1.3x - 1.3x + 2.8$$

$$1.8x - 6.2 = 2.8$$

$$1.8x - 6.2 + 6.2 = 2.8 + 6.2$$

$$1.8x = 9.0$$

$$\frac{1.8x}{1.8} = \frac{9.0}{1.8}$$

$$x = 5$$

$$3.1(x-2) = 1.3x + 2.8$$

$$3.1(5-2) \stackrel{?}{=} 1.3(5) + 2.8$$

$$3.1(3) \stackrel{?}{=} 6.5 + 2.8$$

$$9.3 = 9.3$$

67.

$$2.7(y + 1) = 0.3(3y + 33)$$
$$2.7y + 2.7 = 0.9y + 9.9$$
$$2.7y - 0.9y + 2.7 = 0.9y - 0.9y + 9.9$$
$$1.8y + 2.7 = 9.9$$
$$1.8y + 2.7 - 2.7 = 9.9 - 2.7$$
$$1.8y = 7.2$$
$$\frac{1.8y}{1.8} = \frac{7.2}{1.8}$$
$$y = 4$$

$$2.7(y + 1) = 0.3(3y + 33)$$
$$2.7(4 + 1) = 0.3[3(4) + 33]$$
$$2.7(5) \overset{?}{=} 0.3(12 + 33)$$
$$13.5 \overset{?}{=} 0.3(45)$$
$$13.5 = 13.5$$

69.

$$19.1x - 4(x + 0.3) = -46.5$$
$$19.1x - 4x - 1.2 = -46.5$$
$$15.1x - 1.2 = -46.5$$
$$15.1x - 1.2 + 1.2 = -46.5 + 1.2$$
$$15.1x = -45.3$$
$$\frac{15.1x}{15.1} = \frac{-45.3}{15.1}$$
$$x = -3$$

$$19.1x - 4(x + 0.3) = -46.5$$
$$19.1(-3) - 4(-3 + 0.3) \overset{?}{=} -46.5$$
$$-57.3 - 4(-2.7) \overset{?}{=} -46.5$$
$$-57.3 + 10.8 \overset{?}{=} -46.5$$
$$-46.5 = -46.5$$

71.

$$14.3(x + 2) + 13.7(x - 3) = 15.5$$
$$14.3x + 28.6 + 13.7x - 41.1 = 15.5$$
$$28.0x - 12.5 = 15.5$$
$$28x - 12.5 + 12.5 = 15.5 + 12.5$$
$$28x = 28$$
$$\frac{28x}{28} = \frac{28}{28}$$
$$x = 1$$

$$14.3(x + 2) + 13.7(x - 3) = 15.5$$
$$14.3(1 + 2) + 13.7(1 - 3) \overset{?}{=} 15.5$$
$$14.3(3) + 13.7(-2) \overset{?}{=} 15.5$$
$$42.9 - 27.4 \overset{?}{=} 15.5$$
$$15.5 = 15.5$$

73.

$$8x + 3(2 - x) = 5(x + 2) - 4$$
$$8x + 6 - 3x = 5x + 10 - 4$$
$$5x + 6 = 5x + 6$$
$$5x - 5x + 6 = 5x - 5x + 6$$
$$6 = 6$$

Identity

75.

$$2(s + 2) = 2(s + 1) + 3$$
$$2s + 4 = 2s + 2 + 3$$
$$2s + 4 = 2s + 5$$
$$2s - 2s + 4 = 2s - 2s + 5$$
$$4 \neq 5$$

Contradiction

77.
$$\frac{2(t-1)}{6} - 2 = \frac{t+2}{6}$$
$$6\left[\frac{2(t-1)}{6} - 2\right] = 6 \cdot \frac{t+2}{6}$$
$$6 \cdot \frac{2(t-1)}{6} - 6 \cdot 2 = t + 2$$
$$2(t-1) - 12 = t + 2$$
$$2t - 2 - 12 = t + 2$$
$$2t - 14 = t + 2$$
$$2t - t - 14 = t - t + 2$$
$$t - 14 = 2$$
$$t - 14 + 14 = 2 + 14$$
$$t = 16$$

79.
$$2(3z + 4) = 2(3z - 2) + 13$$
$$6z + 8 = 6z - 4 + 13$$
$$6z + 8 = 6z + 9$$
$$6z - 6z + 8 = 6z - 6z + 9$$
$$8 \neq 9$$
Contradiction

81.
$$2(y - 3) - \frac{y}{2} = \frac{3}{2}(y - 4)$$
$$2\left[2(y - 3) - \frac{y}{2}\right] = 2 \cdot \frac{3}{2}(y - 4)$$
$$2 \cdot 2(y - 3) - 2 \cdot \frac{y}{2} = 3(y - 4)$$
$$4(y - 3) - y = 3y - 12$$
$$4y - 12 - y = 3y - 12$$
$$3y - 12 = 3y - 12$$
$$3y - 3y - 12 = 3y - 3y - 12$$
$$-12 = -12$$
Identity

83.
$$\frac{3x + 14}{2} = x - 2 + \frac{x + 18}{2}$$
$$2 \cdot \frac{3x + 14}{2} = 2\left[x - 2 + \frac{x + 18}{2}\right]$$
$$3x + 14 = 2x - 2 \cdot 2 + 2 \cdot \frac{x + 18}{2}$$
$$3x + 14 = 2x - 4 + x + 18$$
$$3x + 14 = 3x + 14$$
$$3x - 3x + 14 = 3x - 3x + 14$$
$$14 = 14$$
Identity

85. **Answers may vary.**

87. **Answers may vary.**

89. Let x = the number.
$$x = 2x$$
$$x - x = 2x - x$$
$$0 = x$$
The number is 0.

Exercise 2.4 (page 112)

1. $V = \frac{1}{3}Bh = \frac{1}{3}s^2h = \frac{1}{3}(10\,\text{cm})^2(6\,\text{cm}) = \frac{1}{3}(100\,\text{cm}^2)(6\,\text{cm}) = \frac{1}{3}(600\,\text{cm}^3) = 200\,\text{cm}^3$

3. $3(x + 2) + 4(x - 3) = 3 \cdot x + 3 \cdot 2 + 4 \cdot x - 4 \cdot 3 = 3x + 6 + 4x - 12 = 7x - 6$

5. $\frac{1}{2}(x + 1) - \frac{1}{2}(x + 4) = \frac{1}{2} \cdot x + \frac{1}{2} \cdot 1 - \frac{1}{2} \cdot x - \frac{1}{2} \cdot 4 = \frac{1}{2}x + \frac{1}{2} - \frac{1}{2}x - \frac{4}{2} = -\frac{3}{2}$

7. $A = P + Prt = 1200 + 1200(0.08)(3) = 1200 + 288 = \$1488$

9. $2l + 2w$ **11.** vertex   **13.** complementary   **15.** $180°$

17. Let $x =$ the length of one part.
Then $2x =$ the length of the other part.

$$\boxed{\begin{smallmatrix}\text{Length of}\\\text{first part}\end{smallmatrix}} + \boxed{\begin{smallmatrix}\text{Length of}\\\text{second part}\end{smallmatrix}} = 12$$

$$x + 2x = 12$$
$$3x = 12$$
$$\frac{3x}{3} = \frac{12}{3}$$
$$x = 4$$

The parts are 4 feet and 8 feet long.

19.

$$\boxed{\begin{smallmatrix}\text{Sum of 3}\\\text{lengths}\end{smallmatrix}} = 30$$

$$x + 10 + 2x + x = 30$$
$$4x + 10 = 30$$
$$4x + 10 - 10 = 30 - 10$$
$$4x = 20$$
$$\frac{4x}{4} = \frac{20}{4}$$
$$x = 5$$

The sections are 15, 10, and 5 feet long.

21.

$$\boxed{\begin{smallmatrix}\text{Sum of 3}\\\text{lengths}\end{smallmatrix}} = 24$$

$$x + x + 4 + x + 2 = 24$$
$$3x + 6 = 24$$
$$3x + 6 - 6 = 24 - 6$$
$$3x = 18$$
$$\frac{3x}{3} = \frac{18}{3}$$
$$x = 6$$

The sections are 6, 10, and 8 feet long.

23.
$$x + 40° = 50°$$
$$x + 40° - 40° = 50° - 40°$$
$$x = 10°$$

25.
$$x + 21° = 180°$$
$$x + 21° - 21° = 180° - 21°$$
$$x = 159°$$

27.
$$x + 12° = 59°$$
$$x + 12° - 12° = 59° - 12°$$
$$x = 47°$$

29.
$$x + 63° = 90°$$
$$x + 63° - 63° = 90° - 63°$$
$$x = 27°$$

31. Let $x =$ the measure of the complement.
$$x + 37° = 90°$$
$$x + 37° - 37° = 90° - 37°$$
$$x = 53°$$

33. Let $x =$ the measure of the complement.
$$x + 40° = 90°$$
$$x + 40° - 40° = 90° - 40°$$
$$x = 50°$$
Let $x =$ the measure of the supplement of the complement.
$$x + 50° = 180°$$
$$x + 50° - 50° = 180° - 50°$$
$$x = 130°$$

35.

$$\boxed{\begin{smallmatrix}\text{Sum of}\\\text{three sides}\end{smallmatrix}} = 57$$

$$x + x + x = 57$$
$$3x = 57$$
$$\frac{3x}{3} = \frac{57}{3}$$
$$x = 19$$

Each side is 19 feet long.

SECTION 2.4

37. Let L = the length of the pool.
Then $L - 11$ = the width of the pool.

$$\boxed{\text{Perimeter}} = 94$$
$$2L + 2(L - 11) = 94$$
$$2L + 2L - 22 = 94$$
$$4L - 22 = 94$$
$$4L - 22 + 22 = 94 + 22$$
$$4L = 116$$
$$\frac{4L}{4} = \frac{116}{4}$$
$$L = 29$$

The dimensions are 29 m by 18 m.

41. Let a = the measure of each angle.

$$\boxed{\text{Sum of angle measures}} = 180$$
$$a + a + a = 180$$
$$3a = 180$$
$$\frac{3a}{3} = \frac{180}{3}$$
$$a = 60$$

Each angle measures $60°$.

39. Let w = the width of the picture.
Then $2w + 5$ = the length of the picture.

$$\boxed{\text{Perimeter}} = 112$$
$$2w + 2(2w + 5) = 112$$
$$2w + 4w + 10 = 112$$
$$6w + 10 = 112$$
$$6w + 10 - 10 = 112 - 10$$
$$6w = 102$$
$$\frac{6w}{6} = \frac{102}{6}$$
$$w = 17$$

The dimensions are 17 in. by 39 in.

43. Let x = amount invested at 5%.

$$I = PRT$$
$$300 = x(0.05)(1)$$
$$\frac{300}{0.05} = \frac{0.05x}{0.05}$$
$$6000 = x$$

He invested $6,000.

45. Let x = amount in 9% fund. Then $24000 - x$ = amount in 14% fund.

$$\boxed{\text{Interest at 9\%}} + \boxed{\text{Interest at 14\%}} = \boxed{\text{Total interest}}$$
$$0.09x + 0.14(24000 - x) = 3135$$
$$9x + 14(24000 - x) = 313500$$
$$9x + 336000 - 14x = 313500$$
$$-5x + 336000 = 313500$$
$$-5x + 336000 - 336000 = 313500 - 336000$$
$$-5x = -22500$$
$$\frac{-5x}{-5} = \frac{-22500}{-5}$$
$$x = 4500 \Rightarrow \$4,500 \text{ was invested at 9\%, and \$19,500 was invested at 14\%.}$$

47. Let x = amount invested in each account.

$$\boxed{\text{Interest at 8\%}} + \boxed{\text{Interest at 11\%}} = \boxed{\text{Total interest}}$$
$$0.08x + 0.11x = 712.50$$
$$8x + 11x = 71250$$
$$19x = 71250$$
$$\frac{19x}{19} = \frac{71250}{19}$$
$$x = 3750 \Rightarrow \$3,750 \text{ is invested in each account.}$$

49. Let $x =$ amount invested at 7%.

$$\boxed{\begin{array}{c}\text{Interest}\\\text{at 6\%}\end{array}} + \boxed{\begin{array}{c}\text{Interest}\\\text{at 7\%}\end{array}} = \boxed{\begin{array}{c}\text{Total}\\\text{interest}\end{array}}$$

$$0.06(15000) + 0.07x = 1250$$
$$6(15000) + 7x = 125000$$
$$90000 + 7x = 125000$$
$$90000 - 90000 + 7x = 125000 - 90000$$
$$7x = 35000$$
$$\frac{7x}{7} = \frac{35000}{7}$$
$$x = 5000$$

$5,000 should be invested at 7%.

51. Let $r =$ the fund rate and $r + 0.01 =$ the CD rate.
The client invests $21,000 in CDs, and $10,500 in the fund.

$$\boxed{\begin{array}{c}\text{CD}\\\text{interest}\end{array}} = \boxed{\begin{array}{c}\text{Fund}\\\text{interest}\end{array}} + 840$$

$$(r + 0.01)21000 = r(10500) + 840$$
$$21000r + 210 = 10500r + 840$$
$$21000r - 10500r + 210 = 10500r - 10500r + 840$$
$$10500r + 210 = 840$$
$$10500r + 210 - 210 = 840 - 210$$
$$10500r = 630$$
$$\frac{10500r}{10500} = \frac{630}{10500}$$
$$r = 0.06 \Rightarrow \text{The rates are 6\% and 7\%.}$$

53. **Answers may vary.**

55. Pairs of vertical angles have equal measures.

Exercise 2.5 (page 121)

1. $3 + 4(-5) = 3 + (-20) = -17$

3. $2^3 - 3^2 = 2 \cdot 2 \cdot 2 - 3 \cdot 3 = 8 - 9 = -1$

5.
$$-2x + 3 = 9$$
$$-2x + 3 - 3 = 9 - 3$$
$$-2x = 6$$
$$\frac{-2x}{-2} = \frac{6}{-2}$$
$$x = -3$$

7.
$$\frac{2}{3}p + 1 = 5$$
$$\frac{2}{3}p + 1 - 1 = 5 - 1$$
$$\frac{2}{3}p = 4$$
$$\frac{3}{2} \cdot \frac{2}{3}p = \frac{3}{2} \cdot 4$$
$$p = \frac{12}{2}$$
$$p = 6$$

9. $d = rt$

11. $v = pn$

13. Let t = time for cars to meet.

	r	t	d
Car 1 (A to B)	50	t	$50t$
Car 2 (B to A)	55	t	$55t$

$$\boxed{\substack{\text{Distance} \\ \text{for car 1}} + \boxed{\substack{\text{Distance} \\ \text{for car 2}}} = 315$$

$$50t + 55t = 315$$
$$105t = 315$$
$$\frac{105t}{105} = \frac{315}{105}$$
$$t = 3$$

The cars meet after 3 hours.

15. Let t = days for crews to meet.

	r	t	d
Crew 1	1.5	t	$1.5t$
Crew 2	1.2	t	$1.2t$

$$\boxed{\substack{\text{Distance} \\ \text{for crew 1}}} + \boxed{\substack{\text{Distance} \\ \text{for crew 2}}} = 9.45$$

$$1.5t + 1.2t = 9.45$$
$$2.7t = 9.45$$
$$\frac{2.7t}{2.7} = \frac{9.45}{2.7}$$
$$t = 3.5$$

The crews meet after 3.5 days.

17. Let t = time for cars to be 715 miles apart.

	r	t	d
Car 1 (going east)	60	t	$60t$
Car 2 (going west)	50	t	$50t$

$$\boxed{\text{Distance 1}} + \boxed{\text{Distance 2}} = 715$$

$$60t + 50t = 715$$
$$110t = 715$$
$$\frac{110t}{110} = \frac{715}{110}$$
$$t = 6.5$$

They will be 715 miles apart after 6.5 hours.

19. Let t = time for boys to be 2 miles apart.

	r	t	d
Boy 1	3	t	$3t$
Boy 2	4	t	$4t$

$$\boxed{\text{Distance 1}} + \boxed{\text{Distance 2}} = 2$$

$$3t + 4t = 2$$
$$7t = 2$$
$$\frac{7t}{7} = \frac{2}{7}$$
$$t = \frac{2}{7}$$

They will lose contact after $\frac{2}{7}$ hour.

21.

	r	t	d
Car	60 mph	t	$60t$
Bus	50 mph	$t+2$	$50(t+2)$

$$\boxed{\substack{\text{Car} \\ \text{distance}}} = \boxed{\substack{\text{Bus} \\ \text{distance}}}$$

$$60t = 50(t+2)$$
$$60t = 50t + 100$$
$$60t - 50t = 50t - 50t + 100$$
$$10t = 100$$
$$\frac{10t}{10} = \frac{100}{10}$$
$$t = 10$$

The car will overtake the bus after 10 hours.

23. Let t = time for cars to be 82.5 miles apart.

	r	t	d
Car 1	42	t	$42t$
Car 2	53	t	$53t$

$$\boxed{\text{Distance 2}} - \boxed{\text{Distance 1}} = 82.5$$

$$53t - 42t = 82.5$$
$$11t = 82.5$$
$$\frac{11t}{11} = \frac{82.5}{11}$$
$$t = 7.5$$

They will be 82.5 miles apart after 7.5 hours.

25. Let r = rate of slow train. Then
$r + 20$ = rate of fast train.

	r	t	d
Slow train	r	3	$3r$
Fast train	$r + 20$	3	$3(r + 20)$

$\boxed{\text{Slow dist.}} + \boxed{\text{Fast dist.}} = 330$

$$3r + 3(r + 20) = 330$$
$$3r + 3r + 60 = 330$$
$$6r + 60 = 330$$
$$6r + 60 - 60 = 330 - 60$$
$$6r = 270$$
$$\frac{6r}{6} = \frac{270}{6}$$
$$r = 45$$

The rates are 45 mph and 65 mph.

27. Let t = slower time.
Then $5 - t$ = faster time.

	r	t	d
1st part	40	t	$40t$
2nd part	50	$5 - t$	$50(5 - t)$

$\boxed{\text{1st dist.}} + \boxed{\text{2nd dist.}} = 210$

$$40t + 50(5 - t) = 210$$
$$40t + 250 - 50t = 210$$
$$-10t + 250 = 210$$
$$-10t + 250 - 250 = 210 - 250$$
$$-10t = -40$$
$$\frac{-10t}{-10} = \frac{-40}{-10}$$
$$t = 4$$

The car averaged 40 mph for 4 hours.

29. Let T = total number of liters of solution.
12% of the total = liters of acid
$$0.12T = 0.3$$
$$\frac{0.12T}{0.12} = \frac{0.3}{0.12}$$
$$T = 2.5$$

There are 2.5 liters of the solution.

31. Let x = gallons of $1.15 fuel.

$\boxed{\begin{array}{c}\text{Value of}\\\text{\$1.15 fuel}\end{array}} + \boxed{\begin{array}{c}\text{Value of}\\\text{\$1.15 fuel}\end{array}} = \boxed{\begin{array}{c}\text{Value of}\\\text{mixture}\end{array}}$

$$1.15x + 0.85(20) = 1(20 + x)$$
$$115x + 85(20) = 100(20 + x)$$
$$115x + 1700 = 2000 + 100x$$
$$115x - 100x + 1700 = 2000 + 100x - 100x$$
$$15x + 1700 = 2000$$
$$15x + 1700 - 1700 = 2000 - 1700$$
$$15x = 300$$
$$\frac{15x}{15} = \frac{300}{15}$$
$$x = 20 \Rightarrow 20 \text{ gallons of the \$1.15 fuel should be used.}$$

33. Let x = gallons of 3% solution used.

Salt in 3% solution		Salt in 7% solution		Salt in mixture
	+		=	

$$0.03x + 0.07(50) = 0.05(x + 50)$$
$$3x + 7(50) = 5(x + 50)$$
$$3x + 350 = 5x + 250$$
$$3x - 5x + 350 = 5x - 5x + 250$$
$$-2x + 350 = 250$$
$$-2x + 350 - 350 = 250 - 350$$
$$-2x = -100$$
$$\frac{-2x}{-2} = \frac{-100}{-2}$$
$$x = 50$$

50 gallons of the 3% mixture should be used.

35. Let x = ounces of water (0%) added.

Amt. in 10% sol.		Amt. in water		Amt. in 8% sol.
	+		=	

$$0.10(30) + 0(x) = 0.08(30 + x)$$
$$10(30) + 0 = 8(30 + x)$$
$$300 = 240 + 8x$$
$$300 - 240 = 240 - 240 + 8x$$
$$60 = 8x$$
$$\frac{60}{8} = \frac{8x}{8}$$
$$7.5 = x$$

7.5 ounces of water should be added.

37. Let x = pounds of lemondrops. Then $100 - x$ = pounds of jellybeans.

Value of lemondrops		Value of jellybeans		Value of mixture
	+		=	

$$1.90x + 1.20(100 - x) = 1.48(100)$$
$$190x + 120(100 - x) = 148(100)$$
$$190x + 12000 - 120x = 14800$$
$$70x + 12000 = 14800$$
$$70x + 12000 - 12000 = 14800 - 12000$$
$$70x = 2800$$
$$\frac{70x}{70} = \frac{2800}{70}$$
$$x = 40 \Rightarrow 40 \text{ lb of lemondrops and } 60 \text{ lb of jellybeans should be used.}$$

39. Let c = cost of cashews. Then $c - 0.30$ = cost of peanuts. 20 pounds of each are used.

Value of cashews		Value of peanuts		Value of mixture
	+		=	

$$20c + 20(c - 0.30) = 1.05(40)$$
$$20c + 20c - 6 = 42$$
$$40c - 6 = 42$$
$$40c - 6 + 6 = 42 + 6$$
$$40c = 48$$
$$\frac{40c}{40} = \frac{48}{40}$$
$$c = 1.20$$

A bag of cashews is worth $1.20.

41. Let x = pounds of regular coffee used.

Value of regular		Value of gourmet		Value of mixture
	+		=	

$$4(x) + 7(40) = 5(x + 40)$$
$$4x + 280 = 5x + 200$$
$$4x - 5x + 280 = 5x - 5x + 200$$
$$-x + 280 = 200$$
$$-x + 280 - 280 = 200 - 280$$
$$-x = -80$$
$$x = 80$$

80 pounds of regular coffee should be used.

43. Let c = cost of hazelnut beans per pound.

$$\boxed{\begin{array}{c}\text{Value of}\\\text{chocolate}\end{array}} + \boxed{\begin{array}{c}\text{Value of}\\\text{hazelnut}\end{array}} = \boxed{\begin{array}{c}\text{Value of}\\\text{mixture}\end{array}}$$

$$7(2) + c(5) = 6(7)$$
$$14 + 5c = 42$$
$$14 - 14 + 5c = 42 - 14$$
$$5c = 28$$
$$c = \tfrac{28}{5} = \$5.60 \Rightarrow \text{The hazelnut beans cost } \$5.60 \text{ per pound.}$$

45. Answers may vary.   **47.** Answers may vary.

49. Let x and $x + 2$ represent the integers.

$$x + x + 2 = 16$$
$$2x + 2 = 16$$
$$2x = 14$$
$$x = 7$$

This says that the integers are 7 and 9, but these are not **even** integers. The equation has a solution, but not the problem itself.

51. You cannot mix a 10% solution and a 20% solution and end up with a solution that has a greater concentration (30%) than either of the original solutions.

Exercise 2.6 (page 128)

1. $2x - 5y + 3x = 2x + 3x - 5y = (2 + 3)x - 5y = 5x - 5y$

3. $\dfrac{3}{5}(x + 5) - \dfrac{8}{5}(10 + x) = \dfrac{3}{5}x + \dfrac{3}{5} \cdot 5 - \dfrac{8}{5} \cdot 10 - \dfrac{8}{5}x = \dfrac{3}{5}x + 3 - 16 - \dfrac{8}{5}x = \left(\dfrac{3}{5} - \dfrac{8}{5}\right)x - 13$

$$= -\dfrac{5}{5}x - 13$$
$$= -x - 13$$

5. literal   **7.** isolate   **9.** subtract

11. $E = IR$

$$\dfrac{E}{R} = \dfrac{IR}{R}$$
$$\dfrac{E}{R} = I, \text{ or } I = \dfrac{E}{R}$$

13. $V = lwh$

$$\dfrac{V}{lh} = \dfrac{lwh}{lh}$$
$$\dfrac{V}{lh} = w, \text{ or } w = \dfrac{V}{lh}$$

15.
$$P = a + b + c$$
$$P - a = a - a + b + c$$
$$P - a = b + c$$
$$P - a - c = b + c - c$$
$$P - a - c = b, \text{ or } b = P - a - c$$

17.
$$P = 2l + 2w$$
$$P - 2l = 2l - 2l + 2w$$
$$P - 2l = 2w$$
$$\dfrac{P - 2l}{2} = \dfrac{2w}{2}$$
$$\dfrac{P - 2l}{2} = w, \text{ or } w = \dfrac{P - 2l}{2}$$

19.
$$A = P + Prt$$
$$A - P = P - P + Prt$$
$$A - P = Prt$$
$$\frac{A - P}{Pr} = \frac{Prt}{Pr}$$
$$\frac{A - P}{Pr} = t, \text{ or } t = \frac{A - P}{Pr}$$

21.
$$C = 2\pi r$$
$$\frac{C}{2\pi} = \frac{2\pi r}{2\pi}$$
$$\frac{C}{2\pi} = r, \text{ or } r = \frac{C}{2\pi}$$

23.
$$K = \frac{wv^2}{2g}$$
$$2g \cdot K = 2g \cdot \frac{wv^2}{2g}$$
$$2gK = wv^2$$
$$\frac{2gK}{v^2} = \frac{wv^2}{v^2}$$
$$\frac{2gK}{v^2} = w, \text{ or } w = \frac{2gK}{v^2}$$

25.
$$P = I^2 R$$
$$\frac{P}{I^2} = \frac{I^2 R}{I^2}$$
$$\frac{P}{I^2} = R, \text{ or } R = \frac{P}{I^2}$$

27.
$$K = \frac{wv^2}{2g}$$
$$2g \cdot K = 2g \cdot \frac{wv^2}{2g}$$
$$2gK = wv^2$$
$$\frac{2gK}{2K} = \frac{wv^2}{2K}$$
$$g = \frac{wv^2}{2K}$$

29.
$$F = \frac{GMm}{d^2}$$
$$d^2 \cdot F = d^2 \cdot \frac{GMm}{d^2}$$
$$d^2 F = GMm$$
$$\frac{d^2 F}{Gm} = \frac{GMm}{Gm}$$
$$\frac{d^2 F}{Gm} = M, \text{ or } M = \frac{d^2 F}{Gm}$$

31.
$$F = \frac{GMm}{d^2}$$
$$d^2 \cdot F = d^2 \cdot \frac{GMm}{d^2}$$
$$d^2 F = GMm$$
$$\frac{d^2 F}{F} = \frac{GMm}{F}$$
$$d^2 = \frac{GMm}{F}$$

33.
$$G = 2(r - 1)b$$
$$\frac{G}{2b} = \frac{2(r - 1)b}{2b}$$
$$\frac{G}{2b} = r - 1$$
$$\frac{G}{2b} + 1 = r - 1 + 1$$
$$\frac{G}{2b} + 1 = r, \text{ or } r = \frac{G}{2b} + 1$$

35.
$$d = rt \qquad t = \frac{d}{r}$$
$$\frac{d}{r} = \frac{rt}{r} \qquad t = \frac{135}{45}$$
$$\frac{d}{r} = t \qquad t = 3$$

37.
$$i = prt \qquad t = \frac{i}{pr}$$
$$\frac{i}{pr} = \frac{prt}{pr} \qquad t = \frac{12}{100(0.06)}$$
$$\frac{i}{pr} = t \qquad t = \frac{12}{6} = 2$$

39.
$$P = a + b + c \qquad c = P - a - b$$
$$P - a = a - a + b + c \qquad c = 37 - 15 - 19$$
$$P - a = b + c \qquad c = 22 - 19$$
$$P - a - b = b - b + c \qquad c = 3$$
$$P - a - b = c$$

41.
$$K = \frac{1}{2}h(a + b) \qquad h = \frac{2K}{a + b}$$
$$2 \cdot K = 2 \cdot \frac{1}{2}h(a + b) \qquad h = \frac{2(48)}{7 + 5}$$
$$2K = h(a + b)$$
$$\frac{2K}{a + b} = \frac{h(a + b)}{a + b} \qquad h = \frac{96}{12} = 8$$
$$\frac{2K}{a + b} = h$$

43.
$$E = IR \qquad I = \frac{E}{R}$$
$$\frac{E}{R} = \frac{IR}{R} \qquad I = \frac{48}{12}$$
$$\frac{E}{R} = I \qquad I = 4 \text{ amperes}$$

45.
$$C = 2\pi r \qquad r = \frac{C}{2\pi}$$
$$\frac{C}{2\pi} = \frac{2\pi r}{2\pi} \qquad r = \frac{14.32}{2\pi}$$
$$\frac{C}{2\pi} = r \qquad r \approx 2.28 \text{ feet}$$

47.
$$P = I^2 R \qquad R = \frac{P}{I^2}$$
$$\frac{P}{I^2} = \frac{I^2 R}{I^2} \qquad R = \frac{2700}{14^2}$$
$$\frac{P}{I^2} = R \qquad R = \frac{2700}{196} = 13.78 \text{ ohms}$$

49.
$$F = \frac{GMm}{d^2}$$
$$d^2 \cdot F = d^2 \cdot \frac{GMm}{d^2}$$
$$d^2 F = GMm$$
$$\frac{Fd^2}{GM} = \frac{GMm}{GM}$$
$$\frac{Fd^2}{GM} = m, \text{ or } m = \frac{Fd^2}{GM}$$

51.
$$L = 2D + 3.25(r + R)$$
$$L - 3.25(r + R) = 2D + 3.25(r + R) - 3.25(r + R)$$
$$L - 3.25(r + R) = 2D$$
$$\frac{L - 3.25(r + R)}{2} = \frac{2D}{2}$$
$$\frac{L - 3.25r - 3.25R}{2} = D$$

$$D = \frac{L - 3.25r - 3.25R}{2}$$
$$D = \frac{25 - 3.25(1) - 3.25(3)}{2}$$
$$D = \frac{25 - 3.25 - 9.75}{2} = \frac{12}{2} = 6 \text{ ft.}$$

53.
$$C = 0.15(T - C)$$
$$C = 0.15T - 0.15C$$
$$C + 0.15C = 0.15T - 0.15C + 0.15C$$
$$1.15C = 0.15T$$
$$\frac{1.15C}{1.15} = \frac{0.15T}{1.15}$$
$$C \approx 0.1304T$$

The maximum contribution is about 13% of taxable income.

55. **Answers may vary.**

57. $E = mc^2$
$$E = 1(300{,}000)^2$$
$$E = 90{,}000{,}000{,}000 \text{ joules}$$

Exercise 2.7 (page 136)

1. $3x^2 - 2(y^2 - x^2) = 3x^2 + (-2)y^2 - (-2)x^2 = 3x^2 - 2y^2 + 2x^2 = 5x^2 - 2y^2$

3. $\dfrac{1}{3}(x + 6) - \dfrac{4}{3}(x - 9) = \dfrac{1}{3}x + \dfrac{1}{3} \cdot 6 - \dfrac{4}{3}x - \dfrac{4}{3}(-9) = -\dfrac{3}{3}x + 2 + 12 = -x + 14$

5. is less than

7. $\geq$

9. inequality

11.
$$x + 2 > 5$$
$$x + 2 - 2 > 5 - 2$$
$$x > 3$$

3

13.
$$-x - 3 \leq 7$$
$$-x - 3 + 3 \leq 7 + 3$$
$$-x \leq 10$$
$$-1(-x) \geq -1(10)$$
$$x \geq -10$$

-10

15.
$$3 + x < 2$$
$$3 - 3 + x < 2 - 3$$
$$x < -1$$

-1

17.
$$2x - 3 \leq 5$$
$$2x - 3 + 3 \leq 5 + 3$$
$$2x \leq 8$$
$$\frac{2x}{2} \leq \frac{8}{2}$$
$$x \leq 4$$

4

19.
$$-3x - 7 > -1$$
$$-3x - 7 + 7 > -1 + 7$$
$$-3x > 6$$
$$\frac{-3x}{-3} < \frac{6}{-3}$$
$$x < -2$$

-2

21.
$$-4x + 1 > 17$$
$$-4x + 1 - 1 > 17 - 1$$
$$-4x > 16$$
$$\frac{-4x}{-4} < \frac{16}{-4}$$
$$x < -4$$

-4

23.
$$2x + 9 \le x + 8$$
$$2x - x + 9 \le x - x + 8$$
$$x + 9 \le 8$$
$$x + 9 - 9 \le 8 - 9$$
$$x \le -1$$

-1

25.
$$9x + 13 \ge 8x$$
$$9x - 8x + 13 \ge 8x - 8x$$
$$x + 13 \ge 0$$
$$x + 13 - 13 \ge 0 - 13$$
$$x \ge -13$$

-13

27.
$$8x + 4 > 6x - 2$$
$$8x - 6x + 4 > 6x - 6x - 2$$
$$2x + 4 > -2$$
$$2x + 4 - 4 > -2 - 4$$
$$2x > -6$$
$$\frac{2x}{2} > \frac{-6}{2}$$
$$x > -3$$

-3

29.
$$5x + 7 < 2x + 1$$
$$5x - 2x + 7 < 2x - 2x + 1$$
$$3x + 7 < 1$$
$$3x + 7 - 7 < 1 - 7$$
$$3x < -6$$
$$\frac{3x}{3} < \frac{-6}{3}$$
$$x < -2$$

-2

31.
$$7 - x \le 3x - 1$$
$$7 - x - 3x \le 3x - 3x - 1$$
$$7 - 4x \le -1$$
$$7 - 7 - 4x \le -1 - 7$$
$$-4x \le -8$$
$$\frac{-4x}{-4} \ge \frac{-8}{-4}$$
$$x \ge 2$$

2

33.
$$9 - 2x > 24 - 7x$$
$$9 - 2x + 7x > 24 - 7x + 7x$$
$$9 + 5x > 24$$
$$9 - 9 + 5x > 24 - 9$$
$$5x > 15$$
$$\frac{5x}{5} > \frac{15}{5}$$
$$x > 3$$

3

35.
$$3(x - 8) < 5x + 6$$
$$3x - 24 < 5x + 6$$
$$3x - 5x - 24 < 5x - 5x + 6$$
$$-2x - 24 < 6$$
$$-2x - 24 + 24 < 6 + 24$$
$$-2x < 30$$
$$\frac{-2x}{-2} > \frac{30}{-2}$$
$$x > -15$$

-15

37.
$$8(5 - x) \le 10(8 - x)$$
$$40 - 8x \le 80 - 10x$$
$$40 - 8x + 10x \le 80 - 10x + 10x$$
$$40 + 2x \le 80$$
$$40 - 40 + 2x \le 80 - 40$$
$$2x \le 40$$
$$\frac{2x}{2} \le \frac{40}{2}$$
$$x \le 20$$

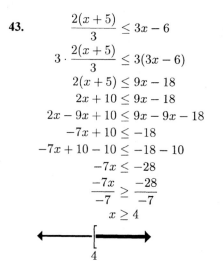

39.
$$\frac{5}{2}(7x - 15) + x \ge \frac{13}{2}x - \frac{3}{2}$$
$$2\left[\frac{5}{2}(7x - 15) + x\right] \ge 2\left[\frac{13}{2}x - \frac{3}{2}\right]$$
$$2 \cdot \frac{5}{2}(7x - 15) + 2x \ge 2 \cdot \frac{13}{2}x - 2 \cdot \frac{3}{2}$$
$$5(7x - 15) + 2x \ge 13x - 3$$
$$35x - 75 + 2x \ge 13x - 3$$
$$37x - 75 \ge 13x - 3$$
$$37x - 13x - 75 \ge 13x - 13x - 3$$
$$24x - 75 \ge -3$$
$$24x - 75 + 75 \ge -3 + 75$$
$$24x \ge 72$$
$$\frac{24x}{24} \ge \frac{72}{24}$$
$$x \ge 3$$

41.
$$\frac{3x - 3}{2} < 2x + 2$$
$$2 \cdot \frac{3x - 3}{2} < 2(2x + 2)$$
$$3x - 3 < 4x + 4$$
$$3x - 4x - 3 < 4x - 4x + 4$$
$$-x - 3 < 4$$
$$-x - 3 + 3 < 4 + 3$$
$$-x < 7$$
$$-1(-x) > -1(7)$$
$$x > -7$$

43.
$$\frac{2(x + 5)}{3} \le 3x - 6$$
$$3 \cdot \frac{2(x + 5)}{3} \le 3(3x - 6)$$
$$2(x + 5) \le 9x - 18$$
$$2x + 10 \le 9x - 18$$
$$2x - 9x + 10 \le 9x - 9x - 18$$
$$-7x + 10 \le -18$$
$$-7x + 10 - 10 \le -18 - 10$$
$$-7x \le -28$$
$$\frac{-7x}{-7} \ge \frac{-28}{-7}$$
$$x \ge 4$$

45.
$$2 < \quad x - 5 \quad < 5$$
$$2 + 5 < x - 5 + 5 < 5 + 5$$
$$7 < \qquad x \qquad < 10$$

47.
$$-5 < \quad x + 4 \quad \le 7$$
$$-5 - 4 < x + 4 - 4 \le 7 - 4$$
$$-9 < \qquad x \qquad \le 3$$

49.
$$0 \le \quad x + 10 \quad \le 10$$
$$0 - 10 \le x + 10 - 10 \le 10 - 10$$
$$-10 \le \quad x \quad \le 0$$

$$-10 \qquad 0$$

51.
$$4 < -2x < 10$$
$$\frac{4}{-2} > \frac{-2x}{2} > \frac{10}{-2}$$
$$-2 > \quad x \quad > -5$$
$$-5 < \quad x \quad < -2$$

$$-5 \qquad -2$$

53.
$$-3 \le \frac{x}{2} \le 5$$
$$2(-3) \le 2 \cdot \frac{x}{2} \le 2(5)$$
$$-6 \le \quad x \quad \le 10$$

$$-6 \qquad 10$$

55.
$$3 \le \quad 2x - 1 \quad < 5$$
$$3 + 1 \le 2x - 1 + 1 < 5 + 1$$
$$4 \le \quad 2x \quad < 6$$
$$\frac{4}{2} \le \quad \frac{2x}{2} \quad < \frac{6}{2}$$
$$2 \le \quad x \quad < 3$$

$$2 \qquad 3$$

57.
$$0 < \quad 10 - 5x \quad \le 15$$
$$0 - 10 < 10 - 10 - 5x \le 15 - 10$$
$$-10 < \quad -5x \quad \le 5$$
$$\frac{-10}{-5} > \quad \frac{-5x}{-5} \quad \ge \frac{5}{-5}$$
$$2 > \quad x \quad \ge -1$$
$$-1 \le \quad x \quad < 2$$

$$-1 \qquad 2$$

59.
$$-6 < \quad 3(x + 2) \quad < 9$$
$$-6 < \quad 3x + 6 \quad < 9$$
$$-6 - 6 < 3x + 6 - 6 < 9 - 6$$
$$-12 < \quad 3x \quad < 3$$
$$\frac{-12}{3} < \quad \frac{3x}{3} \quad < \frac{3}{3}$$
$$-4 < \quad x \quad < 1$$

$$-4 \qquad 1$$

61.
$$-4 < \quad \frac{x - 2}{2} \quad < 6$$
$$2(-4) < 2 \cdot \frac{x - 2}{2} < 2(6)$$
$$-8 < \quad x - 2 \quad < 12$$
$$-8 + 2 < x - 2 + 2 < 12 + 2$$
$$-6 < \quad x \quad < 14$$

$$-6 \qquad 14$$

63. Let $s =$ score on last exam.
$$\text{Average score} \ge 80$$
$$\frac{68 + 75 + 79 + s}{4} \ge 80$$
$$4 \cdot \frac{68 + 75 + 79 + s}{4} \ge 4(80)$$
$$68 + 75 + 79 + s \ge 320$$
$$222 + s \ge 320$$
$$222 - 222 + s \ge 320 - 222$$
$$s \ge 98$$

Her last test score must be at least 98%.

65. Let r = rating of third model. Average rating ≥ 21

$$\frac{17 + 19 + r}{3} \geq 21$$

$$3 \cdot \frac{17 + 19 + r}{3} \geq 3(21)$$

$$17 + 19 + r \geq 63$$

$$36 + r \geq 63$$

$$36 - 36 + r \geq 63 - 36$$

$$r \geq 27 \Rightarrow \text{It must have a rating of at least 27 mpg.}$$

67. Let s = length of each side.

$$0 < \text{Perimeter} \leq 57$$

$$0 < \quad 3s \quad \leq 57$$

$$\frac{0}{3} < \quad \frac{3s}{3} \quad \leq \frac{57}{3}$$

$$0 < \quad s \quad \leq 19$$

Each side can be from 0 feet to 19 feet long.

69. 470 ft $\leq$ range in feet $\leq$ 13,143 ft

$$\frac{470 \text{ ft}}{5280} \leq \text{range in miles} \leq \frac{13143 \text{ ft}}{5280}$$

0.1 miles $\leq$ range in miles $\leq$ 2.5 miles

The range is from 0.1 miles to 2.5 miles.

71. 17500 ft < range in feet < 21700 ft

$$\frac{17500 \text{ ft}}{5280} < \text{range in miles} < \frac{21700 \text{ ft}}{5280}$$

3.3 miles < range in miles < 4.1 miles

The range is between 3.3 miles and 4.1 miles.

73.

$$19° < \quad C \quad < 22°$$

$$19° < \quad \frac{5}{9}(F - 32) \quad < 22°$$

$$\frac{9}{5}(19°) < \frac{9}{5} \cdot \frac{5}{9}(F - 32) < \frac{9}{5}(22°)$$

$$34.2° < \quad F - 32 \quad < 39.6°$$

$$34.2° + 32 < \quad F - 32 + 32 \quad < 39.6° + 32$$

$$66.2° < \quad F \quad < 71.6°$$

The Fahrenheit temperature is between 66.2° and 71.6°.

75.

$$5.9 \text{ in.} < \quad r \quad < 6.1 \text{ in.}$$

$$2\pi(5.9 \text{ in.}) < 2\pi \cdot r < 2\pi(6.1 \text{ in})$$

$$2(3.14)(5.9 \text{ in.}) < \quad 2\pi r \quad < 2(3.14)(6.1 \text{ in.})$$

$$37.052 \text{ in.} < \quad C \quad < 38.308 \text{ in.}$$

The circumference can vary between about 37.052 inches and 38.308 inches.

77. 150 lb < range in lbs < 190 lb

$$\frac{150 \text{ lb}}{2.2} < \text{range in kg} < \frac{190 \text{ lb}}{2.2}$$

68.18 kg < range in kg < 86.36 kg

The weight is between 68.18 and 86.36 kg.

79. Let $w = $ width. Then $2w - 3 = $ length.

$$24 < \text{perimeter} < 48$$
$$24 < 2w + 2(2w - 3) < 48$$
$$24 < 2w + 4w - 6 < 48$$
$$24 < 6w - 6 < 48$$
$$24 + 6 < 6w - 6 + 6 < 48 + 6$$
$$30 < 6w < 54$$
$$\frac{30}{6} < \frac{6w}{6} < \frac{54}{6}$$
$$5 < w < 9$$

The width could be between 5 and 9 feet.

81. Answers may vary.

83. This is not correct because x could represent a negative number, and then multiplying both sides of the equation would require the inequality to change from $<$ to $>$.

Chapter 2 Summary (page 140)

1.
$$3x + 7 = 1$$
$$3(-2) + 7 \stackrel{2}{=} 1$$
$$-6 + 7 \stackrel{2}{=} 1$$
$$1 = 1$$
-2 is a solution.

2.
$$5 - 2x = 3$$
$$5 - 2(-1) \stackrel{2}{=} 3$$
$$5 + 2 \stackrel{2}{=} 3$$
$$7 \neq 3$$
-1 is not a solution.

3.
$$2(x + 3) = x$$
$$2(-3 + 3) \stackrel{2}{=} -3$$
$$2(0) \stackrel{2}{=} -3$$
$$0 \neq -3$$
-3 is not a solution.

4.
$$5(3 - x) = 2 - 4x$$
$$5(3 - 13) \stackrel{2}{=} 2 - 4(13)$$
$$5(-10) \stackrel{2}{=} 2 - 52$$
$$-50 = -50$$
13 is a solution.

5.
$$3(x + 5) = 2(x - 3)$$
$$3(-21 + 5) \stackrel{2}{=} 2(-21 - 3)$$
$$3(-16) \stackrel{2}{=} 2(-24)$$
$$-48 = -48$$
-21 is a solution.

6.
$$2(x - 7) = x + 14$$
$$2(0 - 7) \stackrel{2}{=} 0 + 14$$
$$2(-7) \stackrel{2}{=} 14$$
$$-14 \neq 14$$
0 is not a solution.

7.
$$x - 7 = -6 \qquad x - 7 = -6$$
$$x - 7 + 7 = -6 + 7 \qquad 1 - 7 \stackrel{2}{=} -6$$
$$x = 1 \qquad\qquad -6 = -6$$

8.
$$y - 4 = 5 \qquad y - 4 = 5$$
$$y - 4 + 4 = 5 + 4 \qquad 9 - 4 \stackrel{2}{=} 5$$
$$y = 9 \qquad\qquad 5 = 5$$

9.
$$p + 4 = 20 \qquad p + 4 = 20$$
$$p + 4 - 4 = 20 - 4 \qquad 16 + 4 \stackrel{2}{=} 20$$
$$p = 16 \qquad\qquad 20 = 20$$

10.
$$x + \frac{3}{5} = \frac{3}{5} \qquad\qquad x + \frac{3}{5} = \frac{3}{5}$$
$$x + \frac{3}{5} - \frac{3}{5} = \frac{3}{5} - \frac{3}{5} \qquad 0 + \frac{3}{5} \stackrel{2}{=} \frac{3}{5}$$
$$x = 0 \qquad\qquad\qquad \frac{3}{5} = \frac{3}{5}$$

11.
$$y - \frac{7}{2} = \frac{1}{2} \qquad\qquad y - \frac{7}{2} = \frac{1}{2}$$
$$y - \frac{7}{2} + \frac{7}{2} = \frac{1}{2} + \frac{7}{2} \qquad 4 - \frac{7}{2} \stackrel{?}{=} \frac{1}{2}$$
$$y = \frac{8}{2} = 4 \qquad\qquad \frac{8}{2} - \frac{7}{2} \stackrel{?}{=} \frac{1}{2}$$
$$\frac{1}{2} = \frac{1}{2}$$

12.
$$z + \frac{5}{3} = -\frac{1}{3} \qquad\qquad z + \frac{5}{3} = -\frac{1}{3}$$
$$z + \frac{5}{3} - \frac{5}{3} = -\frac{1}{3} - \frac{5}{3} \qquad -2 + \frac{5}{3} \stackrel{?}{=} -\frac{1}{3}$$
$$z = -\frac{6}{3} \qquad\qquad -\frac{6}{3} + \frac{5}{3} \stackrel{?}{=} -\frac{1}{3}$$
$$z = -2 \qquad\qquad -\frac{1}{3} = -\frac{1}{3}$$

13. Let $r =$ the regular price. Then

$$\boxed{\begin{array}{c}\text{Sale} \\ \text{price}\end{array}} = \boxed{\begin{array}{c}\text{Regular} \\ \text{price}\end{array}} - \boxed{\text{Markdown}}$$

$$69.95 = r - 35.45$$
$$69.95 + 35.45 = r - 35.45 + 35.45$$
$$105.40 = r$$

The regular price is \$105.40.

14. Let $w =$ the wholesale price. Then

$$\boxed{\begin{array}{c}\text{Retail} \\ \text{price}\end{array}} = \boxed{\begin{array}{c}\text{Wholesale} \\ \text{price}\end{array}} + \boxed{\text{Markup}}$$

$$212.95 = w + 115.25$$
$$212.95 - 115.25 = w + 115.25 - 115.25$$
$$97.70 = w$$

The wholesale price is \$97.70.

15.
$$3x = 15 \qquad\qquad 3x = 15$$
$$\frac{3x}{3} = \frac{15}{3} \qquad 3(5) \stackrel{?}{=} 15$$
$$x = 5 \qquad\qquad 15 = 15$$

16.
$$8r = -16 \qquad\qquad 8r = -16$$
$$\frac{8r}{8} = \frac{-16}{8} \qquad 8(-2) \stackrel{?}{=} -16$$
$$r = -2 \qquad\qquad -16 = -16$$

17.
$$10z = 5 \qquad\qquad 10z = 5$$
$$\frac{10z}{10} = \frac{5}{10} \qquad 10\left(\frac{1}{2}\right) \stackrel{?}{=} 5$$
$$z = \frac{1}{2} \qquad\qquad 5 = 5$$

18.
$$14s = 21 \qquad\qquad 14s = 21$$
$$\frac{14s}{14} = \frac{21}{14} \qquad 14\left(\frac{3}{2}\right) \stackrel{?}{=} 21$$
$$s = \frac{3}{2} \qquad\qquad 21 = 21$$

19.
$$\frac{y}{3} = 6 \qquad\qquad \frac{y}{3} = 6$$
$$3 \cdot \frac{y}{3} = 3 \cdot 6 \qquad \frac{18}{3} \stackrel{?}{=} 6$$
$$y = 18 \qquad\qquad 6 = 6$$

20.
$$\frac{w}{7} = -5 \qquad\qquad \frac{w}{7} = -5$$
$$7 \cdot \frac{w}{7} = 7(-5) \qquad \frac{-35}{7} \stackrel{?}{=} -5$$
$$w = -35 \qquad\qquad -5 = -5$$

21.
$$\frac{a}{-7} = \frac{1}{14} \qquad\qquad \frac{a}{-7} = \frac{1}{14}$$
$$-7 \cdot \frac{a}{-7} = -7\left(\frac{1}{14}\right) \qquad \frac{-\frac{1}{2}}{-7} \stackrel{?}{=} \frac{1}{14}$$
$$a = -\frac{7}{14} \qquad\qquad \frac{2\left(-\frac{1}{2}\right)}{2(-7)} \stackrel{?}{=} \frac{1}{14}$$
$$a = -\frac{1}{2} \qquad\qquad \frac{-1}{-14} = \frac{1}{14}$$

22.
$$\frac{t}{12} = \frac{1}{2} \qquad\qquad \frac{t}{12} = \frac{1}{2}$$
$$12 \cdot \frac{t}{12} = 12\left(\frac{1}{2}\right) \qquad \frac{6}{12} \stackrel{?}{=} \frac{1}{2}$$
$$t = \frac{12}{2} \qquad\qquad \frac{1}{2} = \frac{1}{2}$$
$$t = 6$$

23.
$$rb = a$$
$$35\% \cdot 700 = a$$
$$0.35(700) = a$$
$$245 = a$$

24.
$$rb = a$$
$$72\% \cdot b = 936$$
$$0.72b = 936$$
$$\frac{0.72b}{0.72} = \frac{936}{0.72}$$
$$b = 1{,}300$$

25.
$$rb = a$$
$$r \cdot 2300 = 851$$
$$\frac{r \cdot 2300}{2300} = \frac{851}{2300}$$
$$r = 0.37$$
$$r = 37\%$$

26.
$$rb = a$$
$$r \cdot 576 = 72$$
$$\frac{r \cdot 576}{576} = \frac{72}{576}$$
$$r = 0.125$$
$$r = 12.5\%$$

27.
$$5y + 6 = 21$$
$$5y + 6 - 6 = 21 - 6$$
$$5y = 15$$
$$\frac{5y}{5} = \frac{15}{5}$$
$$y = 3$$

$$5y + 6 = 21$$
$$5(3) + 6 \overset{?}{=} 21$$
$$15 + 6 \overset{?}{=} 21$$
$$21 = 21$$

28.
$$5y - 9 = 1$$
$$5y - 9 + 9 = 1 + 9$$
$$5y = 10$$
$$\frac{5y}{5} = \frac{10}{5}$$
$$y = 2$$

$$5y - 9 = 1$$
$$5(2) - 9 \overset{?}{=} 1$$
$$10 - 9 \overset{?}{=} 1$$
$$1 = 1$$

29.
$$-12z + 4 = -8$$
$$-12z + 4 - 4 = -8 - 4$$
$$-12z = -12$$
$$\frac{-12z}{-12} = \frac{-12}{-12}$$
$$z = 1$$

$$-12z + 4 = -8$$
$$-12(1) + 4 \overset{?}{=} -8$$
$$-12 + 4 \overset{?}{=} -8$$
$$-8 = -8$$

30.
$$17z + 3 = 20$$
$$17z + 3 - 3 = 20 - 3$$
$$17z = 17$$
$$\frac{17z}{17} = \frac{17}{17}$$
$$z = 1$$

$$17z + 3 = 20$$
$$17(1) + 3 \overset{?}{=} 20$$
$$17 + 3 \overset{?}{=} 20$$
$$20 = 20$$

31.
$$13 - 13t = 0$$
$$13 - 13 - 13t = 0 - 13$$
$$-13t = -13$$
$$\frac{-13t}{-13} = \frac{-13}{-13}$$
$$t = 1$$

$$13 - 13t = 0$$
$$13 - 13(1) \overset{?}{=} 0$$
$$13 - 13 \overset{?}{=} 0$$
$$0 = 0$$

32.
$$10 + 7t = -4$$
$$10 - 10 + 7t = -4 - 10$$
$$7t = -14$$
$$\frac{7t}{7} = \frac{-14}{7}$$
$$t = -2$$

$$10 + 7t = -4$$
$$10 + 7(-2) \overset{?}{=} -4$$
$$10 + (-14) \overset{?}{=} -4$$
$$-4 = -4$$

33.
$$23a - 43 = 3$$
$$23a - 43 + 43 = 3 + 43$$
$$23a = 46$$
$$\frac{23a}{23} = \frac{46}{23}$$
$$a = 2$$

$$23a - 43 = 3$$
$$23(2) - 43 \overset{?}{=} 3$$
$$46 - 43 \overset{?}{=} 3$$
$$3 = 3$$

34.

$$84 - 21a = -63$$
$$84 - 84 - 21a = -63 - 84$$
$$-21a = -147$$
$$\frac{-21a}{-21} = \frac{-147}{-21}$$
$$a = 7$$

$$84 - 21a = -63$$
$$84 - 21(7) \stackrel{?}{=} -63$$
$$84 - 147 \stackrel{?}{=} -63$$
$$-63 = -63$$

35.

$$3x + 7 = 1$$
$$3x + 7 - 7 = 1 - 7$$
$$3x = -6$$
$$\frac{3x}{3} = \frac{-6}{3}$$
$$x = -2$$

$$3x + 7 = 1$$
$$3(-2) + 7 \stackrel{?}{=} 1$$
$$-6 + 7 \stackrel{?}{=} 1$$
$$1 = 1$$

36.

$$7 - 9x = 16$$
$$7 - 7 - 9x = 16 - 7$$
$$-9x = 9$$
$$\frac{-9x}{-9} = \frac{9}{-9}$$
$$x = -1$$

$$7 - 9x = 16$$
$$7 - 9(-1) \stackrel{?}{=} 16$$
$$7 + 9 \stackrel{?}{=} 16$$
$$16 = 16$$

37.

$$\frac{b + 3}{4} = 2$$
$$4 \cdot \frac{b + 3}{4} = 4 \cdot 2$$
$$b + 3 = 8$$
$$b + 3 - 3 = 8 - 3$$
$$b = 5$$

$$\frac{b + 3}{4} = 2$$
$$\frac{5 + 3}{4} \stackrel{?}{=} 2$$
$$\frac{8}{4} \stackrel{?}{=} 2$$
$$2 = 2$$

38.

$$\frac{b - 7}{2} = -2$$
$$2 \cdot \frac{b - 7}{2} = 2(-2)$$
$$b - 7 = -4$$
$$b - 7 + 7 = -4 + 7$$
$$b = 3$$

$$\frac{b - 7}{2} = -2$$
$$\frac{3 - 7}{2} \stackrel{?}{=} -2$$
$$\frac{-4}{2} \stackrel{?}{=} -2$$
$$-2 = -2$$

39.

$$\frac{x - 8}{5} = 1$$
$$5 \cdot \frac{x - 8}{5} = 5(1)$$
$$x - 8 = 5$$
$$x - 8 + 8 = 5 + 8$$
$$x = 13$$

$$\frac{x - 8}{5} = 1$$
$$\frac{13 - 8}{5} \stackrel{?}{=} 1$$
$$\frac{5}{5} \stackrel{?}{=} 1$$
$$1 = 1$$

40.

$$\frac{x + 10}{2} = -1$$
$$2 \cdot \frac{x + 10}{2} = 2(-1)$$
$$x + 10 = -2$$
$$x + 10 - 10 = -2 - 10$$
$$x = -12$$

$$\frac{x + 10}{2} = -1$$
$$\frac{-12 + 10}{2} \stackrel{?}{=} -1$$
$$\frac{-2}{2} \stackrel{?}{=} -1$$
$$-1 = -1$$

41.

$$\frac{2y - 2}{4} = 2$$
$$4 \cdot \frac{2y - 2}{4} = 4(2)$$
$$2y - 2 = 8$$
$$2y - 2 + 2 = 8 + 2$$
$$2y = 10$$
$$\frac{2y}{2} = \frac{10}{2}$$
$$y = 5$$

$$\frac{2y - 2}{4} = 2$$
$$\frac{2(5) - 2}{4} \stackrel{?}{=} 2$$
$$\frac{10 - 2}{4} \stackrel{?}{=} 2$$
$$\frac{8}{4} \stackrel{?}{=} 2$$
$$2 = 2$$

42.

$$\frac{3y + 12}{11} = 3$$
$$11 \cdot \frac{3y + 12}{11} = 11(3)$$
$$3y + 12 = 33$$
$$3y + 12 - 12 = 33 - 12$$
$$3y = 21$$
$$\frac{3y}{3} = \frac{21}{3}$$
$$y = 7$$

$$\frac{3y + 12}{11} = 3$$
$$\frac{3(7) + 12}{11} \stackrel{?}{=} 3$$
$$\frac{21 + 12}{11} \stackrel{?}{=} 3$$
$$\frac{33}{11} \stackrel{?}{=} 3$$
$$3 = 3$$

43.

$$\frac{x}{2} + 7 = 11 \qquad \frac{x}{2} + 7 = 11$$

$$\frac{x}{2} + 7 - 7 = 11 - 7 \qquad \frac{8}{2} + 7 \overset{?}{=} 11$$

$$\frac{x}{2} = 4 \qquad 4 + 7 \overset{?}{=} 11$$

$$2 \cdot \frac{x}{2} = 2 \cdot 4 \qquad 11 = 11$$

$$x = 8$$

44.

$$\frac{r}{3} - 3 = 7 \qquad \frac{r}{3} - 3 = 7$$

$$\frac{r}{3} - 3 + 3 = 7 + 3 \qquad \frac{30}{3} - 3 \overset{?}{=} 7$$

$$\frac{r}{3} = 10 \qquad 10 - 3 \overset{?}{=} 7$$

$$3 \cdot \frac{r}{3} = 3 \cdot 10 \qquad 7 = 7$$

$$r = 30$$

45.

$$\frac{a}{2} + \frac{9}{4} = 6 \qquad \frac{a}{2} + \frac{9}{4} = 6$$

$$\frac{a}{2} + \frac{9}{4} - \frac{9}{4} = \frac{24}{4} - \frac{9}{4} \qquad \frac{\frac{15}{2}}{2} + \frac{9}{4} \overset{?}{=} 6$$

$$\frac{a}{2} = \frac{15}{4} \qquad \frac{2 \cdot \frac{15}{2}}{2 \cdot 2} + \frac{9}{4} \overset{?}{=} 6$$

$$2 \cdot \frac{a}{2} = 2 \cdot \frac{15}{4} \qquad \frac{15}{4} + \frac{9}{4} \overset{?}{=} 6$$

$$a = \frac{15}{2} \qquad \frac{24}{4} \overset{?}{=} 6$$

$$6 = 6$$

46.

$$\frac{x}{8} - 2.3 = 3.2 \qquad \frac{x}{8} - 2.3 = 3.2$$

$$\frac{x}{8} - 2.3 + 2.3 = 3.2 + 2.3 \qquad \frac{44}{8} - 2.3 \overset{?}{=} 3.2$$

$$\frac{x}{8} = 5.5 \qquad 5.5 - 2.3 \overset{?}{=} 3.2$$

$$8 \cdot \frac{x}{8} = 8(5.5) \qquad 3.2 = 3.2$$

$$x = 44$$

47. Let $x =$ the regular price.

$$\boxed{\text{Sale price}} = \boxed{\text{Regular price}} - \boxed{\text{Markdown}}$$

$$240 = x - 0.25x$$

$$240 = 0.75x$$

$$\frac{240}{0.75} = \frac{0.75x}{0.75}$$

$$320 = x$$

The regular price is $320.

48. Let $r =$ the sales tax rate.

$$\boxed{\text{Total price}} = \boxed{\text{Price before tax}} + \boxed{\text{Sales tax}}$$

$$40.47 = 38 + 38r$$

$$40.47 - 38 = 38 - 38 + 38r$$

$$2.47 = 38r$$

$$\frac{2.47}{38} = \frac{38r}{38}$$

$$0.065 = r$$

The sales tax rate is 6.5%.

49. Let $r =$ the % increase.

$$\boxed{\text{New price}} = \boxed{\text{Original price}} + \boxed{\text{Increase}}$$

$$1100 = 560 + 560r$$

$$1100 - 560 = 560 - 560 + 560r$$

$$540 = 560r$$

$$\frac{540}{560} = \frac{560r}{560}$$

$$0.964 = r$$

The percent increase is 96.4%.

50. Let $r =$ the % discount.

$$\boxed{\text{Markdown}} = \boxed{\text{Percent markdown}} \cdot \boxed{\text{Regular price}}$$

$$465 - 215 = r \cdot 465$$

$$250 = 465r$$

$$\frac{250}{465} = \frac{465r}{465}$$

$$0.538 = r$$

The percent markdown is 53.8%.

51. $5x + 9x = (5 + 9)x = 14x$

52. $7a + 12a = (7 + 12)a = 19a$

53. $18b - 13b = (18 - 13)b = 5b$

54. $21x - 23x = (21 - 23)x = -2x$

55. $5y - 7y = (5 - 7)y = -2y$

56. $19x - 19 \Rightarrow$ not like terms

57. $\begin{aligned}7(x + 2) + 2(x - 7) &= 7x + 14 + 2x - 14 \\ &= 7x + 2x + 14 - 14 \\ &= (7 + 2)x + 0 = 9x\end{aligned}$

58. $\begin{aligned}2(3 - x) + x - 6x &= 6 - 2x + x - 6x \\ &= 6 + (-2 + 1 - 6)x \\ &= 6 - 7x\end{aligned}$

59. $\begin{aligned}y^2 + 3(y^2 - 2) &= y^2 + 3y^2 - 6 \\ &= (1 + 3)y^2 - 6 \\ &= 4y^2 - 6\end{aligned}$

60. $\begin{aligned}2x^2 - 2(x^2 - 2) &= 2x^2 - 2x^2 + 4 \\ &= (2 - 2)x^2 + 4 \\ &= 0x^2 + 4 = 4\end{aligned}$

61.
$$\begin{aligned}2x - 19 &= 2 - x \\ 2x + x - 19 &= 2 - x + x \\ 3x - 19 &= 2 \\ 3x - 19 + 19 &= 2 + 19 \\ 3x &= 21 \\ \frac{3x}{3} &= \frac{21}{3} \\ x &= 7\end{aligned}$$

$$\begin{aligned}2x - 19 &= 2 - x \\ 2(7) - 19 &\overset{?}{=} 2 - 7 \\ 14 - 19 &\overset{?}{=} -5 \\ -5 &= -5\end{aligned}$$

62.
$$\begin{aligned}5b - 19 &= 2b + 20 \\ 5b - 2b - 19 &= 2b - 2b + 20 \\ 3b - 19 &= 20 \\ 3b - 19 + 19 &= 20 + 19 \\ 3b &= 39 \\ \frac{3b}{3} &= \frac{39}{3} \\ b &= 13\end{aligned}$$

$$\begin{aligned}5b - 19 &= 2b + 20 \\ 5(13) - 19 &\overset{?}{=} 2(13) + 20 \\ 65 - 19 &\overset{?}{=} 26 + 20 \\ 46 &= 46\end{aligned}$$

63.
$$\begin{aligned}3x + 20 &= 5 - 2x \\ 3x + 2x + 20 &= 5 - 2x + 2x \\ 5x + 20 &= 5 \\ 5x + 20 - 20 &= 5 - 20 \\ 5x &= -15 \\ \frac{5x}{5} &= \frac{-15}{5} \\ x &= -3\end{aligned}$$

$$\begin{aligned}3x + 20 &= 5 - 2x \\ 3(-3) + 20 &\overset{?}{=} 5 - 2(-3) \\ -9 + 20 &\overset{?}{=} 5 + 6 \\ 11 &= 11\end{aligned}$$

64.

$$0.9x + 10 = 0.7x + 1.8$$
$$0.9x - 0.7x + 10 = 0.7x - 0.7x + 1.8$$
$$0.2x + 10 = 1.8$$
$$0.2x + 10 - 10 = 1.8 - 10$$
$$0.2x = -8.2$$
$$\frac{0.2x}{0.2} = \frac{-8.2}{0.2}$$
$$x = -41$$

$$0.9x + 10 = 0.7x + 1.8$$
$$0.9(-41) + 10 \stackrel{?}{=} 0.7(-41) + 1.8$$
$$-36.9 + 10 \stackrel{?}{=} -28.7 + 1.8$$
$$-26.9 = -26.9$$

65.

$$10(t - 3) = 3(t + 11)$$
$$10t - 30 = 3t + 33$$
$$10t - 3t - 30 = 3t - 3t + 33$$
$$7t - 30 = 33$$
$$7t - 30 + 30 = 33 + 30$$
$$7t = 63$$
$$\frac{7t}{7} = \frac{63}{7}$$
$$t = 9$$

$$10(t - 3) = 3(t + 11)$$
$$10(9 - 3) \stackrel{?}{=} 3(9 + 11)$$
$$10(6) \stackrel{?}{=} 3(20)$$
$$60 = 60$$

66.

$$2(5x - 7) = 2(x - 35)$$
$$10x - 14 = 2x - 70$$
$$10x - 2x - 14 = 2x - 2x - 70$$
$$8x - 14 = -70$$
$$8x - 14 + 14 = -70 + 14$$
$$8x = -56$$
$$\frac{8x}{8} = \frac{-56}{8}$$
$$x = -7$$

$$2(5x - 7) = 2(x - 35)$$
$$2[5(-7) - 7] \stackrel{?}{=} 2(-7 - 35)$$
$$2[-35 - 7] \stackrel{?}{=} 2(-42)$$
$$2(-42) \stackrel{?}{=} 2(-42)$$
$$-84 = -84$$

67.

$$\frac{3u - 6}{5} = 3$$
$$5 \cdot \frac{3u - 6}{5} = 5(3)$$
$$3u - 6 = 15$$
$$3u - 6 + 6 = 15 + 6$$
$$3u = 21$$
$$\frac{3u}{3} = \frac{21}{3}$$
$$u = 7$$

$$\frac{3u - 6}{5} = 3$$
$$\frac{3(7) - 6}{5} \stackrel{?}{=} 3$$
$$\frac{21 - 6}{5} \stackrel{?}{=} 3$$
$$\frac{15}{5} \stackrel{?}{=} 3$$
$$3 = 3$$

68.

$$\frac{5v - 35}{3} = -5$$

$$3 \cdot \frac{5v - 35}{3} = 3(-5)$$

$$5v - 35 = -15$$

$$5v - 35 + 35 = -15 + 35$$

$$5v = 20$$

$$\frac{5v}{5} = \frac{20}{5}$$

$$v = 4$$

$$\frac{5v - 35}{3} = -5$$

$$\frac{5(4) - 35}{3} \stackrel{?}{=} -5$$

$$\frac{20 - 35}{3} \stackrel{?}{=} -5$$

$$\frac{-15}{3} \stackrel{?}{=} -5$$

$$-5 = -5$$

69.

$$\frac{7x - 28}{4} = -21$$

$$4 \cdot \frac{7x - 28}{4} = 4(-21)$$

$$7x - 28 = -84$$

$$7x - 28 + 28 = -84 + 28$$

$$7x = -56$$

$$\frac{7x}{7} = \frac{-56}{7}$$

$$x = -8$$

$$\frac{7x - 28}{4} = -21$$

$$\frac{7(-8) - 28}{4} \stackrel{?}{=} -21$$

$$\frac{-56 - 28}{4} \stackrel{?}{=} -21$$

$$\frac{-84}{4} \stackrel{?}{=} -21$$

$$-21 = -21$$

70.

$$\frac{27 + 9y}{5} = -27$$

$$5 \cdot \frac{27 + 9y}{5} = 5(-27)$$

$$27 + 9y = -135$$

$$27 - 27 + 9y = -135 - 27$$

$$9y = -162$$

$$\frac{9y}{9} = \frac{-162}{9}$$

$$y = -18$$

$$\frac{27 + 9y}{5} = -27$$

$$\frac{27 + 9(-18)}{5} \stackrel{?}{=} -27$$

$$\frac{27 + (-162)}{5} \stackrel{?}{=} -27$$

$$\frac{-135}{5} \stackrel{?}{=} -27$$

$$-27 = -27$$

71.

$$2x - 5 = x - 5 + x$$

$$2x - 5 = 2x - 5$$

$$2x - 2x - 5 = 2x - 2x - 5$$

$$-5 = -5$$

Identity

72.

$$-3(a + 1) - a = -4a + 3$$

$$-3a - 3 - a = -4a + 3$$

$$-4a - 3 = -4a + 3$$

$$-4a + 4a - 3 = -4a + 4a + 3$$

$$-3 \neq 3$$

Contradiction

73. $2(x-1)+4 = 4(1+x)-(2x+2)$
$$2x-2+4 = 4+4x-2x-2$$
$$2x+2 = 2x+2$$
$$2x-2x+2 = 2x-2x+2$$
$$2 = 2$$
Identity

74. Let $x =$ the length of one part. Then $2x-7 =$ the length of the other part.

$$\boxed{\begin{array}{c}\text{Length of}\\\text{first part}\end{array}} + \boxed{\begin{array}{c}\text{Length of}\\\text{second part}\end{array}} = 8$$

$$x+2x-7 = 8$$
$$3x-7 = 8$$
$$3x-7+7 = 8+7$$
$$3x = 15$$
$$\frac{3x}{3} = \frac{15}{3}$$
$$x = 5$$

One part should be cut 5 feet long.

75.
$$x+47° = 62°$$
$$x+47°-47° = 62°-47°$$
$$x = 15°$$

76.
$$x+135° = 180°$$
$$x+135°-135° = 180°-135°$$
$$x = 45°$$

77. Let $x =$ the measure of the complement.
$$x+69° = 90°$$
$$x+69°-69° = 90°-69°$$
$$x = 21°$$

78. Let $x =$ the measure of the supplement.
$$x+69° = 180°$$
$$x+69°-69° = 180°-69°$$
$$x = 111°$$

79. Let $w =$ the width of the picture. Then $2w+3 =$ the length of the picture.

$$\boxed{\text{Perimeter}} = 84$$
$$2w+2(2w+3) = 84$$
$$2w+4w+6 = 84$$
$$6w+6 = 84$$
$$6w+6-6 = 84-6$$
$$6w = 78$$
$$\frac{6w}{6} = \frac{78}{6}$$
$$w = 13 \Rightarrow \text{The width is 13 inches.}$$

80. Let $x =$ amount in 7% CD. Then $27000-x =$ amount in 9% fund.

$$\boxed{\begin{array}{c}\text{Interest}\\\text{at 7\%}\end{array}} + \boxed{\begin{array}{c}\text{Interest}\\\text{at 9\%}\end{array}} = \boxed{\begin{array}{c}\text{Total}\\\text{interest}\end{array}}$$

$$0.07x+0.09(27000-x) = 2110$$
$$7x+9(27000-x) = 211000$$
$$7x+243000-9x = 211000$$
$$-2x+243000 = 211000$$
$$-2x+243000-243000 = 211000-243000$$
$$-2x = -32000$$
$$\frac{-2x}{-2} = \frac{-32000}{-2}$$
$$x = 16000 \Rightarrow \$16{,}000 \text{ is invested at 7\%, and } \$11{,}000 \text{ is invested at 9\%.}$$

81. Let t = time for friends to meet.

	r	t	d
Walk	3	t	$3t$
Bike	12	t	$12t$

$$\boxed{\text{Distance walked}} + \boxed{\text{Distance biked}} = 5$$

$$3t + 12t = 5$$
$$15t = 5$$
$$\frac{15t}{15} = \frac{5}{15}$$
$$t = \frac{1}{3}$$

They meet after $\frac{1}{3}$ hour (or 20 minutes).

82. Let x = liters of 1% butterfat milk used.

$$\boxed{\begin{array}{c}\text{Butterfat in}\\4\% \text{ milk}\end{array}} + \boxed{\begin{array}{c}\text{Butterfat in}\\1\% \text{ milk}\end{array}} = \boxed{\begin{array}{c}\text{Butterfat in}\\2\% \text{ mixture}\end{array}}$$

$$0.04(12) + 0.01x = 0.02(12 + x)$$
$$4(12) + 1x = 2(12 + x)$$
$$48 + x = 24 + 2x$$
$$x - 2x + 48 = 24 + 2x - 2x$$
$$-x + 48 = 24$$
$$-x + 48 - 48 = 24 - 48$$
$$-x = -24$$
$$x = 24$$

24 liters of the 1% milk should be used.

83. Let x = pounds of 90¢ candy.

Then $20 - x$ = pounds of $1.50 candy.

$$\boxed{\begin{array}{c}\text{Value of}\\90¢ \text{ candy}\end{array}} + \boxed{\begin{array}{c}\text{Value of}\\\$1.50 \text{ candy}\end{array}} = \boxed{\begin{array}{c}\text{Value of}\\\text{mixture}\end{array}}$$

$$0.90x + 1.50(20 - x) = 1.20(20)$$
$$90x + 150(20 - x) = 120(20)$$
$$90x + 3000 - 150x = 2400$$
$$-60x + 3000 = 2400$$
$$-60x + 3000 - 3000 = 2400 - 3000$$
$$-60x = -600$$
$$\frac{-60x}{-60} = \frac{-600}{-60}$$
$$x = 10$$

10 lb of each should be used.

84. Let x = # of kilowatt hours used.

$$17.50 + 0.18x = 43.96$$
$$1750 + 18x = 4396$$
$$1750 - 1750 + 18x = 4396 - 1750$$
$$18x = 2646$$
$$\frac{18x}{18} = \frac{2646}{18}$$
$$x = 147$$

147 kilowatt hours were used.

85. Let x = feet of gutter needed.

$$35 + 1.50x = 162.50$$
$$350 + 15x = 1625$$
$$350 - 350 + 15x = 1625 - 350$$
$$15x = 1275$$
$$\frac{15x}{15} = \frac{1275}{15}$$
$$x = 85 \Rightarrow 85 \text{ feet of gutter were needed.}$$

86. $E = IR$

$$\frac{E}{I} = \frac{IR}{I}$$
$$\frac{E}{I} = R, \text{ or } R = \frac{E}{I}$$

87. $i = prt$

$$\frac{i}{pr} = \frac{prt}{pr}$$
$$\frac{i}{pr} = t, \text{ or } t = \frac{i}{pr}$$

88. $P = I^2 R$

$$\frac{P}{I^2} = \frac{I^2 R}{I^2}$$
$$\frac{P}{I^2} = R, \text{ or } R = \frac{P}{I^2}$$

89.
$$d = rt$$
$$\frac{d}{t} = \frac{rt}{t}$$
$$\frac{d}{t} = r, \text{ or } r = \frac{d}{t}$$

90.
$$V = lwh$$
$$\frac{V}{lw} = \frac{lwh}{lw}$$
$$\frac{V}{lw} = h, \text{ or } h = \frac{V}{lw}$$

91.
$$y = mx + b$$
$$y - b = mx + b - b$$
$$y - b = mx$$
$$\frac{y - b}{x} = \frac{mx}{x}$$
$$\frac{y - b}{x} = m, \text{ or } m = \frac{y - b}{x}$$

92.
$$V = \pi r^2 h$$
$$\frac{V}{\pi r^2} = \frac{\pi r^2 h}{\pi r^2}$$
$$\frac{V}{\pi r^2} = h, \text{ or } h = \frac{V}{\pi r^2}$$

93.
$$a = 2\pi rh$$
$$\frac{a}{2\pi h} = \frac{2\pi rh}{2\pi h}$$
$$\frac{a}{2\pi h} = r, \text{ or } r = \frac{a}{2\pi h}$$

94.
$$F = \frac{GMm}{d^2}$$
$$d^2 \cdot F = d^2 \cdot \frac{GMm}{d^2}$$
$$d^2 F = GMm$$
$$\frac{d^2 F}{Mm} = \frac{GMm}{Mm}$$
$$\frac{d^2 F}{Mm} = G, \text{ or } G = \frac{d^2 F}{Mm}$$

95.
$$P = \frac{RT}{mV}$$
$$mV \cdot P = mV \cdot \frac{RT}{mV}$$
$$mVP = RT$$
$$\frac{mVP}{VP} = \frac{RT}{VP}$$
$$m = \frac{RT}{VP}$$

96.
$$3x + 2 < 5$$
$$3x + 2 - 2 < 5 - 2$$
$$3x < 3$$
$$\frac{3x}{3} < \frac{3}{3}$$
$$x < 1$$

1

97.
$$-5x - 8 < 7$$
$$-5x - 8 + 8 < 7 + 8$$
$$-5x < 15$$
$$\frac{-5x}{-5} > \frac{15}{-5}$$
$$x > -3$$

−3

98.
$$5x - 3 \geq 2x + 9$$
$$5x - 2x - 3 \geq 2x - 2x + 9$$
$$3x - 3 \geq 9$$
$$3x - 3 + 3 \geq 9 + 3$$
$$3x \geq 12$$
$$\frac{3x}{3} \geq \frac{12}{3}$$
$$x \geq 4$$

4

99.
$$7x + 1 \leq 8x - 5$$
$$7x - 8x + 1 \leq 8x - 8x - 5$$
$$-x + 1 \leq -5$$
$$-x + 1 - 1 \leq -5 - 1$$
$$-x \leq -6$$
$$-(-x) \geq -(-6)$$
$$x \geq 6$$

6

100.
$$5(3 - x) \le 3(x - 3)$$
$$15 - 5x \le 3x - 9$$
$$15 - 5x - 3x \le 3x - 3x - 9$$
$$15 - 8x \le -9$$
$$15 - 15 - 8x \le -9 - 15$$
$$-8x \le -24$$
$$\frac{-8x}{-8} \ge \frac{-24}{-8}$$
$$x \ge 3$$

101.
$$3(5 - x) \ge 2x$$
$$15 - 3x \ge 2x$$
$$15 - 3x + 3x \ge 2x + 3x$$
$$15 \ge 5x$$
$$\frac{15}{5} \ge \frac{5x}{5}$$
$$3 \ge x, \text{ or } x \le 3$$

102.
$$8 < \quad x + 2 \quad < 13$$
$$8 - 2 < x + 2 - 2 < 13 - 2$$
$$6 < \quad x \quad < 11$$

103.
$$0 \le \quad 2 - 2x \quad < 4$$
$$0 - 2 \le 2 - 2 - 2x < 4 - 2$$
$$-2 \le \quad -2x \quad < 2$$
$$\frac{-2}{-2} \ge \quad \frac{-2x}{-2} \quad > \frac{2}{-2}$$
$$1 \ge \quad x \quad > -1$$
$$-1 < \quad x \quad \le 1$$

104. Let $l =$ length. Then $l - 6 =$ width.

$$\text{perimeter} \quad \le 68$$
$$2l + 2(l - 6) \le 68$$
$$2l + 2l - 12 \le 68$$
$$4l - 12 \quad \le 68$$
$$4l - 12 + 12 \le 68 + 12$$
$$4l \qquad \le 80$$
$$\frac{4l}{4} \qquad \le \frac{80}{4}$$
$$l \qquad \le 20$$

The length must be no more than 20 feet.

Chapter 2 Test (page 144)

1.
$$5x + 3 = -2$$
$$5(-1) + 3 \stackrel{?}{=} -2$$
$$-5 + 3 \stackrel{?}{=} -2$$
$$-2 = -2$$
-1 is a solution.

2.
$$3(x + 2) = 2x$$
$$3(-6 + 2) \stackrel{?}{=} 2(-6)$$
$$3(-4) \stackrel{?}{=} -12$$
$$-12 = -12$$
-6 is a solution.

3.
$$-3(2 - x) = 0$$
$$-3[2 - (-2)] \stackrel{?}{=} 0$$
$$-3[4] \stackrel{?}{=} 0$$
$$-12 \neq 0$$
-2 is not a solution.

CHAPTER 2 TEST

4. $3(x+2) = 2x+7$

$3(1+2) \overset{?}{=} 2(1)+7$

$3(3) \overset{?}{=} 2+7$

$9 = 9$

1 is a solution.

5. $x + 17 = -19$

$x + 17 - 17 = -19 - 17$

$x = -36$

6. $a - 15 = 32$

$a - 15 + 15 = 32 + 15$

$a = 47$

7. $12x = -144$

$\dfrac{12x}{12} = \dfrac{-144}{12}$

$x = -12$

8. $\dfrac{x}{7} = -1$

$7 \cdot \dfrac{x}{7} = 7(-1)$

$x = -7$

9. $8x + 2 = -14$

$8x + 2 - 2 = -14 - 2$

$8x = -16$

$\dfrac{8x}{8} = \dfrac{-16}{8}$

$x = -2$

10. $3 = 5 - 2x$

$3 - 5 = 5 - 5 - 2x$

$-2 = -2x$

$\dfrac{-2}{-2} = \dfrac{-2x}{-2}$

$1 = x$

11. $\dfrac{2x - 5}{3} = 3$

$3 \cdot \dfrac{2x - 5}{3} = 3(3)$

$2x - 5 = 9$

$2x - 5 + 5 = 9 + 5$

$2x = 14$

$\dfrac{2x}{2} = \dfrac{14}{2}$

$x = 7$

12. $\dfrac{3x - 18}{2} = 6x$

$2 \cdot \dfrac{3x - 18}{2} = 2(6x)$

$3x - 18 = 12x$

$3x - 3x - 18 = 12x - 3x$

$-18 = 9x$

$\dfrac{-18}{9} = \dfrac{9x}{9}$

$-2 = x$

13. $23 - 5(x + 10) = -12$

$23 - 5x - 50 = -12$

$-5x - 27 = -12$

$-5x - 27 + 27 = -12 + 27$

$-5x = 15$

$\dfrac{-5x}{-5} = \dfrac{15}{-5}$

$x = -3$

14. $\dfrac{7}{8}(x - 4) = 5x - \dfrac{7}{2}$

$8 \cdot \dfrac{7}{8}(x - 4) = 8\left(5x - \dfrac{7}{2}\right)$

$7(x - 4) = 8(5x) - 8 \cdot \dfrac{7}{2}$

$7x - 28 = 40x - 28$

$7x - 40x - 28 = 40x - 40x - 28$

$-33x - 28 = -28$

$-33x - 28 + 28 = -28 + 28$

$-33x = 0$

$x = 0$

15. $x + 5(x - 3) = x + 5x - 15 = 6x - 15$

16. $3x - 5(2 - x) = 3x - 10 + 5x = 8x - 10$

17. $-3x(x + 3) + 3x(x - 3) = -3x^2 - 9x + 3x^2 - 9x = -3x^2 + 3x^2 - 9x - 9x = -18x$

18. $-4x(2x - 5) - 7x(4x + 1) = -8x^2 + 20x - 28x^2 - 7x = -36x^2 + 13x$

19. $x + 45° = 120°$

$x + 45° - 45° = 120° - 45°$

$x = 75°$

20. Let x = the measure of the supplement.

$x + 105° = 180°$

$x + 105° - 105° = 180° - 105°$

$x = 75°$

21. Let x = amount at 6%. Then $10000 - x$ = amount at 5%.

$$\boxed{\begin{array}{c}\text{Interest}\\\text{at 6\%}\end{array}} + \boxed{\begin{array}{c}\text{Interest}\\\text{at 5\%}\end{array}} = \boxed{\begin{array}{c}\text{Total}\\\text{interest}\end{array}}$$

$$0.06x + 0.05(10000 - x) = 560$$
$$6x + 5(10000 - x) = 56000$$
$$6x + 50000 - 5x = 56000$$
$$x + 50000 = 56000$$
$$x + 50000 - 50000 = 56000 - 50000$$
$$x = 6000 \Rightarrow \$6{,}000 \text{ is invested at 6\%, and \$4{,}000 is invested at 5\%.}$$

22. Let t = time to meet.

	r	t	d
Car	65	t	$65t$
Truck	55	t	$55t$

$$\boxed{\text{Car Dist}} + \boxed{\text{Truck Dist}} = 72$$
$$65t + 55t = 72$$
$$120t = 72$$
$$\frac{120t}{120} = \frac{72}{120}$$
$$t = \tfrac{3}{5}$$

They meet after $\frac{3}{5}$ of an hour.

23. Let x = liters of water used.

$$\boxed{\begin{array}{c}\text{Amt. in}\\\text{10\% sol.}\end{array}} + \boxed{\begin{array}{c}\text{Amt. in}\\\text{water}\end{array}} = \boxed{\begin{array}{c}\text{Amt. in}\\\text{8\% sol.}\end{array}}$$

$$0.10(30) + 0(x) = 0.08(30 + x)$$
$$10(30) + 0 = 8(30 + x)$$
$$300 = 240 + 8x$$
$$300 - 240 = 8x$$
$$60 = 8x$$
$$\frac{60}{8} = \frac{8x}{8}$$
$$\frac{15}{2} = x$$

$7\frac{1}{2}$ liters of water should be used.

24.
$$d = rt$$
$$\frac{d}{r} = \frac{rt}{r}$$
$$\frac{d}{r} = t, \text{ or } t = \frac{d}{r}$$

25.
$$P = 2l + 2w$$
$$P - 2w = 2l + 2w - 2w$$
$$P - 2w = 2l$$
$$\frac{P - 2w}{2} = \frac{2l}{2}$$
$$\frac{P - 2w}{2} = l, \text{ or } l = \frac{P - 2w}{2}$$

26.
$$A = 2\pi rh$$
$$\frac{A}{2\pi r} = \frac{2\pi rh}{2\pi r}$$
$$\frac{A}{2\pi r} = h, \text{ or } h = \frac{A}{2\pi r}$$

27.
$$A = P + Prt$$
$$A - P = P - P + Prt$$
$$A - P = Prt$$
$$\frac{A - P}{Pt} = \frac{Prt}{Pt}$$
$$\frac{A - P}{Pt} = r, \text{ or } r = \frac{A - P}{Pt}$$

28.
$$8x - 20 \geq 4$$
$$8x - 20 + 20 \geq 4 + 20$$
$$8x \geq 24$$
$$\frac{8x}{8} \geq \frac{24}{8}$$
$$x \geq 3$$

29.
$$x - 2(x + 7) > 14$$
$$x - 2x - 14 > 14$$
$$-x - 14 > 14$$
$$-x - 14 + 14 > 14 + 14$$
$$-x > 28$$
$$-(-x) < -(28)$$
$$x < -28$$

30.
$$-4 \le \; 2(x+1) \; < 10$$
$$-4 \le \quad 2x+2 \quad < 10$$
$$-4-2 \le 2x+2-2 < 10-2$$
$$-6 \le \qquad 2x \qquad < 8$$
$$\frac{-6}{2} \le \quad \frac{2x}{2} \quad < \frac{8}{2}$$
$$-3 \le \qquad x \qquad < 4$$

31.
$$-2 < \; 5(x-1) \; \le 10$$
$$-2 < \quad 5x-5 \quad \le 10$$
$$-2+5 < 5x-5+5 \le 10+5$$
$$3 < \qquad 5x \qquad \le 15$$
$$\frac{3}{5} < \quad \frac{5x}{5} \quad \le \frac{15}{5}$$
$$\frac{3}{5} < \qquad x \qquad \le 3$$

Cumulative Review Exercises (page 145)

1. $\frac{27}{9} = 3$: integer, rational number, real number, positive number

2. $-0.25 = -\frac{1}{4}$: rational number, real number, negative number

3.

2 3 4 5 6 7

4.

5. $\frac{|-3|-|3|}{|-3-3|} = \frac{(+3)-(+3)}{|-6|} = \frac{0}{+6} = 0$

6.

7. $2\frac{3}{5} + 5\frac{1}{2} = \frac{13}{5} + \frac{11}{2} = \frac{13 \cdot 2}{5 \cdot 2} + \frac{11 \cdot 5}{2 \cdot 5} = \frac{26}{10} + \frac{55}{10} = \frac{81}{10} = 8\frac{1}{10}$

8. $35.7 - 0.05 = 35.65$

9. $(3x - 2y)z = [3(-5) - 2(3)]0 = 0$

10. $\dfrac{x - 3y + |z|}{2 - x} = \dfrac{-5 - 3(3) + |0|}{2 - (-5)} = \dfrac{-5 - 9 + 0}{2 + 5} = \dfrac{-14}{7} = -2$

11. $x^2 - y^2 + z^2 = (-5)^2 - (3)^2 + 0^2$
$$= 25 - 9 + 0 = 16$$

12. $\dfrac{x}{y} + \dfrac{y + 2}{3 - z} = \dfrac{-5}{3} + \dfrac{3 + 2}{3 - 0} = -\dfrac{5}{3} + \dfrac{5}{3} = 0$

13.
$$rb = a$$
$$0.075(330) = a$$
$$24.75 = a$$

14.
$$rb = a$$
$$0.32b = 1688$$
$$\frac{0.32b}{0.32} = \frac{1688}{0.32}$$
$$b = 5,275$$

15. 2nd term: $5x^2y$; coefficient: 5

16. 3rd term: $37y$; factors: $37, y$

17. $3x - 5x + 2y = -2x + 2y$

18. $3(x - 7) + 2(8 - x) = 3x - 21 + 16 - 2x$
$$= x - 5$$

19. $2x^2y^3 - xy(xy^2) = 2x^2y^3 - x^2y^3 = x^2y^3$

CUMULATIVE REVIEW EXERCISES

20. $x^2(3-y) + x(xy+x) = 3x^2 - x^2y + x^2y + x^2 = 4x^2$

21.
$$3(x-5) + 2 = 2x$$
$$3x - 15 + 2 = 2x$$
$$3x - 13 = 2x$$
$$3x - 3x - 13 = 2x - 3x$$
$$-13 = -x$$
$$13 = x$$

22.
$$\frac{x-5}{3} - 5 = 7$$
$$\frac{x-5}{3} - 5 + 5 = 7 + 5$$
$$\frac{x-5}{3} = 12$$
$$3 \cdot \frac{x-5}{3} = 3(12)$$
$$x - 5 = 36$$
$$x - 5 + 5 = 36 + 5$$
$$x = 41$$

23.
$$\frac{2x-1}{5} = \frac{1}{2}$$
$$10 \cdot \frac{2x-1}{5} = 10 \cdot \frac{1}{2}$$
$$2(2x-1) = 5$$
$$4x - 2 = 5$$
$$4x - 2 + 2 = 5 + 2$$
$$4x = 7$$
$$\frac{4x}{4} = \frac{7}{4}$$
$$x = \frac{7}{4}$$

24.
$$2(a-5) - (3a+1) = 0$$
$$2a - 10 - 3a - 1 = 0$$
$$-a - 11 = 0$$
$$-a + a - 11 = 0 + a$$
$$-11 = a$$

25. Let x = Dealer's invoice.

$$\boxed{\text{Price}} = \boxed{\begin{array}{c}\text{Dealer's}\\\text{invoice}\end{array}} + \boxed{\text{Markup}}$$
$$23499 = x + 0.03x$$
$$23499 = 1.03x$$
$$\frac{23499}{1.03} = \frac{1.03x}{1.03}$$
$$22814.56 = r$$

The dealer's invoice was $22,814.56.

26. Let x = original sofa price.

$$\boxed{\text{Price}} = \boxed{\begin{array}{c}\text{New sofa}\\\text{price}\end{array}} + \boxed{\begin{array}{c}\text{New chair}\\\text{price}\end{array}}$$
$$780 = x - 0.35x + 300 - 0.35(300)$$
$$780 = 0.65x + 300 - 105$$
$$780 = 0.65x + 195$$
$$780 - 195 = 0.65x + 195 - 195$$
$$585 = 0.65x$$
$$\frac{585}{0.65} = \frac{0.65x}{0.65}$$
$$900 = x$$

The original price of the sofa was $900.

27. Let x = original car price.
$$13725.25 = x + 0.085x$$
$$13725.25 = 1.085x$$
$$\frac{13725.25}{1.085} = \frac{1.085x}{1.085}$$
$$12650 = x$$

The original price was $12,650.

28. Let x and $3x$ = the unknown amounts.

$$\boxed{\text{Cement}} + \boxed{\text{Gravel}} = \boxed{\text{Concrete}}$$
$$x + 3x = 500$$
$$4x = 500$$
$$\frac{4x}{4} = \frac{500}{4}$$
$$x = 125 \text{ lbs cement}$$

29. Let x and $2x =$ the lengths.

Total length $= 35$

$14 + x + 2x = 35$

$14 + 3x = 35$

$14 - 14 + 3x = 35 - 14$

$3x = 21$

$\frac{3x}{3} = \frac{21}{3}$

$x = 7$ ft

The section will not span the doorway.

30. Let $x =$ the length of one panel.

$\boxed{\text{Total length}} = 18$

$x + x + 3.4 = 18$

$2x + 3.4 = 18$

$2x + 3.4 - 3.4 = 18 - 3.4$

$2x = 14.6$

$\frac{2x}{2} = \frac{14.6}{2}$

$x = 7.3$

The lengths are 7.3 and 10.7 feet.

31. $A = \frac{1}{2}h(b + B)$

$2A = 2 \cdot \frac{1}{2}h(b + B)$

$2A = h(b + B)$

$\frac{2A}{b+B} = \frac{h(b+B)}{b+B}$

$\frac{2A}{b+B} = h$, or $h = \frac{2A}{b+B}$

32. $y = mx + b$

$y - b = mx + b - b$

$y - b = mx$

$\frac{y-b}{m} = \frac{mx}{m}$

$\frac{y-b}{m} = x$, or $x = \frac{y-b}{m}$

33. $4^2 - 5^2 = 16 - 25 = -9$

34. $(4 - 5)^2 = (-1)^2 = 1$

35. $5(4^3 - 2^3) = 5(64 - 8) = 5(56) = 280$

36. $-2(5^4 - 7^3) = -2(625 - 343)$

$= -2(282) = -564$

37. $8(4 + x) > 10(6 + x)$

$32 + 8x > 60 + 10x$

$32 + 8x - 10x > 60 + 10x - 10x$

$32 - 2x > 60$

$32 - 32 - 2x > 60 - 32$

$-2x > 28$

$\frac{-2x}{-2} < \frac{28}{-2}$

$x < -14$

-14

38. $-9 < 3(x + 2) \leq 3$

$-9 < 3x + 6 \leq 3$

$-9 - 6 < 3x + 6 - 6 \leq 3 - 6$

$-15 < 3x \leq -3$

$\frac{-15}{3} < \frac{3x}{3} \leq \frac{-3}{3}$

$-5 < x \leq -1$

$-5 \quad -1$

Exercise 3.1 (page 156)

1. $-3 - 3(-5) = -3 - (-15) = -3 + 15$
$$= 12$$

3. opposite of -8: 8

5.
$$-4x + 7 = -21$$
$$-4x + 7 - 7 = -21 - 7$$
$$-4x = -28$$
$$\frac{-4x}{-4} = \frac{-28}{-4}$$
$$x = 7$$

7. $(x+1)(x+y)^2 = (-2+1)[-2+(-5)]^2$
$$= (-1)(-7)^2$$
$$= -1(+49) = -49$$

9. ordered pair

11. origin

13. rectangular coordinate

15. no

17. origin, left, up

19. II

21.

x	y
4	3 or -3
0	5 or -5
-3	4 or -4
5 or -5	0
-4	-3 or 3
3	-4 or 4

23. The point $(-10, 60)$ indicates that 10 minutes before the workout started, her heart rate was 60 beats per minute.

25. The point with an x-coordinate of 30 is $(30, 150)$, so her heart rate was 150 beats per minute one half-hour after starting.

27. The points on the graph with a y-coordinate of 100 have x-coordinates of approximately 5 and 50, so her heart rate was 100 beats per minute after about 5 and 50 minutes.

29. Before the workout, her heart rate was 60 beats per minute. After the workout, her heart rate was about 70 beats per minute, or about 10 beats per minute higher.

31.

33.

City	Ordered Pair
Carbondale	(3, J)
Champaign	(4.5, D)
Chicago	(5.5, B)
Peoria	(3.5, C)
Rockford	(3.5, A)
Springfield	(2.25, E)
St. Louis	(2, H)

35. a. The highest point on the graph for the 60° angle is $(-2, 7)$, for a height of 7 ft. The highest point on the graph for the 30° angle is $(0, 3)$, for a height of 3 ft. The difference is 4 ft.

b. The farthest point on the graph for the 60° angle is $(3, 0)$, for a final location of 3 ft. The farthest point on the graph for the 30° angle is $(7, 0)$, for a final location of 7 ft. The difference is 4 ft.

37. a. To find the charge for a 1-day rental, find the y-coordinate of the point with an x-coordinate of 1. This is the point $(1, 2)$. The charge will be $2.

b. To find the charge for a 2-day rental, find the y-coordinate of the point with an x-coordinate of 2. This is the point $(2, 4)$. The charge will be $4.

c. To find the charge for a 5-day rental, find the y-coordinate of the point with an x-coordinate of 5. This is the point $(5, 7)$. The charge will be $7.

d. To find the charge for a 7-day rental, find the y-coordinate of the point with an x-coordinate of 7. This is the point $(7, 9)$. The charge will be $9.

39.

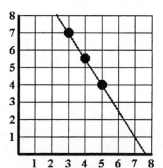

a. $(7, 35) \Rightarrow 35$ miles
b. $(4, 20) \Rightarrow 4$ gallons
c. $(6.5, 32.5) \Rightarrow 32.5$ miles

41.

a. A 3-year old car is worth $7000.
b. $(7, 1) \Rightarrow $1000
c. $(6, 2500) \Rightarrow 6$ years

43. Answers may vary.   **45.** Answers may vary.   **47.** Answers may vary.

Exercise 3.2 (page 170)

1.
$$\frac{x}{8} = -12$$
$$8 \cdot \frac{x}{8} = 8(-12)$$
$$x = -96$$

3. expression

5.
$$rb = a$$
$$0.005(250) = a$$
$$1.25 = a$$

7. $-2.5 - (-2.6) = -2.5 + 2.6 = 0.1$

9. two

11. independent, dependent

13. linear

15. y-intercept

17. $x - 2y = -4$

$4 - 2(4) \overset{?}{=} -4$

$4 - 8 \overset{?}{=} -4$

$-4 = -4 \Rightarrow (4, 4)$ is a solution.

19. $y = \dfrac{2}{3}x + 5$

$12 \overset{?}{=} \dfrac{2}{3}(6) + 5$

$12 \overset{?}{=} 4 + 5$

$12 \neq 9 \Rightarrow (6, 12)$ is not a solution.

21.

$$y = x - 3$$

$x = 0$	$x = 1$	$x = -2$	$x = -4$
$y = 0 - 3$	$y = 1 - 3$	$y = -2 - 3$	$y = -4 - 3$
$y = -3$	$y = -2$	$y = -5$	$y = -7$

x	y
0	-3
1	-2
-2	-5
-4	-7

23.

$$y = -2x$$

$x = 0$	$x = 1$	$x = 3$	$x = -2$
$y = -2(0)$	$y = -2(1)$	$y = -2(3)$	$y = -2(-2)$
$y = 0$	$y = -2$	$y = -6$	$y = 4$

input	output
0	0
1	-2
3	-6
-2	4

25.

$$y = 2x$$

$x = 0$	$x = 1$	$x = -1$
$y = 2(0)$	$y = 2(1)$	$y = 2(-1)$
$y = 0$	$y = 2$	$y = -2$

x	y
0	0
1	2
-1	-2

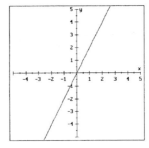

27.

$$y = 2x - 1$$

$x = 0$	$x = 1$	$x = -1$
$y = 2(0) - 1$	$y = 2(1) - 1$	$y = 2(-1) - 1$
$y = 0 - 1$	$y = 2 - 1$	$y = -2 - 1$
$y = -1$	$y = 1$	$y = -3$

x	y
0	-1
1	1
-1	-3

29.
$$y = \frac{x}{2} - 2$$

$$\frac{x = 0}{y = \frac{0}{2} - 2}$$ $$\frac{x = 2}{y = \frac{2}{2} - 2}$$ $$\frac{x = -2}{y = \frac{-2}{2} - 2}$$
$$y = 0 - 2 \qquad y = 1 - 2 \qquad y = -1 - 2$$
$$y = -2 \qquad y = -1 \qquad y = -3$$

x	y
0	-2
2	-1
-2	-3

31.
$$x + y = 7$$

$$\frac{x = 0}{0 + y = 7}$$ $$\frac{y = 0}{x + 0 = 7}$$
$$y = 7 \qquad x = 7$$
$$(0, 7) \qquad (7, 0)$$

33.
$$x - y = 7$$

$$\frac{x = 0}{0 - y = 7}$$ $$\frac{y = 0}{x - 0 = 7}$$
$$-y = 7 \qquad x = 7$$
$$y = -7 \qquad (7, 0)$$
$$(0, -7)$$

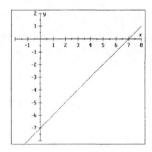

35.
$$y = -2x + 5$$

$$\frac{x = 0}{y = -2(0) + 5}$$ $$\frac{y = 0}{0 = -2x + 5}$$
$$y = 0 + 5 \qquad 2x = 5$$
$$y = 5 \qquad x = \frac{5}{2}$$
$$(0, 5) \qquad \left(\frac{5}{2}, 0\right)$$

37.
$$2x + 3y = 12$$

$x = 0$	$y = 0$
$2(0) + 3y = 12$	$2x + 3(0) = 12$
$0 + 3y = 12$	$2x + 0 = 12$
$3y = 12$	$2x = 12$
$y = 4$	$x = 6$
$(0, 4)$	$(6, 0)$

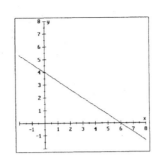

39.
$$y = -5$$
horizontal, y-coordinate $= -5$

41.
$$x = 5$$
vertical, x-coordinate $= 5$

43.
$$y = 0$$
horizontal, y-coordinate $= 0$

45.
$$2x = 5$$
$$x = \frac{5}{2} = 2\frac{1}{2}$$
vertical, x-coordinate $= \frac{5}{2}$

47. **a.** $c = 50 + 25u$

b.

u	c
4	$50 + 25(4) = 150$
8	$50 + 25(8) = 250$
14	$50 + 25(14) = 400$

c. The service fee is $50.

d. cost for 18 units = $500
cost for 12 units = $350
Total cost = $850

49. **a.**

r	h
7	$3.9(7) + 28.9 = 56.2$
8.5	$3.9(8.5) + 28.9 = 62.1$
9	$3.9(9) + 28.9 = 64.0$

b. ...taller the woman is.

c. 58 inches tall

51. Answers may vary. **53.** Answers may vary. **55.** Answers may vary.

57. $x_M = \dfrac{a+c}{2} = \dfrac{5+7}{2} = \dfrac{12}{2} = 6$

$y_M = \dfrac{b+d}{2} = \dfrac{3+9}{2} = \dfrac{12}{2} = 6$

$M(6,6)$

59. $x_M = \dfrac{a+c}{2} = \dfrac{2+(-3)}{2} = \dfrac{-1}{2} = -\dfrac{1}{2}$

$y_M = \dfrac{b+d}{2} = \dfrac{-7+12}{2} = \dfrac{5}{2}$

$M\left(-\dfrac{1}{2}, \dfrac{5}{2}\right)$

61. $x_M = \dfrac{a+c}{2} = \dfrac{4+10}{2} = \dfrac{14}{2} = 7$

$y_M = \dfrac{b+d}{2} = \dfrac{6+6}{2} = \dfrac{12}{2} = 6$

$M(7,6)$

Exercise 3.3 (page 182)

1. $(-2)^4 = (-2)(-2)(-2)(-2) = 16$

3. $3x - x^2 = 3(-3) - (-3)^2 = -9 - (+9)$
$$= -9 - 9 = -18$$

5. system

7. independent

9. inconsistent

11.
$$\begin{array}{ll} x + y = 2 & 2x - y = 1 \\ 1 + 1 \overset{?}{=} 2 & 2(1) - 1 \overset{?}{=} 1 \\ 2 = 2 & 2 - 1 \overset{?}{=} 1 \\ & 1 = 1 \end{array}$$
$(1, 1)$ is a solution to the system.

13.
$$\begin{array}{ll} 2x + y = 4 & x + y = 1 \\ 2(3) + (-2) \overset{?}{=} 4 & 3 + (-2) \overset{?}{=} 1 \\ 6 + (-2) \overset{?}{=} 4 & 1 = 1 \\ 4 = 4 & \end{array}$$
$(3, -2)$ is a solution to the system.

15.
$$\begin{array}{ll} 2x - 3y = -7 & 4x - 5y = 25 \\ 2(4) - 3(5) \overset{?}{=} -7 & 4(4) - 5(5) \overset{?}{=} 25 \\ 8 - 15 \overset{?}{=} -7 & 16 - 25 \overset{?}{=} 25 \\ -7 = -7 & -9 \neq 25 \end{array}$$
$(4, 5)$ is not a solution to the system.

17.
$$\begin{array}{ll} 4x + 5y = -23 & -3x + 2y = 0 \\ 4(-2) + 5(-3) \overset{?}{=} -23 & -3(-2) + 2(-3) \overset{?}{=} 0 \\ -8 + (-15) \overset{?}{=} -23 & 6 + (-6) \overset{?}{=} 0 \\ -23 = -23 & 0 = 0 \end{array}$$
$(-2, -3)$ is a solution to the system.

19.
$$\begin{array}{ll} 2x + y = 4 & 4x - 3y = 11 \\ 2\left(\frac{1}{2}\right) + 3 \overset{?}{=} 4 & 4\left(\frac{1}{2}\right) - 3(3) \overset{?}{=} 11 \\ 1 + 3 \overset{?}{=} 4 & 2 - 9 \overset{?}{=} 11 \\ 4 = 4 & -7 \neq 11 \end{array}$$
$\left(\frac{1}{2}, 3\right)$ is not a solution to the system.

21.
$$\begin{array}{ll} 5x - 4y = -6 & 8y = 10x + 12 \\ 5\left(-\frac{2}{5}\right) - 4\left(\frac{1}{4}\right) \overset{?}{=} -6 & 8\left(\frac{1}{4}\right) \overset{?}{=} 10\left(-\frac{2}{5}\right) + 12 \\ -2 - 1 \overset{?}{=} -6 & 2 \overset{?}{=} -4 + 12 \\ -3 \neq -6 & 2 \neq 8 \end{array}$$
$\left(-\frac{2}{5}, \frac{1}{4}\right)$ is not a solution to the system.

23. $x + y = 2$ $x - y = 0$

x	y
0	2
1	1

x	y
0	0
1	1

$\boxed{(1, 1)}$

25. $x + y = 2$ $x - y = 4$

x	y
0	2
1	1

x	y
0	-4
1	-3

$\boxed{(3, -1)}$

27. $3x + 2y = -8$ $2x - 3y = -1$

x	y
0	-4
2	-7

x	y
1	1
-2	-1

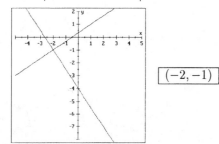

$\boxed{(-2, -1)}$

29. $4x - 2y = 8$ $y = 2x - 4$

x	y
0	-4
1	-2

x	y
0	-4
1	-2

dependent equations

31. $2x - 3y = -18$ $3x + 2y = -1$

x	y
-3	4
-6	2

x	y
1	-2
-1	1

$\boxed{(-3, 4)}$

33. $4x = 3(4 - y)$ $2y = 4(3 - x)$
$4x = 12 - 3y$ $2y = 12 - 4x$

x	y
0	4
3	0

x	y
3	0
0	6

$\boxed{(3, 0)}$

35. $x + 2y = -4$ $x - \frac{1}{2}y = 6$

$2x - y = 12$

x	y
0	-2
2	-3

x	y
4	-4
6	0

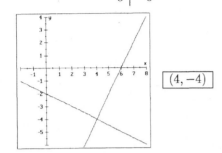

$(4, -4)$

37. $-\frac{3}{4}x + y = 3$ $\frac{1}{4}x + y = -1$

$-3x + 4y = 12$ $x + 4y = -4$

x	y
0	3
-4	0

x	y
0	-1
-4	0

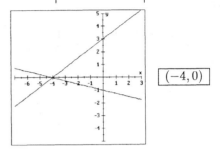

$(-4, 0)$

39. $\frac{1}{2}x + \frac{1}{4}y = 0$ $\frac{1}{4}x - \frac{3}{8}y = -2$

$2x + y = 0$ $2x - 3y = -16$

x	y
0	0
1	-2

x	y
-2	4
-5	2

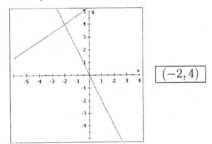

$(-2, 4)$

41. $\frac{1}{3}x - \frac{1}{2}y = \frac{1}{6}$ $\frac{2}{5}x + \frac{1}{2}y = \frac{13}{10}$

$2x - 3y = 1$ $4x + 5y = 13$

x	y
5	3
2	1

x	y
-3	5
2	1

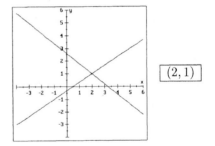

$(2, 1)$

43. $\begin{cases} y = 4 - x \\ y = 2 + x \end{cases}$

$(1, 3)$

45. $\begin{cases} 3x - 6y = 4 & \Rightarrow y = \frac{1}{2}x - \frac{2}{3} \\ 2x + y = 1 & \Rightarrow y = -2x + 1 \end{cases}$

$(0.67, -0.33)$

47. **a.** Donors outnumbered patients.
   b. 1999; about 4100
   c. Patients outnumber donors.

49. **a.** Houston, New Orleans, St. Augustine
   b. St. Louis, Memphis, New Orleans
   c. New Orleans

51. **Answers may vary.**

53. **Answers may vary.**

Exercise 3.4 (page 190)

1. $y^2 - x^2 = (3)^2 - (-2)^2 = 9 - 4 = 5$

3. $\dfrac{3x - 2y}{2x + y} = \dfrac{3(-2) - 2(3)}{2(-2) + 3} = \dfrac{-6 - 6}{-4 + 3}$

$$= \dfrac{-12}{-1} = 12$$

5. $-x(3y - 4) = -(-2)[3(3) - 4] = -(-2)(5) = 10$

7. y, terms

9. remove

11. infinitely many

13. $\begin{cases} (1) & y = 2x \\ (2) & x + y = 6 \end{cases}$
Substitute $y = 2x$ from (1) into (2):
$x + y = 6$
$x + 2x = 6$
$\quad 3x = 6$
$\quad\ \ x = 2$
Substitute and solve for y:
$y = 2x$
$y = 2(2) = 4$
$\boxed{x = 2, y = 4}$

15. $\begin{cases} (1) & y = 2x - 6 \\ (2) & 2x + y = 6 \end{cases}$
Substitute $y = 2x - 6$ from (1) into (2):
$2x + y = 6$
$2x + 2x - 6 = 6$
$\quad 4x = 12$
$\quad\ x = 3$
Substitute and solve for y:
$y = 2x - 6$
$y = 2(3) - 6 = 6 - 6 = 0$
$\boxed{x = 3, y = 0}$

17. $\begin{cases} (1) & y = 2x + 5 \\ (2) & x + 2y = -5 \end{cases}$
Substitute $y = 2x + 5$ from (1) into (2):
$x + 2y = -5$
$x + 2(2x + 5) = -5$
$x + 4x + 10 = -5$
$\quad\quad 5x = -15$
$\quad\quad\ x = -3$
Substitute and solve for y:
$y = 2x + 5$
$y = 2(-3) + 5 = -6 + 5 = -1$
$\boxed{x = -3, y = -1}$

19. $\begin{cases} (1) & 2a + 4b = -24 \\ (2) & a = 20 - 2b \end{cases}$
Substitute $a = 20 - 2b$ from (2) into (1):
$2a + 4b = -24$
$2(20 - 2b) + 4b = -24$
$40 - 4b + 4b = -24$
$40 \neq -24$
$\boxed{\text{inconsistent system}}$

21. $\begin{cases} (1) & 2a = 3b - 13 \\ (2) & b = 2a + 7 \end{cases}$

Substitute $b = 2a + 7$ from (2) into (1):

$2a = 3b - 13$

$2a = 3(2a + 7) - 13$

$2a = 6a + 21 - 13$

$-4a = 8$

$a = -2$

Substitute and solve for b:

$b = 2a + 7$

$b = 2(-2) + 7 = -4 + 7 = 3$

$\boxed{a = -2, b = 3}$

23. $\begin{cases} 2x + y = 4 & \Rightarrow (1) \quad y = -2x + 4 \\ 4x + y = 5 & \Rightarrow (2) \quad 4x + y = 5 \end{cases}$

Substitute $y = -2x + 4$ from (1) into (2):

$4x + y = 5$

$4x + (-2x + 4) = 5$

$4x - 2x + 4 = 5$

$2x = 1$

$x = \dfrac{1}{2}$

Substitute and solve for y:

$y = -2x + 4$

$y = -2\left(\dfrac{1}{2}\right) + 4 = -1 + 4 = 3$

$\boxed{x = \frac{1}{2}, y = 3}$

25. $\begin{cases} r + 3s = 9 & \Rightarrow (1) \quad r = -3s + 9 \\ 3r + 2s = 13 & \Rightarrow (2) \quad 3r + 2s = 13 \end{cases}$

Substitute $r = -3s + 9$ from (1) into (2):

$3r + 2s = 13$

$3(-3s + 9) + 2s = 13$

$-9s + 27 + 2s = 13$

$-7s = -14$

$s = 2$

Substitute and solve for r:

$r = -3s + 9$

$r = -3(2) + 9 = -6 + 9 = 3$

$\boxed{r = 3, s = 2}$

27. $\begin{cases} 4x + 5y = 2 & \Rightarrow (1) \quad 4x + 5y = 2 \\ 3x - y = 11 & \Rightarrow (2) \quad y = 3x - 11 \end{cases}$

Substitute $y = 3x - 11$ from (2) into (1):

$4x + 5y = 2$

$4x + 5(3x - 11) = 2$

$4x + 15x - 55 = 2$

$19x = 57$

$x = 3$

Substitute and solve for y:

$y = 3x - 11$

$y = 3(3) - 11 = 9 - 11 = -2$

$\boxed{x = 3, y = -2}$

29. $\begin{cases} 2x + y = 0 & \Rightarrow (1) \quad y = -2x \\ 3x + 2y = 1 & \Rightarrow (2) \quad 3x + 2y = 1 \end{cases}$

Substitute $y = -2x$ from (1) into (2):

$3x + 2y = 1$

$3x + 2(-2x) = 1$

$3x - 4x = 1$

$-x = 1$

$x = -1$

Substitute and solve for y:

$y = -2x$

$y = -2(-1) = 2$

$\boxed{x = -1, y = 2}$

31. $\begin{cases} 3x + 4y = -7 & \Rightarrow (1) \quad 3x + 4y = -7 \\ 2y - x = -1 & \Rightarrow (2) \quad x = 2y + 1 \end{cases}$

Substitute $x = 2y + 1$ from (2) into (1):

$3x + 4y = -7$

$3(2y + 1) + 4y = -7$

$6y + 3 + 4y = -7$

$10y = -10$

$y = -1$

Substitute and solve for x:

$x = 2y + 1$

$x = 2(-1) + 1 = -2 + 1 = -1$

$\boxed{x = -1, y = -1}$

33. $\begin{cases} 9x = 3y + 12 & \Rightarrow (1) \quad 9x = 3y + 12 \\ 4 = 3x - y & \Rightarrow (2) \quad y = 3x - 4 \end{cases}$

Substitute $y = 3x - 4$ from (2) into (1):

$9x = 3y + 12$

$9x = 3(3x - 4) + 12$

$9x = 9x - 12 + 12$

$0 = 0$ $\boxed{\text{dependent equations}}$

35. $\begin{cases} (1) \quad 2x + 3y = 5 \quad \Rightarrow 2x = 5 - 3y \quad \Rightarrow x = \dfrac{5 - 3y}{2} \\ (2) \quad 3x + 2y = 5 \end{cases}$

Substitute $x = \dfrac{5 - 3y}{2}$ from (1) into (2):

$3x + 2y = 5$

$3 \cdot \dfrac{5 - 3y}{2} + 2y = 5$

$3(5 - 3y) + 4y = 10$

$15 - 9y + 4y = 10$

$-5y = -5$

$y = 1$

Substitute and solve for x:

$x = \dfrac{5 - 3y}{2}$

$x = \dfrac{5 - 3(1)}{2}$

$x = \dfrac{5 - 3}{2} = \dfrac{2}{2} = 1$

Solution:

$\boxed{x = 1, y = 1}$

37. $\begin{cases} (1) \quad a = \dfrac{3}{2}b + 5 \\ (2) \quad 2a - 3b = 8 \end{cases}$

Substitute $a = \dfrac{3}{2}b + 5$ from (1) into (2):

$2a - 3b = 8$

$2\left(\dfrac{3}{2}b + 5\right) - 3b = 8$

$3b + 10 - 3b = 8$

$10 \neq 8$ $\boxed{\text{inconsistent system}}$

39. $\begin{cases} (1) \quad 2x + 5y = -2 \quad \Rightarrow 2x = -2 - 5y \quad \Rightarrow x = \dfrac{-2 - 5y}{2} \\ (2) \quad 4x + 3y = 10 \end{cases}$

Substitute $x = \dfrac{-2 - 5y}{2}$ from (1) into (2):

$4x + 3y = 10$

$4 \cdot \dfrac{-2 - 5y}{2} + 3y = 10$

$4(-2 - 5y) + 6y = 20$

$-8 - 20y + 6y = 20$

$-14y = 28$

$y = -2$

Substitute and solve for x:

$x = \dfrac{-2 - 5y}{2}$

$x = \dfrac{-2 - 5(-2)}{2}$

$x = \dfrac{-2 + 10}{2} = \dfrac{8}{2} = 4$

Solution:

$\boxed{x = 4, y = -2}$

41. $\begin{cases} (1) & 2x - 3y = -3 \quad \Rightarrow 2x = 3y - 3 \quad \Rightarrow x = \dfrac{3y-3}{2} \\ (2) & 3x + 5y = -14 \end{cases}$

Substitute $x = \dfrac{3y-3}{2}$ from (1) into (2):   Substitute and solve for x: Solution:

$$3x + 5y = -14$$

$$3 \cdot \frac{3y-3}{2} + 5y = -14$$

$$3(3y-3) + 10y = -28$$

$$9y - 9 + 10y = -28$$

$$19y = -19$$

$$y = -1$$

$$x = \frac{3y-3}{2}$$

$$x = \frac{3(-1)-3}{2}$$

$$x = \frac{-3-3}{2} = \frac{-6}{2} = -3$$

$\boxed{x = -3, y = -1}$

43. $\begin{cases} (1) & 7x - 2y = -1 \quad \Rightarrow 7x = 2y - 1 \quad \Rightarrow x = \dfrac{2y-1}{7} \\ (2) & -5x + 2y = -1 \end{cases}$

Substitute $x = \dfrac{2y-1}{7}$ from (1) into (2):   Substitute and solve for x: Solution:

$$-5x + 2y = -1$$

$$-5 \cdot \frac{2y-1}{7} + 2y = -1$$

$$-5(2y-1) + 14y = -7$$

$$-10y + 5 + 14y = -7$$

$$4y = -12$$

$$y = -3$$

$$x = \frac{2y-1}{7}$$

$$x = \frac{2(-3)-1}{7}$$

$$x = \frac{-6-1}{7} = \frac{-7}{7} = -1$$

$\boxed{x = -1, y = -3}$

45. $\begin{cases} (1) & 2a + 3b = 2 \quad \Rightarrow 2a = 2 - 3b \quad \Rightarrow a = \dfrac{2-3b}{2} \\ (2) & 8a - 3b = 3 \end{cases}$

Substitute $a = \dfrac{2-3b}{2}$ from (1) into (2):   Substitute and solve for a: Solution:

$$8a - 3b = 3$$

$$8 \cdot \frac{2-3b}{2} - 3b = 3$$

$$4(2-3b) - 3b = 3$$

$$8 - 12b - 3b = 3$$

$$-15b = -5$$

$$b = \frac{1}{3}$$

$$a = \frac{2-3b}{2}$$

$$a = \frac{2-3\left(\frac{1}{3}\right)}{2}$$

$$a = \frac{2-1}{2} = \frac{1}{2}$$

$\boxed{a = \tfrac{1}{2}, b = \tfrac{1}{3}}$

47. $\begin{cases} (1) \quad y - x = 3x \quad \Rightarrow y = 4x \\ (2) \quad 2(x+y) = 14 - y \quad \Rightarrow 2x + 2y = 14 - y \quad \Rightarrow 2x + 3y = 14 \end{cases}$

Substitute $y = 4x$ from (1) into (2):   Substitute and solve for y: Solution:

$2x + 3y = 14$ $y = 4x$ $\boxed{x = 1,\, y = 4}$

$2x + 3(4x) = 14$ $y = 4(1) = 4$

$2x + 12x = 14$

$14x = 14$

$x = 1$

49. $\begin{cases} (1) \quad 3(x-1) + 3 = 8 + 2y \quad \Rightarrow 3x - 3 + 3 = 8 + 2y \quad \Rightarrow 3x = 2y + 8 \quad \Rightarrow x = \dfrac{2y+8}{3} \\ (2) \quad 2(x+1) = 4 + 3y \quad \Rightarrow 2x + 2 = 4 + 3y \quad \Rightarrow 2x - 3y = 2 \end{cases}$

Substitute $x = \dfrac{2y+8}{3}$ from (1) into (2):   Substitute and solve for x: Solution:

$2x - 3y = 2$ $x = \dfrac{2y+8}{3}$ $\boxed{x = 4,\, y = 2}$

$2 \cdot \dfrac{2y+8}{3} - 3y = 2$ $x = \dfrac{2(2)+8}{3}$

$2(2y + 8) - 9y = 6$

$4y + 16 - 9y = 6$ $x = \dfrac{4+8}{3} = \dfrac{12}{3} = 4$

$-5y = -10$

$y = 2$

51. $\begin{cases} (1) \quad 6a = 5(3 + b + a) - a \quad \Rightarrow 6a = 15 + 5b + 5a - a \quad \Rightarrow 2a = 5b + 15 \quad \Rightarrow a = \dfrac{5b+15}{2} \\ (2) \quad 3(a - b) + 4b = 5(1 + b) \quad \Rightarrow 3a - 3b + 4b = 5 + 5b \quad \Rightarrow 3a - 4b = 5 \end{cases}$

Substitute $a = \dfrac{5b+15}{2}$ from (1) into (2):   Substitute and solve for a: Solution:

$3a - 4b = 5$ $a = \dfrac{5b+15}{2}$ $\boxed{a = -5,\, b = -5}$

$3 \cdot \dfrac{5b+15}{2} - 4b = 5$ $a = \dfrac{5(-5)+15}{2}$

$3(5b + 15) - 8b = 10$

$15b + 45 - 8b = 10$ $a = \dfrac{-25+15}{2} = \dfrac{-10}{2} = -5$

$7b = -35$

$b = -5$

53. $\begin{cases} (1) \quad \dfrac{1}{2}x + \dfrac{1}{2}y = -1 \quad \Rightarrow x + y = -2 \quad \Rightarrow x = -y - 2 \\ (2) \quad \dfrac{1}{3}x - \dfrac{1}{2}y = -4 \quad \Rightarrow 2x - 3y = -24 \end{cases}$

Substitute $x = -y - 2$ from (1) into (2):   Substitute and solve for x: Solution:

$2x - 3y = -24$ $x = -y - 2$ $\boxed{x = -6,\, y = 4}$

$2(-y - 2) - 3y = -24$ $x = -4 - 2 = -6$

$-2y - 4 - 3y = -24$

$-5y = -20$

$y = 4$

55. $\begin{cases} (1) \quad 5x = \dfrac{1}{2}y - 1 \quad \Rightarrow 10x = y - 2 \\[2mm] (2) \quad \dfrac{1}{4}y = 10x - 1 \quad \Rightarrow y = 40x - 4 \end{cases}$

Substitute $y = 40x - 4$ from (2) into (1):

$10x = y - 2$
$10x = 40x - 4 - 2$
$-30x = -6$
$x = \dfrac{1}{5}$

Substitute and solve for y:

$y = 40x - 4$
$y = 40\left(\dfrac{1}{5}\right) - 4$
$y = 8 - 4 = 4$

Solution:

$\boxed{x = \tfrac{1}{5}, y = 4}$

57. $\begin{cases} (1) \quad \dfrac{6x - 1}{3} - \dfrac{5}{3} = \dfrac{3y + 1}{2} \quad \Rightarrow [\times 6] \Rightarrow \quad 2(6x - 1) - 2(5) = 3(3y + 1) \\[3mm] \hspace{4.3cm} 12x - 2 - 10 = 9y + 3 \quad \Rightarrow 12x = 9y + 15 \\[3mm] (2) \quad \dfrac{1 + 5y}{4} + \dfrac{x + 3}{4} = \dfrac{17}{2} \quad \Rightarrow [\times 4] \Rightarrow \quad 1 + 5y + x + 3 = 2(17) \Rightarrow x = 30 - 5y \end{cases}$

Substitute $x = 30 - 5y$ from (2) into (1):

$12x = 9y + 15$
$12(30 - 5y) = 9y + 15$
$360 - 60y = 9y + 15$
$-69y = -345$
$y = 5$

Substitute and solve for x:

$x = 30 - 5y$
$x = 30 - 5(5)$
$x = 30 - 25 = 5$

Solution:

$\boxed{x = 5, y = 5}$

59. Answers may vary.

61. Answers may vary.

Exercise 3.5 (page 197)

1.
$8(3x - 5) - 12 = 4(2x + 3)$
$24x - 40 - 12 = 8x + 12$
$24x - 52 = 8x + 12$
$24x - 8x - 52 = 8x - 8x + 12$
$16x - 52 + 52 = 12 + 52$
$16x = 64$
$\dfrac{16x}{16} = \dfrac{64}{16}$
$x = 4$

3.
$x - 2 = \dfrac{x + 2}{3}$
$3(x - 2) = 3 \cdot \dfrac{x + 2}{3}$
$3x - 6 = x + 2$
$3x - x - 6 = x - x + 2$
$2x - 6 + 6 = 2 + 6$
$2x = 8$
$\dfrac{2x}{2} = \dfrac{8}{2}$
$x = 4$

5.
$$7x - 9 \leq 5$$
$$7x - 9 + 9 \leq 5 + 9$$
$$7x \leq 14$$
$$\frac{7x}{7} \leq \frac{14}{7}$$
$$x \leq 2$$

7. coefficient

9. general

11. 15

13.
$$\begin{aligned} x + y &= 5 \\ x - y &= -3 \\ \hline 2x &= 2 \\ x &= 1 \end{aligned}$$
Substitute and solve for y:
$$x + y = 5$$
$$1 + y = 5$$
$$y = 4 \quad \boxed{(1, 4)}$$

15.
$$\begin{aligned} x - y &= -5 \\ x + y &= 1 \\ \hline 2x &= -4 \\ x &= -2 \end{aligned}$$
Substitute and solve for y:
$$x + y = 1$$
$$-2 + y = 1$$
$$y = 3 \quad \boxed{(-2, 3)}$$

17.
$$\begin{aligned} 2x + y &= -1 \\ -2x + y &= 3 \\ \hline 2y &= 2 \\ y &= 1 \end{aligned}$$
Substitute and solve for x:
$$2x + y = -1$$
$$2x + 1 = -1$$
$$2x = -2$$
$$x = -1 \quad \boxed{(-1, 1)}$$

19.
$$\begin{aligned} 2x - 3y &= -11 \\ 3x + 3y &= 21 \\ \hline 5x &= 10 \\ x &= 2 \end{aligned}$$
Substitute and solve for y:
$$3x + 3y = 21$$
$$3(2) + 3y = 21$$
$$6 + 3y = 21$$
$$3y = 15$$
$$y = 5 \quad \boxed{(2, 5)}$$

21.
$$\begin{aligned} 2x + y &= -2 \\ -2x - 3y &= -6 \\ \hline -2y &= -8 \\ y &= 4 \end{aligned}$$
Substitute and solve for x:
$$2x + y = -2$$
$$2x + 4 = -2$$
$$2x = -6$$
$$x = -3 \quad \boxed{(-3, 4)}$$

23.
$$\begin{aligned} 4x + 3y &= 24 \\ 4x - 3y &= -24 \\ \hline 8x &= 0 \\ x &= 0 \end{aligned}$$
Substitute and solve for y:
$$4x + 3y = 24$$
$$4(0) + 3y = 24$$
$$0 + 3y = 24$$
$$3y = 24$$
$$y = 8 \quad \boxed{(0, 8)}$$

25.
$$\begin{aligned} x + y &= 5 \Rightarrow \times (-1) \\ x + 2y &= 8 \end{aligned} \qquad \begin{aligned} -x - y &= -5 \\ x + 2y &= 8 \\ \hline y &= 3 \end{aligned}$$

$$\begin{aligned} x + 2y &= 8 \\ x + 2(3) &= 8 \\ x + 6 &= 8 \\ x &= 2 \end{aligned} \qquad \text{Solution:} \quad \boxed{(2, 3)}$$

27. $2x + \ y = 4 \Rightarrow \times \ (-1) \quad -2x - \ y = -4 \qquad 2x + y = 4 \qquad$ Solution:

$\underline{2x + 3y = 0} \qquad\qquad\qquad \underline{2x + 3y = \ \ \ 0} \quad 2x + (-2) = 4 \qquad \boxed{(3, -2)}$

$2y = -4 \qquad\qquad 2x = 6$

$y = -2 \qquad\qquad\ \ x = 3$

29. $3x + 29 = \ \ 5y \ \Rightarrow \ 3x - 5y = -29 \Rightarrow \times \ (-1) \quad -3x + 5y = 29$

$\underline{4y - 34 = -3x} \ \Rightarrow \ \underline{3x + 4y = \ \ \ 34} \qquad\qquad\qquad \underline{3x + 4y = 34}$

$9y = 63$

$y = \ \ 7$

$3x + 4y = 34 \qquad$ Solution:

$3x + 4(7) = 34 \qquad \boxed{(2, 7)}$

$3x + 28 = 34$

$3x = 6$

$x = 2$

31. $2x = 3(y - 2) \ \Rightarrow \ 2x = 3y - 6 \ \ \Rightarrow \ 2x - 3y = -6 \Rightarrow \times (-1) \ -2x + 3y = \ \ \ 6 \ \boxed{\text{inconsistent}}$

$\underline{2(x + 4) = 3y} \ \Rightarrow \ \underline{2x + 8 = 3y} \ \ \Rightarrow \ \underline{2x - 3y = -8} \qquad\qquad\qquad \underline{2x - 3y = -8} \ \boxed{\text{system}}$

$0 \neq -2$

33. $-2(x + 1) = 3(y - 2) \ \Rightarrow \ \ \ -2x - 2 = 3y - 6 \ \Rightarrow \ -2x - 3y = -4 \ \boxed{\text{dependent}}$

$\underline{3(y + 2) = 6 - 2(x - 2)} \ \Rightarrow \ \underline{3y + 6 = 6 - 2x + 4} \ \Rightarrow \ \ \underline{2x + 3y = 4} \quad \boxed{\text{equations}}$

$0 = 0$

35. $4(x + 1) = 17 - 3(y - 1) \Rightarrow 4x + 4 = 17 - 3y + 3 \ \Rightarrow 4x + 3y = 16$

$\underline{2(x + 2) + 3(y - 1) = 9} \Rightarrow \underline{\ \ 2x + 4 + 3y - 3 = 9} \ \Rightarrow \underline{2x + 3y = 8}$

$4x + 3y = 16 \Rightarrow \times \ (-1) \quad -4x - 3y = -16 \qquad 4x + 3y = 16 \qquad$ Solution:

$\underline{2x + 3y = \ \ 8} \qquad\qquad\qquad \underline{2x + 3y = \ \ \ \ 8} \quad 4(4) + 3y = 16 \qquad \boxed{(4, 0)}$

$-2x \ \ \ \ \ \ = -8 \qquad 16 + 3y = 16$

$x \ \ \ \ \ \ = \ \ \ 4 \qquad\qquad 3y = 0$

$y = 0$

37. $2x + \ y = 10 \Rightarrow \times \ (-2) \quad -4x - 2y = -20 \qquad 2x + y = 10 \qquad$ Solution:

$\underline{x + 2y = 10} \qquad\qquad\qquad \underline{x + 2y = \ \ \ 10} \quad 2\left(\frac{10}{3}\right) + y = 10 \qquad \boxed{\left(\dfrac{10}{3}, \dfrac{10}{3}\right)}$

$-3x \ \ \ \ \ \ = -10 \qquad \ \frac{20}{3} + y = \frac{30}{3}$

$x \ \ \ \ \ \ = \ \ \frac{10}{3} \qquad\qquad\quad y = \frac{10}{3}$

39. $2x - \ y = 16 \Rightarrow \times \ (2) \quad 4x - 2y = 32 \qquad 3x + 2y = 3 \qquad$ Solution:

$\underline{3x + 2y = \ \ 3} \qquad\qquad\quad \underline{3x + 2y = \ \ \ 3} \quad 3(5) + 2y = 3 \qquad \boxed{(5, -6)}$

$7x \ \ \ \ \ \ = 35 \qquad \ 15 + 2y = 3$

$x \ \ \ \ \ \ = \ \ 5 \qquad\qquad 2y = -12$

$y = -6$

41. $2x + 3y = 2 \Rightarrow \times (-2)$ $-4x - 6y = -4$ $2x + 3y = 2$ Solution:

$\underline{4x - 9y = -1}$ $\underline{4x - 9y = -1}$ $2x + 3\left(\frac{1}{3}\right) = 2$ $\boxed{\left(\dfrac{1}{2}, \dfrac{1}{3}\right)}$

 $- 15y = -5$ $2x + 1 = 2$

 $y = \frac{-5}{-15} = \frac{1}{3}$ $2x = 1$

  $x = \frac{1}{2}$

43. $4x + 5y = -20 \Rightarrow \times (4)$ $16x + 20y = -80$ $4x + 5y = -20$ Solution:

$\underline{5x - 4y = -25} \Rightarrow \times (5)$ $\underline{25x - 20y = -125}$ $4(-5) + 5y = -20$ $\boxed{(-5, 0)}$

 $41x = -205$ $-20 + 5y = -20$

 $x = -5$ $5y = 0$

 $y = 0$

45. $6x = -3y \Rightarrow 6x + 3y = 0$ $6x + 3y = 0$ $6x = -3y$ Solution:

$\underline{5y = 2x + 12} \Rightarrow \underline{-2x + 5y = 12} \Rightarrow \times (3)$ $\underline{-6x + 15y = 36}$ $6x = -3(2)$ $\boxed{(-1, 2)}$

$ 18y = 36$ $6x = -6$

$ y = 2$ $x = -1$

47. $4(2x - y) = 18 \Rightarrow 8x - 4y = 18 \Rightarrow 8x - 4y = 18$

$\underline{3(x - 3) = 2y - 1} \Rightarrow \underline{3x - 9 = 2y - 1} \Rightarrow \underline{3x - 2y = 8}$

$8x - 4y = 18$ $8x - 4y = 18$ $8x - 4y = 18$ Solution:

$\underline{3x - 2y = 8} \Rightarrow \times (-2)$ $\underline{-6x + 4y = -16}$ $8(1) - 4y = 18$ $\boxed{\left(1, -\dfrac{5}{2}\right)}$

 $2x = 2$ $8 - 4y = 18$

 $x = 1$ $-4y = 10$

$ y = -\frac{10}{4} = -\frac{5}{2}$

49. $3(x - 2y) = 12 \Rightarrow 3x - 6y = 12$ $3x - 6y = 12$ $3x - 6y = 12$

$\underline{ x = 2(y + 2)} \Rightarrow \underline{x = 2y + 4} \Rightarrow \underline{ x - 2y = 4} \Rightarrow \times (-3)$ $\underline{-3x + 6y = -12}$

$ 0 = 0$

dependent equations

51. $\dfrac{3}{5}x + \dfrac{4}{5}y = 1 \Rightarrow \times (5)$ $3x + 4y = 5 \Rightarrow \times (2)$ $6x + 8y = 10$ $3x + 4y = 5$

$\underline{-\dfrac{1}{4}x + \dfrac{3}{8}y = 1} \Rightarrow \times (8)$ $\underline{-2x + 3y = 8} \Rightarrow \times (3)$ $\underline{-6x + 9y = 24}$ $3x + 4(2) = 5$

$\phantom{-\frac{1}{4}x + \frac{3}{8}y = 1 xxxxxxxxxxxxxxxxxxxxxxxxxxxxxxx} 17y = 34$ $3x + 8 = 5$

$\phantom{-\frac{1}{4}x + \frac{3}{8}y = 1 xxxxxxxxxxxxxxxxxxxxxxxxxxxxxxxxx} y = 2$ $3x = -3$

$\phantom{-\frac{1}{4}x + \frac{3}{8}y = 1 xxxxxxxxxxxxxxxxxxxxxxxxxxxxxxxxxxxxxxxxxxxxx} x = -1$

Solution: $\boxed{(-1, 2)}$

53. $\dfrac{3}{5}x + y = 1 \Rightarrow \times (5)$ $3x + 5y = 5$ $3x + 5y = 5$ Solution:

$\underline{\dfrac{4}{5}x - y = -1} \Rightarrow \times (5)$ $\underline{4x - 5y = -5}$ $3(0) + 5y = 5$ $\boxed{(0, 1)}$

$\phantom{\frac{4}{5}x - y = -1}$ $7x = 0$ $5y = 5$

$\phantom{\frac{4}{5}x - y = -1}$ $x = 0$ $y = 1$

55.

$$\frac{x}{2} - \frac{y}{3} = -2 \Rightarrow \times (6) \qquad\qquad 3x - 2y = -12 \Rightarrow \qquad\qquad 3x - 2y = -12$$

$$\frac{2x-3}{2} + \frac{6y+1}{3} = \frac{17}{6} \Rightarrow \times (6) \quad 3(2x-3) + 2(6y+1) = \quad 17 \Rightarrow 6x - 9 + 12y + 2 = \quad 17$$

$3x - 2y = -12 \Rightarrow \times (-2)$ $-6x + 4y = 24$ $3x - 2y = -12$ Solution:

$6x + 12y = 24$ $\underline{6x + 12y = 24}$ $3x - 2(3) = -12$ $\boxed{(-2, 3)}$

                              $16y = 48$ $3x - 6 = -12$

                                $y = 3$ $3x = -6$

                                         $x = -2$

57.

$$\frac{x-3}{2} + \frac{y+5}{3} = \frac{11}{6} \Rightarrow \times (6) \quad 3(x-3) + 2(y+5) = 11 \Rightarrow 3x - 9 + 2y + 10 = 11$$

$$\frac{x+3}{3} - \frac{5}{12} = \frac{y+3}{4} \Rightarrow \times (12) \quad 4(x+3) - 5 = 3(y+3) \Rightarrow 4x + 12 - 5 = 3y + 9$$

$3x + 2y = 10 \Rightarrow \times (3)$ $9x + 6y = 30$ $3x + 2y = 10$ Solution:

$4x - 3y = 2 \Rightarrow \times (2)$ $\underline{8x - 6y = 4}$ $3(2) + 2y = 10$ $\boxed{(2, 2)}$

                              $17x = 34$ $6 + 2y = 10$

                                $x = 2$ $2y = 4$

                                        $y = 2$

59.   Answers may vary.   **61.** $(1, 4)$

Exercise 3.6   (page 206)

1. $x < 4$

3. $-1 < x \le 2$

5. $8 \cdot 8 \cdot 8 \cdot c = 8^3 c$ **7.** $a \cdot a \cdot b \cdot b = a^2 b^2$ **9.**   variable   **11.**   system

13.   Let x and y represent the integers.   Substitute $x = 2y$ into (2):   Substitute; solve for x:

$\begin{cases} (1) & x = 2y \\ (2) & x + y = 96 \end{cases}$ $\begin{aligned} x + y &= 96 \\ 2y + y &= 96 \\ 3y &= 96 \\ y &= 32 \end{aligned}$ $\begin{aligned} x &= 2y \\ x &= 2(32) = 64 \end{aligned}$

The integers are 32 and 64.

15.   Let x and y represent the integers.

$3x + y = 29 \Rightarrow \times (-2)$ $-6x - 2y = -58$ $x + 2y = 18$ The integers are 5 and 8.

$\underline{x + 2y = 18}$ $\underline{x + 2y = 18}$ $8 + 2y = 18$

                        $-5x = -40$ $2y = 10$

                        $x = 8$ $y = 5$

17. Let x = number of horses and y = number of cows.

$\begin{cases} (1) & y = 5x \\ (2) & x + y = 168 \end{cases}$

Substitute $y = 5x$ from (1) into (2):

$x + y = 168$
$x + 5x = 168$
$6x = 168$
$x = 28$

Substitute and solve for y:

$y = 5x$
$y = 5(28) = 140$
He has 140 cows.

19. Let x = cost of a gallon of paint and y = cost of a brush.

$\begin{array}{l} 8x + 3y = 135 \Rightarrow \times (2) \\ 6x + 2y = 100 \Rightarrow \times (-3) \end{array}$

$\begin{array}{l} 16x + 6y = 270 \\ -18x - 6y = -300 \\ \hline -2x = -30 \\ x = 15 \end{array}$

$6x + 2y = 100$
$6(15) + 2y = 100$
$90 + 2y = 100$
$2y = 10$
$y = 5$

The paint costs $15 per gallon, while each brush costs $5.

21. Let x = cost of contact cleaner and y = cost of soaking solution.

$\begin{array}{l} 2x + 3y = 29.40 \Rightarrow \times (-2) \\ 3x + 2y = 28.60 \Rightarrow \times (3) \end{array}$

$\begin{array}{l} -4x - 6y = -58.80 \\ 9x + 6y = 85.80 \\ \hline 5x = 27 \\ x = 5.40 \end{array}$

$2x + 3y = 29.40$
$2(5.40) + 3y = 29.40$
$10.80 + 3y = 29.40$
$3y = 18.60$
$y = 6.20$

The contact cleaner costs $5.40, while the soaking solution costs $6.20.

23. Let x = the length of one piece.
Let y = the other length.

$\begin{cases} (1) & x = y + 5 \\ (2) & x + y = 25 \end{cases}$

Substitute $x = y + 5$ into (2):

$x + y = 25$
$y + 5 + y = 25$
$2y = 20$
$y = 10$

Substitute; solve for x:

$x = y + 5$
$x = 10 + 5 = 15$
The lengths are 10 ft and 15 ft.

25. Let x = Lena's amount.
Let y = Chayla's amount.

$\begin{cases} (1) & x = y + 50000 \\ (2) & x + y = 250000 \end{cases}$

Substitute $x = y + 50000$ into (2):

$x + y = 250000$
$y + 50000 + y = 250000$
$2y = 200000$
$y = 100000$

Chayla will get $100,000.

27. Let x = time for causes.
Let y = time for outcome.

$\begin{cases} (1) & x = 4y \\ (2) & x + y = 30 \end{cases}$

Substitute $x = 4y$ into (2):

$x + y = 30$
$4y + y = 30$
$5y = 30$
$y = 6$

Substitute; solve for x:

$x = 4y$
$x = 4(6) = 24$
24 minutes will be devoted to the causes, and 6 minutes to the outcome.

29. Let x = deaths from cancer.   Substitute $x = 6y$ into (2): Substitute; solve for x:

Let y = deaths from accidents. $x + y = 630000$ $x = 6y$

$\begin{cases} (1) & x = 6y \\ (2) & x + y = 630000 \end{cases}$ $6y + y = 630000$ $x = 6(90000) = 540000$

$7y = 630000$ 90,000 die from accidents and

$y = 90000$ 540,000 die from cancer.

31. Let w = the width.   Substitute $l = w + 5$ into (2): Substitute; solve for l:

Let l = the length. $2l + 2w = 110$ $l = w + 5$

$\begin{cases} (1) & l = w + 5 \\ (2) & 2l + 2w = 110 \end{cases}$ $\begin{array}{l} 2(w + 5) + 2w = 110 \\ 2w + 10 + 2w = 110 \end{array}$ $l = 25 + 5 = 30$

The dimensions are 25 ft by 30 ft.

$4w = 100$

$w = 25$

33. Let x = the length.   Substitute $x = 2y + 2$ into (2): Substitute; solve for x:

Let y = the width. $2x + 2y = 34$ $x = 2y + 2$

$\begin{cases} (1) & x = 2y + 2 \\ (2) & 2x + 2y = 34 \end{cases}$ $\begin{array}{l} 2(2y + 2) + 2y = 34 \\ 4y + 4 + 2y = 34 \end{array}$ $x = 2(5) + 2 = 12$

The area is $(5)(12) = 60$ ft^2.

$6y = 30$

$y = 5$

35. Let c = the cost.   Substitute: The break point is about 9.9 years.

Let n = number of years. $2250 + 412n = 1715 + 466n$

$\begin{cases} (1) & c = 2250 + 412n \\ (2) & c = 1715 + 466n \end{cases}$ $-54n = -535$

$n \approx 9.9$

37. Since 7 years is less than the break point, choose the furnace with the lower up-front cost (80+).

39. Let x = Bill's principal and y = Janette's principal.

$\begin{array}{r} x + y = 5000 \\ 0.05x + 0.07y = 310 \end{array}$ $\begin{array}{l} \Rightarrow \times (100) \end{array}$ $\begin{array}{r} x + y = 5000 \\ 5x + 7y = 31000 \end{array}$ $\begin{array}{l} \Rightarrow \times (-5) \end{array}$ $\begin{array}{r} -5x - 5y = -25000 \\ 5x + 7y = 31000 \\ \hline 2y = 6000 \\ y = 3000 \end{array}$

$x + y = 5000$ Bill invested $2000.

$x + 3000 = 5000$

$x = 2000$

41. Let x = number of student tickets and y = number of nonstudent tickets.

$\begin{array}{r} x + y = 350 \\ x + 2y = 450 \end{array}$ $\begin{array}{l} \Rightarrow \times (-1) \end{array}$ $\begin{array}{r} -x - y = -350 \\ x + 2y = 450 \\ \hline y = 100 \end{array}$ $\begin{array}{l} x + y = 350 \\ x + 100 = 350 \\ x = 250 \end{array}$ There were 250 student tickets sold.

43. Let b = speed of boat in still water and c = speed of current.

	d	r	t	Equation $(d = r \cdot t)$
Downstream	24	$b+c$	2	$24 = (b+c) \cdot 2 \Rightarrow 2b + 2c = 24$
Upstream	24	$b-c$	3	$24 = (b-c) \cdot 3 \Rightarrow 3b - 3c = 24$

$$\begin{array}{ll} 2b + 2c = 24 \Rightarrow \times (3) & 6b + 6c = 72 \\ 3b - 3c = 24 \Rightarrow \times (2) & 6b - 6c = 48 \\ \hline & 12b \quad\quad = 120 \\ & b \quad\quad = 10 \end{array}$$

The speed of the boat is 10 mph in still water.

45. Let p = airspeed of plane and w = speed of wind.

	d	r	t	Equation $(d = r \cdot t)$
With wind	600	$p+w$	2	$600 = (p+w) \cdot 2 \Rightarrow 2p + 2w = 600$
Against wind	600	$p-w$	3	$600 = (p-w) \cdot 3 \Rightarrow 3p - 3w = 600$

$$\begin{array}{ll} 2p + 2w = 600 \Rightarrow \times (3) & 6p + 6w = 1800 \\ 3p - 3w = 600 \Rightarrow \times (2) & 6p - 6w = 1200 \\ \hline & 12p \quad\quad = 3000 \\ & p \quad\quad = 250 \end{array}$$

$$\begin{array}{l} 2p + 2w = 600 \\ 2(250) + 2w = 600 \\ 500 + 2w = 600 \\ 2w = 100 \\ w = 50 \end{array}$$

The speed of the wind is 50 mph.

47. Let x = liters of first solution and y = liters of second solution.

	Fractional part that is alcohol	Number of liters of solution	Number of liters of alcohol
First solution	0.40	x	$0.40x$
Second solution	0.55	y	$0.55y$
Final solution	0.50	15	$0.50(15) = 7.5$

$$\begin{array}{l} x + y = 15 \\ 0.40x + 0.55y = 7.5 \Rightarrow \times (100) \end{array} \quad \begin{array}{l} x + y = 15 \Rightarrow \times (-40) \\ 40x + 55y = 750 \end{array} \quad \begin{array}{l} -40x - 40y = -600 \\ 40x + 55y = 750 \\ \hline 15y = 150 \\ y = 10 \end{array}$$

$$\begin{array}{l} x + y = 15 \\ x + 10 = 15 \\ x = 5 \end{array}$$

The chemist should use 5 L of the 40% solution and 10 L of the 55% solution.

49. Let x = pounds of peanuts and y = pounds of cashews.

	Cost per pound	Number of pounds	Total value
Peanuts	3	x	$3x$
Cashews	6	y	$6y$
Mixture	4	48	$4(48) = 192$

$$\begin{array}{l} x + y = 48 \Rightarrow \times (-3) \\ 3x + 6y = 192 \end{array} \quad \begin{array}{l} -3x - 3y = -144 \\ 3x + 6y = 192 \\ \hline 3y = 48 \\ y = 16 \end{array}$$

$$\begin{array}{l} x + y = 48 \\ x + 16 = 48 \\ x = 32 \end{array}$$

The merchant should use 32 pounds of peanuts and 16 pounds of cashews.

51. Let x = number of inexpensive radios and y = number of expensive radios.

$$\begin{array}{l} x + y = 25 \Rightarrow \times (-87) \\ 87x + 119y = 2495 \end{array} \qquad \begin{array}{r} -87x - 87y = -2175 \\ \underline{87x + 119y = 2495} \\ 32y = 320 \\ y = 10 \end{array} \qquad \begin{array}{l} x + y = 25 \quad \text{There were 15} \\ x + 10 = 25 \quad \text{inexpensive radios} \\ x = 15 \quad \text{sold.} \end{array}$$

53. Let r = the lower rate. Then $r + 0.01$ = the higher rate.

$$\boxed{\text{Interest on \$950}} + \boxed{\text{Interest on \$1,200}} = \boxed{\text{Total interest}}$$

$$950r + 1200(r + 0.01) = 205.50$$
$$950r + 1200r + 12 = 205.50$$
$$2150r = 193.50$$
$$r = 0.09 \Rightarrow \text{The lower rate is 9\%.}$$

55. $\begin{cases} (1) & p = -\frac{1}{2}q + 1300 \\ (2) & p = \frac{1}{3}q + \frac{1400}{3} \end{cases}$

Substitute:

$$-\frac{1}{2}q + 1300 = \frac{1}{3}q + \frac{1400}{3}$$
$$6\left(-\frac{1}{2}q + 1300\right) = 6\left(\frac{1}{3}q + \frac{1400}{3}\right)$$
$$-3q + 7800 = 2q + 2800$$
$$-5q = -5000$$
$$q = 1000$$

Substitute $q = 1000$ and solve for p:

$$p = -\frac{1}{2}q + 1300$$
$$= -\frac{1}{2}(1000) + 1300$$
$$= -500 + 1300 = 800$$

The equilibrium price is \$800.

57. **Answers may vary.**

Exercise 3.7 (page 218)

1. $3x + 5 = 14$
$$3x = 9$$
$$x = 3$$

3. $A = P + Prt$
$$A - P = Prt$$
$$\frac{A - P}{Pr} = \frac{Prt}{Pr}$$
$$\frac{A - P}{Pr} = t, \text{ or } t = \frac{A - P}{Pr}$$

5. $2a + 5(a - 3) = 2a + 5a - 15 = 7a - 15$

7. $4(b - a) + 3b + 2a = 4b - 4a + 3b + 2a$
$$= -2a + 7b$$

9. inequality

11. boundary

13. inequalities

15. doubly shaded

17. a. $5x - 3y \geq 0$
$$5(1) - 3(1) \overset{?}{\geq} 0$$
$$5 - 3 \overset{?}{\geq} 0$$
$$2 \geq 0$$
$(1, 1)$ is a solution.

b. $5x - 3y \geq 0$
$$5(-2) - 3(-3) \overset{?}{\geq} 0$$
$$-10 + 9 \overset{?}{\geq} 0$$
$$-1 \ngeq 0$$
$(-2, -3)$ is not a solution.

17. **c.**
$$5x - 3y \geq 0$$
$$5(0) - 3(0) \stackrel{?}{\geq} 0$$
$$0 - 0 \stackrel{?}{\geq} 0$$
$$0 \geq 0$$
$(0, 0)$ is a solution.

d.
$$5x - 3y \geq 0$$
$$5\left(\frac{1}{5}\right) - 3\left(\frac{4}{3}\right) \stackrel{?}{\geq} 0$$
$$1 - 4 \stackrel{?}{\geq} 0$$
$$-3 \not\geq 0$$
$\left(\frac{1}{5}, \frac{4}{3}\right)$ is not a solution.

19. **a.** does not include boundary

b. does include boundary

21. Test point: $(0, 0)$
$$y \leq x + 2$$
$$0 \stackrel{?}{\leq} 0 + 2$$
$$0 \leq 2$$
Shade half-plane with test point.

23. Test point: $(0, 0)$
$$y > 2x - 4$$
$$0 \stackrel{?}{>} 2(0) - 4$$
$$0 \stackrel{?}{>} 0 - 4$$
$$0 > -4$$
Shade half-plane with test point.

25. Test point: $(0, 0)$
$$x - 2y \leq 4$$
$$0 - 2(0) \stackrel{?}{\leq} 4$$
$$0 - 0 \stackrel{?}{\leq} 4$$
$$0 \leq 4$$
Shade half-plane with test point.

27. Test point: $(1, 1)$
$$y \leq 4x$$
$$1 \stackrel{?}{\leq} 4(1)$$
$$1 \leq 4$$
Shade half-plane with test point.

29. Boundary   Test point: $(0,0)$
(solid) $y \geq 3 - x$
$y = 3 - x$ $0 \overset{?}{\geq} 3 - 0$

x	y
0	3
3	0

$0 \not\geq 3$

opposite half-plane

31. Boundary   Test point: $(0,0)$
(dotted) $y < 2 - 3x$
$y = 2 - 3x$ $0 \overset{?}{<} 2 - 3(0)$

x	y
0	2
1	-1

$0 \overset{?}{<} 2 - 0$

$0 < 2$

same half-plane

33. Boundary   Test point: $(1,1)$
(solid) $y \geq 2x$
$y = 2x$ $1 \overset{?}{\geq} 2(1)$

x	y
0	0
2	4

$1 \not\geq 2$

opposite half-plane

35. Boundary   Test point: $(0,0)$
(dotted) $2y - x < 8$
$2y - x = 8$ $2(0) - 0 \overset{?}{<} 8$

x	y
0	4
-8	0

$0 - 0 \overset{?}{<} 8$

$0 < 8$

same half-plane

37.   Boundary   Test point: $(0,0)$
        (dotted) $3x - 4y > 12$

$3x - 4y = 12$ $3(0) - 4(0) \overset{?}{>} 12$

x	y
0	-3
4	0

$0 - 0 \overset{?}{>} 12$

$0 \not> 12$

opposite half-plane

39.   Boundary   Test point: $(0,0)$
        (solid) $5x + 4y \geq 20$

$5x + 4y = 20$ $5(0) + 4(0) \overset{?}{\geq} 20$

x	y
0	5
4	0

$0 + 0 \overset{?}{\geq} 20$

$0 \not\geq 20$

opposite half-plane

41.   Boundary   Test point: $(0,0)$
        (dotted) $x < 2$

$x = 2$ $0 < 2$

x	y
2	0
2	4

same half-plane

43.   Boundary   Test point: $(0,0)$
        (solid) $y \leq 1$

$y = 1$ $0 \leq 1$

x	y
0	1
4	1

same half-plane

45. $x + 2y \le 3 \quad 2x - y \ge 1$

x	y
1	1
3	0

x	y
0	-1
2	3

47. $x + y < -1 \quad x - y > -1$

x	y
0	-1
-1	0

x	y
0	1
-1	0

49. $2x - y < 4 \quad x + y \ge -1$

x	y
0	-4
2	0

x	y
0	-1
-1	0

51. $x > 2 \qquad y \le 3$

x	y
2	0
2	2

x	y
0	3
1	3

53. $x + y < 1 \quad x + y > 3$

x	y
0	1
1	0

x	y
0	3
3	0

(no solutions)

55. $3x + 4y > -7 \quad 2x - 3y \ge 1$

x	y
-1	-1
-5	2

x	y
2	1
-1	-1

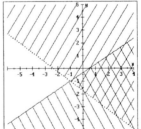

57. $2x - 4y > -6$ $3x + y \geq 5$

x	y
1	2
-3	0

x	y
0	5
1	2

59. $3x - y \leq -4$ $3y > -2(x+5)$

x	y
0	4
-1	1

x	y
1	-4
-5	0

61. $\dfrac{x}{2} + \dfrac{y}{3} \geq 2$ $\dfrac{x}{2} - \dfrac{y}{2} < -1$

$3x + 2y \geq 12$ $x - y < -2$

x	y
0	6
4	0

x	y
0	2
-2	0

63.

$$\boxed{\text{Cake cost}} + \boxed{\text{Pie cost}} \leq 120$$

$$3x + 4y \leq 120$$

Boundary   Test point: $(0,0)$
(solid) $3x + 4y \leq 120$

$3x + 4y = 120$ $3(0) + 4(0) \overset{?}{\leq} 120$

x	y
0	30
40	0

$0 + 0 \overset{?}{\leq} 120$

$0 \leq 120$

same half-plane

Solutions: $(10, 10), (20, 10), (10, 20)$

65. $\boxed{\text{Leather}} + \boxed{\text{Nylon}} \geq 4400$

$100x + 88y \geq 4400$

Boundary Test point: $(0,0)$
(solid) $100x + 88y \geq 4400$

x	y
0	50
44	0

$100(0) + 88(0) \overset{?}{\geq} 4400$

$0 + 0 \overset{?}{\geq} 4400$

$0 \not\geq 4400$

opposite half-plane

Solutions: $(50,50), (30,40), (40,40)$

67. $\boxed{\text{R. stock}} + \boxed{\text{M. stock}} \leq 8000$

$40x + 50y \leq 8000$

Boundary Test point: $(0,0)$
(solid) $40x + 50y \leq 8000$

$40x + 50y = 8000$

x	y
0	160
200	0

$40(0) + 50(0) \overset{?}{\leq} 8000$

$0 + 0 \overset{?}{\leq} 8000$

$0 \leq 8000$

same half-plane

Solutions: $(80,40), (80,80), (120,40)$

69. $10x + 15y \geq 30$ $10x + 15y \leq 60$

x	y
0	2
3	0

x	y
0	4
6	0

Solutions: $(1,2), (4,1)$

71. $150x + 100y \leq 900$ $y > x$

x	y
0	9
6	0

x	y
0	0
1	1

Solutions: $(2,4), (1,5)$

73. **Answers may vary.**

75. **Answers may vary.**

77. **Answers may vary.**

79. **Answers may vary.**

Chapter 3 Summary (page 224)

1-6.

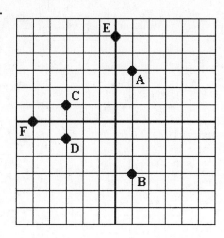

7. $(3, 1)$
8. $(-4, 5)$
9. $(-3, -4)$
10. $(2, -3)$
11. $(0, 0)$
12. $(0, 4)$
13. $(-5, 0)$
14. $(0, -3)$

15.
$$3x - 4y = 12$$
$$3(2) - 4(1) \overset{?}{=} 12$$
$$6 - 4 \overset{?}{=} 12$$
$$2 \neq 12$$
$(2, 1)$ is not a solution.

16.
$$3x - 4y = 12$$
$$3(3) - 4\left(-\tfrac{3}{4}\right) \overset{?}{=} 12$$
$$9 + 3 \overset{?}{=} 12$$
$$12 = 12$$
$\left(3, -\tfrac{3}{4}\right)$ is a solution.

17. $y = x - 5$

x	y
0	-5
2	-3

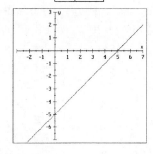

18. $y = 2x + 1$

x	y
0	1
2	5

19. $y = \dfrac{x}{2} + 2$

x	y
0	2
2	3

97

20.

$y = 3$

x	y
0	3
2	3

21.

$x + y = 4$

x	y
0	4
4	0

22.

$x - y = -3$

x	y
0	3
-3	0

23. $3x + 5y = 15$

x	y
0	3
5	0

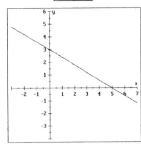

24. $7x - 4y = 28$

x	y
0	-7
4	0

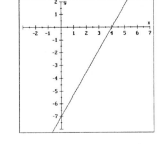

25. $3x - y = -2 \qquad 2x + 3y = 17$

$3(1) - 5 \overset{?}{=} -2 \quad 2(1) + 3(5) \overset{?}{=} 17$

$3 - 5 \overset{?}{=} -2 \qquad 2 + 15 \overset{?}{=} 17$

$-2 = -2 \qquad\qquad 17 = 17$

$(1, 5)$ is a solution to the system.

26. $5x + 3y = 2 \qquad -3x + 2y = 16$

$5(-2) + 3(4) \overset{?}{=} 2 \quad -3(-2) + 2(4) \overset{?}{=} 16$

$-10 + 12 \overset{?}{=} 2 \qquad\qquad 6 + 8 \overset{?}{=} 16$

$2 = 2 \qquad\qquad\qquad 14 \neq 16$

$(-2, 4)$ is not a solution to the system.

27. $2x + 4y = 30 \qquad \dfrac{x}{4} - y = 3$

$2(14) + 4\left(\dfrac{1}{2}\right) \overset{?}{=} 30 \quad \dfrac{14}{4} - \dfrac{1}{2} \overset{?}{=} 3$

$28 + 2 \overset{?}{=} 30 \qquad \dfrac{7}{2} - \dfrac{1}{2} \overset{?}{=} 3$

$30 = 30 \qquad\qquad \dfrac{6}{2} = 3$

$\left(14, \frac{1}{2}\right)$ is a solution to the system.

28. $4x - 6y = 18 \qquad \dfrac{x}{3} + \dfrac{y}{2} = \dfrac{5}{6}$

$4\left(\dfrac{7}{2}\right) - 6\left(-\dfrac{2}{3}\right) \overset{?}{=} 18 \quad \dfrac{\frac{7}{2}}{3} + \dfrac{-\frac{2}{3}}{2} \overset{?}{=} \dfrac{5}{6}$

$14 + 4 \overset{?}{=} 18 \qquad\qquad \dfrac{7}{6} - \dfrac{2}{6} \overset{?}{=} \dfrac{5}{6}$

$18 = 18 \qquad\qquad\qquad \dfrac{5}{6} = \dfrac{5}{6}$

$\left(\frac{7}{2}, -\frac{2}{3}\right)$ is a solution to the system.

29. $\quad x + y = 7 \qquad 2x - y = 5$

x	y
0	7
7	0

x	y
2	-1
3	1

$\boxed{(4, 3)}$

30. $\quad \dfrac{x}{3} + \dfrac{y}{5} = -1 \qquad x - 3y = -3$

x	y
0	-5
-3	0

x	y
0	1
-3	0

$\boxed{(-3, 0)}$

31. $\quad 3x + 6y = 6 \qquad x + 2y = 2$

x	y
0	1
2	0

x	y
0	1
2	0

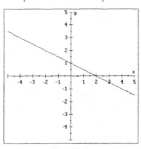

(dependent equations)

32. $\quad 6x + 3y = 12 \qquad 2x + y = 2$

x	y
0	4
2	0

x	y
0	2
1	0

(inconsistent system)

33. $\begin{cases} (1) \quad x = 3y + 5 \\ (2) \quad 5x - 4y = 3 \end{cases}$

Substitute $x = 3y + 5$ from (1):

$$5x - 4y = 3$$
$$5(3y + 5) - 4y = 3$$
$$15y + 25 - 4y = 3$$
$$11y = -22$$
$$y = -2$$

Substitute and solve for x:

$$x = 3y + 5$$
$$x = 3(-2) + 5$$
$$x = -6 + 5 = -1$$

Solution:

$\boxed{x = -1, y = -2}$

34.
$$\begin{cases} 3x - \dfrac{2y}{5} = 2(x-2) \Rightarrow 15x - 2y = 10(x-2) \Rightarrow 5x - 2y = -20 \ \ (1) \\ 2x - 3 = 3 - 2y \quad\quad \Rightarrow 2x + 2y = 6 \quad\quad\quad \Rightarrow 2x + 2y = 6 \ \ (2) \end{cases}$$

Substitute $y = \dfrac{6-2x}{2}$ from (2):   Substitute and solve for y: Solution:

$$5x - 2y = -20$$
$$5x - 2\left(\frac{6-2x}{2}\right) = -20$$
$$5x - (6 - 2x) = -20$$
$$5x - 6 + 2x = -20$$
$$7x = -14$$
$$x = -2$$

$$y = \frac{6-2x}{2}$$
$$y = \frac{6 - 2(-2)}{2}$$
$$y = \frac{6+4}{2} = \frac{10}{2} = 5$$

$$\boxed{x = -2, y = 5}$$

35.
$$\begin{cases} 8x + 5y = 3 \ \ (1) \\ 5x - 8y = 13 \ \ (2) \end{cases}$$

Substitute $y = \dfrac{3-8x}{5}$ from (1):   Substitute and solve for y: Solution:

$$5x - 8y = 13$$
$$5x - 8\left(\frac{3-8x}{5}\right) = 13$$
$$25x - 8(3 - 8x) = 65$$
$$25x - 24 + 64x = 65$$
$$89x = 89$$
$$x = 1$$

$$y = \frac{3-8x}{5}$$
$$y = \frac{3 - 8(1)}{5}$$
$$y = \frac{3-8}{5} = \frac{-5}{5} = -1$$

$$\boxed{x = 1, y = -1}$$

36.
$$\begin{cases} 6(x+2) = y - 1 \Rightarrow 6x + 12 = y - 1 \Rightarrow 6x - y = -13 \ \ (1) \\ 5(y-1) = x + 2 \Rightarrow 5y - 5 = x + 2 \Rightarrow -x + 5y = 7 \ \ (2) \end{cases}$$

Substitute $y = 6x + 13$ from (1):   Substitute and solve for y: Solution:

$$-x + 5y = 7$$
$$-x + 5(6x + 13) = 7$$
$$-x + 30x + 65 = 7$$
$$29x = -58$$
$$x = -2$$

$$y = 6x + 13$$
$$y = 6(-2) + 13$$
$$y = -12 + 13 = 1$$

$$\boxed{x = -2, y = 1}$$

37.
$$\begin{array}{l} 2x + y = 1 \\ \underline{5x - y = 20} \\ 7x \quad\;\; = 21 \\ \;\; x \quad\;\; = 3 \end{array}$$

$$2x + y = 1$$
$$2(3) + y = 1$$
$$6 + y = 1$$
$$y = -5$$

Solution:
$$\boxed{(3, -5)}$$

38.
$$\begin{array}{l} x + 8y = 7 \Rightarrow \times (-1) \\ \underline{x - 4y = 1} \end{array}$$

$$\begin{array}{r} -x - 8y = -7 \\ \underline{x - 4y = 1} \\ -12y = -6 \\ y = \tfrac{1}{2} \end{array}$$

$$x + 8y = 7$$
$$x + 8\left(\tfrac{1}{2}\right) = 7$$
$$x + 4 = 7$$
$$x = 3$$

Solution:
$$\boxed{\left(3, \tfrac{1}{2}\right)}$$

39. $\begin{aligned} 5x + y &= 2 \Rightarrow \times (-2) \\ 3x + 2y &= 11 \end{aligned}$ $\begin{aligned} -10x - 2y &= -4 \\ 3x + 2y &= 11 \\ \hline -7x &= 7 \\ x &= -1 \end{aligned}$ $\begin{aligned} 5x + y &= 2 \\ 5(-1) + y &= 2 \\ -5 + y &= 2 \\ y &= 7 \end{aligned}$ Solution: $\boxed{(-1, 7)}$

40. $\begin{aligned} x + y &= 3 \Rightarrow x + y = 3 \Rightarrow \times (-1) \\ 3x &= 2 - y \Rightarrow 3x + y = 2 \end{aligned}$ $\begin{aligned} -x - y &= -3 \\ 3x + y &= 2 \\ \hline 2x &= -1 \\ x &= -\frac{1}{2} \end{aligned}$ $\begin{aligned} x + y &= 3 \\ -\frac{1}{2} + y &= \frac{6}{2} \\ y &= \frac{7}{2} \end{aligned}$ Solution: $\boxed{\left(-\frac{1}{2}, \frac{7}{2}\right)}$

41. $\begin{aligned} 11x + 3y &= 27 \Rightarrow \times (4) \\ 8x + 4y &= 36 \Rightarrow \times (-3) \end{aligned}$ $\begin{aligned} 44x + 12y &= 108 \\ -24x - 12y &= -108 \\ \hline 20x &= 0 \\ x &= 0 \end{aligned}$ $\begin{aligned} 11x + 3y &= 27 \\ 11(0) + 3y &= 27 \\ 0 + 3y &= 27 \\ 3y &= 27 \\ y &= 9 \end{aligned}$ Solution: $\boxed{(0, 9)}$

42. $\begin{aligned} 9x + 3y &= 5 \Rightarrow 9x + 3y = 5 \\ 3x &= 4 - y \Rightarrow 3x + y = 4 \Rightarrow \times (-3) \end{aligned}$ $\begin{aligned} 9x + 3y &= 5 \\ -9x - 3y &= -12 \\ \hline 0 &\neq -7 \end{aligned}$ $\boxed{\text{inconsistent system}}$

43. $\begin{aligned} 9x + 3y &= 5 \\ 3x + y &= \frac{5}{3} \Rightarrow \times (-3) \end{aligned}$ $\begin{aligned} 9x + 3y &= 5 \\ -9x - 3y &= -5 \\ \hline 0 &= 0 \end{aligned}$ $\boxed{\text{dependent equations}}$

44. $\dfrac{x}{3} + \dfrac{y+2}{2} = 1 \Rightarrow \times (6)$ $2x + 3(y + 2) = 6 \Rightarrow$ $2x + 3y + 6 = 6$

$\dfrac{x+8}{8} + \dfrac{y-3}{3} = 0 \Rightarrow \times (24)$ $3(x + 8) + 8(y - 3) = 0 \Rightarrow 3x + 24 + 8y - 24 = 0$

$\begin{aligned} 2x + 3y &= 0 \Rightarrow \times (3) \\ 3x + 8y &= 0 \Rightarrow \times (-2) \end{aligned}$ $\begin{aligned} 6x + 9y &= 0 \\ -6x - 16y &= 0 \\ \hline -7y &= 0 \\ y &= 0 \end{aligned}$ $\begin{aligned} 2x + 3y &= 0 \\ 2x + 3(0) &= 0 \\ 2x &= 0 \\ x &= 0 \end{aligned}$ Solution: $\boxed{(0, 0)}$

45. Let x and y represent the integers.   Substitute $x = 5y$ into (2):   Substitute; solve for x:

$\begin{cases} (1) \quad x = 5y \\ (2) \quad x + y = 18 \end{cases}$

$\begin{aligned} x + y &= 18 \\ 5y + y &= 18 \\ 6y &= 18 \\ y &= 3 \end{aligned}$

$\begin{aligned} x &= 5y \\ x &= 5(3) = 15 \end{aligned}$

The integers are 3 and 15.

46. Let $w =$ the width. Substitute $l = 3w$ into (2): Substitute; solve for l:

Let $l =$ the length. $\quad 2l + 2w = 24$ $\quad l = 3w$

$\begin{cases} (1) \quad l = 3w \\ (2) \quad 2l + 2w = 24 \end{cases}$ $\quad 2(3w) + 2w = 24$ $\quad l = 3(3) = 9$

$\qquad\qquad\qquad\qquad\quad 6w + 2w = 24$ The dimensions are 3 ft by 9 ft.

$\qquad\qquad\qquad\qquad\qquad\quad 8w = 24$

$\qquad\qquad\qquad\qquad\qquad\quad\; w = 3$

47. Let $x =$ the cost of a grapefruit. Substitute $x = y + 15$ into (2): Substitute; solve for x:

Let $y =$ the cost of an orange. $\quad x + y = 85$ $\quad x = y + 15$

$\begin{cases} (1) \quad x = y + 15 \\ (2) \quad x + y = 85 \end{cases}$ $\quad y + 15 + y = 85$ $\quad x = 35 + 15 = 50$

$\qquad\qquad\qquad\qquad\qquad\quad 2y = 70$ A grapefruit costs 50¢.

$\qquad\qquad\qquad\qquad\qquad\quad\; y = 35$

48. Let $x =$ the electric bill. Substitute $x = y - 23$ into (2): The gas bill was $66.

Let $y =$ the gas bill. $\quad x + y = 109$

$\begin{cases} (1) \quad x = y - 23 \\ (2) \quad x + y = 109 \end{cases}$ $\quad y - 23 + y = 109$

$\qquad\qquad\qquad\qquad\qquad\; 2y = 132$

$\qquad\qquad\qquad\qquad\qquad\;\; y = 66$

49. Let $x =$ cost of a gallon of milk and $y =$ cost of one dozen eggs.

$2x + 3y = 6.80 \Rightarrow \times (-2) \quad -4x - 6y = -13.60$ A gallon of milk costs $1.69.

$3x + 2y = 7.35 \Rightarrow \times (3) \qquad \underline{9x + 6y = \quad 22.05}$

$\qquad\qquad\qquad\qquad\qquad\qquad\quad 5x \quad = \quad 8.45$

$\qquad\qquad\qquad\qquad\qquad\qquad\quad\;\; x \quad = \quad 1.69$

50. Let $x =$ principal at 10% and $y =$ principal at 6%.

$x + \quad y = 3000$ $\qquad\qquad\qquad x + \quad y = \quad 3000 \Rightarrow \times (-6) \quad -6x - 6y = -18000$

$\underline{0.10x + 0.06y = \quad 270} \Rightarrow \times (100) \quad \underline{10x + 6y = 27000}$ $\qquad\qquad\qquad \underline{10x + 6y = \quad 27000}$

$\qquad\qquad\qquad\qquad\qquad\qquad\qquad\qquad\qquad\qquad\qquad\qquad\qquad\qquad\qquad 4x \quad = \quad 9000$

$\qquad\qquad\qquad\qquad\qquad\qquad\qquad\qquad\qquad\qquad\qquad\qquad\qquad\qquad\qquad\;\; x \quad = \quad 2250$

$x + y = 3000$ He invested $750 at the 6% rate.

$2250 + y = 3000$

$\qquad\quad\; y = 750$

51. Boundary   Test point: $(0,0)$
          (solid) $y \geq x + 2$

$y = x + 2$ $0 \overset{?}{\geq} 0 + 2$

x	y
0	2
-2	0

$0 \ngeq 2$

opposite half-plane

52. Boundary   Test point: $(0,0)$
          (dotted) $x < 3$

$x = 3$ $0 < 3$

x	y
3	0
3	4

same half-plane

53. $5x + 3y < 15$ $3x - y > 3$

x	y
0	5
3	0

x	y
0	-3
1	0

54. $5x - 3y \geq 5$ $3x + 2y \geq 3$

x	y
1	0
4	5

x	y
1	0
3	-3

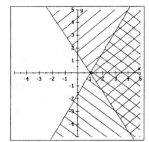

55. $x \geq 3y$ $y < 3x$

x	y
0	0
3	1

x	y
0	0
1	3

56. $x \geq 0$ $x \leq 3$

x	y
0	0
0	4

x	y
3	1
3	0

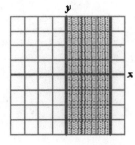

Chapter 3 Test (page 228)

1.
$$y = \frac{x}{2} + 1$$

x	y
0	1
2	3

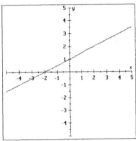

2.
$$2(x+1) - y = 4$$
$$2x + 2 - y = 4$$
$$2x - y = 2$$

x	y
0	-2
1	0

3.
$$x = 1$$

x	y
1	2
1	0

4.
$$2y = 8$$
$$y = 4$$

x	y
0	4
2	4

5.
$$3x - 2y = 12 \qquad 2x + 3y = -5$$
$$3(2) - 2(-3) \overset{?}{=} 12 \quad 2(2) + 3(-3) \overset{?}{=} -5$$
$$6 + 6 \overset{?}{=} 12 \qquad 4 - 9 \overset{?}{=} -5$$
$$12 = 12 \qquad\qquad -5 = -5$$

$(2, -3)$ is a solution to the system.

6.

$$4x + y = -9 \qquad 2x - 3y = -7$$
$$4(-2) + (-1) \overset{?}{=} -9 \quad 2(-2) - 3(-1) \overset{?}{=} -7$$
$$-8 - 1 \overset{?}{=} -9 \qquad -4 + 3 \overset{?}{=} -7$$
$$-9 = -9 \qquad\qquad -1 \neq -7$$

$(-2, -1)$ is not a solution to the system.

7. $3x + y = 7 \qquad x - 2y = 0$

x	y
0	7
2	1

x	y
0	0
4	2

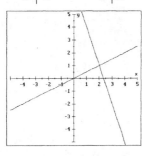

$\boxed{(2, 1)}$

8. $x + \dfrac{y}{2} = 1 \qquad y = 1 - 3x$

x	y
0	2
1	0

x	y
0	1
1	-2

$\boxed{(-1, 4)}$

9.
$$\begin{cases} (1) \quad y = x - 1 \\ (2) \quad x + y = -7 \end{cases}$$

Substitute $y = x - 1$ from (1):   Substitute and solve for y: Solution:

$$x + y = -7 \qquad\qquad y = x - 1 \qquad\qquad \boxed{x = -3, y = -4}$$
$$x + x - 1 = -7 \qquad\qquad y = -3 - 1$$
$$2x = -6 \qquad\qquad y = -4$$
$$x = -3$$

10.
$$\begin{cases} (1) \quad \dfrac{x}{6} + \dfrac{y}{10} = 3 \quad \Rightarrow 5x + 3y = 90 \quad \Rightarrow x = \dfrac{90 - 3y}{5} \\[2mm] (2) \quad \dfrac{5x}{16} - \dfrac{3y}{16} = \dfrac{15}{8} \quad \Rightarrow 5x - 3y = 30 \end{cases}$$

Substitute $x = \dfrac{90 - 3y}{5}$ from (1) into (2):   Substitute and solve for x: Solution:

$$5x - 3y = 30 \qquad\qquad x = \dfrac{90 - 3y}{5} \qquad\qquad \boxed{x = 12, y = 10}$$
$$5\left(\dfrac{90 - 3y}{5}\right) - 3y = 30 \qquad\qquad x = \dfrac{90 - 3(10)}{5}$$
$$90 - 3y - 3y = 30 \qquad\qquad x = \dfrac{90 - 30}{5} = \dfrac{60}{5} = 12$$
$$-6y = -60$$
$$y = 10$$

11. $3x - y = 2 \quad\quad 2x + y = 8$ Solution:

$\dfrac{2x + y = \ 8}{5x \quad\quad = 10} \quad \dfrac{2(2) + y = 8}{4 + y = 8}$ $\boxed{(2, 4)}$

$x \quad\quad = \ 2 \quad\quad\quad y = 4$

12. $4x + \ \ 3 = -3y \quad\quad\quad\quad 4x + 3y = -3 \Rightarrow \times (3) \quad 12x + \ 9y = -9 \quad\quad 4x + 3y = -3$

$\dfrac{-x}{7} + \dfrac{4y}{21} = \ \ 1 \Rightarrow \times (21) \ \underline{-3x + 4y = \ 21 \Rightarrow \times (4)} \ \dfrac{-12x + 16y = \ 84}{25y = 75} \quad\quad \begin{array}{l} 4x + 3(3) = -3 \\ 4x + 9 = -3 \\ 4x = -12 \\ x = -3 \end{array}$

$ y = \ \ 3 \quad\quad\quad \boxed{(-3, 3)}$

13. $2x + 3(y - 2) = 0 \Rightarrow 2x + 3y - 6 = 0 \quad\Rightarrow\quad 2x + 3y = 6$

$\dfrac{-3y = 2(x - 4) \Rightarrow \quad -3y = 2x - 8 \quad \Rightarrow \ -2x - 3y = -8}{ 0 \neq -2 \quad \Rightarrow \text{ inconsistent}}$

14. $\dfrac{x}{3} + y - 4 = 0 \Rightarrow x + 3y - 12 = 0 \quad\Rightarrow\quad x + 3y = 12$

$\dfrac{-3y = x - 12 \Rightarrow \ -x - 3y = -12 \quad \Rightarrow \ -x - 3y = -12}{ 0 = 0 \quad \Rightarrow \text{ dependent (and consistent)}}$

15. Let x and y represent the numbers. Substitute $x = 3y + 2$. Substitute; solve for x:

$\begin{cases} (1) \quad x = 3y + 2 \\ (2) \quad x + y = -18 \end{cases}$
$\quad\quad\quad \begin{array}{l} x + y = -18 \\ 3y + 2 + y = -18 \\ 4y = -20 \\ y = -5 \end{array}$
$\quad\quad \begin{array}{l} x = 3y + 2 \\ x = 3(-5) + 2 \\ x = -15 + 2 = -13 \\ \text{The numbers are } -5 \text{ and } -13. \\ \text{The product is } (-5)(-13) = 65. \end{array}$

16. Let $x =$ principal at 8% and $y =$ principal at 9%.

$\dfrac{x + \quad\ \ y = 10000}{0.08x + 0.09y = \quad 840} \Rightarrow \times (100)$
$\dfrac{x + \ \ y = 10000 \Rightarrow \times (-9)}{8x + 9y = 84000}$
$\dfrac{-9x - 9y = -90000}{\dfrac{8x + 9y = \quad 84000}{\begin{array}{l} -x \quad\quad = \ -6000 \\ x \quad\quad = \quad 6000 \end{array}}}$

$\begin{array}{l} x + y = 10000 \\ 6000 + y = 10000 \\ y = 4000 \end{array}$ She invested \$4000 at the 9% rate.

17. $x + y < 3$ $x - y < 1$

x	y
0	3
3	0

x	y
0	-1
1	0

18. $2x + 3y \le 6$ $x \ge 2$

x	y
0	2
3	0

x	y
2	0
2	3

Exercise 4.1   (page 237)

1.

```
  ◆ ◆ ◆ ◆
 -3  -1  1  3
```

3. the product of 3 and the sum of x and y

5. $|2x| + 3$

7. $-5; 3$

9. $(3x)(3x)(3x)(3x)$

11. $y \cdot y \cdot y \cdot y \cdot y$

13. $x^n y^n$

15. a^{bc}

17. Answers may vary.

19. Base: 4
Exponent: 3

21. Base: x
Exponent: 5

23. Base: $2y$
Exponent: 3

25. Base: x
Exponent: 4

27. Base: x
Exponent: 1

29. Base: x
Exponent: 3

31. $5^3 = 5 \cdot 5 \cdot 5$

33. $x^7 = x \cdot x \cdot x \cdot x \cdot x \cdot x \cdot x$

35. $-4x^5 = -4 \cdot x \cdot x \cdot x \cdot x \cdot x$

37. $(3t)^5 = (3t)(3t)(3t)(3t)(3t)$

39. $2 \cdot 2 \cdot 2 = 2^3$

41. $x \cdot x \cdot x \cdot x = x^4$

43. $(2x)(2x)(2x) = (2x)^3$

45. $-4t \cdot t \cdot t \cdot t = -4t^4$

47. $5^4 = 5 \cdot 5 \cdot 5 \cdot 5 = 625$

49. $2^2 + 3^2 = 4 + 9 = 13$

51. $5^4 - 4^3 = 625 - 64 = 561$

53. $-5(3^4 + 4^3) = -5(81 + 64)$
$= -5(145) = -725$

55. $x^4 x^3 = x^{4+3} = x^7$

57. $x^5 x^5 = x^{5+5} = x^{10}$

59. $tt^2 = t^{1+2} = t^3$

61. $a^3 a^4 a^5 = a^{3+4+5} = a^{12}$

63. $y^3(y^2 y^4) = y^3 y^{2+4} = y^3 y^6 = y^{3+6} = y^9$

65. $4x^2(3x^5) = 4 \cdot 3x^2 x^5 = 12x^{2+5} = 12x^7$

67. $\left(-y^2\right)\left(4y^3\right) = -1 \cdot 4y^2y^3 = -4y^{2+3}$
$$= -4y^5$$

69. $6x^3\left(-x^2\right)\left(-x^4\right) = 6(-1)(-1)x^3x^2x^4$
$$= 6x^{3+2+4} = 6x^9$$

71. $\left(3^2\right)^4 = 3^{2 \cdot 4} = 3^8$

73. $\left(y^5\right)^3 = y^{5 \cdot 3} = y^{15}$

75. $\left(a^3\right)^7 = a^{3 \cdot 7} = a^{21}$

77. $\left(x^2x^3\right)^5 = \left(x^{2+3}\right)^5 = \left(x^5\right)^5 = x^{5 \cdot 5} = x^{25}$

79. $\left(3zz^2z^3\right)^5 = \left(3z^{1+2+3}\right)^5 = \left(3z^6\right)^5 = 3^5\left(z^6\right)^5 = 243z^{6 \cdot 5} = 243z^{30}$

81. $\left(x^5\right)^2\left(x^7\right)^3 = x^{5 \cdot 2}x^{7 \cdot 3} = x^{10}x^{21} = x^{10+21} = x^{31}$

83. $\left(r^3r^2\right)^4\left(r^3r^5\right)^2 = \left(r^{3+2}\right)^4\left(r^{3+5}\right)^2 = \left(r^5\right)^4\left(r^8\right)^2 = r^{5 \cdot 4}r^{8 \cdot 2} = r^{20}r^{16} = r^{20+16} = r^{36}$

85. $\left(s^3\right)^3\left(s^2\right)^2\left(s^5\right)^4 = s^{3 \cdot 3}s^{2 \cdot 2}s^{5 \cdot 4} = s^9s^4s^{20} = s^{9+4+20} = s^{33}$

87. $(xy)^3 = x^3y^3$

89. $\left(r^3s^2\right)^2 = \left(r^3\right)^2\left(s^2\right)^2 = r^{3 \cdot 2}s^{2 \cdot 2} = r^6s^4$

91. $\left(4ab^2\right)^2 = 4^2a^2\left(b^2\right)^2 = 16a^2b^{2 \cdot 2}$
$$= 16a^2b^4$$

93. $\left(-2r^2s^3t\right)^3 = (-2)^3\left(r^2\right)^3\left(s^3\right)^3t^3$
$$= -8r^{2 \cdot 3}s^{3 \cdot 3}t^3 = -8r^6s^9t^3$$

95. $\left(\dfrac{a}{b}\right)^3 = \dfrac{a^3}{b^3}$

97. $\left(\dfrac{x^2}{y^3}\right)^5 = \dfrac{\left(x^2\right)^5}{\left(y^3\right)^5} = \dfrac{x^{2 \cdot 5}}{y^{3 \cdot 5}} = \dfrac{x^{10}}{y^{15}}$

99. $\left(\dfrac{-2a}{b}\right)^5 = \dfrac{(-2a)^5}{b^5} = \dfrac{(-2)^5a^5}{b^5} = \dfrac{-32a^5}{b^5}$

101. $\left(\dfrac{b^2}{3a}\right)^3 = \dfrac{\left(b^2\right)^3}{(3a)^3} = \dfrac{b^{2 \cdot 3}}{3^3a^3} = \dfrac{b^6}{27a^3}$

103. $\dfrac{x^5}{x^3} = x^{5-3} = x^2$

105. $\dfrac{y^3y^4}{yy^2} = \dfrac{y^{3+4}}{y^{1+2}} = \dfrac{y^7}{y^3} = y^{7-3} = y^4$

107. $\dfrac{12a^2a^3a^4}{4\left(a^4\right)^2} = \dfrac{12a^{2+3+4}}{4a^{4 \cdot 2}} = \dfrac{12a^9}{4a^8} = \dfrac{12}{4}a^{9-8} = 3a^1 = 3a$

109. $\dfrac{\left(ab^2\right)^3}{(ab)^2} = \dfrac{a^3\left(b^2\right)^3}{a^2b^2} = \dfrac{a^3b^{2 \cdot 3}}{a^2b^2} = \dfrac{a^3b^6}{a^2b^2} = a^{3-2}b^{6-2} = a^1b^4 = ab^4$

111. $\dfrac{20\left(r^4s^3\right)^4}{6\left(rs^3\right)^3} = \dfrac{20\left(r^4\right)^4\left(s^3\right)^4}{6r^3\left(s^3\right)^3} = \dfrac{20r^{4 \cdot 4}s^{3 \cdot 4}}{6r^3s^{3 \cdot 3}} = \dfrac{20r^{16}s^{12}}{6r^3s^9} = \dfrac{20}{6}r^{16-3}s^{12-9} = \dfrac{10}{3}r^{13}s^3 = \dfrac{10r^{13}s^3}{3}$

113. $\dfrac{17\left(x^4y^3\right)^8}{34\left(x^5y^2\right)^4} = \dfrac{17\left(x^4\right)^8\left(y^3\right)^8}{34\left(x^5\right)^4\left(y^2\right)^4} = \dfrac{17x^{4 \cdot 8}y^{3 \cdot 8}}{34x^{5 \cdot 4}y^{2 \cdot 4}} = \dfrac{17x^{32}y^{24}}{34x^{20}y^8} = \dfrac{17}{34}x^{32-20}y^{24-8} = \dfrac{1}{2}x^{12}y^{16} = \dfrac{x^{12}y^{16}}{2}$

115. $\left(\dfrac{y^3y}{2yy^2}\right)^3 = \left(\dfrac{y^{3+1}}{2y^{1+2}}\right)^3 = \left(\dfrac{y^4}{2y^3}\right)^3 = \left(\dfrac{1}{2}y^{4-3}\right)^3 = \left(\dfrac{1}{2}y^1\right)^3 = \left(\dfrac{1}{2}\right)^3\left(y^1\right)^3 = \dfrac{1}{8}y^{1 \cdot 3} = \dfrac{y^3}{8}$

117. $\left(\dfrac{-2r^3r^3}{3r^4r}\right)^3 = \left(\dfrac{-2r^{3+3}}{3r^{4+1}}\right)^3 = \left(\dfrac{-2r^6}{3r^5}\right)^3 = \left(\dfrac{-2}{3}r^{6-5}\right)^3 = \dfrac{(-2)^3}{3^3}\left(r^1\right)^3 = \dfrac{-8}{27}r^{1\cdot3} = -\dfrac{8r^3}{27}$

119.

Bounce	1	2	3	4
Height	$\frac{1}{2}(32)$	$\frac{1}{2}\left[\frac{1}{2}(32)\right] = 32\left(\frac{1}{2}\right)^2$	$\frac{1}{2}\left[32\left(\frac{1}{2}\right)^2\right] = 32\left(\frac{1}{2}\right)^3$	$\frac{1}{2}\left[32\left(\frac{1}{2}\right)^3\right] = 32\left(\frac{1}{2}\right)^4$

$32\left(\dfrac{1}{2}\right)^4 = 32\left(\dfrac{1^4}{2^4}\right) = 32\left(\dfrac{1}{16}\right) = \dfrac{32}{16} = 2 \text{ ft}$

121.

Years	7	14	21	28
Value	2,000	4,000	8,000	16,000

123. $A = P(1 + r)^t = 8000(1 + 0.06)^{30} = 8000(1.06)^{30} \approx 8000(5.743491) \approx \$45,947.93$

125. Answers may vary.

127. No. $2^3 = 8,\ 3^2 = 9 \Rightarrow 2^3 \neq 3^2$

Exercise 4.2 (page 242)

1. $\dfrac{3a^2 + 4b + 8}{a + 2b^2} = \dfrac{3(-2)^2 + 4(3) + 8}{-2 + 2(3)^2} = \dfrac{3(4) + 12 + 8}{-2 + 2(9)} = \dfrac{12 + 12 + 8}{-2 + 18} = \dfrac{32}{16} = 2$

3.
$$5\left(x - \tfrac{1}{2}\right) = \tfrac{7}{2}$$
$$5x - \tfrac{5}{2} = \tfrac{7}{2}$$
$$5x - \tfrac{5}{2} + \tfrac{5}{2} = \tfrac{7}{2} + \tfrac{5}{2}$$
$$5x = \tfrac{12}{2}$$
$$5x = 6$$
$$\tfrac{5x}{5} = \tfrac{6}{5}$$
$$x = \tfrac{6}{5}$$

5.
$$P = L + \dfrac{s}{f}i$$
$$P - L = \dfrac{s}{f}i$$
$$f(P - L) = f \cdot \dfrac{s}{f}i$$
$$f(P - L) = si$$
$$\dfrac{f(P - L)}{i} = \dfrac{si}{i}$$
$$\dfrac{f(P - L)}{i} = s, \text{ or } s = \dfrac{f(P - L)}{i}$$

7. 1

9. 1

11. $2^5 \cdot 2^{-2} = 2^{5+(-2)} = 2^3 = 8$

13. $4^{-3} \cdot 4^{-2} \cdot 4^5 = 4^{-3+(-2)+5} = 4^0 = 1$

15. $\dfrac{3^5 \cdot 3^{-2}}{3^3} = \dfrac{3^{5+(-2)}}{3^3} = \dfrac{3^3}{3^3} = 3^{3-3} = 3^0 = 1$

17. $\dfrac{2^5 \cdot 2^7}{2^6 \cdot 2^{-3}} = \dfrac{2^{5+7}}{2^{6+(-3)}} = \dfrac{2^{12}}{2^3} = 2^{12-3}$
$$= 2^9 = 512$$

19. $2x^0 = 2 \cdot x^0 = 2 \cdot 1 = 2$

21. $(-x)^0 = 1$

23. $\left(\dfrac{a^2b^3}{ab^4}\right)^0 = 1$

25. $\dfrac{x^0 - 5x^0}{2x^0} = \dfrac{1 - 5(1)}{2(1)} = \dfrac{1 - 5}{2} = \dfrac{-4}{2} = -2$

27. $x^{-2} = \dfrac{1}{x^2}$

29. $b^{-5} = \dfrac{1}{b^5}$

31. $(2y)^{-4} = \dfrac{1}{(2y)^4} = \dfrac{1}{2^4 y^4} = \dfrac{1}{16y^4}$

33. $(ab^2)^{-3} = \dfrac{1}{(ab^2)^3} = \dfrac{1}{a^3(b^2)^3} = \dfrac{1}{a^3 b^6}$

35. $\dfrac{y^4}{y^5} = y^{4-5} = y^{-1} = \dfrac{1}{y}$

37. $\dfrac{(r^2)^3}{(r^3)^4} = \dfrac{r^6}{r^{12}} = r^{6-12} = r^{-6} = \dfrac{1}{r^6}$

39. $\dfrac{y^4 y^3}{y^4 y^{-2}} = \dfrac{y^7}{y^2} = y^{7-2} = y^5$

41. $\dfrac{a^4 a^{-2}}{a^2 a^0} = \dfrac{a^2}{a^2} = a^{2-2} = a^0 = 1$

43. $(ab^2)^{-2} = \dfrac{1}{(ab^2)^2} = \dfrac{1}{a^2(b^2)^2} = \dfrac{1}{a^2 b^4}$

45. $(x^2 y)^{-3} = \dfrac{1}{(x^2 y)^3} = \dfrac{1}{(x^2)^3 y^3} = \dfrac{1}{x^6 y^3}$

47. $(x^{-4} x^3)^3 = (x^{-1})^3 = x^{-3} = \dfrac{1}{x^3}$

49. $(y^3 y^{-2})^{-2} = (y^1)^{-2} = y^{-2} = \dfrac{1}{y^2}$

51. $(a^{-2} b^{-3})^{-4} = (a^{-2})^{-4}(b^{-3})^{-4} = a^8 b^{12}$

53. $(-2x^3 y^{-2})^{-5} = (-2)^{-5}(x^3)^{-5}(y^{-2})^{-5} = \dfrac{1}{(-2)^5} x^{-15} y^{10} = -\dfrac{y^{10}}{32x^{15}}$

55. $\left(\dfrac{a^3}{a^{-4}}\right)^2 = (a^{3-(-4)})^2 = (a^7)^2 = a^{14}$

57. $\left(\dfrac{b^5}{b^{-2}}\right)^{-2} = (b^{5-(-2)})^{-2} = (b^7)^{-2}$
$$= b^{-14} = \dfrac{1}{b^{14}}$$

59. $\left(\dfrac{4x^2}{3x^{-5}}\right)^4 = \left(\dfrac{4}{3} x^{2-(-5)}\right)^4 = \left(\dfrac{4}{3} x^7\right)^4 = \left(\dfrac{4}{3}\right)^4 (x^7)^4 = \dfrac{256}{81} x^{28} = \dfrac{256 x^{28}}{81}$

61. $\left(\dfrac{12y^3 z^{-2}}{3y^{-4} z^3}\right)^2 = \left(\dfrac{12}{3} y^{3-(-4)} z^{-2-3}\right)^2 = (4y^7 z^{-5})^2 = 4^2 (y^7)^2 (z^{-5})^2 = 16 y^{14} z^{-10} = \dfrac{16 y^{14}}{z^{10}}$

63. $\left(\dfrac{2x^3 y^{-2}}{4xy^2}\right)^7 = \left(\dfrac{2}{4} x^{3-1} y^{-2-2}\right)^7 = \left(\dfrac{1}{2} x^2 y^{-4}\right)^7 = \left(\dfrac{1}{2}\right)^7 (x^2)^7 (y^{-4})^7 = \dfrac{1}{128} x^{14} y^{-28} = \dfrac{x^{14}}{128 y^{28}}$

65. $\left(\dfrac{14u^{-2} v^3}{21u^{-3} v}\right)^4 = \left(\dfrac{14}{21} u^{-2-(-3)} v^{3-1}\right)^4 = \left(\dfrac{2}{3} u^1 v^2\right)^4 = \left(\dfrac{2}{3}\right)^4 (u^1)^4 (v^2)^4 = \dfrac{16}{81} u^4 v^8 = \dfrac{16 u^4 v^8}{81}$

67. $\left(\dfrac{6a^2 b^3}{2ab^2}\right)^{-2} = \left(\dfrac{6}{2} a^{2-1} b^{3-2}\right)^{-2} = (3a^1 b^1)^{-2} = \dfrac{1}{(3ab)^2} = \dfrac{1}{3^2 a^2 b^2} = \dfrac{1}{9a^2 b^2}$

69. $\left(\dfrac{18a^2b^3c^{-4}}{3a^{-1}b^2c}\right)^{-3} = \left(\dfrac{18}{3}a^{2-(-1)}b^{3-2}c^{-4-1}\right)^{-3} = \left(6a^3b^1c^{-5}\right)^{-3} = 6^{-3}\left(a^3\right)^{-3}\left(b^1\right)^{-3}\left(c^{-5}\right)^{-3}$

$$= \dfrac{1}{6^3}a^{-9}b^{-3}c^{15} = \dfrac{c^{15}}{216a^9b^3}$$

71. $\dfrac{\left(2x^{-2}y\right)^{-3}}{\left(4x^2y^{-1}\right)^3} = \dfrac{2^{-3}\left(x^{-2}\right)^{-3}y^{-3}}{4^3\left(x^2\right)^3\left(y^{-1}\right)^3} = \dfrac{x^6y^{-3}}{2^3\cdot4^3x^6y^{-3}} = \dfrac{1}{8\cdot64}x^{6-6}y^{-3-(-3)} = \dfrac{1}{512}x^0y^0 = \dfrac{1}{512}$

73. $\dfrac{\left(17x^5y^{-5}z\right)^{-3}}{\left(17x^{-5}y^3z^2\right)^{-4}} = \dfrac{17^{-3}\left(x^5\right)^{-3}\left(y^{-5}\right)^{-3}z^{-3}}{17^{-4}\left(x^{-5}\right)^{-4}\left(y^3\right)^{-4}\left(z^2\right)^{-4}} = \dfrac{17^{-3}x^{-15}y^{15}z^{-3}}{17^{-4}x^{20}y^{-12}z^{-8}}$

$$= 17^{-3-(-4)}x^{-15-20}y^{15-(-12)}z^{-3-(-8)}$$

$$= 17^1x^{-35}y^{27}z^5 = \dfrac{17y^{27}z^5}{x^{35}}$$

75. $x^{2m}x^m = x^{2m+m} = x^{3m}$

77. $u^{2m}v^{3n}u^{3m}v^{-3n} = u^{2m+3m}v^{3n+(-3n)}$

$$= u^{5m}v^0 = u^{5m}$$

79. $y^{3m+2}y^{-m} = y^{3m+2+(-m)} = y^{2m+2}$

81. $\dfrac{y^{3m}}{y^{2m}} = y^{3m-2m} = y^m$

83. $\dfrac{x^{3n}}{x^{6n}} = x^{3n-6n} = x^{-3n} = \dfrac{1}{x^{3n}}$

85. $\left(x^{m+1}\right)^2 = x^{(m+1)\cdot2} = x^{2m+2}$

87. $\left(x^{3-2n}\right)^{-4} = x^{(3-2n)(-4)} = x^{-12+8n} = x^{8n-12}$

89. $\left(y^{2-n}\right)^{-4} = y^{(2-n)(-4)} = y^{-8+4n} = y^{4n-8}$

Problems 91-95 are to be solved using a calculator. The keystrokes needed to solve each problem using a TI-83 graphing calculator appear in each solution. There may be other solutions. Keystrokes for other calculators may be slightly different.

91. $P = A(1+i)^{-n} = 100,000(1+0.07)^{-40}$

| 1 | 0 | 0 | 0 | 0 | 0 | (| 1 | + | . | 0 | 7 |) | ^ | (–) | 4 | 0 | = |

The original deposit should be \$6,678.04.

93. $P = A(1+i)^{-n} = 100,000(1+0.09)^{-40}$

| 1 | 0 | 0 | 0 | 0 | 0 | (| 1 | + | . | 0 | 9 |) | ^ | (–) | 4 | 0 | = |

The original deposit should be \$3,183.76.

95. $P = A(1+i)^{-n} = 1,000,000(1+0.08)^{-60}$

| 1 | 0 | 0 | 0 | 0 | 0 | 0 | (| 1 | + | . | 0 | 8 |) | ^ | (–) | 6 | 0 | = |

The original deposit should be \$9,875.85.

97. **Answers may vary.**

99. If $x > 1$, then x raised to a negative power is less than x. If $x = 1$, then x raised to a negative power is equal to x. If $0 < x < 1$, then x raised to a negative power is greater than x.

Exercise 4.3 (page 249)

1. $-5y^{55} = -5(-1)^{55} = -5(-1) = 5$ **3.** commutative property of addition

5. $3(x-4) - 6 = 0$
$3x - 12 - 6 = 0$
$3x - 18 = 0$
$3x = 18$
$x = 6$

     7. scientific notation   **9.** $23,000 = 2.3 \times 10^4$

11. $1,700,000 = 1.7 \times 10^6$ **13.** $0.062 = 6.2 \times 10^{-2}$ **15.** $0.0000051 = 5.1 \times 10^{-6}$

17. $42.5 \times 10^2 = 4.25 \times 10^1 \times 10^2$
$ = 4.25 \times 10^3$

     19. $0.25 \times 10^{-2} = 2.5 \times 10^{-1} \times 10^{-2}$
$\phantom{0.25 \times 10^{-2}} = 2.5 \times 10^{-3}$

21. $2.3 \times 10^2 = 230$ **23.** $8.12 \times 10^5 = 812,000$ **25.** $1.15 \times 10^{-3} = 0.00115$

27. $9.76 \times 10^{-4} = 0.000976$ **29.** $25 \times 10^6 = 25,000,000$ **31.** $0.51 \times 10^{-3} = 0.00051$

33. $(3.4 \times 10^2)(2.1 \times 10^3) = (3.4)(2.1) \times 10^2 \times 10^3 = 7.14 \times 10^5 = 714,000$

35. $\dfrac{9.3 \times 10^2}{3.1 \times 10^{-2}} = \dfrac{9.3}{3.1} \times \dfrac{10^2}{10^{-2}} = 3 \times 10^4 = 30,000$

37. $\dfrac{96,000}{(12,000)(0.00004)} = \dfrac{9.6 \times 10^4}{(1.2 \times 10^4)(4 \times 10^{-5})} = \dfrac{9.6 \times 10^4}{4.8 \times 10^{-1}} = \dfrac{9.6}{4.8} \times \dfrac{10^4}{10^{-1}} = 2 \times 10^5 = 200,000$

39. $25,700,000,000,000 = 2.57 \times 10^{13}$ mi   **41.** $1.14 \times 10^8 = 114,000,000$ mi

43. $0.00622 = 6.22 \times 10^{-3}$ mi

45. 3.6×10^7 mi $\times 5280$ ft/mi $= 19008 \times 10^7$ ft $= 1.9008 \times 10^{11}$ ft

47. $\dfrac{3.3 \times 10^4 \text{ cm}}{\text{sec}} = \dfrac{3.3 \times 10^4 \text{ cm}}{\text{sec}} \cdot \dfrac{1 \text{ m}}{100 \text{ cm}} \cdot \dfrac{1 \text{ km}}{1,000 \text{ m}} = \dfrac{3.3 \times 10^4}{10^5}$ km/sec $= 3.3 \times 10^{-1}$ km/sec

49. 1.5×10^{-4} in.; $25,000,000,000,000 = 2.5 \times 10^{13}$

51. **Answers may vary.**   **53.** **Answers may vary.**

Exercise 4.4 (page 258)

1. $5(u - 5) + 9 = 2(u + 4)$
$5u - 25 + 9 = 2u + 8$
$5u - 16 = 2u + 8$
$3u - 16 = 8$
$3u = 24$
$u = 8$

3. $-4(3y + 2) \leq 28$
$-12y - 8 \leq 28$
$-12y \leq 36$
$\dfrac{-12y}{-12} \geq \dfrac{36}{-12}$
$y \geq -3$

5. $\left(x^2 x^4\right)^3 = \left(x^6\right)^3 = x^{18}$

7. $\left(\dfrac{y^2 y^5}{y^4}\right)^3 = \left(\dfrac{y^7}{y^4}\right)^3 = \left(y^3\right)^3 = y^9$

9. algebraic

11. polynomial

13. trinomial

15. degree

17. function

19. domain

21. yes

23. yes

25. binomial

27. trinomial

29. monomial

31. binomial

33. trinomial

35. none of these (4 terms)

37. 4th

39. 3rd

41. 8th $(3 + 5 = 8)$

43. 6th $(3 + 1 + 2 = 6)$

45. 12th

47. 0th

49. $5x - 3 = 5(2) - 3 = 10 - 3 = 7$

51. $5x - 3 = 5(-1) - 3 = -5 - 3 = -8$

53. $-x^2 - 4 = -(0)^2 - 4 = -0 - 4 = -4$

55. $-x^2 - 4 = -(-1)^2 - 4 = -1 - 4 = -5$

57. $f(0) = 5(0) + 1 = 0 + 1 = 1$

59. $f(-2) = 5(-2) + 1 = -10 + 1 = -9$

61. $f(0) = (0)^2 - 2(0) + 3 = 0 - 0 + 3 = 3$

63. $f(-2) = (-2)^2 - 2(-2) + 3$
$= 4 + 4 + 3 = 11$

65.

x	$x^2 - 3$
-2	$(-2)^2 - 3 = 4 - 3 = 1$
-1	$(-1)^2 - 3 = 1 - 3 = -2$
0	$(0)^2 - 3 = 0 - 3 = -3$
1	$(1)^2 - 3 = 1 - 3 = -2$
2	$(2)^2 - 3 = 4 - 3 = 1$

67.

x	$x^3 + 2$
-2	$(-2)^3 + 2 = -8 + 2 = -6$
-1	$(-1)^3 + 2 = -1 + 2 = 1$
0	$(0)^3 + 2 = 0 + 2 = 2$
1	$(1)^3 + 2 = 1 + 2 = 3$
2	$(2)^3 + 2 = 8 + 2 = 10$

69. $f(x) = x^2 - 1$

x	$f(x)$
-2	3
-1	0
0	-1
1	0
2	3

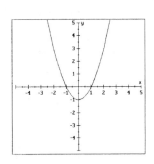

71. $f(x) = x^3 + 2$

x	$f(x)$
-2	-6
-1	1
0	2
1	3
2	10

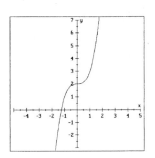

73. $h = -16t^2 + 64t$
$h = -16(2)^2 + 64(2)$
$h = -16(4) + 128$
$h = -64 + 128 = 64$ ft

75. $f(d) = -0.08d^2 + 100d$
$f(815) = -0.08(815)^2 + 100(815)$
$\qquad = -0.08(664225) + 81500$
$\qquad = -53138 + 81500$
$\qquad = \$28,362$

77. $d = 0.04v^2 + 0.9v$
$d = 0.04(30)^2 + 0.9(30)$
$d = 0.04(900) + 27$
$d = 36 + 27 = 63$ ft

79. **Answers may vary.**

81. There are many possible polynomials. One is $2x - 2$.

Exercise 4.5 (page 264)

1. $ab + cd = (3)(-2) + (-1)(2)$
$\qquad = -6 + (-2) = -8$

3. $a(b + c) = 3[-2 + (-1)] = 3(-3) = -9$

5. $-4(2x - 9) \geq 12$
$-8x + 36 \geq 12$
$\qquad -8x \geq -24$
$\qquad \dfrac{-8x}{-8} \leq \dfrac{-24}{-8}$
$\qquad x \leq 3$

7. monomial

9. coefficients, variables

11. like terms

13. like terms; $3y + 4y = 7y$

15. unlike terms

17. like terms; $3x^3 + 4x^3 + 6x^3 = 13x^3$

19. like terms; $-5x^3y^2 + 13x^3y^2 = 8x^3y^2$

21. like terms; $-23t^6 + 32t^6 + 56t^6 = 65t^6$

23. unlike terms

25. $4y + 5y = 9y$

27. $-8t^2 - 4t^2 = -12t^2$

29. $32u^3 - 16u^3 = 16u^3$

31. $18x^5y^2 - 11x^5y^2 = 7x^5y^2$

33. $3rst + 4rst + 7rst = 14rst$

35. $-4a^2bc + 5a^2bc - 7a^2bc = -6a^2bc$

37. $(3x)^2 - 4x^2 + 10x^2 = 9x^2 - 4x^2 + 10x^2 = 15x^2$

39. $5x^2y^2 + 2(xy)^2 - (3x^2)y^2 = 5x^2y^2 + 2x^2y^2 - 3x^2y^2 = 4x^2y^2$

41. $(-3x^2y)^4 + (4x^4y^2)^2 - 2x^8y^4 = 81x^8y^4 + 16x^8y^4 - 2x^8y^4 = 95x^8y^4$

43. $(3x + 7) + (4x - 3) = 3x + 7 + 4x - 3$
$$= 7x + 4$$

45. $(4a + 3) - (2a - 4) = 4a + 3 - 2a + 4$
$$= 2a + 7$$

47. $(2x + 3y) + (5x - 10y) = 2x + 3y + 5x - 10y = 7x - 7y$

49. $(-8x - 3y) - (11x + y) = -8x - 3y - 11x - y = -19x - 4y$

51. $(3x^2 - 3x - 2) + (3x^2 + 4x - 3) = 3x^2 - 3x - 2 + 3x^2 + 4x - 3 = 6x^2 + x - 5$

53. $(2b^2 + 3b - 5) - (2b^2 - 4b - 9) = 2b^2 + 3b - 5 - 2b^2 + 4b + 9 = 7b + 4$

55. $(2x^2 - 3x + 1) - (4x^2 - 3x + 2) + (2x^2 + 3x + 2)$
$$= 2x^2 - 3x + 1 - 4x^2 + 3x - 2 + 2x^2 + 3x + 2 = 3x + 1$$

57. $2(x + 3) + 3(x + 3) = 2x + 6 + 3x + 9 = 5x + 15$

59. $-8(x - y) + 11(x - y) = -8x + 8y + 11x - 11y = 3x - 3y$

61. $2(x^2 - 5x - 4) - 3(x^2 - 5x - 4) + 6(x^2 - 5x - 4)$
$$= 2x^2 - 10x - 8 - 3x^2 + 15x + 12 + 6x^2 - 30x - 24 = 5x^2 - 25x - 20$$

63.
$$\begin{array}{r} 3x^2 + 4x + \ 5 \\ +\ \underline{2x^2 - 3x + \ 6} \\ 5x^2 + \ x + 11 \end{array}$$

65.
$$\begin{array}{r} 2x^3 - 3x^2 + 4x - 7 \\ +\ \underline{-9x^3 - 4x^2 - 5x + 6} \\ -7x^3 - 7x^2 - \ x - 1 \end{array}$$

67.
$$\begin{array}{r} -3x^2y + 4xy + 25y^2 \\ +\ \underline{5x^2y - 3xy - 12y^2} \\ 2x^2y + \ xy + 13y^2 \end{array}$$

69.
$$\begin{array}{r} 3x^2 + 4x - 5 \\ -\ \underline{-2x^2 - 2x + 3} \\ 3x^2 + 4x - 5 \\ +\ \underline{2x^2 + 2x - 3} \\ 5x^2 + 6x - 8 \end{array}$$

71.
$$\begin{array}{r} 4x^3 + 4x^2 - 3x + 10 \\ -\ \underline{5x^3 - 2x^2 - 4x - \ 4} \\ 4x^3 + 4x^2 - 3x + 10 \\ +\ \underline{-5x^3 + 2x^2 + 4x + \ 4} \\ -x^3 + 6x^2 + \ x + 14 \end{array}$$

73.
$$\begin{array}{r} -2x^2y^2 - \ 4xy + 12y^2 \\ -\ \underline{10x^2y^2 + \ 9xy - 24y^2} \\ -2x^2y^2 - \ 4xy + 12y^2 \\ +\ \underline{-10x^2y^2 - \ 9xy + 24y^2} \\ -12x^2y^2 - 13xy + 36y^2 \end{array}$$

SECTION 4.5

75. $\left[(2x^2 - 3x + 4) + (3x^2 - 2)\right] + (x^2 + x - 3)$
$$= \left[2x^2 - 3x + 4 + 3x^2 - 2\right] + x^2 + x - 3$$
$$= 5x^2 - 3x + 2 + x^2 + x - 3 = 6x^2 - 2x - 1$$

77. $\left[(3t^3 + t^2) + (-t^3 + 6t - 3)\right] - (t^3 - 2t^2 + 2)$
$$= \left[3t^3 + t^2 - t^3 + 6t - 3\right] - t^3 + 2t^2 - 2$$
$$= 2t^3 + t^2 + 6t - 3 - t^3 + 2t^2 - 2 = t^3 + 3t^2 + 6t - 5$$

79. $\left[(-2x^2 - 7x + 1) + (-4x^2 + 8x - 1)\right] + (3x^2 + 4x - 7)$
$$= \left[-2x^2 - 7x + 1 - 4x^2 + 8x - 1\right] + 3x^2 + 4x - 7$$
$$= -6x^2 + x + 3x^2 + 4x - 7 = -3x^2 + 5x - 7$$

81. $2(x + 3) + 4(x - 2) = 2x + 6 + 4x - 8 = 6x - 2$

83. $-2(x^2 + 7x - 1) - 3(x^2 - 2x + 7) = -2x^2 - 14x + 2 - 3x^2 + 6x - 21 = -5x^2 - 8x - 19$

85. $2(2y^2 - 2y + 2) - 4(3y^2 - 4y - 1) + 4(y^3 - y^2 - y)$
$$= 4y^2 - 4y + 4 - 12y^2 + 16y + 4 + 4y^3 - 4y^2 - 4y = 4y^3 - 12y^2 + 8y + 8$$

87. $2(a^2b^2 - ab) - 3(ab + 2ab^2) + (b^2 - ab + a^2b^2) = 2a^2b^2 - 2ab - 3ab - 6ab^2 + b^2 - ab + a^2b^2$
$$= 3a^2b^2 - 6ab + b^2 - 6ab^2$$

89. $-4(x^2y^2 + xy^3 + xy^2z) - 2(x^2y^2 - 4xy^2z) - 2(8xy^3 - y)$
$$= -4x^2y^2 - 4xy^3 - 4xy^2z - 2x^2y^2 + 8xy^2z - 16xy^3 + 2y$$
$$= -6x^2y^2 + 4xy^2z - 20xy^3 + 2y$$

91. $y = 900x + 105,000$
$y = 900(10) + 105,000$
$y = 9,000 + 105,000 = \$114,000$

93. $y = 1,000x + 120,000$
$y = 1,000(12) + 120,000$
$y = 12,000 + 120,000 = \$132,000$

95. a. $\quad y = 900x + 105,000 \qquad\qquad y = 1,000x + 120,000$
$\qquad y = 900(20) + 105,000 \qquad\quad y = 1,000(20) + 120,000$
$\qquad y = 18,000 + 105,000 = \$123,000 \quad y = 20,000 + 120,000 = \$140,000$
$$\$123,000 + \$140,000 = \$263,000$$

b. $\quad y = 1,900x + 225,000 = 1,900(20) + 225,000 = 38,000 + 225,000 = \$263,000$

97. $y = -1,100x + 6,600$

99. $y = (-1,100x + 6,600) + (-1,700x + 9,200) = -2,800x + 15,800$

101. Answers may vary.

103. $P(x + h) + P(x) = [3(x + h) - 5] + [3x - 5] = 3x + 3h - 5 + 3x - 5 = 6x + 3h - 10$

105. $P(x) - Q(x) = (x^{23} + 5x^2 + 73) - (x^{23} + 4x^2 + 73) = x^{23} + 5x^2 + 73 - x^{23} - 4x^2 - 73 = x^2$

$P(7) - Q(7) = 7^2 = 49$

Exercise 4.6 (page 274)

1. distributive property

3. commutative property of multiplication

5. $\dfrac{5}{3}(5y + 6) - 10 = 0$

$3\left[\dfrac{5}{3}(5y + 6) - 10\right] = 3(0)$

$3 \cdot \dfrac{5}{3}(5y + 6) - 3 \cdot 10 = 0$

$5(5y + 6) - 30 = 0$

$25y + 30 - 30 = 0$

$25y = 0$

$y = 0$

7. monomial

9. trinomial

11. $(2x)(3x) = 6x^2$

13. $(5)(3x) = 15x$

15. $(3x^2)(4x^3) = (3)(4)x^2 x^3 = 12x^5$

17. $(3b^2)(-2b)(4b^3) = (3)(-2)(4)b^2 b b^3$
$= -24b^6$

19. $(2x^2 y^3)(3x^3 y^2) = (2)(3)x^2 x^3 y^3 y^2 = 6x^5 y^5$

21. $(x^2 y^5)(x^2 z^5)(-3y^2 z^3) = (1)(1)(-3)x^2 x^2 y^5 y^2 z^5 x^3 = -3x^4 y^7 z^8$

23. $(x^2 y^3)^5 = (x^2)^5 (y^3)^5 = x^{10} y^{15}$

25. $(a^3 b^2 c)(abc^3)^2 = (a^3 b^2 c)\left[a^2 b^2 (c^3)^2\right] = (a^3 b^2 c)(a^2 b^2 c^6) = a^3 a^2 b^2 b^2 c c^6 = a^5 b^4 c^7$

27. $3(x + 4) = 3(x) + 3(4) = 3x + 12$

29. $-4(t + 7) = -4(t) + (-4)(7) = -4t - 28$

31. $3x(x - 2) = 3x(x) - 3x(2) = 3x^2 - 6x$

33. $-2x^2(3x^2 - x) = -2x^2(3x^2) - (-2x^2)(x)$
$= -6x^4 + 2x^3$

35. $3xy(x + y) = 3xy(x) + 3xy(y) = 3x^2 y + 3xy^2$

37. $2x^2(3x^2 + 4x - 7) = 2x^2(3x^2) + 2x^2(4x) - 2x^2(7) = 6x^4 + 8x^3 - 14x^2$

39. $\frac{1}{4}x^2(8x^5 - 4) = \frac{1}{4}x^2(8x^5) - \frac{1}{4}x^2(4) = 2x^7 - x^2$

41. $-\frac{2}{3}r^2 t^2(9r - 3t) = -\frac{2}{3}r^2 t^2(9r) - \left(-\frac{2}{3}r^2 t^2\right)(3t) = -6r^3 t^2 + 2r^2 t^3$

43. $(3xy)(-2x^2 y^3)(x + y) = (-6x^3 y^4)(x + y) = -6x^3 y^4(x) + (-6x^3 y^4)(y) = -6x^4 y^4 - 6x^3 y^5$

45. $(a+4)(a+5) = a(a) + a(5) + 4(a) + 4(5) = a^2 + 5a + 4a + 20 = a^2 + 9a + 20$

47. $(3x-2)(x+4) = 3x(x) + 3x(4) + (-2)(x) + (-2)(4) = 3x^2 + 12x - 2x - 8 = 3x^2 + 10x - 8$

49. $(2a+4)(3a-5) = 2a(3a) - 2a(5) + 4(3a) - 4(5) = 6a^2 - 10a + 12a - 20 = 6a^2 + 2a - 20$

51. $(3x-5)(2x+1) = 3x(2x) + 3x(1) + (-5)(2x) + (-5)(1) = 6x^2 + 3x - 10x - 5$
$$= 6x^2 - 7x - 5$$

53. $(x+3)(2x-3) = x(2x) - x(3) + 3(2x) - 3(3) = 2x^2 - 3x + 6x - 9 = 2x^2 + 3x - 9$

55. $(2s+3t)(3s-t) = 2s(3s) - 2s(t) + 3t(3s) - 3t(t) = 6s^2 - 2st + 9st - 3t^2 = 6s^2 + 7st - 3t^2$

57. $(x+y)(x+z) = x(x) + x(z) + y(x) + y(z) = x^2 + xz + xy + yz$

59. $(u+v)(u+2t) = u(u) + u(2t) + v(u) + v(2t) = u^2 + 2tu + uv + 2tv$

61. $(-2r-3s)(2r+7s) = (-2r)(2r) + (-2r)(7s) + (-3s)(2r) + (-3s)(7s)$
$$= -4r^2 - 14rs - 6rs - 21s^2 = -4r^2 - 20rs - 21s^2$$

63.

$$
\begin{array}{r}
4x + 3 \\
x + 2 \\
\hline
4x^2 + 3x \\
8x + 6 \\
\hline
4x^2 + 11x + 6
\end{array}
$$

65.

$$
\begin{array}{r}
4x - 2y \\
3x + 5y \\
\hline
12x^2 - 6xy \\
20xy - 10y^2 \\
\hline
12x^2 + 14xy - 10y^2
\end{array}
$$

67.

$$
\begin{array}{r}
x^2 + x + 1 \\
x - 1 \\
\hline
x^3 + x^2 + x \\
-x^2 - x - 1 \\
\hline
x^3 - 1
\end{array}
$$

69. $(2x+1)(x^2+3x-1) = (2x+1)(x^2) + (2x+1)(3x) - (2x+1)(1)$
$$= 2x(x^2) + 1(x^2) + 2x(3x) + 1(3x) - 2x(1) - 1(1)$$
$$= 2x^3 + x^2 + 6x^2 + 3x - 2x - 1 = 2x^3 + 7x^2 + x - 1$$

71. $(x+4)(x+4) = x(x) + x(4) + 4(x) + 4(4) = x^2 + 4x + 4x + 16 = x^2 + 8x + 16$

73. $(t-3)(t-3) = t(t) - t(3) + (-3)(t) - (-3)(3) = t^2 - 3t - 3t + 9 = t^2 - 6t + 9$

75. $(r+4)(r-4) = r(r) - r(4) + 4(r) - 4(4) = r^2 - 4r + 4r - 16 = r^2 - 16$

77. $(x+5)^2 = (x+5)(x+5) = x(x) + x(5) + 5(x) + 5(5) = x^2 + 5x + 5x + 25 = x^2 + 10x + 25$

79. $(2s+1)(2s+1) = 2s(2s) + 2s(1) + 1(2s) + 1(1) = 4s^2 + 2s + 2s + 1 = 4s^2 + 4s + 1$

81. $(4x+5)(4x-5) = 4x(4x) - 4x(5) + 5(4x) - 5(5) = 16x^2 - 20x + 20x - 25 = 16x^2 - 25$

83. $(x-2y)^2 = (x-2y)(x-2y) = x(x) - x(2y) + (-2y)(x) - (-2y)(2y)$
$$= x^2 - 2xy - 2xy + 4y^2 = x^2 - 4xy + 4y^2$$

85. $(2a - 3b)^2 = (2a - 3b)(2a - 3b) = 2a(2a) - 2a(3b) + (-3b)(2a) - (-3b)(3b)$
$$= 4a^2 - 6ab - 6ab + 9b^2 = 4a^2 - 12ab + 9b^2$$

87. $(4x + 5y)(4x - 5y) = 4x(4x) - 4x(5y) + 5y(4x) - 5y(5y)$
$$= 16x^2 - 20xy + 20xy - 25y^2 = 16x^2 - 25y^2$$

89. $2(x - 4)(x + 1) = 2[x(x) + x(1) + (-4)(x) + (-4)(1)] = 2\left[x^2 + x - 4x - 4\right]$
$$= 2\left[x^2 - 3x - 4\right] = 2x^2 - 6x - 8$$

91. $3a(a + b)(a - b) = 3a[a(a) - a(b) + b(a) - b(b)] = 3a\left[a^2 - ab + ab - b^2\right]$
$$= 3a\left[a^2 - b^2\right] = 3a^3 - 3ab^2$$

93. $(4t + 3)\left(t^2 + 2t + 3\right) = 4t\left(t^2\right) + 4t(2t) + 4t(3) + 3\left(t^2\right) + 3(2t) + 3(3)$
$$= 4t^3 + 8t^2 + 12t + 3t^2 + 6t + 9 = 4t^3 + 11t^2 + 18t + 9$$

95. $(-3x + y)\left(x^2 - 8xy + 16y^2\right)$
$$= -3x\left(x^2\right) - (-3x)(8xy) + (-3x)\left(16y^2\right) + y\left(x^2\right) - y(8xy) + y\left(16y^2\right)$$
$$= -3x^3 + 24x^2y - 48xy^2 + x^2y - 8xy^2 + 16y^3 = -3x^3 + 25x^2y - 56xy^2 + 16y^3$$

97. $(x - 2y)\left(x^2 + 2xy + 4y^2\right)$
$$= x\left(x^2\right) + x(2xy) + x\left(4y^2\right) + (-2y)\left(x^2\right) + (-2y)(2xy) + (-2y)\left(4y^2\right)$$
$$= x^3 + 2x^2y + 4xy^2 - 2x^2y - 4xy^2 - 8y^3 = x^3 - 8y^3$$

99. $2t(t + 2) + 3t(t - 5) = [2t(t) + 2t(2)] + [3t(t) - 3t(5)] = \left[2t^2 + 4t\right] + \left[3t^2 - 15t\right]$
$$= 2t^2 + 4t + 3t^2 - 15t = 5t^2 - 11t$$

101. $3xy(x + y) - 2x(xy - x) = [3xy(x) + 3xy(y)] - [(2x)(xy) - (2x)(x)]$
$$= \left[3x^2y + 3xy^2\right] - \left[2x^2y - 2x^2\right]$$
$$= 3x^2y + 3xy^2 - 2x^2y + 2x^2 = x^2y + 3xy^2 + 2x^2$$

103. $(x + y)(x - y) + x(x + y) = [x(x) - x(y) + y(x) - y(y)] + [x(x) + x(y)]$
$$= \left[x^2 - xy + xy - y^2\right] + \left[x^2 + xy\right]$$
$$= \left[x^2 - y^2\right] + \left[x^2 + xy\right] = x^2 - y^2 + x^2 + xy = 2x^2 + xy - y^2$$

105. $(x + 2)^2 - (x - 2)^2 = [(x + 2)(x + 2)] - [(x - 2)(x - 2)]$
$$= [x(x) + x(2) + 2(x) + 2(2)] - [x(x) - x(2) + (-2)(x) - (-2)(2)]$$
$$= \left[x^2 + 2x + 2x + 4\right] - \left[x^2 - 2x - 2x + 4\right]$$
$$= \left[x^2 + 4x + 4\right] - \left[x^2 - 4x + 4\right] = x^2 + 4x + 4 - x^2 + 4x - 4 = 8x$$

107. $(2s - 3)(s + 2) + (3s + 1)(s - 3)$
$$= [2s(s) + 2s(2) + (-3)(s) + (-3)(2)] + [3s(s) - 3s(3) + 1(s) - 1(3)]$$
$$= [2s^2 + 4s - 3s - 6] + [3s^2 - 9s + s - 3]$$
$$= [2s^2 + s - 6] + [3s^2 - 8s - 3] = 2s^2 + s - 6 + 3s^2 - 8s - 3 = 5s^2 - 7s - 9$$

109. $(s - 4)(s + 1) = s^2 + 5$
$$s^2 + s - 4s - 4 = s^2 + 5$$
$$s^2 - 3s - 4 = s^2 + 5$$
$$-3s - 4 = 5$$
$$-3s = 9$$
$$s = -3$$

111. $z(z + 2) = (z + 4)(z - 4)$
$$z^2 + 2z = z^2 - 4z + 4z - 16$$
$$z^2 + 2z = z^2 - 16$$
$$2z = -16$$
$$z = -8$$

113. $(x + 4)(x - 4) = (x - 2)(x + 6)$
$$x^2 - 4x + 4x - 16 = x^2 + 6x - 2x - 12$$
$$x^2 - 16 = x^2 + 4x - 12$$
$$-16 = 4x - 12$$
$$-4 = 4x$$
$$-1 = x$$

115. $(a - 3)^2 = (a + 3)^2$
$$(a - 3)(a - 3) = (a + 3)(a + 3)$$
$$a^2 - 3a - 3a + 9 = a^2 + 3a + 3a + 9$$
$$a^2 - 6a + 9 = a^2 + 6a + 9$$
$$-6a + 9 = 6a + 9$$
$$-6a = 6a$$
$$12a = 0$$
$$a = 0$$

117. $4 + (2y - 3)^2 = (2y - 1)(2y + 3)$
$$4 + (2y - 3)(2y - 3) = (2y - 1)(2y + 3)$$
$$4 + 4y^2 - 6y - 6y + 9 = 4y^2 + 6y - 2y - 3$$
$$4y^2 - 12y + 13 = 4y^2 + 4y - 3$$
$$-12y + 13 = 4y - 3$$
$$-12y = 4y - 16$$
$$-16y = -16$$
$$y = 1$$

119. Let $r =$ the smaller radius.
Then $r + 3 =$ the larger radius.

$$\boxed{\begin{array}{c}\text{Larger}\\\text{area}\end{array}} - \boxed{\begin{array}{c}\text{Smaller}\\\text{area}\end{array}} = 15\pi$$

$$\pi(r + 3)^2 - \pi r^2 = 15\pi$$
$$\pi(r + 3)(r + 3) - \pi r^2 = 15\pi$$
$$\pi\left(r^2 + 3r + 3r + 9\right) - \pi r^2 = 15\pi$$
$$\pi\left(r^2 + 6r + 9\right) - \pi r^2 = 15\pi$$
$$\pi r^2 + 6\pi r + 9\pi - \pi r^2 = 15\pi$$
$$6\pi r + 9\pi = 15\pi$$
$$6\pi r = 6\pi$$
$$\frac{6\pi r}{6\pi} = \frac{6\pi}{6\pi}$$
$$r = 1$$

The larger radius $= 1 + 3 = 4$ m.

121. Let $s =$ the side of the softball field. Then
$s + 30 =$ the side of the baseball field.

$$\boxed{\begin{array}{c}\text{Larger}\\\text{area}\end{array}} - \boxed{\begin{array}{c}\text{Smaller}\\\text{area}\end{array}} = 4500$$

$$(s + 30)^2 - s^2 = 4500$$
$$(s + 30)(s + 30) - s^2 = 4500$$
$$s^2 + 30s + 30s + 900 - s^2 = 4500$$
$$60s + 900 = 4500$$
$$60s = 3600$$
$$s = 60$$

The baseball field has a side of
$60 + 30 = 90$ feet.

123. Answers may vary.

125. Answers may vary.

Exercise 4.7 (page 281)

1. binomial

3. none of these

5. 2

7. polynomial

9. two

11. $\dfrac{a}{b}$

13. $\dfrac{5}{15} = \dfrac{5(1)}{5(3)} = \dfrac{1}{3}$

15. $\dfrac{-125}{75} = -\dfrac{25(5)}{25(3)} = -\dfrac{5}{3}$

17. $\dfrac{120}{160} = \dfrac{40(3)}{40(4)} = \dfrac{3}{4}$

19. $\dfrac{-3612}{-3612} = 1$

21. $\dfrac{-90}{360} = -\dfrac{90(1)}{90(4)} = -\dfrac{1}{4}$

23. $\dfrac{5880}{2660} = \dfrac{140(42)}{140(19)} = \dfrac{42}{19}$

25. $\dfrac{xy}{yz} = \dfrac{x}{z}$

27. $\dfrac{r^3 s^2}{rs^3} = r^{3-1}s^{2-3} = r^2 s^{-1} = \dfrac{r^2}{s}$

29. $\dfrac{8x^3 y^2}{4xy^3} = \dfrac{8}{4}x^{3-1}y^{2-3} = 2x^2 y^{-1} = \dfrac{2x^2}{y}$

31. $\dfrac{12u^5 v}{-4u^2 v^3} = \dfrac{12}{-4}u^{5-2}v^{1-3} = -3u^3 v^{-2} = \dfrac{-3u^3}{v^2}$

33. $\dfrac{-16r^3 y^2}{-4r^2 y^4} = \dfrac{-16}{-4}r^{3-2}y^{2-4} = 4ry^{-2}$
$$= \dfrac{4r}{y^2}$$

35. $\dfrac{-65rs^2 t}{15r^2 s^3 t} = \dfrac{-65}{15}r^{1-2}s^{2-3}t^{1-1}$
$$= -\dfrac{13}{3}r^{-1}s^{-1}t^0 = -\dfrac{13}{3rs}$$

37. $\dfrac{x^2 x^3}{xy^6} = \dfrac{x^5}{xy^6} = \dfrac{x^{5-1}}{y^6} = \dfrac{x^4}{y^6}$

39. $\dfrac{(a^3 b^4)^3}{ab^4} = \dfrac{a^9 b^{12}}{ab^4} = a^{9-1}b^{12-4} = a^8 b^8$

41. $\dfrac{15(r^2 s^3)^2}{-5(rs^5)^3} = \dfrac{15r^4 s^6}{-5r^3 s^{15}} = \dfrac{15}{-5}r^{4-3}s^{6-15} = -3r^1 s^{-9} = -\dfrac{3r}{s^9}$

43. $\dfrac{-32(x^3y)^3}{128(x^2y^2)^3} = \dfrac{-32x^9y^3}{128x^6y^6} = \dfrac{-32}{128}x^{9-6}y^{3-6} = -\dfrac{1}{4}x^3y^{-3} = -\dfrac{x^3}{4y^3}$

45. $\dfrac{(5a^2b)^3}{(2a^2b^2)^3} = \dfrac{125a^6b^3}{8a^6b^6} = \dfrac{125}{8}a^{6-6}b^{3-6} = \dfrac{125}{8}a^0b^{-3} = \dfrac{125}{8b^3}$

47. $\dfrac{-(3x^3y^4)^3}{-(9x^4y^5)^2} = \dfrac{-27x^9y^{12}}{-81x^8y^{10}} = \dfrac{-27}{-81}x^{9-8}y^{12-10} = \dfrac{1}{3}xy^2 = \dfrac{xy^2}{3}$

49. $\dfrac{(a^2a^3)^4}{(a^4)^3} = \dfrac{(a^5)^4}{a^{12}} = \dfrac{a^{20}}{a^{12}} = a^{20-12} = a^8$

51. $\dfrac{(z^3z^{-4})^3}{(z^{-3})^2} = \dfrac{(z^{-1})^3}{z^{-6}} = \dfrac{z^{-3}}{z^{-6}} = z^{-3-(-6)} = z^3$

53. $\dfrac{6x+9y}{3xy} = \dfrac{6x}{3xy} + \dfrac{9y}{3xy} = \dfrac{2}{y} + \dfrac{3}{x}$

55. $\dfrac{5x-10y}{25xy} = \dfrac{5x}{25xy} - \dfrac{10y}{25xy} = \dfrac{1}{5y} - \dfrac{2}{5x}$

57. $\dfrac{3x^2+6y^3}{3x^2y^2} = \dfrac{3x^2}{3x^2y^2} + \dfrac{6y^3}{3x^2y^2} = \dfrac{1}{y^2} + \dfrac{2y}{x^2}$

59. $\dfrac{15a^3b^2 - 10a^2b^3}{5a^2b^2} = \dfrac{15a^3b^2}{5a^2b^2} - \dfrac{10a^2b^3}{5a^2b^2}$
$= 3a - 2b$

61. $\dfrac{4x-2y+8z}{4xy} = \dfrac{4x}{4xy} - \dfrac{2y}{4xy} + \dfrac{8z}{4xy} = \dfrac{1}{y} - \dfrac{1}{2x} + \dfrac{2z}{xy}$

63. $\dfrac{12x^3y^2 - 8x^2y - 4x}{4xy} = \dfrac{12x^3y^2}{4xy} - \dfrac{8x^2y}{4xy} - \dfrac{4x}{4xy} = 3x^2y - 2x - \dfrac{1}{y}$

65. $\dfrac{-25x^2y + 30xy^2 - 5xy}{-5xy} = \dfrac{-25x^2y}{-5xy} + \dfrac{30xy^2}{-5xy} - \dfrac{5xy}{-5xy} = 5x - 6y + 1$

67. $\dfrac{5x(4x-2y)}{2y} = \dfrac{20x^2 - 10xy}{2y} = \dfrac{20x^2}{2y} - \dfrac{10xy}{2y} = \dfrac{10x^2}{y} - 5x$

69. $\dfrac{(-2x)^3 + (3x^2)^2}{6x^2} = \dfrac{-8x^3 + 9x^4}{6x^2} = \dfrac{-8x^3}{6x^2} + \dfrac{9x^4}{6x^2} = -\dfrac{4x}{3} + \dfrac{3x^2}{2}$

71. $\dfrac{4x^2y^2 - 2(x^2y^2 + xy)}{2xy} = \dfrac{4x^2y^2 - 2x^2y^2 - 2xy}{2xy} = \dfrac{2x^2y^2 - 2xy}{2xy} = \dfrac{2x^2y^2}{2xy} - \dfrac{2xy}{2xy} = xy - 1$

73. $\dfrac{(3x-y)(2x-3y)}{6xy} = \dfrac{6x^2 - 9xy - 2xy + 3y^2}{6xy} = \dfrac{6x^2 - 11xy + 3y^2}{6xy} = \dfrac{6x^2}{6xy} - \dfrac{11xy}{6xy} + \dfrac{3y^2}{6xy}$
$= \dfrac{x}{y} - \dfrac{11}{6} + \dfrac{y}{2x}$

75.
$$\frac{(a+b)^2 - (a-b)^2}{2ab} = \frac{(a+b)(a+b) - (a-b)(a-b)}{2ab}$$
$$= \frac{[a^2 + ab + ab + b^2] - [a^2 - ab - ab + b^2]}{2ab}$$
$$= \frac{[a^2 + 2ab + b^2] - [a^2 - 2ab + b^2]}{2ab}$$
$$= \frac{a^2 + 2ab + b^2 - a^2 + 2ab - b^2}{2ab} = \frac{4ab}{2ab} = 2$$

77.
$$l = \frac{P - 2w}{2}$$
$$l = \frac{P}{2} - \frac{2w}{2}$$
$$l = \frac{P}{2} - w$$
They are the same.

79.
$$C = \frac{0.15x + 12}{x}$$
$$C = \frac{0.15x}{x} + \frac{12}{x}$$
$$C = 0.15 + \frac{12}{x}$$
They are the same.

81. Answers may vary.

83.
$$\frac{x^{500} - x^{499}}{x^{499}} = \frac{x^{500}}{x^{499}} - \frac{x^{499}}{x^{499}} = x - 1$$
Let $x = 501$: $x - 1 = 501 - 1 = 500$

Exercise 4.8 (page 287)

1. $21, 22, 24, 25, 26, 27, 28$

3. $|a - b| = |-2 - 3| = |-5| = 5$

5. $-|a^2 - b^2| = -|(-2)^2 - 3^2| = -|4 - 9| = -|-5| = -(+5) = -5$

7. $3(2x^2 - 4x + 5) + 2(x^2 + 3x - 7) = 6x^2 - 12x + 15 + 2x^2 + 6x - 14 = 8x^2 - 6x + 1$

9. divisor; dividend

11. remainder

13. $4x^3 - 2x^2 + 7x + 6$

15. $6x^4 - x^3 + 2x^2 + 9x$

17. $0x^3$ and $0x$

19.
$$\begin{array}{r} x + 2 \\ x+2\overline{\smash{\big)}\,x^2 + 4x + 4} \\ \underline{x^2 + 2x} \\ 2x + 4 \\ \underline{2x + 4} \\ 0 \end{array}$$

21.
$$\begin{array}{r} y + 12 \\ y+1\overline{\smash{\big)}\,y^2 + 13y + 12} \\ \underline{y^2 + \ y} \\ 12y + 12 \\ \underline{12y + 12} \\ 0 \end{array}$$

23.
$$\begin{array}{r} a + b \\ a+b\overline{\smash{\big)}\,a^2 + 2ab + b^2} \\ \underline{a^2 + \ ab} \\ ab + b^2 \\ \underline{ab + b^2} \\ 0 \end{array}$$

25.
$$\begin{array}{r} 3a - 2 \\ 2a+3\overline{\smash{\big)}\,6a^2 + 5a - 6} \\ \underline{6a^2 + 9a} \\ -4a - 6 \\ \underline{-4a - 6} \\ 0 \end{array}$$

27.
$$\begin{array}{r} b + 3 \\ 3b+2\overline{\smash{\big)}\,3b^2 + 11b + 6} \\ \underline{3b^2 + \ 2b} \\ 9b + 6 \\ \underline{9b + 6} \\ 0 \end{array}$$

29.
$$\begin{array}{r} x - 3y \\ 2x-y\overline{\smash{\big)}\,2x^2 - 7xy + 3y^2} \\ \underline{2x^2 - \ xy} \\ -6xy + 3y^2 \\ \underline{-6xy + 3y^2} \\ 0 \end{array}$$

31.
$$\begin{array}{r} 2x + 1 \\ 5x+3\,\overline{\big)\,10x^2 + 11x + 3} \\ \underline{10x^2 + 6x} \\ 5x + 3 \\ \underline{5x + 3} \\ 0 \end{array}$$

33.
$$\begin{array}{r} x - 7 \\ 2x+4\,\overline{\big)\,2x^2 - 10x - 28} \\ \underline{2x^2 + 4x} \\ -\,14x - 28 \\ \underline{-\,14x - 28} \\ 0 \end{array}$$

35.
$$\begin{array}{r} 3x + 2y \\ 2x-y\,\overline{\big)\,6x^2 + xy - 2y^2} \\ \underline{6x^2 - 3xy} \\ 4xy - 2y^2 \\ \underline{4xy - 2y^2} \\ 0 \end{array}$$

37.
$$\begin{array}{r} 2x - y \\ x+3y\,\overline{\big)\,2x^2 + 5xy - 3y^2} \\ \underline{2x^2 + 6xy} \\ -\,xy - 3y^2 \\ \underline{-\,xy - 3y^2} \\ 0 \end{array}$$

39.
$$\begin{array}{r} x + 5y \\ 3x-2y\,\overline{\big)\,3x^2 + 13xy - 10y^2} \\ \underline{3x^2 - 2xy} \\ 15xy - 10y^2 \\ \underline{15xy - 10y^2} \\ 0 \end{array}$$

41.
$$\begin{array}{r} x - 5y \\ 4x+y\,\overline{\big)\,4x^2 - 19xy - 5y^2} \\ \underline{4x^2 + xy} \\ -\,20xy - 5y^2 \\ \underline{-\,20xy - 5y^2} \\ 0 \end{array}$$

43.
$$\begin{array}{r} x^2 + 2x - 1 \\ 2x+3\,\overline{\big)\,2x^3 + 7x^2 + 4x - 3} \\ \underline{2x^3 + 3x^2} \\ 4x^2 + 4x \\ \underline{4x^2 + 6x} \\ -\,2x - 3 \\ \underline{-\,2x - 3} \\ 0 \end{array}$$

45.
$$\begin{array}{r} 2x^2 + 2x + 1 \\ 3x+2\,\overline{\big)\,6x^3 + 10x^2 + 7x + 2} \\ \underline{6x^3 + 4x^2} \\ 6x^2 + 7x \\ \underline{6x^2 + 4x} \\ 3x + 2 \\ \underline{3x + 2} \\ 0 \end{array}$$

47.
$$\begin{array}{r} x^2 + xy + y^2 \\ 2x+y\,\overline{\big)\,2x^3 + 3x^2y + 3xy^2 + y^3} \\ \underline{2x^3 + x^2y} \\ 2x^2y + 3xy^2 \\ \underline{2x^2y + xy^2} \\ 2xy^2 + y^3 \\ \underline{2xy^2 + y^3} \\ 0 \end{array}$$

49.
$$\begin{array}{r} x + 1 + \frac{-1}{2x+3} \\ 2x+3\,\overline{\big)\,2x^2 + 5x + 2} \\ \underline{2x^2 + 3x} \\ 2x + 2 \\ \underline{2x + 3} \\ -1 \end{array}$$

51.
$$\begin{array}{r} 2x + 2 + \frac{-3}{2x+1} \\ 2x+1\,\overline{\big)\,4x^2 + 6x - 1} \\ \underline{4x^2 + 2x} \\ 4x - 1 \\ \underline{4x + 2} \\ -3 \end{array}$$

53.
$$\begin{array}{r} x^2 + 2x + 1 \\ x+1\,\overline{\big)\,x^3 + 3x^2 + 3x + 1} \\ \underline{x^3 + x^2} \\ 2x^2 + 3x \\ \underline{2x^2 + 2x} \\ x + 1 \\ \underline{x + 1} \\ 0 \end{array}$$

55.
$$\begin{array}{r} x^2 + 2x - 1 + \frac{6}{2x+3} \\ 2x+3\,\overline{\big)\,2x^3 + 7x^2 + 4x + 3} \\ \underline{2x^3 + 3x^2} \\ 4x^2 + 4x \\ \underline{4x^2 + 6x} \\ -\,2x + 3 \\ \underline{-\,2x - 3} \\ 6 \end{array}$$

57.
$$\begin{array}{r} 2x^2 + 8x + 14 + \frac{31}{x-2} \\ x-2\,\overline{\big)\,2x^3 + 4x^2 - 2x + 3} \\ \underline{2x^3 - 4x^2} \\ 8x^2 - 2x \\ \underline{8x^2 - 16x} \\ 14x + 3 \\ \underline{14x - 28} \\ 31 \end{array}$$

59.

$$\begin{array}{r}
x + 1 \\
x - 1 \,\overline{\big)\, x^2 + 0x - 1} \\
\underline{x^2 - x\phantom{{}-1}} \\
x - 1 \\
\underline{x - 1} \\
0
\end{array}$$

61.

$$\begin{array}{r}
2x - 3 \\
2x + 3 \,\overline{\big)\, 4x^2 + 0x - 9} \\
\underline{4x^2 + 6x\phantom{{}-9}} \\
-6x - 9 \\
\underline{-6x - 9} \\
0
\end{array}$$

63.

$$\begin{array}{r}
x^2 - x + 1 \\
x + 1 \,\overline{\big)\, x^3 + 0x^2 + 0x + 1} \\
\underline{x^3 + x^2\phantom{{}+0x+1}} \\
-x^2 + 0x \\
\underline{-x^2 - x} \\
x + 1 \\
\underline{x + 1} \\
0
\end{array}$$

65.

$$\begin{array}{r}
a^2 - 3a + 10 + \dfrac{-30}{a+3} \\
a + 3 \,\overline{\big)\, a^3 + 0a^2 + a + 0} \\
\underline{a^3 + 3a^2} \\
-3a^2 + a \\
\underline{-3a^2 - 9a} \\
10a + 0 \\
\underline{10a + 30} \\
-30
\end{array}$$

67.

$$\begin{array}{r}
5x^2 - x + 4 + \dfrac{16}{3x-4} \\
3x - 4 \,\overline{\big)\, 15x^3 - 23x^2 + 16x + 0} \\
\underline{15x^3 - 20x^2} \\
-3x^2 + 16x \\
\underline{-3x^2 + 4x} \\
12x + 0 \\
\underline{12x - 16} \\
16
\end{array}$$

69. Answers may vary.

71. $x^2 - 2x$ is added to the dividend when it should be subtracted.

Chapter 4 Summary (page 289)

1. $(-3x)^4 = (-3x)(-3x)(-3x)(-3x)$

2. $\left(\tfrac{1}{2}pq\right)^3 = \left(\tfrac{1}{2}pq\right)\left(\tfrac{1}{2}pq\right)\left(\tfrac{1}{2}pq\right)$

3. $5^3 = 5 \cdot 5 \cdot 5 = 125$

4. $3^5 = 3 \cdot 3 \cdot 3 \cdot 3 \cdot 3 = 243$

5. $(-8)^2 = (-8)(-8) = 64$

6. $-8^2 = -1 \cdot 8 \cdot 8 = -64$

7. $3^2 + 2^2 = 9 + 4 = 13$

8. $(3+2)^2 = (5)^2 = 25$

9. $x^3 x^2 = x^{3+2} = x^5$

10. $x^2 x^7 = x^{2+7} = x^9$

11. $\left(y^7\right)^3 = y^{7\cdot 3} = y^{21}$

12. $\left(x^{21}\right)^2 = x^{21\cdot 2} = x^{42}$

13. $(ab)^3 = a^3 b^3$

14. $(3x)^4 = 3^4 x^4 = 81x^4$

15. $b^3 b^4 b^5 = b^{3+4+5} = b^{12}$

16. $-z^2(z^3 y^2) = -z^{2+3}y^2 = -y^2 z^5$

17. $(16s)^2 s = 16^2 s^2 s = 256 s^3$

18. $-3y(y^5) = -3y^6$

19. $\left(x^2 x^3\right)^3 = \left(x^5\right)^3 = x^{15}$

20. $\left(2x^2 y\right)^2 = 2^2 \left(x^2\right)^2 y^2 = 4x^4 y^2$

21. $\dfrac{x^7}{x^3} = x^{7-3} = x^4$

22. $\left(\dfrac{x^2 y}{xy^2}\right)^2 = \dfrac{x^4 y^2}{x^2 y^4} = x^2 y^{-2} = \dfrac{x^2}{y^2}$

23. $\dfrac{8(y^2x)^2}{4(yx^2)^2} = \dfrac{8y^4x^2}{4y^2x^4} = \dfrac{8}{4}y^2x^{-2} = \dfrac{2y^2}{x^2}$

24. $\dfrac{(5y^2z^3)^3}{25(yz)^5} = \dfrac{125y^6z^9}{25y^5z^5} = 5yz^4$

25. $x^0 = 1$

26. $(3x^2y^2)^0 = 1$

27. $(3x^0)^2 = (3 \cdot 1)^2 = 3^2 = 9$

28. $(3x^2y^0)^2 = (3x^2 \cdot 1)^2 = (3x^2)^2 = 9x^4$

29. $x^{-3} = \dfrac{1}{x^3}$

30. $x^{-2}x^3 = x^1 = x$

31. $y^4y^{-3} = y^1 = y$

32. $\dfrac{x^3}{x^{-7}} = x^{3-(-7)} = x^{10}$

33. $\left(x^{-3}x^4\right)^{-2} = \left(x^1\right)^{-2} = x^{-2} = \dfrac{1}{x^2}$

34. $\left(a^{-2}b\right)^{-3} = a^6b^{-3} = \dfrac{a^6}{b^3}$

35. $\left(\dfrac{x^2}{x}\right)^{-5} = (x)^{-5} = \dfrac{1}{x^5}$

36. $\left(\dfrac{15z^4}{5z^3}\right)^{-2} = (3z)^{-2} = \dfrac{1}{(3z)^2} = \dfrac{1}{9z^2}$

37. $728 = 7.28 \times 10^2$

38. $9{,}370 = 9.37 \times 10^3$

39. $0.0136 = 1.36 \times 10^{-2}$

40. $0.00942 = 9.42 \times 10^{-3}$

41. $7.73 = 7.73 \times 10^0$

42. $753 \times 10^3 = 7.53 \times 10^2 \times 10^3$
$= 7.53 \times 10^5$

43. $0.018 \times 10^{-2} = 1.8 \times 10^{-2} \times 10^{-2}$
$= 1.8 \times 10^{-4}$

44. $600 \times 10^2 = 6.00 \times 10^2 \times 10^2$
$= 6.00 \times 10^4$

45. $7.26 \times 10^5 = 726{,}000$

46. $3.91 \times 10^{-4} = 0.000391$

47. $2.68 \times 10^0 = 2.68$

48. $5.76 \times 10^1 = 57.6$

49. $739 \times 10^{-2} = 7.39$

50. $0.437 \times 10^{-3} = 0.000437$

51. $\dfrac{(0.00012)(0.00004)}{0.00000016} = \dfrac{(1.2 \times 10^{-4})(4 \times 10^{-5})}{1.6 \times 10^{-7}} = \dfrac{4.8 \times 10^{-9}}{1.6 \times 10^{-7}} = 3 \times 10^{-2} = 0.03$

52. $\dfrac{(4{,}800)(20{,}000)}{600{,}000} = \dfrac{(4.8 \times 10^3)(2 \times 10^4)}{6 \times 10^5} = \dfrac{9.6 \times 10^7}{6 \times 10^5} = 1.6 \times 10^2 = 160$

53. $13x^7$: monomial, degree $= 7$

54. $5^3x + x^2$: binomial, degree $= 2$

55. $-3x^5 + x - 1$: trinomial, degree $= 5$

56. $9x + 21x^3y^2$: binomial,
degree $= 3 + 2 = 5$

57. $3x + 2 = 3(3) + 2 = 9 + 2 = 11$

58. $3x + 2 = 3(0) + 2 = 0 + 2 = 2$

59. $3x + 2 = 3(-2) + 2 = -6 + 2 = -4$

60. $3x + 2 = 3\left(\frac{2}{3}\right) + 2 = 2 + 2 = 4$

61. $5x^4 - x = 5(3)^4 - 3 = 5(81) - 3$
$$= 405 - 3 = 402$$

62. $5x^4 - x = 5(0)^4 - 0 = 5(0) - 0$
$$= 0 - 0 = 0$$

63. $5x^4 - x = 5(-2)^4 - (-2) = 5(16) + 2$
$$= 80 + 2 = 82$$

64. $5x^4 - x = 5(-0.3)^4 - (-0.3)$
$$= 5(0.0081) + 0.3$$
$$= 0.0405 + 3 = 0.3405$$

65. $f(0) = 0^2 - 4 = 0 - 4 = -4$

66. $f(5) = 5^2 - 4 = 25 - 4 = 21$

67. $f(-2) = (-2)^2 - 4 = 4 - 4 = 0$

68. $f\left(\frac{1}{2}\right) = \left(\frac{1}{2}\right)^2 - 4 = \frac{1}{4} - 4 = -\frac{15}{4}$

69. $f(x) = x^2 - 5$

x	-2	-1	0	1	2
$f(x)$	-1	-4	-5	-4	-1

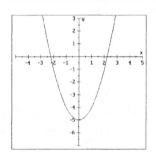

70. $f(x) = x^3 - 2$

x	-2	-1	0	1	2
$f(x)$	-10	-3	-2	-1	6

71. $3x + 5x - x = 7x$

72. $3x + 2y$: not like terms

73. $(xy)^2 + 3x^2y^2 = x^2y^2 + 3x^2y^2 = 4x^2y^2$

74. $-2x^2yz + 3x^2yz = x^2yz$

75. $(3x^2 + 2x) + (5x^2 - 8x) = 3x^2 + 2x + 5x^2 - 8x = 8x^2 - 6x$

76. $(7a^2 + 2a - 5) - (3a^2 - 2a + 1) = 7a^2 + 2a - 5 - 3a^2 + 2a - 1 = 4a^2 + 4a - 6$

77. $3(9x^2 + 3x + 7) - 2(11x^2 - 5x + 9) = 27x^2 + 9x + 21 - 22x^2 + 10x - 18 = 5x^2 + 19x + 3$

78. $4(4x^3 + 2x^2 - 3x - 8) - 5(2x^3 - 3x + 8) = 16x^3 + 8x^2 - 12x - 32 - 10x^3 + 15x - 40$
$$= 6x^3 + 8x^2 + 3x - 72$$

79. $(2x^2y^3)(5xy^2) = 10x^3y^5$

80. $(xyz^3)(x^3z)^2 = xyz^3x^6z^2 = x^7yz^5$

81. $5(x + 3) = 5x + 15$

82. $3(2x + 4) = 6x + 12$

83. $x^2(3x^2 - 5) = 3x^4 - 5x^2$

84. $2y^2(y^2 + 5y) = 2y^4 + 10y^3$

85. $-x^2y(y^2 - xy) = -x^2y^3 + x^3y^2$

86. $-3xy(xy - x) = -3x^2y^2 + 3x^2y$

87. $(x + 3)(x + 2) = x^2 + 2x + 3x + 6$
$$= x^2 + 5x + 6$$

88. $(2x + 1)(x - 1) = 2x^2 - 2x + x - 1$
$$= 2x^2 - x - 1$$

89. $(3a - 3)(2a + 2) = 6a^2 + 6a - 6a - 6$
$$= 6a^2 - 6$$

90. $6(a - 1)(a + 1) = 6\left(a^2 + a - a - 1\right)$
$$= 6\left(a^2 - 1\right) = 6a^2 - 6$$

91. $(a - b)(2a + b) = 2a^2 + ab - 2ab - b^2$
$$= 2a^2 - ab - b^2$$

92. $(3x - y)(2x + y) = 6x^2 + 3xy - 2xy - y^2$
$$= 6x^2 + xy - y^2$$

93. $(x + 3)(x + 3) = x^2 + 3x + 3x + 9$
$$= x^2 + 6x + 9$$

94. $(x + 5)(x - 5) = x^2 - 5x + 5x - 25$
$$= x^2 - 25$$

95. $(y - 2)(y + 2) = y^2 + 2y - 2y - 4$
$$= y^2 - 4$$

96. $(x + 4)^2 = (x + 4)(x + 4)$
$$= x^2 + 4x + 4x + 16$$
$$= x^2 + 8x + 16$$

97. $(x - 3)^2 = (x - 3)(x - 3) = x^2 - 3x - 3x + 9 = x^2 - 6x + 9$

98. $(y - 1)^2 = (y - 1)(y - 1) = y^2 - y - y + 1 = y^2 - 2y + 1$

99. $(2y + 1)^2 = (2y + 1)(2y + 1) = 4y^2 + 2y + 2y + 1 = 4y^2 + 4y + 1$

100. $(y^2 + 1)(y^2 - 1) = y^4 - y^2 + y^2 - 1 = y^4 - 1$

101. $(3x + 1)\left(x^2 + 2x + 1\right) = 3x^3 + 6x^2 + 3x + x^2 + 2x + 1 = 3x^3 + 7x^2 + 5x + 1$

102. $(2a - 3)\left(4a^2 + 6a + 9\right) = 8a^3 + 12a^2 + 18a - 12a^2 - 18a - 27 = 8a^3 - 27$

103. $x^2 + 3 = x(x + 3)$
$$x^2 + 3 = x^2 + 3x$$
$$3 = 3x$$
$$1 = x$$

104. $x^2 + x = (x + 1)(x + 2)$
$$x^2 + x = x^2 + 2x + x + 2$$
$$x^2 + x = x^2 + 3x + 2$$
$$x = 3x + 2$$
$$-2x = 2$$
$$x = -1$$

105.
$$(x+2)(x-5) = (x-4)(x-1)$$
$$x^2 - 5x + 2x - 10 = x^2 - x - 4x + 4$$
$$x^2 - 3x - 10 = x^2 - 5x + 4$$
$$-3x - 10 = -5x + 4$$
$$-3x = -5x + 14$$
$$2x = 14$$
$$x = 7$$

106.
$$(x-1)(x-2) = (x-3)(x+1)$$
$$x^2 - 2x - x + 2 = x^2 + x - 3x - 3$$
$$x^2 - 3x + 2 = x^2 - 2x - 3$$
$$-3x + 2 = -2x - 3$$
$$-3x = -2x - 5$$
$$-x = -5$$
$$x = 5$$

107.
$$x^2 + x(x+2) = x(2x+1) + 1$$
$$x^2 + x^2 + 2x = 2x^2 + x + 1$$
$$2x^2 + 2x = 2x^2 + x + 1$$
$$2x = x + 1$$
$$x = 1$$

108.
$$(x+5)(3x+1) = x^2 + (2x-1)(x-5)$$
$$3x^2 + 16x + 5 = 3x^2 - 11x + 5$$
$$16x + 5 = -11x + 5$$
$$16x = -11x$$
$$27x = 0$$
$$x = 0$$

109. $\dfrac{3x + 6y}{2xy} = \dfrac{3x}{2xy} + \dfrac{6y}{2xy} = \dfrac{3}{2y} + \dfrac{3}{x}$

110. $\dfrac{14xy - 21x}{7xy} = \dfrac{14xy}{7xy} - \dfrac{21x}{7xy} = 2 - \dfrac{3}{y}$

111. $\dfrac{15a^2bc + 20ab^2c - 25abc^2}{-5abc} = \dfrac{15a^2bc}{-5abc} + \dfrac{20ab^2c}{-5abc} - \dfrac{25abc^2}{-5abc} = -3a - 4b + 5c$

112.
$$\dfrac{(x+y)^2 + (x-y)^2}{-2xy} = \dfrac{(x+y)(x+y) + (x-y)(x-y)}{-2xy}$$
$$= \dfrac{x^2 + xy + xy + y^2 + x^2 - xy - xy + y^2}{-2xy}$$
$$= \dfrac{2x^2 + 2y^2}{-2xy} = \dfrac{2x^2}{-2xy} + \dfrac{2y^2}{-2xy} = -\dfrac{x}{y} - \dfrac{y}{x}$$

113.
$$\begin{array}{r}
x + 1 + \frac{3}{x+2} \\
x+2 \overline{\smash{\big)}\, x^2 + 3x + 5} \\
\underline{x^2 + 2x} \\
x + 5 \\
\underline{x + 2} \\
3
\end{array}$$

114.
$$\begin{array}{r}
x - 5 \\
x-1 \overline{\smash{\big)}\, x^2 - 6x + 5} \\
\underline{x^2 - x} \\
-5x + 5 \\
\underline{-5x + 5} \\
0
\end{array}$$

115.
$$\begin{array}{r}
2x + 1 \\
x+3 \overline{\smash{\big)}\, 2x^2 + 7x + 3} \\
\underline{2x^2 + 6x} \\
x + 3 \\
\underline{x + 3} \\
0
\end{array}$$

116.
$$\begin{array}{r}
x + 5 + \frac{3}{3x-1} \\
3x-1 \overline{\smash{\big)}\, 3x^2 + 14x - 2} \\
\underline{3x^2 - x} \\
15x - 2 \\
\underline{15x - 5} \\
3
\end{array}$$

117.

$$3x^2 + 2x + 1 + \frac{2}{2x-1}$$

$$2x - 1 \overline{)6x^3 + \ x^2 + 0x + \quad 1}$$
$$\underline{6x^3 - 3x^2}$$
$$4x^2 + 0x$$
$$\underline{4x^2 - 2x}$$
$$2x + \quad 1$$
$$\underline{2x - \quad 1}$$
$$2$$

118.

$$3x^2 - \ x - \ 4$$

$$3x + 1 \overline{)9x^3 + 0x^2 - 13x - 4}$$
$$\underline{9x^3 + 3x^2}$$
$$-3x^2 - 13x$$
$$\underline{-3x^2 - \quad x}$$
$$-12x - 4$$
$$\underline{-12x - 4}$$
$$0$$

Chapter 4 Test   (page 293)

1. $2xxxyyyy = 2x^3y^4$
2. $3^2 + 5^3 = 9 + 125 = 134$
3. $y^2(yy^3) = y^2y^4 = y^6$

4. $(-3b^2)(2b^3)(-b^2) = (-3)(2)(-1)b^2b^3b^2 = 6b^7$

5. $(2x^3)^5(x^2)^3 = 32x^{15}x^6 = 32x^{21}$
6. $(2rr^2r^3)^3 = (2r^6)^3 = 8r^{18}$

7. $3x^0 = 3(1) = 3$
8. $2y^{-5}y^2 = 2y^{-3} = \dfrac{2}{y^3}$
9. $\dfrac{y^2}{yy^{-2}} = \dfrac{y^2}{y^{-1}} = y^3$

10. $\left(\dfrac{a^2b^{-1}}{4a^3b^{-2}}\right)^{-3} = \left(\dfrac{1}{4}a^{-1}b^1\right)^{-3} = \left(\dfrac{1}{4}\right)^{-3}a^3b^{-3} = \dfrac{64a^3}{b^3}$

11. $28{,}000 = 2.8 \times 10^4$
12. $0.0025 = 2.5 \times 10^{-3}$
13. $7.4 \times 10^3 = 7{,}400$

14. $9.3 \times 10^{-5} = 0.000093$
15. $3x^2 + 2$: binomial

16. $2 + 3 + 5 = 10$
17. $x^2 + x + 2 = (-2)^2 + (-2) - 2$
$$= 4 - 2 - 2 = 0$$

18. $f(x) = x^2 + 2$

x	$f(x)$
-2	6
-1	3
0	2
1	3
2	6

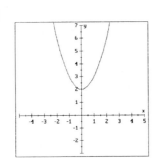

19. $-6(x - y) + 2(x + y) - 3(x + 2y) = -6x + 6y + 2x + 2y - 3x - 6y = -7x + 2y$

20. $-2(x^2 + 3x - 1) - 3(x^2 - x + 2) + 5(x^2 + 2) = -2x^2 - 6x + 2 - 3x^2 + 3x - 6 + 5x^2 + 10$
$$= -3x + 6$$

21.
$$3x^3 + 4x^2 - x - 7$$
$$+ \ 2x^3 - 2x^2 + 3x + 2$$
$$\overline{5x^3 + 2x^2 + 2x - 5}$$

22.
$$2x^2 - 7x + 3 \quad \Rightarrow \qquad 2x^2 - 7x + 3$$
$$- \ 3x^2 - 2x - 1 \qquad + \ -3x^2 + 2x + 1$$
$$\overline{} \qquad\qquad \overline{-x^2 - 5x + 4}$$

23. $(-2x^3)(2x^2y) = -4x^5y$

24. $3y^2(y^2 - 2y + 3) = 3y^4 - 6y^3 + 9y^2$

25. $(2x - 5)(3x + 4) = 6x^2 + 8x - 15x - 20 = 6x^2 - 7x - 20$

26. $(2x - 3)(x^2 - 2x + 4) = 2x^3 - 4x^2 + 8x - 3x^2 + 6x - 12 = 2x^3 - 7x^2 + 14x - 12$

27.
$$(a + 2)^2 = (a - 3)^2$$
$$(a + 2)(a + 2) = (a - 3)(a - 3)$$
$$a^2 + 4a + 4 = a^2 - 6a + 9$$
$$4a + 4 = -6a + 9$$
$$10a = 5$$
$$a = \frac{5}{10} = \frac{1}{2}$$

28. $\dfrac{8x^2y^3z^4}{16x^3y^2z^4} = \dfrac{1}{2}x^{-1}y^1z^0 = \dfrac{y}{2x}$

29.
$$\frac{6a^2 - 12b^2}{24ab} = \frac{6a^2}{24ab} - \frac{12b^2}{24ab}$$
$$= \frac{a}{4b} - \frac{b}{2a}$$

30.
$$\begin{array}{r} x - 2 \\ 2x+3 \overline{)\,2x^2 - x - 6} \\ \underline{2x^2 + 3x} \\ -4x - 6 \\ \underline{-4x - 6} \\ 0 \end{array}$$

Cumulative Review Exercises (page 294)

1. $5 + 3 \cdot 2 = 5 + 6 = 11$

2. $3 \cdot 5^2 - 4 = 3 \cdot 25 - 4 = 75 - 4 = 71$

3. $\dfrac{3x - y}{xy} = \dfrac{3(2) - (-5)}{2(-5)} = \dfrac{6 + 5}{-10} = -\dfrac{11}{10}$

4. $\dfrac{x^2 - y^2}{x + y} = \dfrac{2^2 - (-5)^2}{2 + (-5)} = \dfrac{4 - 25}{-3} = \dfrac{-21}{-3} = 7$

5.
$$\frac{4}{5}x + 6 = 18$$
$$\frac{4}{5}x = 12$$
$$5 \cdot \frac{4}{5}x = 5(12)$$
$$4x = 60$$
$$x = 15$$

6.
$$x - 2 = \frac{x + 2}{3}$$
$$3(x - 2) = 3 \cdot \frac{x + 2}{3}$$
$$3x - 6 = x + 2$$
$$2x - 6 = 2$$
$$2x = 8$$
$$x = 4$$

7.
$$2(5x + 2) = 3(3x - 2)$$
$$10x + 4 = 9x - 6$$
$$x + 4 = -6$$
$$x = -10$$

8. $4(y+1) = -2(4-y)$
$4y + 4 = -8 + 2y$
$2y + 4 = -8$
$2y = -12$
$y = -6$

9. $5x - 3 > 7$
$5x > 10$
$x > 2$

10. $7x - 9 < 5$
$7x < 14$
$x < 2$

11. $-2 < -x + 3 < 5$
$-5 < \quad -x \quad < 2$
$\dfrac{-5}{-1} > \dfrac{-x}{-1} \quad > \dfrac{2}{-1}$
$5 > \quad x \quad > -2$
$-2 < \quad x \quad < 5$

12. $0 \le \dfrac{4-x}{3} \le 2$
$3(0) \le 3 \cdot \dfrac{4-x}{3} \le 3(2)$
$0 \le \quad 4 - x \quad \le 6$
$-4 \le \quad -x \quad \le 2$
$\dfrac{-4}{-1} \ge \dfrac{-x}{-1} \quad \ge \dfrac{2}{-1}$
$4 \ge \quad x \quad \ge -2$
$-2 \le \quad x \quad \le 4$

13. $A = p + prt$
$A - p = p - p + prt$
$A - p = prt$
$\dfrac{A-p}{pt} = \dfrac{prt}{pt}$
$\dfrac{A-p}{pt} = r, \text{ or } r = \dfrac{A-p}{pt}$

14. $A = \dfrac{1}{2}bh$
$2A = 2 \cdot \dfrac{1}{2}bh$
$2A = bh$
$\dfrac{2A}{b} = \dfrac{bh}{b}$
$\dfrac{2A}{b} = h, \text{ or } h = \dfrac{2A}{b}$

15. $3x - 4y = 12$

$x = 0$	$y = 0$
$3(0) - 4y = 12$	$3x - 4(0) = 12$
$0 - 4y = 12$	$3x - 0 = 12$
$-4y = 12$	$3x = 12$
$y = -3$	$x = 4$

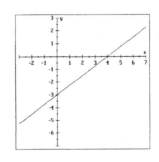

16.
$$y - 2 = \frac{1}{2}(x - 4)$$

$x = 0$	$x = 4$
$y - 2 = \frac{1}{2}(0 - 4)$	$y - 2 = \frac{1}{2}(4 - 4)$
$y - 2 = \frac{1}{2}(-4)$	$y - 2 = \frac{1}{2}(0)$
$y - 2 = -2$	$y - 2 = 0$
$y = 0$	$y = 2$

17. $f(0) = 5(0) - 2 = 0 - 2 = -2$

18. $f(3) = 5(3) - 2 = 15 - 2 = 13$

19. $f(-2) = 5(-2) - 2 = -10 - 2 = -12$

20. $f\left(\frac{1}{5}\right) = 5\left(\frac{1}{5}\right) - 2 = 1 - 2 = -1$

21. $(y^3 y^5) y^6 = y^8 y^6 = y^{14}$

22. $\dfrac{x^3 y^4}{x^2 y^3} = x^{3-2} y^{4-3} = x^1 y^1 = xy$

23. $\dfrac{a^4 b^{-3}}{a^{-3} b^3} = a^{4-(-3)} b^{-3-3} = a^7 b^{-6} = \dfrac{a^7}{b^6}$

24. $\left(\dfrac{-x^{-2} y^3}{x^{-3} y^2}\right)^2 = \left(-x^{-2-(-3)} y^{3-2}\right)^2$
$$= \left(-x^1 y^1\right)^2 = x^2 y^2$$

25. $(3x^2 + 2x - 7) - (2x^2 - 2x + 7) = 3x^2 + 2x - 7 - 2x^2 + 2x - 7 = x^2 + 4x - 14$

26. $(3x - 7)(2x + 8) = 6x^2 + 24x - 14x - 56 = 6x^2 + 10x - 56$

27. $(x - 2)(x^2 + 2x + 4) = x^3 + 2x^2 + 4x - 2x^2 - 4x - 8 = x^3 - 8$

28.
$$\begin{array}{r} 2x + 1 \\ x - 3 \overline{\smash)2x^2 - 5x - 3} \\ \underline{2x^2 - 6x} \\ x - 3 \\ \underline{x - 3} \\ 0 \end{array}$$

29. $(1.6 \times 10^2)(3 \times 10^{16}) = 4.8 \times 10^{18}$ m

30. $A = 2lw + 2wd + 2ld$
$202 = 2(9)(5) + 2(5)d + 2(9)d$
$202 = 90 + 28d$
$112 = 28d \Rightarrow d = 4$ inches

31. $A = \pi(R + r)(R - r)$
$= \pi(17 + 3)(17 - 3)$
$= \pi(20)(14)$
$= 280\pi \approx 879.6$ in.^{2}

32. Let $r =$ the regular price. Then an employee can purchase the TV for $0.75r$.
Purchase price + Sales tax = Total price
$$0.75r + 0.08(0.75r) = 414.72$$
$$0.81r = 414.72$$
$$\frac{0.81r}{0.81} = \frac{414.72}{0.81}$$
$$r = 512 \Rightarrow \text{The regular price is } \$512.$$

Exercise 5.1 (page 303)

1. $3x - 2(x+1) = 5$
$\quad\ 3x - 2x - 2 = 5$
$\qquad\qquad\quad x = 7$

3. $\dfrac{2x-7}{5} = 3$

$\quad 5 \cdot \dfrac{2x-7}{5} = 5(3)$
$\qquad\ 2x - 7 = 15$
$\qquad\qquad 2x = 22$
$\qquad\qquad\ x = 11$

5. prime

7. largest (or greatest)

9. $a + b$

11. $12 = 2 \cdot 6 = 2 \cdot (2 \cdot 3) = 2^2 \cdot 3$

13. $15 = 3 \cdot 5$

15. $40 = 4 \cdot 10 = (2 \cdot 2) \cdot (2 \cdot 5) = 2^3 \cdot 5$

17. $98 = 2 \cdot 49 = 2 \cdot (7 \cdot 7) = 2 \cdot 7^2$

19. $225 = 15 \cdot 15 = (3 \cdot 5) \cdot (3 \cdot 5) = 3^2 \cdot 5^2$

21. $288 = 24 \cdot 12 = (4 \cdot 6) \cdot (4 \cdot 3) = (2^2 \cdot 2 \cdot 3) \cdot (2^2 \cdot 3) = 2^5 \cdot 3^2$

23. $4a + 12 = \boxed{4}(a + 3)$

25. $r^4 + r^2 = r^2(\boxed{r^2} + 1)$

27. $4y^2 + 8y - 2xy = 2y(2y + \boxed{4} - \boxed{x})$

29. $a(x+y) + b(x+y) = (x+y)\boxed{(a+b)}$

31. $(r-s)p - (r-s)q = (r-s)\boxed{(p-q)}$

33. $3x + 6 = 3(x + 2)$

35. $xy - xz = x(y - z)$

37. $t^3 + 2t^2 = t^2(t + 2)$

39. $r^4 - r^2 = r^2(r^2 - 1)$

41. $a^3b^3z^3 - a^2b^3z^2 = a^2b^3z^2(az - 1)$

43. $24x^2y^3z^4 + 8xy^2z^3 = 8xy^2z^3(3xyz + 1)$

45. $12uvw^3 - 18uv^2w^2 = 6uvw^2(2w - 3v)$

47. $3x + 3y - 6z = 3(x + y - 2z)$

49. $ab + ac - ad = a(b + c - d)$

51. $4y^2 + 8y - 2xy = 2y(2y + 4 - x)$

53. $12r^2 - 3rs + 9r^2s^2 = 3r(4r - s + 3rs^2)$

55. $abx - ab^2x + abx^2 = abx(1 - b + x)$

57. $4x^2y^2z^2 - 6xy^2z^2 + 12xyz^2 = 2xyz^2(2xy - 3y + 6)$

59. $70a^3b^2c^2 + 49a^2b^3c^3 - 21a^2b^2c^2 = 7a^2b^2c^2(10a + 7bc - 3)$

61. $-a - b = -(a + b)$

63. $-2x + 5y = -(2x - 5y)$

65. $-2a + 3b = -(2a - 3b)$

67. $-3m - 4n + 1 = -(3m + 4n - 1)$

69. $-3xy + 2z + 5w = -(3xy - 2z - 5w)$ **71.** $-3ab - 5ac + 9bc = -(3ab + 5ac - 9bc)$

73. $-3x^2y - 6xy^2 = -3xy(x + 2y)$ **75.** $-4a^2b^3 + 12a^3b^2 = -4a^2b^2(b - 3a)$

77. $-4a^2b^2c^2 + 14a^2b^2c - 10ab^2c^2 = -2ab^2c(2ac - 7a + 5c)$

79. $-14a^6b^6 + 49a^2b^3 - 21ab = -7ab(2a^5b^5 - 7ab^2 + 3)$

81. $-5a^2b^3c + 15a^3b^4c^2 - 25a^4b^3c = -5a^2b^3c(1 - 3abc + 5a^2)$

83. $(x + y)2 + (x + y)b = (x + y)(2 + b)$ **85.** $3(x + y) - a(x + y) = (x + y)(3 - a)$

87. $3(r - 2s) - x(r - 2s) = (r - 2s)(3 - x)$

89. $(x - 3)^2 + (x - 3) = (x - 3)(x - 3) + 1(x - 3) = (x - 3)(x - 3 + 1) = (x - 3)(x - 2)$

91. $2x(a^2 + b) + 2y(a^2 + b) = (a^2 + b)(2x + 2y) = 2(a^2 + b)(x + y)$

93. $3x^2(r + 3s) - 6y^2(r + 3s) = (r + 3s)(3x^2 - 6y^2) = 3(r + 3s)(x^2 - 2y^2)$

95. $3x(a + b + c) - 2y(a + b + c) = (a + b + c)(3x - 2y)$

97. $14x^2y(r + 2s - t) - 21xy(r + 2s - t) = (r + 2s - t)(14x^2y - 21xy) = 7xy(r + 2s - t)(2x - 3)$

99. $(x + 3)(x + 1) - y(x + 1) = (x + 1)(x + 3 - y)$

101. $(3x - y)(x^2 - 2) + (x^2 - 2) = (3x - y)(x^2 - 2) + 1(x^2 - 2) = (x^2 - 2)(3x - y + 1)$

103. $2x + 2y + ax + ay = 2(x + y) + a(x + y) = (x + y)(2 + a)$

105. $7r + 7s - kr - ks = 7(r + s) - k(r + s) = (r + s)(7 - k)$

107. $xr + xs + yr + ys = x(r + s) + y(r + s) = (r + s)(x + y)$

109. $2ax + 2bx + 3a + 3b = 2x(a + b) + 3(a + b) = (a + b)(2x + 3)$

111. $2ab + 2ac + 3b + 3c = 2a(b + c) + 3(b + c) = (b + c)(2a + 3)$

113. $2x^2 + 2xy - 3x - 3y = 2x(x + y) - 3(x + y) = (x + y)(2x - 3)$

115. $3tv - 9tw + uv - 3uw = 3t(v - 3w) + u(v - 3w) = (v - 3w)(3t + u)$

117. $9mp + 3mq - 3np - nq = 3m(3p + q) - n(3p + q) = (3p + q)(3m - n)$

119. $mp - np - m + n = p(m - n) - 1(m - n) = (m - n)(p - 1)$

121. $x(a - b) + y(b - a) = x(a - b) - y(a - b) = (a - b)(x - y)$

123. $ax^3 + bx^3 + 2ax^2y + 2bx^2y = x^2[ax + bx + 2ay + 2by] = x^2[x(a+b) + 2y(a+b)]$
$$= x^2(a+b)(x+2y)$$

125. $4a^2b + 12a^2 - 8ab - 24a = 4a[ab + 3a - 2b - 6] = 4a[a(b+3) - 2(b+3)]$
$$= 4a(b+3)(a-2)$$

127. $x^3 + 2x^2 + x + 2 = x^2(x+2) + 1(x+2) = (x+2)(x^2+1)$

129. $x^3y - x^2y - xy^2 + y^2 = y[x^3 - x^2 - xy + y] = y[x^2(x-1) - y(x-1)]$
$$= y(x-1)(x^2 - y)$$

131. $2r - bs - 2s + br = 2r - 2s + br - bs = 2(r-s) + b(r-s) = (r-s)(2+b)$

133. $ax + by + bx + ay = ax + ay + bx + by = a(x+y) + b(x+y) = (x+y)(a+b)$

135. $ac + bd - ad - bc = ac - bc - ad + bd = c(a-b) - d(a-b) = (a-b)(c-d)$

137. $ar^2 - brs + ars - br^2 = r[ar - bs + as - br] = r[ar - br + as - bs] = r[r(a-b) + s(a-b)]$
$$= r(a-b)(r+s)$$

139. $ba + 3 + a + 3b = ba + a + 3b + 3 = a(b+1) + 3(b+1) = (b+1)(a+3)$

141. $pr + qs - ps - qr = pr - qr - ps + qs = r(p-q) - s(p-q) = (p-q)(r-s)$

143. Answers may vary. 　　　　　　　　　**145. Answers may vary.**

147. $ax + ay + bx + by = a(x+y) + b(x+y) = (x+y)(a+b)$
$ax + bx + ay + by = x(a+b) + y(a+b) = (a+b)(x+y)$

Exercise 5.2   (page 309)

1. $\dfrac{p}{w} + \dfrac{v^2}{2g} + h = k$

$\dfrac{p}{w} = k - h - \dfrac{v^2}{2g}$

$w \cdot \dfrac{p}{w} = w\left(k - h - \dfrac{v^2}{2g}\right)$

$p = w\left(k - h - \dfrac{v^2}{2g}\right)$

3. difference of two squares

5. $(p-q)$

7. $x^2 - 9 = (x+3)\boxed{(x-3)}$

9. $4m^2 - 9n^2 = (2m+3n)\boxed{(2m-3n)}$

11. $x^2 - 16 = x^2 - 4^2 = (x+4)(x-4)$

13. $y^2 - 49 = y^2 - 7^2 = (y+7)(y-7)$

15. $4y^2 - 49 = (2y)^2 - 7^2 = (2y+7)(2y-7)$

17. $9x^2 - y^2 = (3x)^2 - y^2 = (3x+y)(3x-y)$

19. $25t^2 - 36u^2 = (5t)^2 - (6u)^2$
$$= (5t + 6u)(5t - 6u)$$

21. $16a^2 - 25b^2 = (4a)^2 - (5b)^2$
$$= (4a + 5b)(4a - 5b)$$

23. $a^2 + b^2$: sum of squares $\Rightarrow$ prime

25. $a^4 - 4b^2 = (a^2)^2 - (2b)^2$
$$= (a^2 + 2b)(a^2 - 2b)$$

27. $49y^2 - 225z^4 = (7y)^2 - (15z^2)^2 = (7y + 15z^2)(7y - 15z^2)$

29. $196x^4 - 169y^2 = (14x^2)^2 - (13y)^2 = (14x^2 + 13y)(14x^2 - 13y)$

31. $8x^2 - 32y^2 = 8[x^2 - 4y^2] = 8[x^2 - (2y)^2] = 8(x + 2y)(x - 2y)$

33. $2a^2 - 8y^2 = 2[a^2 - 4y^2] = 2[a^2 - (2y)^2] = 2(a + 2y)(a - 2y)$

35. $3r^2 - 12s^2 = 3[r^2 - 4s^2] = 3[r^2 - (2s)^2] = 3(r + 2s)(r - 2s)$

37. $x^3 - xy^2 = x(x^2 - y^2) = x(x + y)(x - y)$

39. $4a^2x - 9b^2x = x[4a^2 - 9b^2] = x[(2a)^2 - (3b)^2] = x(2a + 3b)(2a - 3b)$

41. $3m^3 - 3mn^2 = 3m(m^2 - n^2) = 3m(m + n)(m - n)$

43. $4x^4 - x^2y^2 = x^2(4x^2 - y^2) = x^2[(2x)^2 - y^2] = x^2(2x + y)(2x - y)$

45. $2a^3b - 242ab^3 = 2ab[a^2 - 121b^2] = 2ab[a^2 - (11b)^2] = 2ab(a + 11b)(a - 11b)$

47. $x^4 - 81 = (x^2 + 9)(x^2 - 9)$
$$= (x^2 + 9)(x + 3)(x - 3)$$

49. $a^4 - 16 = (a^2 + 4)(a^2 - 4)$
$$= (a^2 + 4)(a + 2)(a - 2)$$

51. $a^4 - b^4 = (a^2 + b^2)(a^2 - b^2) = (a^2 + b^2)(a + b)(a - b)$

53. $81r^4 - 256s^4 = (9r^2 + 16s^2)(9r^2 - 16s^2) = (9r^2 + 16s^2)(3r + 4s)(3r - 4s)$

55. $a^4 - b^8 = (a^2 + b^4)(a^2 - b^4) = (a^2 + b^4)(a + b^2)(a - b^2)$

57. $x^8 - y^8 = (x^4 + y^4)(x^4 - y^4) = (x^4 + y^4)(x^2 + y^2)(x^2 - y^2) = (x^4 + y^4)(x^2 + y^2)(x + y)(x - y)$

59. $2x^4 - 2y^4 = 2(x^4 - y^4) = 2(x^2 + y^2)(x^2 - y^2) = 2(x^2 + y^2)(x + y)(x - y)$

61. $a^4b - b^5 = b(a^4 - b^4) = b(a^2 + b^2)(a^2 - b^2) = b(a^2 + b^2)(a + b)(a - b)$

63. $48m^4n - 243n^5 = 3n(16m^4 - 81n^4) = 3n(4m^2 + 9n^2)(4m^2 - 9n^2)$
$$= 3n(4m^2 + 9n^2)(2m + 3n)(2m - 3n)$$

65. $3a^5y + 6ay^5 = 3ay(a^4 + 2y^4)$

67. $3a^{10} - 3a^2b^4 = 3a^2(a^8 - b^4) = 3a^2(a^4 + b^2)(a^4 - b^2) = 3a^2(a^4 + b^2)(a^2 + b)(a^2 - b)$

69. $2x^8y^2 - 32y^6 = 2y^2(x^8 - 16y^4) = 2y^2(x^4 + 4y^2)(x^4 - 4y^2) = 2y^2(x^4 + 4y^2)(x^2 + 2y)(x^2 - 2y)$

71. $a^6b^2 - a^2b^6c^4 = a^2b^2(a^4 - b^4c^4) = a^2b^2(a^2 + b^2c^2)(a^2 - b^2c^2)$
$$= a^2b^2(a^2 + b^2c^2)(a + bc)(a - bc)$$

73. $a^2b^7 - 625a^2b^3 = a^2b^3(b^4 - 625) = a^2b^3(b^2 + 25)(b^2 - 25) = a^2b^3(b^2 + 25)(b + 5)(b - 5)$

75. $243r^5s - 48rs^5 = 3rs(81r^4 - 16s^4) = 3rs(9r^2 + 4s^2)(9r^2 - 4s^2)$
$$= 3rs(9r^2 + 4s^2)(3r + 2s)(3r - 2s)$$

77. $16(x - y)^2 - 9 = [4(x - y)]^2 - 3^2 = [4(x - y) + 3][4(x - y) - 3] = (4x - 4y + 3)(4x - 4y - 3)$

79. $a^3 - 9a + 3a^2 - 27 = a(a^2 - 9) + 3(a^2 - 9) = (a^2 - 9)(a + 3) = (a + 3)(a - 3)(a + 3)$

81. $y^3 - 16y - 3y^2 + 48 = y(y^2 - 16) - 3(y^2 - 16) = (y^2 - 16)(y - 3) = (y + 4)(y - 4)(y - 3)$

83. $3x^3 - 12x + 3x^2 - 12 = 3[x^3 - 4x + x^2 - 4] = 3[x(x^2 - 4) + 1(x^2 - 4)]$
$$= 3(x^2 - 4)(x + 1) = 3(x + 2)(x - 2)(x + 1)$$

85. $3m^3 - 3mn^2 + 3am^2 - 3an^2 = 3[m^3 - mn^2 + am^2 - an^2] = 3[m(m^2 - n^2) + a(m^2 - n^2)]$
$$= 3(m^2 - n^2)(m + a)$$
$$= 3(m + n)(m - n)(m + a)$$

87. $2m^3n^2 - 32mn^2 + 8m^2 - 128 = 2[m^3n^2 - 16mn^2 + 4m^2 - 64]$
$$= 2[mn^2(m^2 - 16) + 4(m^2 - 16)]$$
$$= 2(m^2 - 16)(mn^2 + 4) = 2(m + 4)(m - 4)(mn^2 + 4)$$

89. **Answers may vary.**

91. $399 \cdot 401 = (400 - 1)(400 + 1) = 400^2 - 1^2 = 160,000 - 1 = 159,999$

Exercise 5.3 (page 318)

1. $x - 3 > 5$
$x > 8$

3. $-3x - 5 \geq 4$
$-3x \geq 9$
$\frac{-3x}{-3} \leq \frac{9}{-3}$
$x \leq -3$

5. $\dfrac{3(x-1)}{4} < 12$

$4 \cdot \dfrac{3x-3}{4} < 4(12)$

$3x - 3 < 48$

$3x < 51$

$x < 17$

7. $-2 < x \le 4$

9. $x^2 + 2xy + y^2 = (x+y)^2$

11. $y^2 + 6y + 8 = \left(y + \boxed{4}\right)\left(y + \boxed{2}\right)$

13. $x^2 - xy - 2y^2 = \left(x + \boxed{y}\right)\left(x - \boxed{2y}\right)$

15. $x^2 + 3x + 2 = (x+2)(x+1)$

17. $z^2 + 12z + 11 = (z+1)(z+11)$

19. $a^2 - 4a - 5 = (a-5)(a+1)$

21. $t^2 - 9t + 14 = (t-2)(t-7)$

23. $u^2 + 10u + 15$: prime

25. $y^2 - y - 30 = (y-6)(y+5)$

27. $a^2 + 6a - 16 = (a+8)(a-2)$

29. $t^2 - 5t - 50 = (t-10)(t+5)$

31. $r^2 - 9r - 12$: prime

33. $y^2 + 2yz + z^2 = (y+z)(y+z)$

35. $x^2 + 4xy + 4y^2 = (x+2y)(x+2y)$

37. $m^2 + 3mn - 10n^2 = (m+5n)(m-2n)$

39. $a^2 - 4ab - 12b^2 = (a-6b)(a+2b)$

41. $u^2 + 2uv - 15v^2 = (u+5v)(u-3v)$

43. $-x^2 - 7x - 10 = -\left(x^2 + 7x + 10\right)$
$= -(x+5)(x+2)$

45. $-y^2 - 2y + 15 = -\left(y^2 + 2y - 15\right)$
$= -(y+5)(y-3)$

47. $-t^2 - 15t + 34 = -\left(t^2 + 15t - 34\right)$
$= -(t+17)(t-2)$

49. $-r^2 + 14r - 40 = -\left(r^2 - 14r + 40\right)$
$= -(r-4)(r-10)$

51. $-a^2 - 4ab - 3b^2 = -\left(a^2 + 4ab + 3b^2\right)$
$= -(a+3b)(a+b)$

53. $-x^2 + 6xy + 7y^2 = -\left(x^2 - 6xy - 7y^2\right)$
$= -(x-7y)(x+y)$

55. $4 - 5x + x^2 = x^2 - 5x + 4$
$= (x-4)(x-1)$

57. $10y + 9 + y^2 = y^2 + 10y + 9$
$= (y+9)(y+1)$

59. $c^2 - 5 + 4c = c^2 + 4c - 5$
$= (c+5)(c-1)$

61. $-r^2 + 2s^2 + rs = -r^2 + rs + 2s^2$
$= -\left(r^2 - rs - 2s^2\right)$
$= -(r-2s)(r+s)$

63. $4rx + r^2 + 3x^2 = r^2 + 4rx + 3x^2$
$= (r+3x)(r+x)$

65. $-3ab + a^2 + 2b^2 = a^2 - 3ab + 2b^2$
$$= (a - 2b)(a - b)$$

67. $2x^2 + 10x + 12 = 2(x^2 + 5x + 6)$
$$= 2(x + 2)(x + 3)$$

69. $3y^3 + 6y^2 + 3y = 3y(y^2 + 2y + 1)$
$$= 3y(y + 1)(y + 1)$$

71. $-5a^2 + 25a - 30 = -5(a^2 - 5a + 6)$
$$= -5(a - 3)(a - 2)$$

73. $3z^2 - 15tz + 12t^2 = 3(z^2 - 5tz + 4t^2) = 3(z - 4t)(z - t)$

75. $12xy + 4x^2y - 72y = 4y(3x + x^2 - 18) = 4y(x^2 + 3x - 18) = 4y(x + 6)(x - 3)$

77. $-4x^2y - 4x^3 + 24xy^2 = -4x(xy + x^2 - 6y^2) = -4x(x^2 + xy - 6y^2) = -4x(x + 3y)(x - 2y)$

79. $x^2 + 4x + 4 - y^2 = (x + 2)(x + 2) - y^2 = (x + 2)^2 - y^2 = (x + 2 + y)(x + 2 - y)$

81. $b^2 - 6b + 9 - c^2 = (b - 3)(b - 3) - c^2 = (b - 3)^2 - c^2 = (b - 3 + c)(b - 3 - c)$

83. $a^2 + 2ab + b^2 - 4 = (a + b)(a + b) - 4 = (a + b)^2 - 2^2 = (a + b + 2)(a + b - 2)$

85. $b^2 - y^2 - 4y - 4 = b^2 - (y^2 + 4y + 4) = b^2 - (y + 2)(y + 2) = b^2 - (y + 2)^2$
$$= [b + (y + 2)][b - (y + 2)]$$
$$= (b + y + 2)(b - y - 2)$$

87. $x^2 + 6x + 9 = (x + 3)(x + 3)$

89. $y^2 - 8y + 16 = (y - 4)(y - 4)$

91. $t^2 + 20t + 100 = (t + 10)(t + 10)$

93. $u^2 - 18u + 81 = (u - 9)(u - 9)$

95. $x^2 + 4xy + 4y^2 = (x + 2y)(x + 2y)$

97. $r^2 - 10rs + 25s^2 = (r - 5s)(r - 5s)$

99. Answers may vary.

101. Both answers check. Both may be factored more completely:
$$(2x + 6)(x + 7) = 2(x + 3)(x + 7)$$
$$(x + 3)(2x + 14) = (x + 3) \cdot 2(x + 7)$$

Exercise 5.4 (page 326)

1. $$l = f + (n - 1)d$$
$$l = f + nd - d$$
$$l - f + d = nd$$
$$\frac{l - f + d}{d} = \frac{nd}{d}$$
$$\frac{l - f + d}{d} = n$$

3. descending

5. opposites

7. $6x^2 + x - 2 = \left(3x + \boxed{2}\right)\left(2x - \boxed{1}\right)$

9. $12x^2 - 7xy + y^2 = \left(3x - \boxed{y}\right)\left(4x - \boxed{y}\right)$

11. $2x^2 - 3x + 1 = (2x - 1)(x - 1)$

13. $3a^2 + 13a + 4 = (3a + 1)(a + 4)$

15. $4z^2 + 13z + 3 = (z + 3)(4z + 1)$

17. $6y^2 + 7y + 2 = (3y + 2)(2y + 1)$

19. $6x^2 - 7x + 2 = (2x - 1)(3x - 2)$

21. $3a^2 - 4a - 4 = (3a + 2)(a - 2)$

23. $2x^2 - 3x - 2 = (2x + 1)(x - 2)$

25. $2m^2 + 5m - 12 = (2m - 3)(m + 4)$

27. $10y^2 - 3y - 1 = (5y + 1)(2y - 1)$

29. $12y^2 - 5y - 2 = (4y + 1)(3y - 2)$

31. $5t^2 + 13t + 6 = (5t + 3)(t + 2)$

33. $16m^2 - 14m + 3 = (8m - 3)(2m - 1)$

35. $3x^2 - 4xy + y^2 = (3x - y)(x - y)$

37. $2u^2 + uv - 3v^2 = (2u + 3v)(u - v)$

39. $4a^2 - 4ab + b^2 = (2a - b)(2a - b)$

41. $6r^2 + rs - 2s^2 = (2r - s)(3r + 2s)$

43. $4x^2 + 8xy + 3y^2 = (2x + y)(2x + 3y)$

45. $4a^2 - 15ab + 9b^2 = (4a - 3b)(a - 3b)$

47. $-13x + 3x^2 - 10 = 3x^2 - 13x - 10$
$$= (3x + 2)(x - 5)$$

49. $15 + 8a^2 - 26a = 8a^2 - 26a + 15$
$$= (4a - 3)(2a - 5)$$

51. $12y^2 + 12 - 25y = 12y^2 - 25y + 12$
$$= (3y - 4)(4y - 3)$$

53. $3x^2 + 6 + x = 3x^2 + x + 6 \Rightarrow$ prime

55. $2a^2 + 3b^2 + 5ab = 2a^2 + 5ab + 3b^2$
$$= (2a + 3b)(a + b)$$

57. $pq + 6p^2 - q^2 = 6p^2 + pq - q^2$
$$= (2p + q)(3p - q)$$

59. $b^2 + 4a^2 + 16ab = 4a^2 + 16ab + b^2$
PRIME

61. $12x^2 + 10y^2 - 23xy = 12x^2 - 23xy + 10y^2$
$$= (3x - 2y)(4x - 5y)$$

63. $4x^2 + 10x - 6 = 2(2x^2 + 5x - 3)$
$$= 2(2x - 1)(x + 3)$$

65. $y^3 + 13y^2 + 12y = y(y^2 + 13y + 12)$
$$= y(y + 12)(y + 1)$$

67. $6x^3 - 15x^2 - 9x = 3x(2x^2 - 5x - 3)$
$$= 3x(2x + 1)(x - 3)$$

69. $30r^5 + 63r^4 - 30r^3 = 3r^3(10r^2 + 21r - 10)$
$$= 3r^3(5r - 2)(2r + 5)$$

71. $4a^2 - 4ab - 8b^2 = 4(a^2 - ab - 2b^2)$
$$= 4(a - 2b)(a + b)$$

73. $8x^2 - 12xy - 8y^2 = 4(2x^2 - 3xy - 2y^2)$
$$= 4(2x + y)(x - 2y)$$

75. $-16m^3n - 20m^2n^2 - 6mn^3 = -2mn(8m^2 + 10mn + 3n^2) = -2mn(4m + 3n)(2m + n)$

77. $-28u^3v^3 + 26u^2v^4 - 6uv^5 = -2uv^3(14u^2 - 13uv + 3v^2) = -2uv^3(7u - 3v)(2u - v)$

79. $4x^2 + 12x + 9 = (2x + 3)(2x + 3)$
$$= (2x + 3)^2$$

81. $9x^2 + 12x + 4 = (3x + 2)(3x + 2)$
$$= (3x + 2)^2$$

83. $16x^2 - 8xy + y^2 = (4x - y)(4x - y) = (4x - y)^2$

85. $4x^2 + 4xy + y^2 - 16 = (4x^2 + 4xy + y^2) - 16 = (2x + y)(2x + y) - 16$
$$= (2x + y)^2 - 4^2 = (2x + y + 4)(2x + y - 4)$$

87. $9 - a^2 - 4ab - 4b^2 = 9 - (a^2 + 4ab + 4b^2) = 9 - (a + 2b)(a + 2b)$
$$= 3^2 - (a + 2b)^2$$
$$= (3 + (a + 2b))(3 - (a + 2b))$$
$$= (3 + a + 2b)(3 - a - 2b)$$

89. $9p^2 + 1 + 6p - q^2 = 9p^2 + 6p + 1 - q^2 = (3p + 1)(3p + 1) - q^2 = (3p + 1)^2 - q^2$
$$= (3p + 1 + q)(3p + 1 - q)$$

91. Answers may vary.

93. $(6x + 1)(x + 6) = 6x^2 + 37x + 6; (6x - 1)(x - 6) = 6x^2 - 37x + 6 \Rightarrow b = \pm 37$
$(6x + 2)(x + 3) = 6x^2 + 20x + 6; (6x - 2)(x - 3) = 6x^2 - 20x + 6 \Rightarrow b = \pm 20$
$(6x + 3)(x + 2) = 6x^2 + 15x + 6; (6x - 3)(x - 2) = 6x^2 - 15x + 6 \Rightarrow b = \pm 15$
$(6x + 6)(x + 1) = 6x^2 + 12x + 6; (6x - 6)(x - 1) = 6x^2 - 12x + 6 \Rightarrow b = \pm 12$
$(2x + 1)(3x + 6) = 6x^2 + 15x + 6; (2x - 1)(3x - 6) = 6x^2 - 15x + 6 \Rightarrow b = \pm 15$
$(2x + 2)(3x + 3) = 6x^2 + 12x + 6; (2x - 2)(3x - 3) = 6x^2 - 12x + 6 \Rightarrow b = \pm 12$
$(2x + 3)(3x + 2) = 6x^2 + 13x + 6; (2x - 3)(3x - 2) = 6x^2 - 13x + 6 \Rightarrow b = \pm 13$
$(2x + 6)(3x + 1) = 6x^2 + 20x + 6; (2x - 6)(3x - 1) = 6x^2 - 20x + 6 \Rightarrow b = \pm 20$
$b = \pm 12, \pm 13, \pm 15, \pm 20, \pm 37$

Exercise 5.5 (page 331)

1. $1 \times 10^{-13} = 0.0000000000001$ cm

3. $x^3 + y^3 = (x + y)\boxed{(x^2 - xy + y^2)}$

5. $y^3 + 1 = y^3 + 1^3$
$$= (y + 1)(y^2 - 1y + 1^2)$$
$$= (y + 1)(y^2 - y + 1)$$

7. $a^3 - 27 = a^3 - 3^3$
$$= (a - 3)(a^2 + 3a + 3^2)$$
$$= (a - 3)(a^2 + 3a + 9)$$

9. $8 + x^3 = 2^3 + x^3$
$$= (2 + x)(2^2 - 2x + x^2)$$
$$= (2 + x)(4 - 2x + x^2)$$

11. $s^3 - t^3 = (s - t)(s^2 + st + t^2)$

13. $27x^3 + y^3 = (3x)^3 + y^3$
$$= (3x + y)((3x)^2 - 3xy + y^2)$$
$$= (3x + y)(9x^2 - 3xy + y^2)$$

15. $a^3 + 8b^3 = a^3 + (2b)^3$
$$= (a + 2b)(a^2 - 2ab + (2b)^2)$$
$$= (a + 2b)(a^2 - 2ab + 4b^2)$$

17. $64x^3 - 27 = (4x)^3 - 3^3 = (4x - 3)((4x)^2 + 3(4x) + 3^2) = (4x - 3)(16x^2 + 12x + 9)$

19. $27x^3 - 125y^3 = (3x)^3 - (5y)^3 = (3x - 5y)\left((3x)^2 + (3x)(5y) + (5y)^2\right)$
$$= (3x - 5y)\left(9x^2 + 15xy + 25y^2\right)$$

21. $a^6 - b^3 = \left(a^2\right)^3 - b^3$
$$= \left(a^2 - b\right)\left(\left(a^2\right)^2 + a^2b + b^2\right)$$
$$= \left(a^2 - b\right)\left(a^4 + a^2b + b^2\right)$$

23. $x^9 + y^6 = \left(x^3\right)^3 + \left(y^2\right)^3$
$$= \left(x^3 + y^2\right)\left(\left(x^3\right)^2 - x^3y^2 + \left(y^2\right)^2\right)$$
$$= \left(x^3 + y^2\right)\left(x^6 - x^3y^2 + y^4\right)$$

25. $2x^3 + 54 = 2\left(x^3 + 27\right)$
$$= 2\left(x^3 + 3^3\right)$$
$$= 2(x + 3)\left(x^2 - 3x + 3^2\right)$$
$$= 2(x + 3)\left(x^2 - 3x + 9\right)$$

27. $-x^3 + 216 = -\left(x^3 - 216\right)$
$$= -\left(x^3 - 6^3\right)$$
$$= -(x - 6)\left(x^2 + 6x + 6^2\right)$$
$$= -(x - 6)\left(x^2 + 6x + 36\right)$$

29. $64m^3x - 8n^3x = 8x\left(8m^3 - n^3\right) = 8x\left((2m)^3 - n^3\right) = 8x(2m - n)\left((2m)^2 + 2mn + n^2\right)$
$$= 8x(2m - n)\left(4m^2 + 2mn + n^2\right)$$

31. $x^4y + 216xy^4 = xy\left(x^3 + 216y^3\right) = xy\left(x^3 + (6y)^3\right) = xy(x + 6y)\left(x^2 - 6xy + (6y)^2\right)$
$$= xy(x + 6y)\left(x^2 - 6xy + 36y^2\right)$$

33. $81r^4s^2 - 24rs^5 = 3rs^2\left(27r^3 - 8s^3\right) = 3rs^2\left((3r)^3 - (2s)^3\right)$
$$= 3rs^2(3r - 2s)\left((3r)^2 + (3r)(2s) + (2s)^2\right)$$
$$= 3rs^2(3r - 2s)\left(9r^2 + 6rs + 4s^2\right)$$

35. $125a^6b^2 + 64a^3b^5 = a^3b^2\left(125a^3 + 64b^3\right) = a^3b^2\left((5a)^3 + (4b)^3\right)$
$$= a^3b^2(5a + 4b)\left((5a)^2 - (5a)(4b) + (4b)^2\right)$$
$$= a^3b^2(5a + 4b)\left(25a^2 - 20ab + 16b^2\right)$$

37. $y^7z - yz^4 = yz\left(y^6 - z^3\right) = yz\left(\left(y^2\right)^3 - z^3\right) = yz\left(y^2 - z\right)\left(\left(y^2\right)^2 + y^2z + z^2\right)$
$$= yz\left(y^2 - z\right)\left(y^4 + y^2z + z^2\right)$$

39. $2mp^4 + 16mpq^3 = 2mp\left(p^3 + 8q^3\right) = 2mp\left(p^3 + (2q)^3\right) = 2mp(p + 2q)\left(p^2 - 2pq + (2q)^2\right)$
$$= 2mp(p + 2q)\left(p^2 - 2pq + 4q^2\right)$$

41. $x^6 - 1 = \left(x^3\right)^2 - 1^2 = \left(x^3 + 1\right)\left(x^3 - 1\right) = \left(x^3 + 1^3\right)\left(x^3 - 1^3\right)$
$$= (x + 1)\left(x^2 - 1x + 1^2\right)(x - 1)\left(x^2 + 1x + 1^2\right)$$
$$= (x + 1)\left(x^2 - x + 1\right)(x - 1)\left(x^2 + x + 1\right)$$

43. $x^{12} - y^6 = (x^6)^2 - (y^3)^2 = (x^6 + y^3)(x^6 - y^3)$
$$= \left((x^2)^3 + y^3\right)\left((x^2)^3 - y^3\right)$$
$$= (x^2 + y)\left((x^2)^2 - x^2 y + y^2\right)(x^2 - y)\left((x^2)^2 + x^2 y + y^2\right)$$
$$= (x^2 + y)(x^4 - x^2 y + y^2)(x^2 - y)(x^4 + x^2 y + y^2)$$

45. $3(x^3 + y^3) - z(x^3 + y^3) = (x^3 + y^3)(3 - z) = (x + y)(x^2 - xy + y^2)(3 - z)$

47. $(m^3 + 8n^3) + (m^3 x + 8n^3 x) = 1(m^3 + 8n^3) + x(m^3 + 8n^3)$
$$= (m^3 + 8n^3)(1 + x)$$
$$= (m + 2n)(m^2 - 2mn + (2n)^2)(1 + x)$$
$$= (m + 2n)(m^2 - 2mn + 4n^2)(1 + x)$$

49. $(a^4 + 27a) - (a^3 b + 27b) = a(a^3 + 27) - b(a^3 + 27) = (a^3 + 27)(a - b)$
$$= (a^3 + 3^3)(a - b)$$
$$= (a + 3)(a^2 - 3a + 3^2)(a - b)$$
$$= (a + 3)(a^2 - 3a + 9)(a - b)$$

51. $y^3(y^2 - 1) - 27(y^2 - 1) = (y^2 - 1)(y^3 - 27) = (y^2 - 1^2)(y^3 - 3^3)$
$$= (y + 1)(y - 1)(y - 3)(y^2 + 3y + 9)$$

53. **Answers may vary.**

55. $a^3 - b^3 = 11^3 - 7^3 = 1331 - 343 = \boxed{988}$
$(a - b)(a^2 + ab + b^2) = (11 - 7)(11^2 + 11 \cdot 7 + 7^2) = 4(121 + 77 + 49) = 4(247) = \boxed{988}$

Exercise 5.6 (page 335)

1. $2(t - 5) + t = 3(2 - t)$
$2t - 10 + t = 6 - 3t$
$3t - 10 = 6 - 3t$
$6t = 16$
$t = \dfrac{16}{6} = \dfrac{8}{3}$

3. $5 - 3(t + 1) = t + 2$
$5 - 3t - 3 = t + 2$
$-3t + 2 = t + 2$
$-4t = 0$
$t = \dfrac{0}{-4} = 0$

5. factors

7. binomials

9. $6x + 3 = 3(2x + 1)$

11. $x^2 - 6x - 7 = (x - 7)(x + 1)$

13. $6t^2 + 7t - 3 = (2t + 3)(3t - 1)$

15. $4x^2 - 25 = (2x)^2 - 5^2 = (2x + 5)(2x - 5)$

17. $t^2 - 2t + 1 = (t - 1)(t - 1)$

19. $a^3 - 8 = a^3 - 2^3 = (a - 2)(a^2 + 2a + 2^2) = (a - 2)(a^2 + 2a + 4)$

21. $x^2 y^2 - 2x^2 - y^2 + 2 = x^2 (y^2 - 2) - 1 (y^2 - 2) = (y^2 - 2)(x^2 - 1) = (y^2 - 2)(x^2 - 1^2)$
$$= (y^2 - 2)(x + 1)(x - 1)$$

23. $70p^4 q^3 - 35p^4 q^2 + 49p^5 q^2 = 7p^4 q^2 (10q - 5 + 7p)$

25. $2ab^2 + 8ab - 24a = 2a(b^2 + 4b - 12) = 2a(b + 6)(b - 2)$

27. $-8p^3 q^7 - 4p^2 q^3 = -4p^2 q^3 (2pq^4 + 1)$

29. $4a^2 - 4ab + b^2 - 9 = (4a^2 - 4ab + b^2) - 9 = (2a - b)^2 - 3^2 = (2a - b + 3)(2a - b - 3)$

31. $x^2 + 7x + 1 \Rightarrow$ prime

33. $-2x^5 + 128x^2 = -2x^2 (x^3 - 64) = -2x^2 (x^3 - 4^3) = -2x^2 (x - 4)(x^2 + 4x + 16)$

35. $14t^3 - 40t^2 + 6t^4 = 2t^2 (7t - 20 + 3t^2) = 2t^2 (3t^2 + 7t - 20) = 2t^2 (3t - 5)(t + 4)$

37. $a^2 (x - a) - b^2 (x - a) = (x - a)(a^2 - b^2) = (x - a)(a + b)(a - b)$

39. $8p^6 - 27q^6 = (2p^2)^3 - (3q^2)^3 = (2p^2 - 3q^2)\left((2p^2)^2 + (2p^2)(3q^2) + (3q^2)^2\right)$
$$= (2p^2 - 3q^2)(4p^4 + 6p^2 q^2 + 9q^4)$$

41. $125p^3 - 64y^3 = (5p)^3 - (4y)^3 = (5p - 4y)\left((5p)^2 + (5p)(4y) + (4y)^2\right)$
$$= (5p - 4y)(25p^2 + 20py + 16y^2)$$

43. $-16x^4 y^2 z + 24x^5 y^3 z^4 - 15x^2 y^3 z^7 = -x^2 y^2 z(16x^2 - 24x^3 yz^3 + 15yz^6)$

45. $81p^4 - 16q^4 = (9p^2)^2 - (4q^2)^2 = (9p^2 + 4q^2)(9p^2 - 4q^2) = (9p^2 + 4q^2)(3p + 2q)(3p - 2q)$

47. $4x^2 + 9y^2 \Rightarrow$ prime

49. $54x^3 + 250y^6 = 2(27x^3 + 125y^6) = 2\left((3x)^3 + (5y^2)^3\right)$
$$= 2(3x + 5y^2)\left((3x)^2 - (3x)(5y^2) + (5y^2)^2\right)$$
$$= 2(3x + 5y^2)(9x^2 - 15xy^2 + 25y^4)$$

51. $10r^2 - 13r - 4 \Rightarrow$ prime

53. $21t^3 - 10t^2 + t = t(21t^2 - 10t + 1) = t(3t - 1)(7t - 1)$

55. $x^5 - x^3 y^2 + x^2 y^3 - y^5 = x^3 (x^2 - y^2) + y^3 (x^2 - y^2) = (x^2 - y^2)(x^3 + y^3)$
$$= (x + y)(x - y)(x + y)(x^2 - xy + y^2)$$

57. $2a^2c - 2b^2c + 4a^2d - 4b^2d = 2\left(a^2c - b^2c + 2a^2d - 2b^2d\right)$
$$= 2\left(c\left(a^2 - b^2\right) + 2d\left(a^2 - b^2\right)\right)$$
$$= 2\left(a^2 - b^2\right)(c + 2d) = 2(a + b)(a - b)(c + 2d)$$

59. **Answers may vary.**   **61.** **Answers may vary.**

Exercise 5.7   (page 340)

1. $u^3 u^2 u^4 = u^{3+2+4} = u^9$ **3.** $\dfrac{a^3 b^4}{a^2 b^5} = a^1 b^{-1} = \dfrac{a}{b}$

5.   quadratic   **7.**   second

9. $(x - 2)(x + 3) = 0$ **11.** $(x - 4)(x + 1) = 0$
$x - 2 = 0$ **or** $x + 3 = 0$ $x - 4 = 0$ **or** $x + 1 = 0$
$x = 2 x = -3$ $x = 4 x = -1$

13. $(2x - 5)(3x + 6) = 0$
$2x - 5 = 0$ **or** $3x + 6 = 0$
$2x = 5 3x = -6$
$x = \frac{5}{2} x = -2$

15. $(x - 1)(x + 2)(x - 3) = 0$
$x - 1 = 0$ **or** $x + 2 = 0$ **or** $x - 3 = 0$
$x = 1 x = -2 x = 3$

17. $x^2 - 3x = 0$ **19.** $2x^2 - 5x = 0$
$x(x - 3) = 0$ $x(2x - 5) = 0$
$x = 0$ **or** $x - 3 = 0$ $x = 0$ **or** $2x - 5 = 0$
$x = 0 x = 3$ $x = 0 2x = 5$
$ x = 0 x = \frac{5}{2}$

21. $x^2 - 7x = 0$ **23.** $3x^2 + 8x = 0$
$x(x - 7) = 0$ $x(3x + 8) = 0$
$x = 0$ **or** $x - 7 = 0$ $x = 0$ **or** $3x + 8 = 0$
$x = 0 x = 7$ $x = 0 3x = -8$
$ x = 0 x = -\frac{8}{3}$

25. $8x^2 - 16x = 0$ **27.** $10x^2 + 2x = 0$
$8x(x - 2) = 0$ $2x(5x + 1) = 0$
$8x = 0$ **or** $x - 2 = 0$ $2x = 0$ **or** $5x + 1 = 0$
$x = 0 x = 2$ $x = 0 5x = -1$
$ x = 0 x = -\frac{1}{5}$

29.
$$x^2 - 25 = 0$$
$$(x + 5)(x - 5) = 0$$
$$x + 5 = 0 \quad \textbf{or} \quad x - 5 = 0$$
$$x = -5 \qquad\qquad x = 5$$

31.
$$y^2 - 49 = 0$$
$$(y + 7)(y - 7) = 0$$
$$y + 7 = 0 \quad \textbf{or} \quad y - 7 = 0$$
$$y = -7 \qquad\qquad y = 7$$

33.
$$4x^2 - 1 = 0$$
$$(2x + 1)(2x - 1) = 0$$
$$2x + 1 = 0 \quad \textbf{or} \quad 2x - 1 = 0$$
$$2x = -1 \qquad\qquad 2x = 1$$
$$x = -\tfrac{1}{2} \qquad\qquad x = \tfrac{1}{2}$$

35.
$$9y^2 - 4 = 0$$
$$(3y + 2)(3y - 2) = 0$$
$$3y + 2 = 0 \quad \textbf{or} \quad 3y - 2 = 0$$
$$3y = -2 \qquad\qquad 3y = 2$$
$$y = -\tfrac{2}{3} \qquad\qquad y = \tfrac{2}{3}$$

37.
$$x^2 = 49$$
$$x^2 - 49 = 0$$
$$(x + 7)(x - 7) = 0$$
$$x + 7 = 0 \quad \textbf{or} \quad x - 7 = 0$$
$$x = -7 \qquad\qquad x = 7$$

39.
$$4x^2 = 81$$
$$4x^2 - 81 = 0$$
$$(2x + 9)(2x - 9) = 0$$
$$2x + 9 = 0 \quad \textbf{or} \quad 2x - 9 = 0$$
$$2x = -9 \qquad\qquad 2x = 9$$
$$x = -\tfrac{9}{2} \qquad\qquad x = \tfrac{9}{2}$$

41.
$$x^2 - 13x + 12 = 0$$
$$(x - 1)(x - 12) = 0$$
$$x - 1 = 0 \quad \textbf{or} \quad x - 12 = 0$$
$$x = 1 \qquad\qquad x = 12$$

43.
$$x^2 - 2x - 15 = 0$$
$$(x + 3)(x - 5) = 0$$
$$x + 3 = 0 \quad \textbf{or} \quad x - 5 = 0$$
$$x = -3 \qquad\qquad x = 5$$

45.
$$x^2 - 4x - 21 = 0$$
$$(x + 3)(x - 7) = 0$$
$$x + 3 = 0 \quad \textbf{or} \quad x - 7 = 0$$
$$x = -3 \qquad\qquad x = 7$$

47.
$$x^2 + 8 - 9x = 0$$
$$x^2 - 9x + 8 = 0$$
$$(x - 8)(x - 1) = 0$$
$$x - 8 = 0 \quad \textbf{or} \quad x - 1 = 0$$
$$x = 8 \qquad\qquad x = 1$$

49.
$$a^2 + 8a = -15$$
$$a^2 + 8a + 15 = 0$$
$$(a + 3)(a + 5) = 0$$
$$a + 3 = 0 \quad \textbf{or} \quad a + 5 = 0$$
$$a = -3 \qquad\qquad a = -5$$

51.
$$2y - 8 = -y^2$$
$$y^2 + 2y - 8 = 0$$
$$(y + 4)(y - 2) = 0$$
$$y + 4 = 0 \quad \textbf{or} \quad y - 2 = 0$$
$$y = -4 \qquad\qquad y = 2$$

53.
$$2x^2 - 5x + 2 = 0$$
$$(2x - 1)(x - 2) = 0$$
$$2x - 1 = 0 \quad \textbf{or} \quad x - 2 = 0$$
$$2x = 1 \qquad\qquad x = 2$$
$$x = \tfrac{1}{2} \qquad\qquad x = 2$$

55.
$$6x^2 + x - 2 = 0$$
$$(2x - 1)(3x + 2) = 0$$
$$2x - 1 = 0 \quad \textbf{or} \quad 3x + 2 = 0$$
$$2x = 1 \qquad\qquad 3x = -2$$
$$x = \tfrac{1}{2} \qquad\qquad x = -\tfrac{2}{3}$$

57.
$$5p^2 - 6p + 1 = 0$$
$$(5p - 1)(p - 1) = 0$$
$$5p - 1 = 0 \quad \textbf{or} \quad p - 1 = 0$$
$$5p = 1 \qquad\qquad p = 1$$
$$p = \tfrac{1}{5} \qquad\qquad p = 1$$

59.
$$14m^2 + 23m + 3 = 0$$
$$(7m + 1)(2m + 3) = 0$$
$$7m + 1 = 0 \quad \textbf{or} \quad 2m + 3 = 0$$
$$7m = -1 \qquad\qquad 2m = -3$$
$$m = -\tfrac{1}{7} \qquad\qquad m = -\tfrac{3}{2}$$

61.
$$(x - 1)\left(x^2 + 5x + 6\right) = 0$$
$$(x - 1)(x + 2)(x + 3) = 0$$
$$x - 1 = 0 \quad \textbf{or} \quad x + 2 = 0 \quad \textbf{or} \quad x + 3 = 0$$
$$x = 1 \qquad\qquad x = -2 \qquad\qquad x = -3$$

63.
$$(x + 3)\left(x^2 + 2x - 15\right) = 0$$
$$(x + 3)(x + 5)(x - 3) = 0$$
$$x + 3 = 0 \quad \textbf{or} \quad x + 5 = 0 \quad \textbf{or} \quad x - 3 = 0$$
$$x = -3 \qquad\qquad x = -5 \qquad\qquad x = 3$$

65.
$$(x - 5)\left(2x^2 + x - 3\right) = 0$$
$$(x - 5)(2x + 3)(x - 1) = 0$$
$$x - 5 = 0 \quad \textbf{or} \quad 2x + 3 = 0 \quad \textbf{or} \quad x - 1 = 0$$
$$x = 5 \qquad\qquad 2x = -3 \qquad\qquad x = 1$$
$$x = -\tfrac{3}{2}$$

67.
$$\left(p^2 - 81\right)(p + 2) = 0$$
$$(p + 9)(p - 9)(p + 2) = 0$$
$$p + 9 = 0 \quad \textbf{or} \quad p - 9 = 0 \quad \textbf{or} \quad p + 2 = 0$$
$$p = -9 \qquad\qquad p = 9 \qquad\qquad p = -2$$

69.
$$x^3 + 3x^2 + 2x = 0$$
$$x\left(x^2 + 3x + 2\right) = 0$$
$$x(x + 2)(x + 1) = 0$$
$$x = 0 \quad \textbf{or} \quad x + 2 = 0 \quad \textbf{or} \quad x + 1 = 0$$
$$x = 0 \qquad\qquad x = -2 \qquad\qquad x = -1$$

71.
$$x^3 - 27x - 6x^2 = 0$$
$$x\left(x^2 - 27 - 6x\right) = 0$$
$$x\left(x^2 - 6x - 27\right) = 0$$
$$x(x - 9)(x + 3) = 0$$
$$x = 0 \quad \textbf{or} \quad x - 9 = 0 \quad \textbf{or} \quad x + 3 = 0$$
$$x = 0 \qquad\qquad x = 9 \qquad\qquad x = -3$$

73.
$$3x^2 - 8x = 3$$
$$3x^2 - 8x - 3 = 0$$
$$(3x + 1)(x - 3) = 0$$
$$3x + 1 = 0 \quad \textbf{or} \quad x - 3 = 0$$
$$3x = -1 \qquad\qquad x = 3$$
$$x = -\tfrac{1}{3} \qquad\qquad x = 3$$

75.
$$15x^2 - 2 = 7x$$
$$15x^2 - 7x - 2 = 0$$
$$(3x - 2)(5x + 1) = 0$$
$$3x - 2 = 0 \quad \textbf{or} \quad 5x + 1 = 0$$
$$3x = 2 \qquad\qquad 5x = -1$$
$$x = \tfrac{2}{3} \qquad\qquad x = -\tfrac{1}{5}$$

77.
$$x(6x + 5) = 6$$
$$6x^2 + 5x = 6$$
$$6x^2 + 5x - 6 = 0$$
$$(3x - 2)(2x + 3) = 0$$
$$3x - 2 = 0 \quad \textbf{or} \quad 2x + 3 = 0$$
$$3x = 2 \qquad\qquad 2x = -3$$
$$x = \tfrac{2}{3} \qquad\qquad x = -\tfrac{3}{2}$$

79.
$$(x + 1)(8x + 1) = 18x$$
$$8x^2 + 9x + 1 = 18x$$
$$8x^2 - 9x + 1 = 0$$
$$(x - 1)(8x - 1) = 0$$
$$x - 1 = 0 \quad \textbf{or} \quad 8x - 1 = 0$$
$$x = 1 \qquad\qquad 8x = 1$$
$$x = 1 \qquad\qquad x = \tfrac{1}{8}$$

81.
$$2x(3x^2 + 10x) = -6x$$
$$6x^3 + 20x^2 = -6x$$
$$6x^3 + 20x^2 + 6x = 0$$
$$2x(3x^2 + 10x + 3) = 0$$
$$2x(3x + 1)(x + 3) = 0$$
$$2x = 0 \quad \textbf{or} \quad 3x + 1 = 0 \quad \textbf{or} \quad x + 3 = 0$$
$$x = 0 \qquad\quad 3x = -1 \qquad\quad x = -3$$
$$x = 0 \qquad\quad x = -\tfrac{1}{3} \qquad\quad x = -3$$

83.
$$x^3 + 7x^2 = x^2 - 9x$$
$$x^3 + 6x^2 + 9x = 0$$
$$x(x^2 + 6x + 9) = 0$$
$$x(x + 3)(x + 3) = 0$$
$$x = 0 \quad \textbf{or} \quad x + 3 = 0 \quad \textbf{or} \quad x + 3 = 0$$
$$x = 0 \qquad\quad x = -3 \qquad\quad x = -3$$

85. Answers may vary.

87. Answers may vary.

89.
$$3a^2 + 9a - 2a - 6 = 0$$
$$3a(a + 3) - 2(a + 3) = 0$$
$$(a + 3)(3a - 2) = 0$$
$$a + 3 = 0 \quad \textbf{or} \quad 3a - 2 = 0$$
$$a = -3 \qquad\qquad 3a = 2$$
$$a = \tfrac{2}{3}$$

$$3a^2 + 9a - 2a - 6 = 0$$
$$3a^2 + 7a - 6 = 0$$
$$(a + 3)(3a - 2) = 0$$
$$a + 3 = 0 \quad \textbf{or} \quad 3a - 2 = 0$$
$$a = -3 \qquad\qquad 3a = 2$$
$$a = \tfrac{2}{3}$$

Exercise 5.8 (page 345)

1.
$$-2(5z + 2) = 3(2 - 3z)$$
$$-10z - 4 = 6 - 9z$$
$$-4 = 6 + z$$
$$-10 = z$$

3. Let $w =$ the width and $3w =$ the length.
$$\text{Perimeter} = 120$$
$$2(w) + 2(3w) = 120$$
$$2w + 6w = 120$$
$$8w = 120$$
$$w = 15$$
The area $= 15(45) = 675 \text{ cm}^2$.

5. analyze

7. Let $x =$ the first positive integer. Then $x + 2 =$ the other positive integer.

$$x(x+2) = 35$$
$$x^2 + 2x = 35$$
$$x^2 + 2x - 35 = 0$$
$$(x+7)(x-5) = 0$$
$$x + 7 = 0 \quad \textbf{or} \quad x - 5 = 0$$
$$x = -7 \qquad\qquad x = 5$$

The answer of -7 is not a positive integer, so the integers are 5 and 7.

9. Let $x =$ the composite integer.

$$x^2 + 4 = 10x - 5$$
$$x^2 - 10x + 9 = 0$$
$$(x-9)(x-1) = 0$$
$$x - 9 = 0 \quad \textbf{or} \quad x - 1 = 0$$
$$x = 9 \qquad\qquad x = 1$$

The answer of 1 is not a composite integer, so the integer is 9.

11. Let $v = 144$ and $h = 0$:

$$h = vt - 16t^2$$
$$0 = 144t - 16t^2$$
$$0 = 16t(9 - t)$$
$$16t = 0 \quad \textbf{or} \quad 9 - t = 0$$
$$t = 0 \qquad\qquad 9 = t$$

Since $t = 0$ is when the object was first thrown, it will hit the ground in 9 seconds.

13. Let $v = 220$ and $h = 600$:

$$h = vt - 16t^2$$
$$600 = 220t - 16t^2$$
$$16t^2 - 220t + 600 = 0$$
$$4(4t^2 - 55t + 150) = 0$$
$$4(4t - 15)(t - 10) = 0$$
$$4t - 15 = 0 \quad \textbf{or} \quad t - 10 = 0$$
$$4t = 15 \qquad\qquad t = 10$$
$$t = \tfrac{15}{4} \qquad\qquad t = 10$$

The cannonball will be at a height of 600 feet after $\frac{15}{4}$ seconds and after 10 seconds.

15.
$$h = -16t^2 + 64$$
$$0 = -16t^2 + 64$$
$$16t^2 - 64 = 0$$
$$16(t^2 - 4) = 0$$
$$16(t + 2)(t - 2) = 0$$
$$t + 2 = 0 \quad \textbf{or} \quad t - 2 = 0$$
$$t = -2 \qquad\qquad t = 2$$

The value of $t = -2$ does not make sense, so the dive lasts 2 seconds.

17.
$$h^2 + 72^2 = 75^2$$
$$h^2 + 5184 = 5625$$
$$h^2 - 441 = 0$$
$$(h + 21)(h - 21) = 0$$
$$h + 21 = 0 \quad \textbf{or} \quad h - 21 = 0$$
$$h = -21 \qquad\qquad h = 21$$

The value of $h = -21$ does not make sense, so the camper started at a height of 21 feet.

19. Area = Length · Width
$$36 = (2w + 1)w$$
$$36 = 2w^2 + w$$
$$0 = 2w^2 + w - 36$$
$$0 = (2w + 9)(w - 4)$$
$$2w + 9 = 0 \quad \textbf{or} \quad w - 4 = 0$$
$$2w = -9 \qquad\qquad w = 4$$
$$w = -\tfrac{9}{2} \qquad\qquad w = 4$$

Since the answer $w = -\frac{9}{2}$ does not make sense, the dimensions are 4 m by 9 m.

21. Let w = the width and $w + 2$ = the length.
Area = Length · Width
$$143 = (w + 2)w$$
$$143 = w^2 + 2w$$
$$0 = w^2 + 2w - 143$$
$$0 = (w + 13)(w - 11)$$
$$w + 13 = 0 \quad \textbf{or} \quad w - 11 = 0$$
$$w = -13 \qquad\qquad w = 11$$

Since the answer $w = -13$ does not make sense, the dimensions are 11 ft by 13 ft and the perimeter is 48 ft.

23. Let b = the base and $5b - 2$ = the height.
$$A = \frac{1}{2}bh$$
$$36 = \frac{1}{2}b(5b - 2)$$
$$72 = b(5b - 2)$$
$$72 = 5b^2 - 2b$$
$$0 = 5b^2 - 2b - 72$$
$$0 = (5b + 18)(b - 4)$$
$$5b + 18 = 0 \quad \textbf{or} \quad b - 4 = 0$$
$$5b = -18 \qquad\qquad b = 4$$
$$h = -\tfrac{18}{5} \qquad\qquad b = 4$$

Since the answer $b = -\frac{18}{5}$ does not make sense, the base is 4 in. and the height is 18 in.

25. Let x = the base and the height.

| Base | + | Height | = | Area | − 6 |

$$x + x = \tfrac{1}{2}(x)(x) - 6$$
$$2x = \tfrac{1}{2}x^2 - 6$$
$$0 = \tfrac{1}{2}x^2 - 2x - 6$$
$$0 = x^2 - 4x - 12$$
$$0 = (x - 6)(x + 2)$$
$$x - 6 = 0 \quad \textbf{or} \quad x + 2 = 0$$
$$x = 6 \qquad\qquad x = -2$$

The answer $x = -2$ does not make sense, so the area is $\frac{1}{2}(6)(6) = 18$ square units.

27.

| Large area | − | Small area | = | Border area |

$$(10 + 2w)(25 + 2w) - (10)(25) = 74$$
$$250 + 70w + 4w^2 - 250 = 74$$
$$4w^2 + 70w - 74 = 0$$
$$2(2w^2 + 35w - 37) = 0$$
$$2(2w + 37)(w - 1) = 0$$
$$2w + 37 = 0 \quad \textbf{or} \quad w - 1 = 0$$
$$2w = -37 \qquad\qquad w = 1$$
$$w = -\tfrac{37}{2} \qquad\qquad w = 1$$

Since the answer $w = -\frac{37}{2}$ does not make sense, the width should be 1 meter.

29. Let w = the width and $2w + 1$ = the height.
$$V = lwh$$
$$210 = 10w(2w + 1)$$
$$210 = 20w^2 + 10w$$
$$0 = 20w^2 + 10w - 210$$
$$0 = 10(2w^2 + w - 21)$$
$$0 = 10(2w + 7)(w - 3)$$
$$2w + 7 = 0 \quad \textbf{or} \quad w - 3 = 0$$
$$2w = -7 \qquad\qquad w = 3$$
$$w = -\tfrac{7}{2} \qquad\qquad w = 3$$

Since the answer $w = -\frac{7}{2}$ does not make sense, the width is 3 cm.

31. Let $x =$ one edge and $x - 3 =$ the other.

$$V = \frac{Bh}{3}$$

$$84 = \frac{x(x-3)(9)}{3}$$

$$252 = 9x^2 - 27x$$

$$0 = 9x^2 - 27x - 252$$

$$0 = 9(x + 4)(x - 7)$$

$$x + 4 = 0 \quad \textbf{or} \quad x - 7 = 0$$

$$x = -4 \qquad\qquad x = 7$$

Since the answer $x = -4$ does not make sense, the base is 7 cm by 4 cm.

33.

$$C = \tfrac{1}{2}(n^2 - n)$$

$$6 = \tfrac{1}{2}(n^2 - n)$$

$$2(66) = 2 \cdot \tfrac{1}{2}(n^2 - n)$$

$$132 = n^2 - n$$

$$0 = n^2 - n - 132$$

$$0 = (n + 11)(n - 12)$$

$$n + 11 = 0 \quad \textbf{or} \quad n - 12 = 0$$

$$n = -11 \qquad\qquad n = 12$$

The value of $n = -11$ does not make sense, so 12 telephones are needed.

35. Area #1 $= \pi r^2 = \pi(20 \text{ m})^2 = 400\pi \text{ m}^2$

Area #2 $= \pi r^2 = \pi(21 \text{ m})^2 = 441\pi \text{ m}^2$

Total Area $= 400\pi + 441\pi \text{ m}^2 = 841\pi \text{ m}^2$

37. **Answers may vary.**

39. Let $w =$ the width and $w + 2 =$ the length.

$$\text{Area} = 18$$

$$w(w + 2) = 18$$

$$w^2 + 2w = 18$$

$$w^2 + 2w - 18 = 0 \Rightarrow \text{This is a prime trinomial, so it cannot be factored in order to solve.}$$

Chapter 5 Summary (page 349)

1. $35 = 5 \cdot 7$

2. $45 = 9 \cdot 5 = 3^2 \cdot 5$

3. $96 = 12 \cdot 8 = 4 \cdot 3 \cdot 2^3 = 2^2 \cdot 3 \cdot 2^3 = 2^5 \cdot 3$

4. $102 = 2 \cdot 51 = 2 \cdot 3 \cdot 17$

5. $87 = 3 \cdot 29$

6. $99 = 9 \cdot 11 = 3^2 \cdot 11$

7. $2{,}050 = 50 \cdot 41 = 25 \cdot 2 \cdot 41 = 2 \cdot 5^2 \cdot 41$

8. $4{,}096 = 64 \cdot 64 = 2^6 \cdot 2^6 = 2^{12}$

9. $3x + 9y = 3(x + 3y)$

10. $5ax^2 + 15a = 5a(x^2 + 3)$

11. $7x^2 + 14x = 7x(x + 2)$

12. $3x^2 - 3x = 3x(x - 1)$

13. $2x^3 + 4x^2 - 8x = 2x(x^2 + 2x - 4)$

14. $ax + ay - az = a(x + y - z)$

15. $ax + ay - a = a(x + y - 1)$

16. $x^2yz + xy^2z = xyz(x + y)$

17. $(x + y)a + (x + y)b = (x + y)(a + b)$

18. $(x + y)^2 + (x + y) = (x + y)(x + y) + 1(x + y) = (x + y)(x + y + 1)$

19. $2x^2(x+2) + 6x(x+2) = (x+2)(2x^2+6x) = (x+2)(2x)(x+3) = 2x(x+2)(x+3)$

20. $3x(y+z) - 9x(y+z)^2 = 3x[(y+z) - 3(y+z)(y+z)] = 3x(y+z)(1 - 3(y+z))$
$$= 3x(y+z)(1 - 3y - 3z)$$

21. $3p + 9q + ap + 3aq = 3(p+3q) + a(p+3q) = (p+3q)(3+a)$

22. $ar - 2as + 7r - 14s = a(r-2s) + 7(r-2s) = (r-2s)(a+7)$

23. $x^2 + ax + bx + ab = x(x+a) + b(x+a)$ **24.** $xy + 2x - 2y - 4 = x(y+2) - 2(y+2)$
$$= (x+a)(x+b)$$
$$= (y+2)(x-2)$$

25. $xa + yb + ya + xb = xa + ya + xb + yb = a(x+y) + b(x+y) = (x+y)(a+b)$

26. $x^2 - 9 = x^2 - 3^2 = (x+3)(x-3)$ **27.** $x^2y^2 - 16 = (xy+4)(xy-4)$

28. $(x+2)^2 - y^2 = (x+2+y)(x+2-y)$

29. $z^2 - (x+y)^2 = [z+(x+y)][z-(x+y)] = (z+x+y)(z-x-y)$

30. $6x^2y - 24y^3 = 6y(x^2 - 4y^2) = 6y[x^2 - (2y)^2] = 6y(x+2y)(x-2y)$

31. $(x+y)^2 - z^2 = [(x+y)+z][(x+y)-z] = (x+y+z)(x+y-z)$

32. $x^2 + 10x + 21 = (x+3)(x+7)$ **33.** $x^2 + 4x - 21 = (x+7)(x-3)$

34. $x^2 + 2x - 24 = (x+6)(x-4)$ **35.** $x^2 - 4x - 12 = (x-6)(x+2)$

36. $2x^2 - 5x - 3 = (2x+1)(x-3)$ **37.** $3x^2 - 14x - 5 = (3x+1)(x-5)$

38. $6x^2 + 7x - 3 = (2x+3)(3x-1)$ **39.** $6x^2 + 3x - 3 = 3(2x^2 + x - 1)$
$$= 3(2x-1)(x+1)$$

40. $6x^3 + 17x^2 - 3x = x(6x^2 + 17x - 3)$ **41.** $4x^3 - 5x^2 - 6x = x(4x^2 - 5x - 6)$
$$= x(x+3)(6x-1)$$
$$= x(x-2)(4x+3)$$

42. $c^3 - 27 = c^3 - 3^3 = (c-3)(c^2 + 3c + 9)$ **43.** $d^3 + 8 = d^3 + 2^3 = (d+2)(d^2 - 2d + 4)$

44. $2x^3 + 54 = 2(x^3 + 27)$ **45.** $2ab^4 - 2ab = 2ab(b^3 - 1)$
$$= 2(x^3 + 3^3)$$
$$= 2ab(b^3 - 1^3)$$
$$= 2(x+3)(x^2 - 3x + 9)$$
$$= 2ab(b-1)(b^2 + b + 1)$$

46. $3x^2y - xy^2 - 6xy + 2y^2 = y(3x^2 - xy - 6x + 2y) = y[x(3x-y) - 2(3x-y)]$
$$= y(3x-y)(x-2)$$

47. $5x^2 + 10x - 15xy - 30y = 5(x^2 + 2x - 3xy - 6y) = 5[x(x+2) - 3y(x+2)]$
$$= 5(x+2)(x-3y)$$

48. $2a^2x + 2abx + a^3 + a^2b = a(2ax + 2bx + a^2 + ab) = a[2x(a+b) + a(a+b)]$
$$= a(a+b)(2x+a)$$

49. $x^2 + 2ax + a^2 - y^2 = (x^2 + 2ax + a^2) - y^2 = (x+a)(x+a) - y^2$
$$= (x+a)^2 - y^2 = (x+a+y)(x+a-y)$$

50. $x^2 - 4 + bx + 2b = x^2 - 2^2 + bx + 2b = (x+2)(x-2) + b(x+2) = (x+2)(x-2+b)$

51. $ax^6 - ay^6 = a(x^6 - y^6) = a\left[(x^3)^2 - (y^3)^2\right] = a(x^3 + y^3)(x^3 - y^3)$
$$= a(x+y)(x^2 - xy + y^2)(x-y)(x^2 + xy + y^2)$$

52. $x^2 + 2x = 0$
$x(x+2) = 0$
$x = 0$ **or** $x + 2 = 0$
$x = 0 \qquad x = -2$

53. $2x^2 - 6x = 0$
$2x(x-3) = 0$
$2x = 0$ **or** $x - 3 = 0$
$x = 0 \qquad x = 3$

54. $3x^2 = 2x$
$3x^2 - 2x = 0$
$x(3x - 2) = 0$
$x = 0$ **or** $3x - 2 = 0$
$3x = 2$
$x = \frac{2}{3}$

55. $5x^2 + 25x = 0$
$5x(x+5) = 0$
$5x = 0$ **or** $x + 5 = 0$
$x = \frac{0}{5} \qquad x = -5$
$x = 0$

56. $x^2 - 9 = 0$
$(x+3)(x-3) = 0$
$x + 3 = 0$ **or** $x - 3 = 0$
$x = -3 \qquad x = 3$

57. $x^2 - 25 = 0$
$(x+5)(x-5) = 0$
$x + 5 = 0$ **or** $x - 5 = 0$
$x = -5 \qquad x = 5$

58. $a^2 - 7a + 12 = 0$
$(a-3)(a-4) = 0$
$a - 3 = 0$ **or** $a - 4 = 0$
$a = 3 \qquad a = 4$

59. $x^2 - 2x - 15 = 0$
$(x+3)(x-5) = 0$
$x + 3 = 0$ **or** $x - 5 = 0$
$x = -3 \qquad x = 5$

60. $2x - x^2 + 24 = 0$
$x^2 - 2x - 24 = 0$
$(x+4)(x-6) = 0$
$x + 4 = 0$ **or** $x - 6 = 0$
$x = -4 \qquad x = 6$

61. $16 + x^2 - 10x = 0$
$x^2 - 10x + 16 = 0$
$(x-2)(x-8) = 0$
$x - 2 = 0$ **or** $x - 8 = 0$
$x = 2 \qquad x = 8$

62.
$$2x^2 - 5x - 3 = 0$$
$$(2x + 1)(x - 3) = 0$$
$$2x + 1 = 0 \quad \text{or} \quad x - 3 = 0$$
$$2x = -1 \qquad\qquad x = 3$$
$$x = -\tfrac{1}{2} \qquad\qquad x = 3$$

63.
$$2x^2 + x - 3 = 0$$
$$(2x + 3)(x - 1) = 0$$
$$2x + 3 = 0 \quad \text{or} \quad x - 1 = 0$$
$$2x = -3 \qquad\qquad x = 1$$
$$x = -\tfrac{3}{2} \qquad\qquad x = 1$$

64.
$$4x^2 = 1$$
$$4x^2 - 1 = 0$$
$$(2x + 1)(2x - 1) = 0$$
$$2x + 1 = 0 \quad \text{or} \quad 2x - 1 = 0$$
$$2x = -1 \qquad\qquad 2x = 1$$
$$x = -\tfrac{1}{2} \qquad\qquad x = \tfrac{1}{2}$$

65.
$$9x^2 = 4$$
$$9x^2 - 4 = 0$$
$$(3x + 2)(3x - 2) = 0$$
$$3x + 2 = 0 \quad \text{or} \quad 3x - 2 = 0$$
$$3x = -2 \qquad\qquad 3x = 2$$
$$x = -\tfrac{2}{3} \qquad\qquad x = \tfrac{2}{3}$$

66.
$$x^3 - 7x^2 + 12x = 0$$
$$x(x^2 - 7x + 12) = 0$$
$$x(x - 3)(x - 4) = 0$$
$$x = 0 \quad \text{or} \quad x - 3 = 0 \quad \text{or} \quad x - 4 = 0$$
$$x = 0 \qquad\qquad x = 3 \qquad\qquad x = 4$$

67.
$$x^3 + 5x^2 + 6x = 0$$
$$x(x^2 + 5x + 6) = 0$$
$$x(x + 2)(x + 3) = 0$$
$$x = 0 \quad \text{or} \quad x + 2 = 0 \quad \text{or} \quad x + 3 = 0$$
$$x = 0 \qquad\qquad x = -2 \qquad\qquad x = -3$$

68.
$$2x^3 + 5x^2 = 3x$$
$$2x^3 + 5x^2 - 3x = 0$$
$$x(2x^2 + 5x - 3) = 0$$
$$x(2x - 1)(x + 3) = 0$$
$$x = 0 \quad \text{or} \quad 2x - 1 = 0 \quad \text{or} \quad x + 3 = 0$$
$$x = 0 \qquad\qquad 2x = 1 \qquad\qquad x = -3$$
$$x = 0 \qquad\qquad x = \tfrac{1}{2} \qquad\qquad x = -3$$

69.
$$3x^3 - 2x = x^2$$
$$3x^3 - x^2 - 2x = 0$$
$$x(3x^2 - x - 2) = 0$$
$$x(3x + 2)(x - 1) = 0$$
$$x = 0 \quad \text{or} \quad 3x + 2 = 0 \quad \text{or} \quad x - 1 = 0$$
$$x = 0 \qquad\qquad 3x = -2 \qquad\qquad x = 1$$
$$x = 0 \qquad\qquad x = -\tfrac{2}{3} \qquad\qquad x = 1$$

70. Let $x =$ one number. Then
$12 - x =$ the other number.
$$x(12 - x) = 35$$
$$12x - x^2 = 35$$
$$0 = x^2 - 12x + 35$$
$$0 = (x - 5)(x - 7)$$
$x - 5 = 0$ **or** $x - 7 = 0$
$\quad x = 5 \qquad\qquad x = 7$
The numbers are 5 and 7.

71. Let $x =$ the positive number.
$$3x^2 + 5x = 2$$
$$3x^2 + 5x - 2 = 0$$
$$(3x - 1)(x + 2) = 0$$
$3x - 1 = 0$ **or** $x + 2 = 0$
$\quad 3x = 1 \qquad\qquad x = -2$
$\quad x = \frac{1}{3} \qquad\qquad x = -2$
The answer of -2 is not positive,
so the number is $\frac{1}{3}$.

72. Let $w =$ the width and $w + 2 =$ the length.
$$\text{Area} = \text{Length} \cdot \text{Width}$$
$$48 = (w + 2)w$$
$$0 = w^2 + 2w - 48$$
$$0 = (w + 8)(w - 6)$$
$w + 8 = 0$ **or** $w - 6 = 0$
$\quad w = -8 \qquad\qquad w = 6$
Since the answer $w = -8$ does not make
sense, the dimensions are 6 ft by 8 ft.

73. Let $w =$ the width and $2w + 3 =$ the length.
$$\text{Area} = \text{Length} \cdot \text{Width}$$
$$27 = (2w + 3)w$$
$$0 = 2w^2 + 3w - 27$$
$$0 = (2w + 9)(w - 3)$$
$2w + 9 = 0$ **or** $w - 3 = 0$
$\quad 2w = -9 \qquad\qquad w = 3$
$\quad w = -\frac{9}{2} \qquad\qquad w = 3$
Since the answer $w = -\frac{9}{2}$ does not make
sense, the dimensions are 3 ft by 9 ft.

74. Let $w =$ the width and $w + 3 =$ the length.
$$\text{Area} = \text{Perimeter}$$
$$(w + 3)w = 2w + 2(w + 3)$$
$$w^2 + 3w = 2w + 2w + 6$$
$$w^2 + 3w = 4w + 6$$
$$w^2 - w - 6 = 0$$
$$(w + 2)(w - 3) = 0$$
$w + 2 = 0$ **or** $w - 3 = 0$
$\quad w = -2 \qquad\qquad w = 3$
Since the answer $w = -2$ does not make sense, the dimensions are 3 ft by 6 ft.

Chapter 5 Test (page 352)

1. $196 = 14 \cdot 14 = 2 \cdot 7 \cdot 2 \cdot 7 = 2^2 \cdot 7^2$

2. $111 = 3 \cdot 37$

3. $60ab^2c^3 + 30a^3b^2c - 25a = 5a(12b^2c^3 + 6a^2b^2c - 5)$

4. $3x^2(a + b) - 6xy(a + b) = 3x[x(a + b) - 2y(a + b)] = 3x(a + b)(x - 2y)$

5. $ax + ay + bx + by = a(x + y) + b(x + y)$
$\qquad\qquad\qquad\qquad = (x + y)(a + b)$

6. $x^2 - 25 = x^2 - 5^2 = (x + 5)(x - 5)$

7. $3a^2 - 27b^2 = 3(a^2 - 9b^2) = 3\left(a^2 - (3b)^2\right) = 3(a + 3b)(a - 3b)$

8. $16x^4 - 81y^4 = \left(4x^2\right)^2 - \left(9y^2\right)^2 = \left(4x^2 + 9y^2\right)\left(4x^2 - 9y^2\right)$
$$= \left(4x^2 + 9y^2\right)\left((2x)^2 - (3y)^2\right)$$
$$= \left(4x^2 + 9y^2\right)(2x + 3y)(2x - 3y)$$

9. $x^2 + 4x + 3 = (x + 3)(x + 1)$ **10.** $x^2 - 9x - 22 = (x - 11)(x + 2)$

11. $x^2 + 10xy + 9y^2 = (x + y)(x + 9y)$ **12.** $6x^2 - 30xy + 24y^2 = 6\left(x^2 - 5xy + 4y^2\right)$
$$= 6(x - 4y)(x - y)$$

13. $3x^2 + 13x + 4 = (3x + 1)(x + 4)$ **14.** $2a^2 + 5a - 12 = (2a - 3)(a + 4)$

15. $2x^2 + 3xy - 2y^2 = (2x - y)(x + 2y)$ **16.** $12 - 25x + 12x^2 = 12x^2 - 25x + 12$
$$= (4x - 3)(3x - 4)$$

17. $12a^2 + 6ab - 36b^2 = 6\left(2a^2 + ab - 6b^2\right)$ **18.** $x^3 - 64 = x^3 - 4^3 = (x - 4)\left(x^2 + 4x + 16\right)$
$$= 6(2a - 3b)(a + 2b)$$

19. $216 + 8a^3 = 8(27 + a^3) = 8(3^3 + a^3) = 8(3 + a)(9 - 3a + a^2)$

20. $x^9z^3 - y^3z^6 = z^3\left(x^9 - y^3z^3\right) = z^3\left(\left(x^3\right)^3 - (yz)^3\right) = z^3\left(x^3 - yz\right)\left(\left(x^3\right)^2 + x^3yz + (yz)^2\right)$
$$= z^3\left(x^3 - yz\right)\left(x^6 + x^3yz + y^2z^2\right)$$

21. $x^2 + 3x = 0$ **22.** $2x^2 + 5x + 3 = 0$
$x(x + 3) = 0$ $(2x + 3)(x + 1) = 0$
$x = 0 \quad \text{or} \quad x + 3 = 0$ $2x + 3 = 0 \quad \text{or} \quad x + 1 = 0$
$x = 0 \qquad\qquad x = -3$ $2x = -3 \qquad\qquad x = -1$
$\qquad\qquad\qquad x = -\frac{3}{2} \qquad\qquad x = -1$

23. $9y^2 - 81 = 0$ **24.** $-3(y - 6) + 2 = y^2 + 2$
$9\left(y^2 - 9\right) = 0$ $-3y + 18 + 2 = y^2 + 2$
$9(y + 3)(y - 3) = 0$ $-3y + 20 = y^2 + 2$
$y + 3 = 0 \quad \text{or} \quad y - 3 = 0$ $0 = y^2 + 3y - 18$
$y = -3 \qquad\qquad y = 3$ $0 = (y + 6)(y - 3)$
$\qquad\qquad\qquad\qquad y + 6 = 0 \quad \text{or} \quad y - 3 = 0$
$\qquad\qquad\qquad\qquad y = -6 \qquad\qquad y = 3$

25.
$$10x^2 - 13x = 9$$
$$10x^2 - 13x - 9 = 0$$
$$(2x + 1)(5x - 9) = 0$$
$$2x + 1 = 0 \quad \text{or} \quad 5x - 9 = 0$$
$$2x = -1 \qquad\qquad 5x = 9$$
$$x = -\tfrac{1}{2} \qquad\qquad x = \tfrac{9}{5}$$

26.
$$10x^2 - x = 9$$
$$10x^2 - x - 9 = 0$$
$$(10x + 9)(x - 1) = 0$$
$$10x + 9 = 0 \quad \text{or} \quad x - 1 = 0$$
$$10x = -9 \qquad\qquad x = 1$$
$$x = -\tfrac{9}{10} \qquad\qquad x = 1$$

27.
$$10x^2 + 43x = 9$$
$$10x^2 + 43x - 9 = 0$$
$$(2x + 9)(5x - 1) = 0$$
$$2x + 9 = 0 \quad \text{or} \quad 5x - 1 = 0$$
$$2x = -9 \qquad\qquad 5x = 1$$
$$x = -\tfrac{9}{2} \qquad\qquad x = \tfrac{1}{5}$$

28.
$$10x^2 - 89x = 9$$
$$10x^2 - 89x - 9 = 0$$
$$(10x + 1)(x - 9) = 0$$
$$10x + 1 = 0 \quad \text{or} \quad x - 9 = 0$$
$$10x = -1 \qquad\qquad x = 9$$
$$x = -\tfrac{1}{10} \qquad\qquad x = 9$$

29. Let $v = 192$ and $h = 0$:
$$h = vt - 16t^2$$
$$0 = 192t - 16t^2$$
$$0 = 16t(12 - t)$$
$$16t = 0 \quad \text{or} \quad 12 - t = 0$$
$$t = 0 \qquad\qquad 12 = t$$

Since $t = 0$ is when the cannonball was fired, it will hit the ground in 12 seconds.

30. Let $h = $ the height and $h + 2 = $ the base.
$$A = \tfrac{1}{2}bh$$
$$40 = \tfrac{1}{2}(h + 2)h$$
$$80 = (h + 2)h$$
$$0 = h^2 + 2h - 80$$
$$0 = (h + 10)(h - 8)$$
$$h + 10 = 0 \quad \text{or} \quad h - 8 = 0$$
$$h = -10 \qquad\qquad h = 8$$

Since the answer $h = -10$ does not make sense, the height is 8 m and the base is 10 m.

Exercise 6.1 (page 359)

1.
$$2x + 4 = 38$$
$$2x = 34$$
$$x = 17$$

3.
$$3(x + 2) = 24$$
$$3x + 6 = 24$$
$$3x = 18$$
$$x = 6$$

5. $2x + 6 = 2(x + 3)$

7. $2x^2 - x - 6 = (2x + 3)(x - 2)$

9. quotient

11. equal

13. Answers may vary.

15. $\dfrac{5}{7}$

17. $\dfrac{17}{34} = \dfrac{1}{2}$

19. $\dfrac{22}{33} = \dfrac{2}{3}$

21. $\dfrac{7}{24.5} = \dfrac{14}{49} = \dfrac{2}{7}$

23. $\dfrac{4 \text{ oz}}{12 \text{ oz}} = \dfrac{1}{3}$

25. $\dfrac{12 \text{ min}}{1 \text{ hr}} = \dfrac{12 \text{ min}}{60 \text{ min}} = \dfrac{1}{5}$

27. $\dfrac{3 \text{ days}}{1 \text{ week}} = \dfrac{3 \text{ days}}{7 \text{ days}} = \dfrac{3}{7}$

29. $\dfrac{18 \text{ months}}{2 \text{ years}} = \dfrac{18 \text{ months}}{24 \text{ months}} = \dfrac{3}{4}$

31. $750 + 652 + 188 + 125 + 110 = \$1{,}825$

33. $\dfrac{110}{1825} = \dfrac{22}{365}$

35. $995 + 1245 + 1680 + 4580 + 225 = \$8{,}725$

37. $\dfrac{1680}{8725} = \dfrac{336}{1745}$

39. $\dfrac{125}{2000} = \dfrac{1}{16}$

41. $\dfrac{\$30.09}{17 \text{ gal}} = \$1.77/\text{gal}$

43. $\dfrac{84\text{¢}}{12 \text{ oz}} = 7\text{¢}/\text{oz}$

45. $\dfrac{89\text{¢}}{6 \text{ oz}} \approx 14.873\text{¢}/\text{oz}; \dfrac{119\text{¢}}{8 \text{ oz}} = 14.875\text{¢}/\text{oz}$
The 6-oz can is the better buy.

47. $\dfrac{54 \text{ pg}}{40 \text{ min}} = 1.35 \dfrac{\text{pg}}{\text{min}}; \dfrac{80 \text{ pg}}{62 \text{ min}} \approx 1.29 \dfrac{\text{pg}}{\text{min}}$
The 1st student read faster.

49. $\dfrac{11880 \text{ gal}}{27 \text{ min}} = 440 \text{ gal/min}$

51. $\dfrac{36.75 - 35}{35} = \dfrac{1.75}{35} = 0.05 = 5\%$

53. $\dfrac{325 \text{ mi}}{5 \text{ hr}} = 65 \text{ mi/hr}$

55. $\dfrac{1235 \text{ mi}}{51.3 \text{ gal}} \approx 24 \dfrac{\text{mi}}{\text{gal}}; \dfrac{1456 \text{ mi}}{55.78 \text{ gal}} \approx 26 \dfrac{\text{mi}}{\text{gal}}$
The 2nd car had the better mpg rating.

57. **Answers may vary.**

59. $\dfrac{17}{19} = \dfrac{17 \cdot 21}{19 \cdot 21} = \dfrac{357}{399}; \dfrac{19}{21} = \dfrac{19 \cdot 19}{21 \cdot 19} = \dfrac{361}{399}; \dfrac{19}{21}$ is larger.

Exercise 6.2 (page 369)

1. $\dfrac{9}{10} = 0.9 = 90\%$

3. $33\dfrac{1}{3}\% = \dfrac{100}{3}\% = \dfrac{100}{3} \cdot \dfrac{1}{100} = \dfrac{1}{3}$

5.
$$rb = a$$
$$0.30(1600) = a$$
$$480 = a$$

7.
$$rb = a$$
$$0.25(98) = a$$
$$24.50 = a$$
$$98 - 24.50 = \$73.50$$

9. proportion; ratios

11. means

13. shape

15. $ad; bc$

17. triangle

19.
$$\dfrac{9}{7} \overset{?}{=} \dfrac{81}{70}$$
$$9 \cdot 70 \overset{?}{=} 7 \cdot 81$$
$$630 \neq 567$$
not a proportion

21.
$$\dfrac{-7}{3} \overset{?}{=} \dfrac{14}{-6}$$
$$-7(-6) \overset{?}{=} 3 \cdot 14$$
$$42 = 42$$
proportion

23.
$$\dfrac{9}{19} \overset{?}{=} \dfrac{38}{80}$$
$$9 \cdot 80 \overset{?}{=} 19 \cdot 38$$
$$720 \neq 722$$
not a proportion

25.
$$\frac{10.4}{3.6} \stackrel{?}{=} \frac{41.6}{14.4}$$
$$10.4(14.4) \stackrel{?}{=} 3.6(41.6)$$
$$149.76 = 149.76$$
proportion

27.
$$\frac{2}{3} = \frac{x}{6}$$
$$2(6) = 3x$$
$$12 = 3x$$
$$4 = x$$

29.
$$\frac{5}{10} = \frac{3}{c}$$
$$5c = 10(3)$$
$$5c = 30$$
$$c = 6$$

31.
$$\frac{-6}{x} = \frac{8}{4}$$
$$-6(4) = 8x$$
$$-24 = 8x$$
$$-3 = x$$

33.
$$\frac{x}{3} = \frac{9}{3}$$
$$x(3) = 3(9)$$
$$3x = 27$$
$$x = 9$$

35.
$$\frac{x+1}{5} = \frac{3}{15}$$
$$15(x+1) = 5(3)$$
$$15x + 15 = 15$$
$$15x = 0$$
$$x = 0$$

37.
$$\frac{x+3}{12} = \frac{-7}{6}$$
$$6(x+3) = 12(-7)$$
$$6x + 18 = -84$$
$$6x = -102$$
$$x = -17$$

39.
$$\frac{4-x}{13} = \frac{11}{26}$$
$$26(4-x) = 13(11)$$
$$104 - 26x = 143$$
$$-26x = 39$$
$$x = -\frac{39}{26} = -\frac{3}{2}$$

41.
$$\frac{2x+1}{18} = \frac{14}{3}$$
$$3(2x+1) = 18(14)$$
$$6x + 3 = 252$$
$$6x = 249$$
$$x = \frac{249}{6} = \frac{83}{2}$$

43.
$$\frac{3p-2}{12} = \frac{p+1}{3}$$
$$3(3p-2) = 12(p+1)$$
$$9p - 6 = 12p + 12$$
$$-3p = 18$$
$$p = -\frac{18}{3} = -6$$

45. Let $c =$ the cost of 51 pints of yogurt.
$$\frac{3 \text{ pints}}{51 \text{ pints}} = \frac{\text{cost of 3 pints}}{\text{cost of 51 pints}}$$
$$\frac{3}{51} = \frac{1}{c}$$
$$c \cdot 3 = 51 \cdot 1$$
$$3c = 51$$
$$c = 17$$
The 51 pints will cost $17.

47. Let $c =$ the cost of 39 packets.
$$\frac{3 \text{ packets}}{39 \text{ packets}} = \frac{\text{cost of 3 packets}}{\text{cost of 39 packets}}$$
$$\frac{3}{39} = \frac{50}{c}$$
$$c \cdot 3 = 39 \cdot 50$$
$$3c = 1{,}950$$
$$c = 650 \text{ cents} = \$6.50$$
The 39 packets will cost $6.50.

49. Let $d =$ the # of drops of pure essence needed with 56 drops of alcohol.
$$\frac{3 \text{ pure}}{7 \text{ alcohol}} = \frac{\text{drops of pure}}{56 \text{ drops of alcohol}}$$
$$\frac{3}{7} = \frac{d}{56}$$
$$3 \cdot 56 = 7d$$
$$168 = 7d$$
$$24 = d$$
24 drops of pure essence will be needed.

51. Let f = the cups of flour needed for 12 dozen cookies.

$$\frac{1\frac{1}{4} \text{ C. flour}}{3\frac{1}{2} \text{ dozen}} = \frac{\text{C. flour}}{12 \text{ dozen}}$$

$$\frac{1\frac{1}{4}}{3\frac{1}{2}} = \frac{f}{12}$$

$$1\frac{1}{4} \cdot 12 = 3\frac{1}{2}f$$

$$\frac{5}{4} \cdot 12 = \frac{7}{2}f$$

$$15 = \frac{7}{2}f$$

$$30 = 7f$$

$$\frac{30}{7} = f$$

$$4\frac{2}{7} = f$$

$4\frac{2}{7}\left(\approx 4\frac{1}{4}\right)$ cups of flour will be needed.

53. Let n = the number defective.

$$\frac{5 \text{ defective}}{100 \text{ parts}} = \frac{\text{number defective}}{940 \text{ parts}}$$

$$\frac{5}{100} = \frac{n}{940}$$

$$5 \cdot 940 = 100n$$

$$4700 = 100n$$

$$47 = n$$

There will be 47 defective parts.

55. Let g = gallons of gas needed for 315 miles.

$$\frac{42 \text{ miles}}{315 \text{ miles}} = \frac{\text{gas for } 42 \text{ miles}}{\text{gas for } 315 \text{ miles}}$$

$$\frac{42}{315} = \frac{1}{g}$$

$$g \cdot 42 = 315 \cdot 1$$

$$42g = 315$$

$$g = \frac{315}{42} = \frac{15}{2} = 7\frac{1}{2}$$

To go 315 miles, $7\frac{1}{2}$ gallons of gas are needed.

57. Let p = the amount he was paid last week.

$$\frac{30 \text{ hours}}{40 \text{ hours}} = \frac{\text{pay for } 30 \text{ hours}}{\text{pay for } 40 \text{ hours}}$$

$$\frac{30}{40} = \frac{p}{412}$$

$$412 \cdot 30 = 40 \cdot p$$

$$12{,}360 = 40p$$

$$309 = p$$

He was paid $309 last week.

59. Let l = the length of a real caboose.

$$\frac{169 \text{ feet}}{1 \text{ foot}} = \frac{\text{length of a real caboose}}{\text{length of a model caboose}}$$

$$\frac{169}{1} = \frac{l}{3.5}$$

$$169(3.5) = 1 \cdot l$$

$$591.5 = l$$

A real engine is 591.5 inches ($49'\ 3.5''$) long.

61. Let t = the number of teachers required.

$$\frac{3 \text{ teachers}}{50 \text{ students}} = \frac{\text{number of teachers}}{\text{number of students}}$$

$$\frac{3}{50} = \frac{t}{2{,}700}$$

$$2{,}700 \cdot 3 = 50 \cdot t$$

$$8{,}100 = 50t$$

$$162 = t$$

A total of 162 teachers are needed.

63. Let x = the amount of oil to be added.

$$\frac{50}{1} = \frac{\text{ounces of gasoline}}{\text{ounces of oil}}$$

$$\frac{50}{1} = \frac{6 \cdot 128}{x}$$

$$\frac{50}{1} = \frac{768}{x}$$

$$x \cdot 50 = 1 \cdot 768$$

$$50x = 768$$

$$x = \frac{768}{50} \approx 15.36 \text{ ounces}$$

The directions are close.

65.
$$\frac{\text{height of man}}{\text{man's shadow}} = \frac{\text{height of tree}}{\text{tree's shadow}}$$

$$\frac{6}{4} = \frac{h}{26}$$

$$6 \cdot 26 = 4h$$

$$156 = 4h$$

$$39 = h$$

The tree is 39 feet tall.

67.
$$\frac{75}{32} = \frac{w}{20}$$

$$75 \cdot 20 = 32w$$

$$1500 = 32w$$

$$\frac{1500}{32} = w, \text{ or } w = 46\frac{7}{8}$$

The river is $46\frac{7}{8}$ feet wide.

69.
$$\frac{x}{1350} = \frac{5}{1}$$

$$1 \cdot x = 1350 \cdot 5$$

$$x = 6750$$

The plane will lose 6,750 feet in altitude.

71.
$$\frac{x}{750} = \frac{52800}{2500}$$

$$2500 \cdot x = 750 \cdot 52800$$

$$2500x = 39,600,000$$

$$x = 15,840$$

The road will rise 15,840 ft.

73. Answers may vary.

75. If $\dfrac{a}{b} = \dfrac{c}{d}$, then $ad = bc$. Thus, $\dfrac{a}{b} = \dfrac{a(b+d)}{b(b+d)} = \dfrac{ab+ad}{b(b+d)} = \dfrac{ab+bc}{b(b+d)} = \dfrac{b(a+c)}{b(b+d)} = \dfrac{a+c}{b+d}$.

Exercise 6.3 (page 378)

1. If a, b and c are real numbers, then $(a + b) + c = a + (b + c)$.

3. 0 is the additive identity.

5. $\frac{5}{3}$ is the additive inverse of $-\frac{5}{3}$.

7. numerator

9. 0

11. negatives (or opposites)

13. $\frac{a}{b}$

15. factor; common

17. The expression is undefined when the denominator equals 0:
$$y - 2 = 0$$
$$y = 2$$

19. The expression is undefined when the denominator equals 0:
$$x^2 - x - 2 = 0$$
$$(x + 1)(x - 2) = 0$$
$$x + 1 = 0 \quad \text{or} \quad x - 2 = 0$$
$$x = -1 \qquad x = 2$$

21. The expression is undefined when the denominator equals 0:
$$2m^2 - m - 3 = 0$$
$$(m + 1)(2m - 3) = 0$$
$$m + 1 = 0 \quad \text{or} \quad 2m - 3 = 0$$
$$m = -1 \qquad 2m = 3$$
$$m = \tfrac{3}{2}$$

23. $\dfrac{8}{10} = \dfrac{2 \cdot 4}{2 \cdot 5} = \dfrac{4}{5}$

25. $\dfrac{28}{35} = \dfrac{7 \cdot 4}{7 \cdot 5} = \dfrac{4}{5}$

27. $\dfrac{8}{52} = \dfrac{4 \cdot 2}{4 \cdot 13} = \dfrac{2}{13}$

29. $\dfrac{10}{45} = \dfrac{5 \cdot 2}{5 \cdot 9} = \dfrac{2}{9}$

31. $\dfrac{-18}{54} = -\dfrac{18 \cdot 1}{18 \cdot 3} = -\dfrac{1}{3}$

33. $\dfrac{4x}{2} = \dfrac{2 \cdot 2x}{2 \cdot 1} = 2x$

35. $\dfrac{-6x}{18} = -\dfrac{6 \cdot x}{6 \cdot 3} = -\dfrac{x}{3}$

37. $\dfrac{45}{9a} = \dfrac{9 \cdot 5}{9 \cdot a} = \dfrac{5}{a}$

39. $\dfrac{7 + 3}{5z} = \dfrac{10}{5z} = \dfrac{5 \cdot 2}{5 \cdot z} = \dfrac{2}{z}$

41. $\dfrac{(3 + 4)a}{24 - 3} = \dfrac{7a}{21} = \dfrac{7 \cdot a}{7 \cdot 3} = \dfrac{a}{3}$

43. $\dfrac{2x}{3x} = \dfrac{x \cdot 2}{x \cdot 3} = \dfrac{2}{3}$

45. $\dfrac{6x^2}{4x^2} = \dfrac{2x^2 \cdot 3}{2x^2 \cdot 2} = \dfrac{3}{2}$

47. $\dfrac{2x^2}{3y} \Rightarrow$ simplest form

49. $\dfrac{15x^2y}{5xy^2} = \dfrac{5xy \cdot 3x}{5xy \cdot y} = \dfrac{3x}{y}$

51. $\dfrac{28x}{32y} = \dfrac{4 \cdot 7x}{4 \cdot 8y} = \dfrac{7x}{8y}$

53. $\dfrac{x + 3}{3(x + 3)} = \dfrac{1(x + 3)}{3(x + 3)} = \dfrac{1}{3}$

55. $\dfrac{5x + 35}{x + 7} = \dfrac{5(x + 7)}{1(x + 7)} = 5$

57. $\dfrac{x^2 + 3x}{2x + 6} = \dfrac{x(x + 3)}{2(x + 3)} = \dfrac{x}{2}$

59. $\dfrac{15x - 3x^2}{25y - 5xy} = \dfrac{3x(5 - x)}{5y(5 - x)} = \dfrac{3x}{5y}$

61. $\dfrac{6a - 6b + 6c}{9a - 9b + 9c} = \dfrac{6(a - b + c)}{9(a - b + c)} = \dfrac{6}{9} = \dfrac{2}{3}$

63. $\dfrac{x - 7}{7 - x} = \dfrac{-(7 - x)}{7 - x} = -1$

65. $\dfrac{6x - 3y}{3y - 6x} = \dfrac{-(3y - 6x)}{3y - 6x} = -1$

67. $\dfrac{a + b - c}{c - a - b} = \dfrac{-(c - a - b)}{c - a - b} = -1$

69. $\dfrac{x^2 + 3x + 2}{x^2 + x - 2} = \dfrac{(x + 2)(x + 1)}{(x + 2)(x - 1)} = \dfrac{x + 1}{x - 1}$

71. $\dfrac{x^2 - 8x + 15}{x^2 - x - 6} = \dfrac{(x - 5)(x - 3)}{(x + 2)(x - 3)} = \dfrac{x - 5}{x + 2}$

73. $\dfrac{2x^2 - 8x}{x^2 - 6x + 8} = \dfrac{2x(x - 4)}{(x - 2)(x - 4)} = \dfrac{2x}{x - 2}$

75. $\dfrac{xy + 2x^2}{2xy + y^2} = \dfrac{x(y + 2x)}{y(2x + y)} = \dfrac{x}{y}$

77. $\dfrac{x^2 + 3x + 2}{x^3 + x^2} = \dfrac{(x + 2)(x + 1)}{x^2(x + 1)} = \dfrac{x + 2}{x^2}$

79. $\dfrac{x^2 - 8x + 16}{x^2 - 16} = \dfrac{(x - 4)(x - 4)}{(x + 4)(x - 4)} = \dfrac{x - 4}{x + 4}$

81. $\dfrac{2x^2 - 8}{x^2 - 3x + 2} = \dfrac{2(x^2 - 4)}{(x - 1)(x - 2)} = \dfrac{2(x + 2)(x - 2)}{(x - 1)(x - 2)} = \dfrac{2(x + 2)}{x - 1}$

83. $\dfrac{x^2 - 2x - 15}{x^2 + 2x - 15} = \dfrac{(x - 5)(x + 3)}{(x + 5)(x - 3)}$

85. $\dfrac{x^2 - 3(2x - 3)}{9 - x^2} = \dfrac{x^2 - 6x + 9}{3^2 - x^2} = \dfrac{(x - 3)(x - 3)}{(3 + x)(3 - x)} = -\dfrac{x - 3}{x + 3}$

87. $\dfrac{4(x + 3) + 4}{3(x + 2) + 6} = \dfrac{4x + 12 + 4}{3x + 6 + 6} = \dfrac{4x + 16}{3x + 12} = \dfrac{4(x + 4)}{3(x + 4)} = \dfrac{4}{3}$

89. $\dfrac{x^2 - 9}{(2x + 3) - (x + 6)} = \dfrac{x^2 - 9}{2x + 3 - x - 6} = \dfrac{x^2 - 9}{x - 3} = \dfrac{(x + 3)(x - 3)}{x - 3} = x + 3$

91. $\dfrac{x^3 + 1}{x^2 - x + 1} = \dfrac{x^3 + 1^3}{x^2 - x + 1} = \dfrac{(x + 1)(x^2 - 1x + 1^2)}{x^2 - x + 1} = \dfrac{(x + 1)(x^2 - x + 1)}{x^2 - x + 1} = x + 1$

93. $\dfrac{2a^3 - 16}{2a^2 + 4a + 8} = \dfrac{2(a^3 - 8)}{2(a^2 + 2a + 4)} = \dfrac{a^3 - 2^3}{a^2 + 2a + 4} = \dfrac{(a - 2)(a^2 + 2a + 2^2)}{a^2 + 2a + 4}$

$$= \dfrac{(a - 2)(a^2 + 2a + 4)}{a^2 + 2a + 4} = a - 2$$

95. $\dfrac{ab + b + 2a + 2}{ab + a + b + 1} = \dfrac{b(a + 1) + 2(a + 1)}{a(b + 1) + 1(b + 1)} = \dfrac{(a + 1)(b + 2)}{(b + 1)(a + 1)} = \dfrac{b + 2}{b + 1}$

97. $\dfrac{xy + 3y + 3x + 9}{x^2 - 9} = \dfrac{y(x + 3) + 3(x + 3)}{(x + 3)(x - 3)} = \dfrac{(x + 3)(y + 3)}{(x + 3)(x - 3)} = \dfrac{y + 3}{x - 3}$

99. Answers may vary.

101. $\dfrac{3 - x}{3 + x} = \dfrac{-(x - 3)}{x + 3} = -\dfrac{x - 3}{x + 3}$

Exercise 6.4 (page 386)

1. $2x^3y^2(-3x^2y^4z) = -6x^5y^6z$

3. $(3y)^{-4} = \dfrac{1}{(3y)^4} = \dfrac{1}{81y^4}$

5. $\dfrac{x^{3m}}{x^{4m}} = x^{-m} = \dfrac{1}{x^m}$

7. $-4\left(y^3 - 4y^2 + 3y - 2\right) + 6\left(-2y^2 + 4\right) - 4\left(-2y^3 - y\right)$
$$= -4y^3 + 16y^2 - 12y + 8 - 12y^2 + 24 + 8y^3 + 4y$$
$$= 4y^3 + 4y^2 - 8y + 32$$

9. numerator

11. numerators; denominators

13. 1

15. divisor; multiply

17. $\dfrac{5}{7} \cdot \dfrac{9}{13} = \dfrac{5 \cdot 9}{7 \cdot 13} = \dfrac{45}{91}$

19. $\dfrac{25}{35} \cdot \dfrac{-21}{55} = \dfrac{-5 \cdot 5 \cdot 3 \cdot 7}{7 \cdot 5 \cdot 5 \cdot 11} = -\dfrac{3}{11}$

21. $\dfrac{2}{3} \cdot \dfrac{15}{2} \cdot \dfrac{1}{7} = \dfrac{2 \cdot 3 \cdot 5}{3 \cdot 2 \cdot 7} = \dfrac{5}{7}$

23. $\dfrac{3x}{y} \cdot \dfrac{y}{2} = \dfrac{3x \cdot y}{y \cdot 2} = \dfrac{3x}{2}$

25. $\dfrac{5y}{7} \cdot \dfrac{7x}{5z} = \dfrac{5 \cdot y \cdot 7 \cdot x}{7 \cdot 5 \cdot z} = \dfrac{xy}{z}$

27. $\dfrac{7z}{9z} \cdot \dfrac{4z}{2z} = \dfrac{7 \cdot z \cdot 2 \cdot 2 \cdot z}{9 \cdot z \cdot 2 \cdot z} = \dfrac{14}{9}$

29. $\dfrac{2x^2y}{3xy} \cdot \dfrac{3xy^2}{2} = \dfrac{6x^3y^3}{6xy} = x^2y^2$

31. $\dfrac{8x^2y^2}{4x^2} \cdot \dfrac{2xy}{2y} = \dfrac{16x^3y^3}{8x^2y} = 2xy^2$

33. $\dfrac{-2xy}{x^2} \cdot \dfrac{3xy}{2} = -\dfrac{6x^2y^2}{2x^2} = -3y^2$

35. $\dfrac{ab^2}{a^2b} \cdot \dfrac{b^2c^2}{abc} \cdot \dfrac{abc^2}{a^3c^2} = \dfrac{a^2b^5c^4}{a^6b^2c^3} = \dfrac{b^3c}{a^4}$

37. $\dfrac{10r^2st^3}{6rs^2} \cdot \dfrac{3r^3t}{2rst} \cdot \dfrac{2s^3t^4}{5s^2t^3} = \dfrac{60r^5s^4t^8}{60r^2s^5t^4} = \dfrac{r^3t^4}{s}$

39. $\dfrac{z+7}{7} \cdot \dfrac{z+2}{z} = \dfrac{(z+7)(z+2)}{7z}$

41. $\dfrac{x-2}{2} \cdot \dfrac{2x}{x-2} = \dfrac{2x(x-2)}{2(x-2)} = x$

43. $\dfrac{x+5}{5} \cdot \dfrac{x}{x+5} = \dfrac{x(x+5)}{5(x+5)} = \dfrac{x}{5}$

45. $\dfrac{(x+1)^2}{x+1} \cdot \dfrac{x+2}{x+1} = \dfrac{(x+2)(x+1)^2}{(x+1)^2}$
$$= x+2$$

47. $\dfrac{2x+6}{x+3} \cdot \dfrac{3}{4x} = \dfrac{2(x+3)}{x+3} \cdot \dfrac{3}{4x}$
$$= \dfrac{6(x+3)}{4x(x+3)} = \dfrac{3}{2x}$$

49. $\dfrac{x^2 - x}{x} \cdot \dfrac{3x - 6}{3x - 3} = \dfrac{x(x-1)}{x} \cdot \dfrac{3(x-2)}{3(x-1)} = \dfrac{3x(x-1)(x-2)}{3x(x-1)} = x - 2$

51. $\dfrac{7y - 14}{y - 2} \cdot \dfrac{x^2}{7x} = \dfrac{7(y-2)}{y-2} \cdot \dfrac{x^2}{7x} = \dfrac{7x^2(y-2)}{7x(y-2)} = x$

53. $\dfrac{x^2 + x - 6}{5x} \cdot \dfrac{5x - 10}{x + 3} = \dfrac{(x+3)(x-2)}{5x} \cdot \dfrac{5(x-2)}{x+3} = \dfrac{5(x+3)(x-2)^2}{5x(x+3)} = \dfrac{(x-2)^2}{x}$

55. $\dfrac{m^2 - 2m - 3}{2m + 4} \cdot \dfrac{m^2 - 4}{m^2 + 3m + 2} = \dfrac{(m-3)(m+1)}{2(m+2)} \cdot \dfrac{(m+2)(m-2)}{(m+2)(m+1)}$

$$= \dfrac{(m-3)(m+1)(m+2)(m-2)}{2(m+2)(m+2)(m+1)} = \dfrac{(m-3)(m-2)}{2(m+2)}$$

57. $\dfrac{x^2 + 7xy + 12y^2}{x^2 + 2xy - 8y^2} \cdot \dfrac{x^2 - xy - 2y^2}{x^2 + 4xy + 3y^2} = \dfrac{(x+4y)(x+3y)}{(x+4y)(x-2y)} \cdot \dfrac{(x-2y)(x+y)}{(x+3y)(x+y)}$

$$= \dfrac{(x+4y)(x+3y)(x-2y)(x+y)}{(x+4y)(x-2y)(x+3y)(x+y)} = 1$$

59. $\dfrac{ax + bx + ay + by}{x^3 - y^3} \cdot \dfrac{x^2 + xy + y^2}{ax + bx} = \dfrac{x(a+b) + y(a+b)}{(x-y)(x^2 + xy + y^2)} \cdot \dfrac{x^2 + xy + y^2}{x(a+b)}$

$$= \dfrac{(a+b)(x+y)}{(x-y)(x^2 + xy + y^2)} \cdot \dfrac{x^2 + xy + y^2}{x(a+b)}$$

$$= \dfrac{(a+b)(x+y)(x^2 + xy + y^2)}{(x-y)(x^2 + xy + y^2) \cdot x(a+b)} = \dfrac{x+y}{x(x-y)}$$

61. $\dfrac{x^2 - y^2}{y^2 - xy} \cdot \dfrac{yx^3 - y^4}{ax + ay + bx + by} = \dfrac{(x+y)(x-y)}{y(y-x)} \cdot \dfrac{y(x^3 - y^3)}{a(x+y) + b(x+y)}$

$$= -\dfrac{(x+y)}{y} \cdot \dfrac{y(x-y)(x^2 + xy + y^2)}{(x+y)(a+b)}$$

$$= -\dfrac{(x+y) \cdot y(x-y)(x^2 + xy + y^2)}{y(x+y)(a+b)}$$

$$= -\dfrac{(x-y)(x^2 + xy + y^2)}{a+b}$$

63. $\dfrac{abc^2}{a+1} \cdot \dfrac{c}{a^2 b^2} \cdot \dfrac{a^2 + a}{ac} = \dfrac{abc^2}{a+1} \cdot \dfrac{c}{a^2 b^2} \cdot \dfrac{a(a+1)}{ac} = \dfrac{a^2 bc^3(a+1)}{a^3 b^2 c(a+1)} = \dfrac{c^2}{ab}$

65. $\dfrac{3x^2 + 5x + 2}{x^2 - 9} \cdot \dfrac{x - 3}{x^2 - 4} \cdot \dfrac{x^2 + 5x + 6}{6x + 4} = \dfrac{(3x+2)(x+1)}{(x+3)(x-3)} \cdot \dfrac{x-3}{(x+2)(x-2)} \cdot \dfrac{(x+2)(x+3)}{2(3x+2)}$

$$= \dfrac{(3x+2)(x+1)(x-3)(x+2)(x+3)}{2(x+3)(x-3)(x+2)(x-2)(3x+2)} = \dfrac{x+1}{2(x-2)}$$

67. $\dfrac{1}{3} \div \dfrac{1}{2} = \dfrac{1}{3} \cdot \dfrac{2}{1} = \dfrac{2}{3}$

69. $\dfrac{21}{14} \div \dfrac{5}{2} = \dfrac{21}{14} \cdot \dfrac{2}{5} = \dfrac{3 \cdot 7 \cdot 2}{2 \cdot 7 \cdot 5} = \dfrac{3}{5}$

71. $\dfrac{2}{y} \div \dfrac{4}{3} = \dfrac{2}{y} \cdot \dfrac{3}{4} = \dfrac{2 \cdot 3}{y \cdot 2 \cdot 2} = \dfrac{3}{2y}$

73. $\dfrac{3x}{2} \div \dfrac{x}{2} = \dfrac{3x}{2} \cdot \dfrac{2}{x} = \dfrac{3 \cdot x \cdot 2}{2 \cdot x} = 3$

75. $\dfrac{3x}{y} \div \dfrac{2x}{4} = \dfrac{3x}{y} \cdot \dfrac{4}{2x} = \dfrac{3 \cdot x \cdot 2 \cdot 2}{y \cdot 2 \cdot x} = \dfrac{6}{y}$

77. $\dfrac{4x}{3x} \div \dfrac{2y}{9y} = \dfrac{4x}{3x} \cdot \dfrac{9y}{2y} = \dfrac{2 \cdot 2 \cdot x \cdot 3 \cdot 3 \cdot y}{3 \cdot x \cdot 2 \cdot y}$
$= 6$

79. $\dfrac{x^2}{3} \div \dfrac{2x}{4} = \dfrac{x^2}{3} \cdot \dfrac{4}{2x} = \dfrac{x \cdot x \cdot 2 \cdot 2}{3 \cdot 2 \cdot x} = \dfrac{2x}{3}$

81. $\dfrac{x^2 y}{3xy} \div \dfrac{xy^2}{6y} = \dfrac{x^2 y}{3xy} \cdot \dfrac{6y}{xy^2} = \dfrac{6x^2 y^2}{3x^2 y^3} = \dfrac{2}{y}$

83. $\dfrac{x+2}{3x} \div \dfrac{x+2}{2} = \dfrac{x+2}{3x} \cdot \dfrac{2}{x+2} = \dfrac{2(x+2)}{3x(x+2)} = \dfrac{2}{3x}$

85. $\dfrac{(z-2)^2}{3z^2} \div \dfrac{z-2}{6z} = \dfrac{(z-2)^2}{3z^2} \cdot \dfrac{6z}{z-2} = \dfrac{6z(z-2)^2}{3z^2(z-2)} = \dfrac{2(z-2)}{z}$

87. $\dfrac{(z-7)^2}{z+2} \div \dfrac{z(z-7)}{5z^2} = \dfrac{(z-7)^2}{z+2} \cdot \dfrac{5z^2}{z(z-7)} = \dfrac{5z^2(z-7)^2}{z(z+2)(z-7)} = \dfrac{5z(z-7)}{z+2}$

89. $\dfrac{x^2-4}{3x+6} \div \dfrac{x-2}{x+2} = \dfrac{x^2-4}{3x+6} \cdot \dfrac{x+2}{x-2} = \dfrac{(x+2)(x-2)(x+2)}{3(x+2)(x-2)} = \dfrac{x+2}{3}$

91. $\dfrac{x^2-1}{3x-3} \div \dfrac{x+1}{3} = \dfrac{x^2-1}{3x-3} \cdot \dfrac{3}{x+1} = \dfrac{(x+1)(x-1)3}{3(x-1)(x+1)} = 1$

93. $\dfrac{5x^2+13x-6}{x+3} \div \dfrac{5x^2-17x+6}{x-2} = \dfrac{5x^2+13x-6}{x+3} \cdot \dfrac{x-2}{5x^2-17x+6}$
$= \dfrac{(5x-2)(x+3)(x-2)}{(x+3)(5x-2)(x-3)} = \dfrac{x-2}{x-3}$

95. $\dfrac{2x^2+8x-42}{x-3} \div \dfrac{2x^2+14x}{x^2+5x} = \dfrac{2(x^2+4x-21)}{x-3} \cdot \dfrac{x^2+5x}{2x^2+14x}$
$= \dfrac{2(x+7)(x-3)(x)(x+5)}{(x-3)(2x)(x+7)} = x+5$

97. $\dfrac{ab+4a+2b+8}{b^2+4b+16} \div \dfrac{b^2-16}{b^3-64} = \dfrac{ab+4a+2b+8}{b^2+4b+16} \cdot \dfrac{b^3-64}{b^2-16}$
$= \dfrac{a(b+4)+2(b+4)}{b^2+4b+16} \cdot \dfrac{b^3-4^3}{(b+4)(b-4)}$
$= \dfrac{(b+4)(a+2)}{b^2+4b+16} \cdot \dfrac{(b-4)(b^2+4b+16)}{(b+4)(b-4)}$
$= \dfrac{(b+4)(a+2)(b-4)(b^2+4b+16)}{(b^2+4b+16)(b+4)(b-4)} = a+2$

99.
$$\frac{p^3 - p^2q + pq^2}{mp - mq + np - nq} \div \frac{q^3 + p^3}{q^2 - p^2} = \frac{p^3 - p^2q + pq^2}{mp - mq + np - nq} \cdot \frac{q^2 - p^2}{q^3 + p^3}$$
$$= \frac{p(p^2 - pq + q^2)}{m(p - q) + n(p - q)} \cdot \frac{(q + p)(q - p)}{(q + p)(q^2 - pq + p^2)}$$
$$= \frac{p(p^2 - pq + q^2)}{(p - q)(m + n)} \cdot \frac{(q + p)(q - p)}{(q + p)(q^2 - pq + p^2)}$$
$$= \frac{p(p^2 - pq + q^2)(q + p)(q - p)}{(p - q)(m + n)(q + p)(q^2 - pq + p^2)} = -\frac{p}{m + n}$$

101.
$$\frac{x}{3} \cdot \frac{9}{4} \div \frac{x^2}{6} = \frac{x}{3} \cdot \frac{9}{4} \cdot \frac{6}{x^2}$$
$$= \frac{x \cdot 3 \cdot 3 \cdot 2 \cdot 3}{3 \cdot 2 \cdot 2 \cdot x \cdot x} = \frac{9}{2x}$$

103.
$$\frac{x^2}{18} \div \frac{x^3}{6} \div \frac{12}{x^2} = \frac{x^2}{18} \cdot \frac{6}{x^3} \cdot \frac{x^2}{12}$$
$$= \frac{6x^4}{216x^3} = \frac{x}{36}$$

105.
$$\frac{x^2 - 1}{x^2 - 9} \cdot \frac{x + 3}{x + 2} \div \frac{5}{x + 2} = \frac{x^2 - 1}{x^2 - 9} \cdot \frac{x + 3}{x + 2} \cdot \frac{x + 2}{5} = \frac{(x + 1)(x - 1)(x + 3)(x + 2)}{5(x + 3)(x - 3)(x + 2)}$$
$$= \frac{(x + 1)(x - 1)}{5(x - 3)}$$

107.
$$\frac{x^2 - 4}{2x + 6} \div \frac{x + 2}{4} \cdot \frac{x + 3}{x - 2} = \frac{x^2 - 4}{2x + 6} \cdot \frac{4}{x + 2} \cdot \frac{x + 3}{x - 2} = \frac{4(x + 2)(x - 2)(x + 3)}{2(x + 3)(x + 2)(x - 2)} = 2$$

109.
$$\frac{x - x^2}{x^2 - 4}\left(\frac{2x + 4}{x + 2} \div \frac{5}{x + 2}\right) = \frac{x - x^2}{x^2 - 4}\left(\frac{2x + 4}{x + 2} \cdot \frac{x + 2}{5}\right) = \frac{x - x^2}{x^2 - 4}\left(\frac{2(x + 2)(x + 2)}{5(x + 2)}\right)$$
$$= \frac{x(1 - x)}{(x + 2)(x - 2)} \cdot \frac{2(x + 2)}{5}$$
$$= \frac{2x(1 - x)}{5(x - 2)}$$

111.
$$\frac{y^2}{x + 1} \cdot \frac{x^2 + 2x + 1}{x^2 - 1} \div \frac{3y}{xy - y} = \frac{y^2}{x + 1} \cdot \frac{(x + 1)(x + 1)}{(x + 1)(x - 1)} \cdot \frac{xy - y}{3y} = \frac{y^2(x + 1)(x + 1)(y)(x - 1)}{(x + 1)(x + 1)(x - 1)(3y)}$$
$$= \frac{y^2}{3}$$

113.
$$\frac{x^2 + x - 6}{x^2 - 4} \cdot \frac{x^2 + 2x}{x - 2} \div \frac{x^2 + 3x}{x + 2} = \frac{x^2 + x - 6}{x^2 - 4} \cdot \frac{x^2 + 2x}{x - 2} \cdot \frac{x + 2}{x^2 + 3x}$$
$$= \frac{(x + 3)(x - 2)(x)(x + 2)(x + 2)}{(x + 2)(x - 2)(x - 2)(x)(x + 3)} = \frac{x + 2}{x - 2}$$

115. Answers may vary.

117. Answers may vary.

119. You always get the original value of x after simplifying.

Exercise 6.5 (page 398)

1. $49 = 7 \cdot 7 = 7^2$

3. $136 = 4 \cdot 34 = 2 \cdot 2 \cdot 2 \cdot 17 = 2^3 \cdot 17$

5. $102 = 2 \cdot 51 = 2 \cdot 3 \cdot 17$

7. $144 = 16 \cdot 9 = 2 \cdot 2 \cdot 2 \cdot 2 \cdot 3 \cdot 3 = 2^4 \cdot 3^2$

9. LCD

11. numerators; common denominator

13. $\dfrac{1}{3} + \dfrac{1}{3} = \dfrac{1+1}{3} = \dfrac{2}{3}$

15. $\dfrac{2}{9} + \dfrac{1}{9} = \dfrac{2+1}{9} = \dfrac{3}{9} = \dfrac{1}{3}$

17. $\dfrac{2x}{y} + \dfrac{2x}{y} = \dfrac{2x+2x}{y} = \dfrac{4x}{y}$

19. $\dfrac{4}{7y} + \dfrac{10}{7y} = \dfrac{4+10}{7y} = \dfrac{14}{7y} = \dfrac{2}{y}$

21. $\dfrac{y+2}{5z} + \dfrac{y+4}{5z} = \dfrac{y+2+y+4}{5z} = \dfrac{2y+6}{5z}$

23. $\dfrac{3x-5}{x-2} + \dfrac{6x-13}{x-2} = \dfrac{3x-5+6x-13}{x-2} = \dfrac{9x-18}{x-2} = \dfrac{9(x-2)}{x-2} = 9$

25. $\dfrac{5}{7} - \dfrac{4}{7} = \dfrac{5-4}{7} = \dfrac{1}{7}$

27. $\dfrac{35}{72} - \dfrac{44}{72} = \dfrac{35-44}{72} = \dfrac{-9}{72} = -\dfrac{1}{8}$

29. $\dfrac{2x}{y} - \dfrac{x}{y} = \dfrac{2x-x}{y} = \dfrac{x}{y}$

31. $\dfrac{9y}{3x} - \dfrac{6y}{3x} = \dfrac{9y-6y}{3x} = \dfrac{3y}{3x} = \dfrac{y}{x}$

33. $\dfrac{6x-5}{3xy} - \dfrac{3x-5}{3xy} = \dfrac{6x-5-(3x-5)}{3xy} = \dfrac{6x-5-3x+5}{3xy} = \dfrac{3x}{3xy} = \dfrac{1}{y}$

35. $\dfrac{3y-2}{y+3} - \dfrac{2y-5}{y+3} = \dfrac{3y-2-(2y-5)}{y+3} = \dfrac{3y-2-2y+5}{y+3} = \dfrac{y+3}{y+3} = 1$

37. $\dfrac{13x}{15} + \dfrac{12x}{15} - \dfrac{5x}{15} = \dfrac{13x+12x-5x}{15}$
$$= \dfrac{20x}{15} = \dfrac{4x}{3}$$

39. $\dfrac{x}{3y} + \dfrac{2x}{3y} - \dfrac{x}{3y} = \dfrac{x+2x-x}{3y} = \dfrac{2x}{3y}$

41. $\dfrac{3x}{y+2} - \dfrac{3y}{y+2} + \dfrac{x+y}{y+2} = \dfrac{3x-3y+x+y}{y+2} = \dfrac{4x-2y}{y+2} = \dfrac{2(2x-y)}{y+2}$

43. $\dfrac{x+1}{x-2} - \dfrac{2(x-3)}{x-2} + \dfrac{3(x+1)}{x-2} = \dfrac{x+1-2(x-3)+3(x+1)}{x-2} = \dfrac{x+1-2x+6+3x+3}{x-2}$
$$= \dfrac{2x+10}{x-2} = \dfrac{2(x+5)}{x-2}$$

45. $\dfrac{25}{4} = \dfrac{25 \cdot 5}{4 \cdot 5} = \dfrac{125}{20}$

47. $\dfrac{8}{x} = \dfrac{8 \cdot xy}{x \cdot xy} = \dfrac{8xy}{x^2 y}$

49. $\dfrac{3x}{x+1} = \dfrac{3x(x+1)}{(x+1)(x+1)} = \dfrac{3x(x+1)}{(x+1)^2}$

51. $\dfrac{2y}{x} = \dfrac{2y(x+1)}{x(x+1)} = \dfrac{2y(x+1)}{x^2+x}$

53. $\dfrac{z}{z-1} = \dfrac{z(z+1)}{(z-1)(z+1)} = \dfrac{z(z+1)}{z^2-1}$

55. $\dfrac{2}{x+1} = \dfrac{2(x+2)}{(x+1)(x+2)} = \dfrac{2(x+2)}{x^2+3x+2}$

57. $2x = 2 \cdot x$
$6x = 2 \cdot 3 \cdot x$
LCD $= 2 \cdot 3 \cdot x = 6x$

59. $3x = 3 \cdot x$
$6y = 2 \cdot 3 \cdot y$
$9xy = 3^2 \cdot x \cdot y$
LCD $= 2 \cdot 3^2 \cdot x \cdot y = 18xy$

61. $x^2 - 1 = (x+1)(x-1)$
$x + 1 = x + 1$
LCD $= (x+1)(x-1) = x^2 - 1$

63. $x^2 + 6x = x(x+6)$
$x + 6 = x + 6$
$x = x$
LCD $= x(x+6) = x^2 + 6x$

65. $x^2 - 4x - 5 = (x+1)(x-5)$
$x^2 - 25 = (x+5)(x-5)$
LCD $= (x+1)(x-5)(x+5)$

67. $\dfrac{1}{2} + \dfrac{2}{3} = \dfrac{1 \cdot 3}{2 \cdot 3} + \dfrac{2 \cdot 2}{3 \cdot 2} = \dfrac{3}{6} + \dfrac{4}{6} = \dfrac{7}{6}$

69. $\dfrac{2y}{9} + \dfrac{y}{3} = \dfrac{2y}{9} + \dfrac{y \cdot 3}{3 \cdot 3} = \dfrac{2y}{9} + \dfrac{3y}{9} = \dfrac{5y}{9}$

71. $\dfrac{21x}{14} - \dfrac{5x}{21} = \dfrac{21x \cdot 3}{14 \cdot 3} - \dfrac{5x \cdot 2}{21 \cdot 2}$
$= \dfrac{63x}{42} - \dfrac{10x}{42} = \dfrac{53x}{42}$

73. $\dfrac{4x}{3} + \dfrac{2x}{y} = \dfrac{4x \cdot y}{3 \cdot y} + \dfrac{2x \cdot 3}{y \cdot 3}$
$= \dfrac{4xy}{3y} + \dfrac{6x}{3y} = \dfrac{4xy + 6x}{3y}$

75. $\dfrac{2}{x} - 3x = \dfrac{2}{x} - \dfrac{3x}{1} = \dfrac{2}{x} - \dfrac{3x \cdot x}{1 \cdot x}$
$= \dfrac{2}{x} - \dfrac{3x^2}{x} = \dfrac{2 - 3x^2}{x}$

77. $\dfrac{y+2}{5y} + \dfrac{y+4}{15y} = \dfrac{(y+2)3}{5y \cdot 3} + \dfrac{y+4}{15y} = \dfrac{3y+6}{15y} + \dfrac{y+4}{15y} = \dfrac{4y+10}{15y}$

79. $\dfrac{x+5}{xy} - \dfrac{x-1}{x^2y} = \dfrac{(x+5)x}{xy \cdot x} - \dfrac{x-1}{x^2y} = \dfrac{x^2+5x}{x^2y} - \dfrac{x-1}{x^2y} = \dfrac{x^2+4x+1}{x^2y}$

81. $\dfrac{x}{x+1} + \dfrac{x-1}{x} = \dfrac{x \cdot x}{(x+1)x} + \dfrac{(x-1)(x+1)}{x(x+1)} = \dfrac{x^2}{x(x+1)} + \dfrac{x^2-1}{x(x+1)} = \dfrac{2x^2-1}{x(x+1)}$

83. $\dfrac{x-1}{x} + \dfrac{y+1}{y} = \dfrac{(x-1)y}{x \cdot y} + \dfrac{(y+1)x}{y \cdot x} = \dfrac{xy-y}{xy} + \dfrac{xy+x}{xy} = \dfrac{2xy+x-y}{xy}$

85. $\dfrac{x}{x-2} + \dfrac{4+2x}{x^2-4} = \dfrac{x}{x-2} + \dfrac{2(x+2)}{(x+2)(x-2)} = \dfrac{x}{x-2} + \dfrac{2}{x-2} = \dfrac{x+2}{x-2}$

87. $\dfrac{x+1}{x-1} + \dfrac{x-1}{x+1} = \dfrac{(x+1)(x+1)}{(x-1)(x+1)} + \dfrac{(x-1)(x-1)}{(x+1)(x-1)} = \dfrac{x^2+2x+1}{(x+1)(x-1)} + \dfrac{x^2-2x+1}{(x+1)(x-1)}$

$$= \dfrac{2x^2+2}{(x+1)(x-1)}$$

89. $\dfrac{2x+2}{x-2} - \dfrac{2x}{2-x} = \dfrac{2x+2}{x-2} + \dfrac{2x}{x-2} = \dfrac{4x+2}{x-2} = \dfrac{2(2x+1)}{x-2}$

91. $\dfrac{2x}{x^2-3x+2} + \dfrac{2x}{x-1} - \dfrac{x}{x-2} = \dfrac{2x}{(x-2)(x-1)} + \dfrac{2x}{x-1} - \dfrac{x}{x-2}$

$$= \dfrac{2x}{(x-2)(x-1)} + \dfrac{2x(x-2)}{(x-1)(x-2)} - \dfrac{x(x-1)}{(x-2)(x-1)}$$

$$= \dfrac{2x+2x(x-2)-x(x-1)}{(x-2)(x-1)}$$

$$= \dfrac{2x+2x^2-4x-x^2+x}{(x-2)(x-1)}$$

$$= \dfrac{x^2-x}{(x-2)(x-1)} = \dfrac{x(x-1)}{(x-2)(x-1)} = \dfrac{x}{x-2}$$

93. $\dfrac{2x}{x-1} + \dfrac{3x}{x+1} - \dfrac{x+3}{x^2-1} = \dfrac{2x}{x-1} + \dfrac{3x}{x+1} - \dfrac{x+3}{(x+1)(x-1)}$

$$= \dfrac{2x(x+1)}{(x-1)(x+1)} + \dfrac{3x(x-1)}{(x+1)(x-1)} - \dfrac{x+3}{(x+1)(x-1)}$$

$$= \dfrac{2x(x+1)+3x(x-1)-(x+3)}{(x+1)(x-1)}$$

$$= \dfrac{2x^2+2x+3x^2-3x-x-3}{(x+1)(x-1)}$$

$$= \dfrac{5x^2-2x-3}{(x+1)(x-1)} = \dfrac{(5x+3)(x-1)}{(x+1)(x-1)} = \dfrac{5x+3}{x+1}$$

95. $\dfrac{x+1}{2x+4} - \dfrac{x^2}{2x^2-8} = \dfrac{x+1}{2(x+2)} - \dfrac{x^2}{2(x+2)(x-2)} = \dfrac{(x+1)(x-2)}{2(x+2)(x-2)} - \dfrac{x^2}{2(x+2)(x-2)}$

$$= \dfrac{x^2-2x+x-2-x^2}{2(x+2)(x-2)}$$

$$= \dfrac{-x-2}{2(x+2)(x-2)}$$

$$= \dfrac{-(x+2)}{2(x+2)(x-2)} = -\dfrac{1}{2(x-2)}$$

97. Answers may vary.

99. Answers may vary.

101. The subtraction needs to be distributed in the numerator of the 2nd fraction.

103. $\dfrac{a}{b} + \dfrac{c}{d} = \dfrac{a(d)}{b(d)} + \dfrac{c(b)}{d(b)} = \dfrac{ad+bc}{bd}$

Exercise 6.6 (page 405)

1. $t^3 t^4 t^2 = t^{3+4+2} = t^9$

3. $-2r(r^3)^2 = -2rr^6 = -2r^7$

5. $\left(\dfrac{3r}{4r^3}\right)^4 = \left(\dfrac{3}{4r^2}\right)^4 = \dfrac{3^4}{(4r^2)^4} = \dfrac{81}{256r^8}$

7. $\left(\dfrac{6r^{-2}}{2r^3}\right)^{-2} = (3r^{-5})^{-2} = \dfrac{r^{10}}{9}$

9. complex fraction

11. single; divide

13. $\dfrac{\frac{2}{3}}{\frac{3}{4}} = \dfrac{2}{3} \div \dfrac{3}{4} = \dfrac{2}{3} \cdot \dfrac{4}{3} = \dfrac{8}{9}$

15. $\dfrac{\frac{4}{5}}{\frac{32}{15}} = \dfrac{4}{5} \div \dfrac{32}{15} = \dfrac{4}{5} \cdot \dfrac{15}{32} = \dfrac{3}{8}$

17. $\begin{aligned}\dfrac{\frac{2}{3}+1}{\frac{1}{3}+1} &= \dfrac{\left(\frac{2}{3}+1\right)3}{\left(\frac{1}{3}+1\right)3} = \dfrac{\frac{2}{3}\cdot 3 + 1\cdot 3}{\frac{1}{3}\cdot 3 + 1 \cdot 3}\\ &= \dfrac{2+3}{1+3} = \dfrac{5}{4}\end{aligned}$

19. $\begin{aligned}\dfrac{\frac{1}{2}+\frac{3}{4}}{\frac{3}{2}+\frac{1}{4}} &= \dfrac{\left(\frac{1}{2}+\frac{3}{4}\right)4}{\left(\frac{3}{2}+\frac{1}{4}\right)4} = \dfrac{\frac{1}{2}\cdot 4 + \frac{3}{4}\cdot 4}{\frac{3}{2}\cdot 4 + \frac{1}{4}\cdot 4}\\ &= \dfrac{2+3}{6+1} = \dfrac{5}{7}\end{aligned}$

21. $\dfrac{\frac{x}{y}}{\frac{1}{x}} = \dfrac{x}{y} \div \dfrac{1}{x} = \dfrac{x}{y} \cdot \dfrac{x}{1} = \dfrac{x^2}{y}$

23. $\begin{aligned}\dfrac{\frac{5t^2}{9x^2}}{\frac{3t}{x^2t}} &= \dfrac{5t^2}{9x^2} \div \dfrac{3t}{x^2t} = \dfrac{5t^2}{9x^2} \cdot \dfrac{x^2t}{3t}\\ &= \dfrac{5x^2t^3}{27x^2t} = \dfrac{5t^2}{27}\end{aligned}$

25. $\dfrac{\frac{1}{x}-3}{\frac{5}{x}+2} = \dfrac{\left(\frac{1}{x}-3\right)x}{\left(\frac{5}{x}+2\right)x} = \dfrac{\frac{1}{x}\cdot x - 3\cdot x}{\frac{5}{x}\cdot x + 2\cdot x} = \dfrac{1-3x}{5+2x}$

27. $\dfrac{\frac{2}{x}+2}{\frac{4}{x}+2} = \dfrac{\left(\frac{2}{x}+2\right)x}{\left(\frac{4}{x}+2\right)x} = \dfrac{\frac{2}{x}\cdot x + 2\cdot x}{\frac{4}{x}\cdot x + 2\cdot x} = \dfrac{2+2x}{4+2x} = \dfrac{2(1+x)}{2(2+x)} = \dfrac{1+x}{2+x}$

29. $\dfrac{\frac{3y}{x}-y}{y-\frac{y}{x}} = \dfrac{\left(\frac{3y}{x}-y\right)x}{\left(y-\frac{y}{x}\right)x} = \dfrac{\frac{3y}{x}\cdot x - y\cdot x}{y\cdot x - \frac{y}{x}\cdot x} = \dfrac{3y-xy}{xy-y} = \dfrac{y(3-x)}{y(x-1)} = \dfrac{3-x}{x-1}$

31. $\dfrac{\frac{1}{x+1}}{1+\frac{1}{x+1}} = \dfrac{\left(\frac{1}{x+1}\right)(x+1)}{\left(1+\frac{1}{x+1}\right)(x+1)} = \dfrac{\frac{1}{x+1}(x+1)}{1(x+1)+\frac{1}{x+1}(x+1)} = \dfrac{1}{x+1+1} = \dfrac{1}{x+2}$

33. $\begin{aligned}\dfrac{\frac{x}{x+2}}{\frac{x}{x+2}+x} &= \dfrac{\left(\frac{x}{x+2}\right)(x+2)}{\left(\frac{x}{x+2}+x\right)(x+2)} = \dfrac{\frac{x}{x+2}(x+2)}{\frac{x}{x+2}(x+2)+x(x+2)} = \dfrac{x}{x+x^2+2x} = \dfrac{x}{x^2+3x}\\ &= \dfrac{x}{x(x+3)} = \dfrac{1}{x+3}\end{aligned}$

35. $\dfrac{1}{\frac{1}{x}+\frac{1}{y}} = \dfrac{1(xy)}{\left(\frac{1}{x}+\frac{1}{y}\right)xy} = \dfrac{xy}{\frac{1}{x}\cdot xy + \frac{1}{y}\cdot xy} = \dfrac{xy}{y+x}$

37. $\dfrac{\frac{2}{x}}{\frac{2}{y}-\frac{4}{x}} = \dfrac{\frac{2}{x}(xy)}{\left(\frac{2}{y}-\frac{4}{x}\right)(xy)} = \dfrac{\frac{2}{x}(xy)}{\frac{2}{y}(xy)-\frac{4}{x}(xy)} = \dfrac{2y}{2x-4y} = \dfrac{2y}{2(x-2y)} = \dfrac{y}{x-2y}$

39. $\dfrac{3+\frac{3}{x-1}}{3-\frac{3}{x}} = \dfrac{\left(3+\frac{3}{x-1}\right)(x)(x-1)}{\left(3-\frac{3}{x}\right)(x)(x-1)} = \dfrac{3x(x-1)+\frac{3}{x-1}(x)(x-1)}{3x(x-1)-\frac{3}{x}(x)(x-1)} = \dfrac{3x^2-3x+3x}{3x^2-3x-3(x-1)}$

$$= \dfrac{3x^2}{3x^2-6x+3}$$

$$= \dfrac{3x^2}{3(x^2-2x+1)}$$

$$= \dfrac{x^2}{x^2-2x+1} = \dfrac{x^2}{(x-1)^2}$$

41. $\dfrac{\frac{3}{x}+\frac{4}{x+1}}{\frac{2}{x+1}-\frac{3}{x}} = \dfrac{\left(\frac{3}{x}+\frac{4}{x+1}\right)(x)(x+1)}{\left(\frac{2}{x+1}-\frac{3}{x}\right)(x)(x+1)} = \dfrac{\frac{3}{x}(x)(x+1)+\frac{4}{x+1}(x)(x+1)}{\frac{2}{x+1}(x)(x+1)-\frac{3}{x}(x)(x+1)}$

$$= \dfrac{3(x+1)+4x}{2x-3(x+1)} = \dfrac{3x+3+4x}{2x-3x-3} = \dfrac{7x+3}{-x-3}$$

43. $\dfrac{\frac{2}{x}-\frac{3}{x+1}}{\frac{2}{x+1}-\frac{3}{x}} = \dfrac{\left(\frac{2}{x}-\frac{3}{x+1}\right)(x)(x+1)}{\left(\frac{2}{x+1}-\frac{3}{x}\right)(x)(x+1)} = \dfrac{2(x+1)-3x}{2x-3(x+1)}$

$$= \dfrac{2x+2-3x}{2x-3x-3} = \dfrac{-x+2}{-x-3} = \dfrac{-(x-2)}{-(x+3)} = \dfrac{x-2}{x+3}$$

45. $\dfrac{\frac{1}{y^2+y}-\frac{1}{xy+x}}{\frac{1}{xy+x}-\frac{1}{y^2+y}} = \dfrac{\frac{1}{y(y+1)}-\frac{1}{x(y+1)}}{\frac{1}{x(y+1)}-\frac{1}{y(y+1)}} = \dfrac{\left(\frac{1}{y(y+1)}-\frac{1}{x(y+1)}\right)(x)(y)(y+1)}{\left(\frac{1}{x(y+1)}-\frac{1}{y(y+1)}\right)(x)(y)(y+1)}$

$$= \dfrac{1x-1y}{1y-1x} = \dfrac{x-y}{-(x-y)} = -1$$

47. $\dfrac{x^{-2}}{y^{-1}} = \dfrac{\frac{1}{x^2}}{\frac{1}{y}} = \dfrac{1}{x^2} \div \dfrac{1}{y} = \dfrac{1}{x^2} \cdot \dfrac{y}{1} = \dfrac{y}{x^2}$

49. $\dfrac{1+x^{-1}}{x^{-1}-1} = \dfrac{1+\frac{1}{x}}{\frac{1}{x}-1} = \dfrac{\left(1+\frac{1}{x}\right)x}{\left(\frac{1}{x}-1\right)x} = \dfrac{1x+\frac{1}{x}\cdot x}{\frac{1}{x}\cdot x-1x}$

$$= \dfrac{x+1}{1-x}$$

51. $\dfrac{a^{-2}+a}{a+1} = \dfrac{\frac{1}{a^2}+a}{a+1} = \dfrac{\left(\frac{1}{a^2}+a\right)a^2}{(a+1)a^2} = \dfrac{1+a^3}{a^3+a^2} = \dfrac{(1+a)(1-a+a^2)}{a^2(a+1)} = \dfrac{a^2-a+1}{a^2}$

53. $\dfrac{2x^{-1}+4x^{-2}}{2x^{-2}+x^{-1}} = \dfrac{\frac{2}{x}+\frac{4}{x^2}}{\frac{2}{x^2}+\frac{1}{x}} = \dfrac{\left(\frac{2}{x}+\frac{4}{x^2}\right)x^2}{\left(\frac{2}{x^2}+\frac{1}{x}\right)x^2} = \dfrac{2x+4}{2+x} = \dfrac{2(x+2)}{x+2} = 2$

55. $\dfrac{1-25y^{-2}}{1+10y^{-1}+25y^{-2}} = \dfrac{1-\frac{25}{y^2}}{1+\frac{10}{y}+\frac{25}{y^2}} = \dfrac{\left(1-\frac{25}{y^2}\right)y^2}{\left(1+\frac{10}{y}+\frac{25}{y^2}\right)y^2} = \dfrac{y^2-25}{y^2+10y+25}$

$$= \dfrac{(y+5)(y-5)}{(y+5)(y+5)} = \dfrac{y-5}{y+5}$$

57. Answers may vary.

59. $\dfrac{1}{1+1} = \dfrac{1}{2}; \dfrac{1}{1+\frac{1}{2}} = \dfrac{1}{\frac{3}{2}} = \dfrac{2}{3}; \dfrac{1}{1+\frac{1}{1+\frac{1}{2}}} = \dfrac{1}{1+\frac{1}{\frac{3}{2}}} = \dfrac{1}{1+\frac{2}{3}} = \dfrac{1}{\frac{5}{3}} = \dfrac{3}{5}$

$\dfrac{1}{1+\frac{1}{1+\frac{1}{1+\frac{1}{2}}}} = \dfrac{1}{1+\frac{1}{1+\frac{1}{\frac{3}{2}}}} = \dfrac{1}{1+\frac{1}{1+\frac{2}{3}}} = \dfrac{1}{1+\frac{1}{\frac{5}{3}}} = \dfrac{1}{1+\frac{3}{5}} = \dfrac{1}{\frac{8}{5}} = \dfrac{5}{8}$

Exercise 6.7 (page 412)

1. $x^2 + 4x = x(x+4)$

3. $2x^2 + x - 3 = (2x+3)(x-1)$

5. $x^4 - 16 = (x^2)^2 - 4^2 = (x^2+4)(x^2-4) = (x^2+4)(x+2)(x-2)$

7. extraneous

9. LCD

11. xy

13. $\dfrac{x}{2} + 4 = \dfrac{3x}{2}$

$2\left(\dfrac{x}{2}+4\right) = 2\left(\dfrac{3x}{2}\right)$

$x + 8 = 3x$

$8 = 2x$

$4 = x$

15. $\dfrac{2y}{5} - 8 = \dfrac{4y}{5}$

$5\left(\dfrac{2y}{5}-8\right) = 5\left(\dfrac{4y}{5}\right)$

$2y - 40 = 4y$

$-40 = 2y$

$-20 = y$

17. $\dfrac{x}{3} + 1 = \dfrac{x}{2}$

$6\left(\dfrac{x}{3}+1\right) = 6\left(\dfrac{x}{2}\right)$

$2x + 6 = 3x$

$6 = x$

19. $\dfrac{x}{5} - \dfrac{x}{3} = -8$

$15\left(\dfrac{x}{5}-\dfrac{x}{3}\right) = 15(-8)$

$3x - 5x = -120$

$-2x = -120$

$x = 60$

21. $\dfrac{3a}{2} + \dfrac{a}{3} = -22$

$6\left(\dfrac{3a}{2}+\dfrac{a}{3}\right) = 6(-22)$

$9a + 2a = -132$

$11a = -132$

$a = -12$

23. $\dfrac{x-3}{3} + 2x = -1$

$3\left(\dfrac{x-3}{3}+2x\right) = 3(-1)$

$x - 3 + 6x = -3$

$7x = 0$

$x = 0$

25.
$$\frac{z-3}{2} = z+2$$
$$2\left(\frac{z-3}{2}\right) = 2(z+2)$$
$$z-3 = 2z+4$$
$$-7 = z$$

27.
$$\frac{5(x+1)}{8} = x+1$$
$$8\left(\frac{5(x+1)}{8}\right) = 8(x+1)$$
$$5(x+1) = 8(x+1)$$
$$5x+5 = 8x+8$$
$$-3x = 3$$
$$x = -1$$

29.
$$\frac{c-4}{4} = \frac{c+4}{8}$$
$$8\left(\frac{c-4}{4}\right) = 8\left(\frac{c+4}{8}\right)$$
$$2(c-4) = c+4$$
$$2c-8 = c+4$$
$$c = 12$$

31.
$$\frac{x+1}{3} + \frac{x-1}{5} = \frac{2}{15}$$
$$15\left(\frac{x+1}{3} + \frac{x-1}{5}\right) = 15\left(\frac{2}{15}\right)$$
$$5(x+1) + 3(x-1) = 2$$
$$5x+5+3x-3 = 2$$
$$8x = 0$$
$$x = 0$$

33.
$$\frac{3x-1}{6} - \frac{x+3}{2} = \frac{3x+4}{3}$$
$$12\left(\frac{3x-1}{6} - \frac{x+3}{2}\right) = 12\left(\frac{3x+4}{3}\right)$$
$$2(3x-1) - 6(x+3) = 4(3x+4)$$
$$6x-2-6x-18 = 12x+16$$
$$-36 = 12x$$
$$-3 = x$$

35.
$$\frac{3}{x} + 2 = 3$$
$$x\left(\frac{3}{x} + 2\right) = x(3)$$
$$3+2x = 3x$$
$$3 = x \Rightarrow \text{The answer checks.}$$

37.
$$\frac{5}{a} - \frac{4}{a} = 8 + \frac{1}{a}$$
$$a\left(\frac{5}{a} - \frac{4}{a}\right) = a\left(8 + \frac{1}{a}\right)$$
$$5-4 = 8a+1$$
$$0 = 8a$$
$$0 = a$$
0 does not check. $\Rightarrow$ no solutions

39.
$$\frac{2}{y+1} + 5 = \frac{12}{y+1}$$
$$(y+1)\left(\frac{2}{y+1} + 5\right) = (y+1)\left(\frac{12}{y+1}\right)$$
$$2+5(y+1) = 12$$
$$2+5y+5 = 12$$
$$5y = 5$$
$$y = 1$$
The answer checks.

41.
$$\frac{1}{x-1} + \frac{3}{x-1} = 1$$
$$(x-1)\left(\frac{1}{x-1} + \frac{3}{x-1}\right) = (x-1)(1)$$
$$1+3 = x-1$$
$$5 = x$$
The answer checks.

43.
$$\frac{a^2}{a+2} - \frac{4}{a+2} = a$$
$$(a+2)\left(\frac{a^2}{a+2} - \frac{4}{a+2}\right) = (a+2)a$$
$$a^2 - 4 = a^2 + 2a$$
$$-4 = 2a$$
$$-2 = a$$

-2 does not check $\Rightarrow$ no solutions

45.
$$\frac{x}{x-5} - \frac{5}{x-5} = 3$$
$$(x-5)\left(\frac{x}{x-5} - \frac{5}{x-5}\right) = (x-5)3$$
$$x - 5 = 3x - 15$$
$$-2x = -10$$
$$x = 5$$

5 does not check $\Rightarrow$ no solutions

47.
$$\frac{3r}{2} - \frac{3}{r} = \frac{3r}{2} + 3$$
$$2r\left(\frac{3r}{2} - \frac{3}{r}\right) = 2r\left(\frac{3r}{2} + 3\right)$$
$$3r^2 - 6 = 3r^2 + 6r$$
$$-6 = 6r$$
$$-1 = r \Rightarrow \text{The answer checks.}$$

49.
$$\frac{1}{3} + \frac{2}{x-3} = 1$$
$$3(x-3)\left(\frac{1}{3} + \frac{2}{x-3}\right) = 3(x-3)(1)$$
$$x - 3 + 3(2) = 3x - 9$$
$$x - 3 + 6 = 3x - 9$$
$$12 = 2x$$
$$6 = x$$

The answer checks.

51.
$$\frac{u}{u-1} + \frac{1}{u} = \frac{u^2+1}{u^2-u}$$
$$\frac{u}{u-1} + \frac{1}{u} = \frac{u^2+1}{u(u-1)}$$
$$u(u-1)\left(\frac{u}{u-1} + \frac{1}{u}\right) = u(u-1)\left(\frac{u^2+1}{u(u-1)}\right)$$
$$u^2 + u - 1 = u^2 + 1$$
$$u - 1 = 1$$
$$u = 2 \Rightarrow \text{The answer checks.}$$

53.
$$\frac{3}{x-2} + \frac{1}{x} = \frac{2(3x+2)}{x^2-2x}$$
$$\frac{3}{x-2} + \frac{1}{x} = \frac{6x+4}{x(x-2)}$$
$$x(x-2)\left(\frac{3}{x-2} + \frac{1}{x}\right) = x(x-2)\left(\frac{6x+4}{x(x-2)}\right)$$
$$3x + x - 2 = 6x + 4$$
$$4x - 2 = 6x + 4$$
$$-2x = 6$$
$$x = -3 \Rightarrow \text{The answer checks.}$$

55.
$$\frac{7}{q^2 - q - 2} + \frac{1}{q + 1} = \frac{3}{q - 2}$$

$$\frac{7}{(q - 2)(q + 1)} + \frac{1}{q + 1} = \frac{3}{q - 2}$$

$$(q - 2)(q + 1)\left(\frac{7}{(q - 2)(q + 1)} + \frac{1}{q + 1}\right) = (q - 2)(q + 1)\left(\frac{3}{q - 2}\right)$$

$$7 + q - 2 = 3(q + 1)$$

$$q + 5 = 3q + 3$$

$$-2q = -2$$

$$q = 1 \Rightarrow \text{The answer checks.}$$

57.
$$\frac{3y}{3y - 6} + \frac{8}{y^2 - 4} = \frac{2y}{2y + 4}$$

$$\frac{3y}{3(y - 2)} + \frac{8}{(y + 2)(y - 2)} = \frac{2y}{2(y + 2)}$$

$$\frac{y}{y - 2} + \frac{8}{(y + 2)(y - 2)} = \frac{y}{y + 2}$$

$$(y + 2)(y - 2)\left(\frac{y}{y - 2} + \frac{8}{(y + 2)(y - 2)}\right) = (y + 2)(y - 2)\left(\frac{y}{y + 2}\right)$$

$$y(y + 2) + 8 = y(y - 2)$$

$$y^2 + 2y + 8 = y^2 - 2y$$

$$8 = -4y$$

$$-2 = y: \quad -2 \text{ does not check} \Rightarrow \text{no solutions}$$

59.
$$y + \frac{2}{3} = \frac{2y - 12}{3y - 9}$$

$$y + \frac{2}{3} = \frac{2(y - 6)}{3(y - 3)}$$

$$3(y - 3)\left(y + \frac{2}{3}\right) = 3(y - 3)\left(\frac{2y - 12}{3(y - 3)}\right)1$$

$$3y(y - 3) + 2(y - 3) = 2y - 12$$

$$3y^2 - 9y + 2y - 6 = 2y - 12$$

$$3y^2 - 9y + 6 = 0$$

$$3(y^2 - 3y + 2) = 0$$

$$3(y - 2)(y - 1) = 0$$

$$y - 2 = 0 \quad \textbf{or} \quad y - 1 = 0$$

$$y = 2 \qquad\qquad y = 1$$

Both answers check.

61.

$$\frac{5}{4y+12} - \frac{3}{4} = \frac{5}{4y+12} - \frac{y}{4}$$

$$\frac{5}{4(y+3)} - \frac{3}{4} = \frac{5}{4(y+3)} - \frac{y}{4}$$

$$4(y+3)\left(\frac{5}{4(y+3)} - \frac{3}{4}\right) = 4(y+3)\left(\frac{5}{4(y+3)} - \frac{y}{4}\right)$$

$$5 - 3(y+3) = 5 - y(y+3)$$

$$5 - 3y - 9 = 5 - y^2 - 3y$$

$$y^2 - 9 = 0$$

$$(y+3)(y-3) = 0$$

$$y + 3 = 0 \quad \textbf{or} \quad y - 3 = 0 \quad \Rightarrow y = -3 \text{ does not check, so the only solution is } y = 3.$$
$$y = -3 \qquad\qquad y = 3$$

63.

$$\frac{x}{x-1} - \frac{12}{x^2 - x} = \frac{-1}{x-1}$$

$$\frac{x}{x-1} - \frac{12}{x(x-1)} = \frac{-1}{x-1}$$

$$x(x-1)\left(\frac{x}{x-1} - \frac{12}{x(x-1)}\right) = x(x-1)\left(\frac{-1}{x-1}\right)$$

$$x^2 - 12 = -x$$

$$x^2 + x - 12 = 0$$

$$(x+4)(x-3) = 0$$

$$x + 4 = 0 \quad \textbf{or} \quad x - 3 = 0 \quad \Rightarrow \text{ Both answers check.}$$
$$x = -4 \qquad\qquad x = 3$$

65.

$$\frac{z-4}{z-3} = \frac{z+2}{z+1}$$

$$(z-3)(z+1)\left(\frac{z-4}{z-3}\right) = (z-3)(z+1)\left(\frac{z+2}{z+1}\right)$$

$$(z+1)(z-4) = (z-3)(z+2)$$

$$z^2 - 3z - 4 = z^2 - z - 6$$

$$-2z = -2$$

$$z = 1 \Rightarrow \text{ The answer checks.}$$

67.

$$\frac{n}{n^2 - 9} + \frac{n+8}{n+3} = \frac{n-8}{n-3}$$

$$\frac{n}{(n+3)(n-3)} + \frac{n+8}{n+3} = \frac{n-8}{n-3}$$

$$(n+3)(n-3)\left(\frac{n}{(n+3)(n-3)} + \frac{n+8}{n+3}\right) = (n+3)(n-3)\left(\frac{n-8}{n-3}\right)$$

$$n + (n-3)(n+8) = (n+3)(n-8)$$

$$n + n^2 + 5n - 24 = n^2 - 5n - 24$$

$$11n = 0$$

$$n = 0 \Rightarrow \text{The answer checks.}$$

69.

$$\frac{b+2}{b+3} + 1 = \frac{-7}{b-5}$$

$$(b+3)(b-5)\left(\frac{b+2}{b+3} + 1\right) = (b+3)(b-5)\left(\frac{-7}{b-5}\right)$$

$$(b-5)(b+2) + (b+3)(b-5) = -7(b+3)$$

$$b^2 - 3b - 10 + b^2 - 2b - 15 = -7b - 21$$

$$2b^2 + 2b - 4 = 0$$

$$2(b^2 + b - 2) = 0$$

$$2(b+2)(b-1) = 0$$

$$b + 2 = 0 \quad \textbf{or} \quad b - 1 = 0 \Rightarrow \text{Both answers check.}$$

$$b = -2 \qquad\qquad b = 1$$

71.

$$\frac{1}{a} + \frac{1}{b} = 1$$

$$ab\left(\frac{1}{a} + \frac{1}{b}\right) = ab(1)$$

$$b + a = ab$$

$$b = ab - a$$

$$b = a(b-1)$$

$$\frac{b}{b-1} = a$$

73.

$$\frac{1}{f} = \frac{1}{d_1} + \frac{1}{d_2}$$

$$fd_1d_2 \cdot \frac{1}{f} = fd_1d_2\left(\frac{1}{d_1} + \frac{1}{d_2}\right)$$

$$d_1d_2 = fd_2 + fd_1$$

$$d_1d_2 = f(d_2 + d_1)$$

$$\frac{d_1d_2}{d_2 + d_1} = f$$

75. Answers may vary.

77.
$$x = \frac{1}{x}$$
$$x(x) = x\left(\frac{1}{x}\right)$$
$$x^2 = 1$$
$$x^2 - 1 = 0$$
$$(x + 1)(x - 1) = 0$$
$$x + 1 = 0 \quad \text{or} \quad x - 1 = 0$$
$$x = -1 \qquad\qquad x = 1$$

The numbers 1 and -1 are equal to their own reciprocals.

Exercise 6.8 (page 417)

1.
$$x^2 - 5x - 6 = 0$$
$$(x - 6)(x + 1) = 0$$
$$x - 6 = 0 \quad \text{or} \quad x + 1 = 0$$
$$x = 6 \qquad\qquad x = -1$$

3.
$$(t + 2)\left(t^2 + 7t + 12\right) = 0$$
$$(t + 2)(t + 4)(t + 3) = 0$$
$$t + 2 = 0 \quad \text{or} \quad t + 4 = 0 \quad \text{or} \quad t + 3 = 0$$
$$t = -2 \qquad\qquad t = -4 \qquad\qquad t = -3$$

5.
$$y^3 - y^2 = 0$$
$$y^2(y - 1) = 0$$
$$y = 0 \quad \text{or} \quad y = 0 \quad \text{or} \quad y - 1 = 0$$
$$y = 0 \qquad\quad y = 0 \qquad\qquad y = 1$$

7.
$$\left(x^2 - 1\right)\left(x^2 - 4\right) = 0$$
$$(x + 1)(x - 1)(x + 2)(x - 2) = 0$$
$$x + 1 = 0 \quad \text{or} \quad x - 1 = 0 \quad \text{or} \quad x + 2 = 0 \quad \text{or} \quad x - 2 = 0$$
$$x = -1 \qquad\qquad x = 1 \qquad\qquad x = -2 \qquad\qquad x = 2$$

9. Answers may vary.

11. Let $x =$ the number.
$$\frac{2(3)}{4 + x} = 1$$
$$(4 + x)\left(\frac{6}{4 + x}\right) = (4 + x)(1)$$
$$6 = 4 + x$$
$$2 = x \Rightarrow \text{The number is 2.}$$

13. Let x = the number.

$$\frac{3+x}{4+2x} = \frac{4}{7}$$

$$7(4+2x)\left(\frac{3+x}{4+2x}\right) = 7(4+2x)\left(\frac{4}{7}\right)$$

$$7(x+3) = 4(4+2x)$$

$$7x+21 = 16+8x$$

$$5 = x$$

The number is 5.

15. Let x = the number. $\Rightarrow x + \dfrac{1}{x} = \dfrac{13}{6}$

$$6x\left(x + \frac{1}{x}\right) = 6x\left(\frac{13}{6}\right)$$

$$6x^2 + 6 = 13x$$

$$6x^2 - 13x + 6 = 0$$

$$(2x-3)(3x-2) = 0$$

$$2x - 3 = 0 \quad \textbf{or} \quad 3x - 2 = 0$$

$$2x = 3 \qquad\qquad 3x = 2$$

$$x = \tfrac{3}{2} \qquad\qquad x = \tfrac{2}{3}$$

The numbers are $\frac{3}{2}$ and $\frac{2}{3}$.

17. Let x = hours for both pipes to fill the pool.

$$\boxed{\begin{array}{c}\text{1st pipe}\\\text{in 1 hour}\end{array}} + \boxed{\begin{array}{c}\text{2nd pipe}\\\text{in 1 hour}\end{array}} = \boxed{\begin{array}{c}\text{Total in}\\\text{1 hour}\end{array}}$$

$$\frac{1}{5} + \frac{1}{4} = \frac{1}{x}$$

$$20x\left(\frac{1}{5} + \frac{1}{4}\right) = 20x\left(\frac{1}{x}\right)$$

$$4x + 5x = 20$$

$$9x = 20$$

$$x = \frac{20}{9}$$

The pool can be filled in $2\frac{2}{9}$ hours.

19. Let x = days for both working together.

$$\boxed{\begin{array}{c}\text{Roofer}\\\text{in 1 day}\end{array}} + \boxed{\begin{array}{c}\text{Owner}\\\text{in 1 day}\end{array}} = \boxed{\begin{array}{c}\text{Total in}\\\text{1 day}\end{array}}$$

$$\frac{1}{4} + \frac{1}{7} = \frac{1}{x}$$

$$28x\left(\frac{1}{4} + \frac{1}{7}\right) = 28x\left(\frac{1}{x}\right)$$

$$7x + 4x = 28$$

$$11x = 28$$

$$x = \frac{28}{11}$$

The pool can be filled in $2\frac{6}{11}$ days.

21. Let r = the rate at which he walks.

$$\boxed{\begin{array}{c}\text{Time he}\\\text{walks}\end{array}} = \boxed{\begin{array}{c}\text{Time he}\\\text{rides}\end{array}}$$

$$\frac{8}{r} = \frac{28}{r+10}$$

$$r(r+10)\left(\frac{8}{r}\right) = r(r+10)\left(\frac{28}{r+10}\right)$$

$$8(r+10) = 28r$$

$$8r + 80 = 28r$$

$$80 = 20r$$

$$4 = r$$

	d	r	t
Walks	8	r	$\dfrac{8}{r}$
Rides	28	$r+10$	$\dfrac{28}{r+10}$

He walks 4 miles per hour, so it will take him $\frac{30}{4} = 7\frac{1}{2}$ hours to walk 30 miles.

23. Let $r =$ the speed of the current.

$$\boxed{\text{Time downstream}} = \boxed{\text{Time upstream}}$$

$$\frac{22}{18+r} = \frac{14}{18-r}$$

	d	r	t
Downstream	22	$18+r$	$\dfrac{22}{18+r}$
Upstream	14	$18-r$	$\dfrac{14}{18-r}$

$$(18+r)(18-r)\left(\frac{22}{18+r}\right) = (18+r)(18-r)\left(\frac{14}{18-r}\right)$$

$$22(18-r) = 14(18+r)$$

$$396 - 22r = 252 + 14r$$

$$144 = 36r$$

$$4 = r$$

The current has a speed of 4 miles per hour.

25. Let $r =$ the lower interest rate.

	I	P	r
Lower rate CD	175	$\dfrac{175}{r}$	r
Higher rate CD	200	$\dfrac{200}{r+.01}$	$r+.01$

$$\boxed{\text{Lower rate principal}} = \boxed{\text{Higher rate principal}}$$

$$\frac{175}{r} = \frac{200}{r+.01}$$

$$r(r+.01)\left(\frac{175}{r}\right) = r(r+.01)\left(\frac{200}{r+.01}\right)$$

$$175(r+.01) = 200r$$

$$175r + 1.75 = 200r$$

$$1.75 = 25r$$

$$0.07 = r$$

The rates are 7% and 8%.

27. Let $x =$ the number who contributed.

$$\boxed{\text{Original share}} = \boxed{\text{Share with more workers}} + 2$$

$$\frac{35}{x} = \frac{35}{x+2} + 2$$

$$x(x+2)\left(\frac{35}{x}\right) = x(x+2)\left(\frac{35}{x+2} + 2\right)$$

$$35(x+2) = 35x + 2x(x+2)$$

$$35x + 70 = 35x + 2x^2 + 4x$$

$$0 = 2x^2 + 4x - 70$$

$$0 = 2\left(x^2 + 2x - 35\right)$$

$$0 = 2(x+7)(x-5)$$

$$x + 7 = 0 \quad \textbf{or} \quad x - 5 = 0$$

$$x = -7 \qquad\qquad x = 5$$

Since the answer cannot be negative, 5 workers must have contributed.

29. Let $x =$ the number bought.

Then each cost $\frac{120}{x}$.

$$\boxed{\text{Number bought}} \cdot \boxed{\begin{array}{c}\text{Amount charged}\\\text{for each}\end{array}} = 120$$

$$(x+10) \cdot \left(\frac{120}{x} - 1\right) = 120$$

$$120 - x + \frac{1200}{x} - 10 = 120$$

$$\frac{1200}{x} - x - 10 = 0$$

$$\frac{1200}{x} - x - 10 = 0$$

$$x\left(\frac{1200}{x} - x - 10\right) = x(0)$$

$$-x^2 - 10x + 1200 = 0$$

$$-\left(x^2 + 10x - 1200\right) = 0$$

$$-(x+40)(x-30) = 0$$

$$x + 40 = 0 \quad \textbf{or} \quad x - 30 = 0$$

$$x = -40 \qquad\qquad x = 30$$

The store can buy 30 at the regular price.

31. Let $r =$ the still-water speed.

$$\boxed{\begin{array}{c}\text{Time}\\\text{upstream}\end{array}} + \boxed{\begin{array}{c}\text{Time}\\\text{downstream}\end{array}} = 5$$

$$\frac{60}{r-5} + \frac{60}{r+5} = 5$$

$$(r+5)(r-5)\left(\frac{60}{r-5} + \frac{60}{r+5}\right) = 5(r+5)(r-5)$$

$$60(r+5) + 60(r-5) = 5(r^2 - 25)$$

$$60r + 300 + 60r - 300 = 5r^2 - 125$$

$$0 = 5r^2 - 120r - 125$$

$$0 = 5(r^2 - 24r - 25)$$

$$0 = 5(r+1)(r-25)$$

$$r + 1 = 0 \quad \text{or} \quad r - 25 = 0$$

$$r = -1 \qquad\qquad r = 25$$

The still-water speed should be 25 miles per hour.

	d	r	t
Upstream	60	$r - 5$	$\dfrac{60}{r-5}$
Downstream	60	$r + 5$	$\dfrac{60}{r+5}$

33. Answers may vary.

35. Answers may vary.

Chapter 6 Summary (page 420)

1. $\dfrac{3}{6} = \dfrac{1}{2}$

2. $\dfrac{12x}{15x} = \dfrac{4}{5}$

3. $\dfrac{2 \text{ ft}}{1 \text{ yd}} = \dfrac{2 \text{ ft}}{3 \text{ ft}} = \dfrac{2}{3}$

4. $\dfrac{5 \text{ pt}}{3 \text{ qt}} = \dfrac{5 \text{ pt}}{6 \text{ pt}} = \dfrac{5}{6}$

5. $\dfrac{\$8.79}{3 \text{ lb}} = \$2.93/\text{lb}$

6. $\dfrac{2275 \text{ kwh}}{4 \text{ weeks}} = 568.75 \text{ kwh/week}$

7. $\dfrac{4}{7} \stackrel{?}{=} \dfrac{20}{34}$

$4 \cdot 34 \stackrel{?}{=} 7 \cdot 20$

$136 \neq 140$

not a proportion

8. $\dfrac{5}{7} \stackrel{?}{=} \dfrac{30}{42}$

$5 \cdot 42 \stackrel{?}{=} 7 \cdot 30$

$210 = 210$

proportion

9. $\dfrac{3}{x} = \dfrac{6}{9}$

$3(9) = 6x$

$27 = 6x$

$\dfrac{27}{6} = x$

$\dfrac{9}{2} = x$

10. $\dfrac{x}{3} = \dfrac{x}{5}$

$5x = 3x$

$2x = 0$

$x = 0$

11. $\dfrac{x-2}{5} = \dfrac{x}{7}$

$7(x-2) = 5x$

$7x - 14 = 5x$

$2x = 14$

$x = 7$

12. $\dfrac{4x-1}{18} = \dfrac{x}{6}$

$6(4x-1) = 18x$

$24x - 6 = 18x$

$6x = 6$

$x = 1$

13. Let h = the height of the pole.

$$\frac{\text{height of pole}}{\text{shadow of pole}} = \frac{\text{height of man}}{\text{shadow of man}}$$

$$\frac{h}{12} = \frac{6}{3.6}$$

$$3.6h = 12(6)$$

$$3.6h = 72$$

$$h = \frac{72}{3.6} = 20 \text{ ft}$$

14. The expression is undefined when the denominator equals 0:

$$(x+3)(x-3) = 0$$

$$x+3=0 \quad \text{or} \quad x-3=0$$

$$x=-3 \qquad\qquad x=3$$

15. $\dfrac{10}{25} = \dfrac{5 \cdot 2}{5 \cdot 5} = \dfrac{2}{5}$

16. $\dfrac{-12}{18} = -\dfrac{6 \cdot 2}{6 \cdot 3} = -\dfrac{2}{3}$

17. $\dfrac{-51}{153} = -\dfrac{51 \cdot 1}{51 \cdot 3} = -\dfrac{1}{3}$

18. $\dfrac{105}{45} = \dfrac{15 \cdot 7}{15 \cdot 3} = \dfrac{7}{3}$

19. $\dfrac{3x^2}{6x^3} = \dfrac{1}{2x}$

20. $\dfrac{5xy^2}{2x^2y^2} = \dfrac{5}{2x}$

21. $\dfrac{x^2}{x^2+x} = \dfrac{x^2}{x(x+1)} = \dfrac{x}{x+1}$

22. $\dfrac{x+2}{x^2+2x} = \dfrac{x+2}{x(x+2)} = \dfrac{1}{x}$

23. $\dfrac{6xy}{3xy} = 2$

24. $\dfrac{8x^2y}{2x(4xy)} = \dfrac{8x^2y}{8x^2y} = 1$

25. $\dfrac{3p-2}{2-3p} = -1$

26. $\dfrac{x^2-x-56}{x^2-5x-24} = \dfrac{(x-8)(x+7)}{(x-8)(x+3)} = \dfrac{x+7}{x+3}$

27. $\dfrac{2x^2-16x}{2x^2-18x+16} = \dfrac{2x(x-8)}{2(x^2-9x+8)} = \dfrac{2x(x-8)}{2(x-8)(x-1)} = \dfrac{x}{x-1}$

28. $\dfrac{a^2+2a+ab+2b}{a^2+2ab+b^2} = \dfrac{a(a+2)+b(a+2)}{(a+b)(a+b)} = \dfrac{(a+2)(a+b)}{(a+b)(a+b)} = \dfrac{a+2}{a+b}$

29. $\dfrac{3xy}{2x} \cdot \dfrac{4x}{2y^2} = \dfrac{12x^2y}{4xy^2} = \dfrac{3x}{y}$

30. $\dfrac{3x}{x^2-x} \cdot \dfrac{2x-2}{x^2} = \dfrac{3x(2x-2)}{(x^2-x)x^2} = \dfrac{6x(x-1)}{x^3(x-1)}$
$$= \dfrac{6}{x^2}$$

31. $\dfrac{x^2+3x+2}{x^2+2x} \cdot \dfrac{x}{x+1} = \dfrac{(x^2+3x+2)x}{(x^2+2x)(x+1)} = \dfrac{x(x+2)(x+1)}{x(x+2)(x+1)} = 1$

32. $\dfrac{x^2+x}{3x-15} \cdot \dfrac{6x-30}{x^2+2x+1} = \dfrac{(x^2+x)(6x-30)}{(3x-15)(x^2+2x+1)} = \dfrac{x(x+1) \cdot 6(x-5)}{3(x-5) \cdot (x+1)(x+1)} = \dfrac{2x}{x+1}$

33. $\dfrac{3x^2}{5x^2y} \div \dfrac{6x}{15xy^2} = \dfrac{3x^2}{5x^2y} \cdot \dfrac{15xy^2}{6x} = \dfrac{45x^3y^2}{30x^3y} = \dfrac{3y}{2}$

34. $\dfrac{x^2+5x}{x^2+4x-5} \div \dfrac{x^2}{x-1} = \dfrac{x^2+5x}{x^2+4x-5} \cdot \dfrac{x-1}{x^2} = \dfrac{x(x+5)(x-1)}{x^2(x+5)(x-1)} = \dfrac{1}{x}$

35. $\dfrac{x^2-x-6}{2x-1} \div \dfrac{x^2-2x-3}{2x^2+x-1} = \dfrac{x^2-x-6}{2x-1} \cdot \dfrac{2x^2+x-1}{x^2-2x-3} = \dfrac{(x-3)(x+2)(2x-1)(x+1)}{(2x-1)(x-3)(x+1)}$

$$= x+2$$

36. $\dfrac{x^2-3x}{x^2-x-6} \div \dfrac{x^2-x}{x^2+x-2} = \dfrac{x^2-3x}{x^2-x-6} \cdot \dfrac{x^2+x-2}{x^2-x} = \dfrac{x(x-3)\cdot(x+2)(x-1)}{(x+2)(x-3)\cdot x(x-1)} = 1$

37. $\dfrac{x^2+4x+4}{x^2+x-6}\left(\dfrac{x-2}{x-1} \div \dfrac{x+2}{x^2+2x-3}\right) = \dfrac{x^2+4x+4}{x^2+x-6}\left(\dfrac{x-2}{x-1} \cdot \dfrac{x^2+2x-3}{x+2}\right)$

$$= \dfrac{(x+2)(x+2)}{(x+3)(x-2)}\left(\dfrac{(x-2)(x+3)(x-1)}{(x-1)(x+2)}\right)$$

$$= \dfrac{(x+2)(x+2)(x-2)(x+3)(x-1)}{(x+3)(x-2)(x-1)(x+2)} = x+2$$

38. $\dfrac{x}{x+y} + \dfrac{y}{x+y} = \dfrac{x+y}{x+y} = 1$

39. $\dfrac{3x}{x-7} - \dfrac{x-2}{x-7} = \dfrac{3x-x+2}{x-7} = \dfrac{2x+2}{x-7}$

40. $\dfrac{x}{x-1} + \dfrac{1}{x} = \dfrac{(x)(x)}{(x-1)(x)} + \dfrac{1(x-1)}{x(x-1)} = \dfrac{x^2}{x(x-1)} + \dfrac{x-1}{x(x-1)} = \dfrac{x^2+x-1}{x(x-1)}$

41. $\dfrac{1}{7} - \dfrac{1}{x} = \dfrac{1(x)}{7(x)} - \dfrac{1(7)}{x(7)} = \dfrac{x}{7x} - \dfrac{7}{7x} = \dfrac{x-7}{7x}$

42. $\dfrac{3}{x+1} - \dfrac{2}{x} = \dfrac{3(x)}{(x+1)(x)} - \dfrac{2(x+1)}{x(x+1)} = \dfrac{3x}{x(x+1)} - \dfrac{2x+2}{x(x+1)} = \dfrac{3x-2x-2}{x(x+1)} = \dfrac{x-2}{x(x+1)}$

43. $\dfrac{x+2}{2x} - \dfrac{2-x}{x^2} = \dfrac{(x+2)x}{2x(x)} - \dfrac{(2-x)2}{x^2(2)} = \dfrac{x^2+2x}{2x^2} - \dfrac{4-2x}{2x^2} = \dfrac{x^2+2x-4+2x}{2x^2}$

$$= \dfrac{x^2+4x-4}{2x^2}$$

44. $\dfrac{x}{x+2} + \dfrac{3}{x} - \dfrac{4}{x^2+2x} = \dfrac{x}{x+2} + \dfrac{3}{x} - \dfrac{4}{x(x+2)} = \dfrac{x(x)}{x(x+2)} + \dfrac{3(x+2)}{x(x+2)} - \dfrac{4}{x(x+2)}$

$$= \dfrac{x^2}{x(x+2)} + \dfrac{3x+6}{x(x+2)} - \dfrac{4}{x(x+2)}$$

$$= \dfrac{x^2+3x+2}{x(x+2)} = \dfrac{(x+2)(x+1)}{x(x+2)} = \dfrac{x+1}{x}$$

45. $\dfrac{2}{x-1} - \dfrac{3}{x+1} + \dfrac{x-5}{x^2-1} = \dfrac{2}{x-1} - \dfrac{3}{x+1} + \dfrac{x-5}{(x+1)(x-1)}$

$$= \dfrac{2(x+1)}{(x-1)(x+1)} - \dfrac{3(x-1)}{(x+1)(x-1)} + \dfrac{x-5}{(x+1)(x-1)}$$

$$= \dfrac{2(x+1) - 3(x-1) + x - 5}{(x+1)(x-1)}$$

$$= \dfrac{2x + 2 - 3x + 3 + x - 5}{(x+1)(x-1)} = \dfrac{0}{(x+1)(x-1)} = 0$$

46. $\dfrac{\frac{3}{2}}{\frac{2}{3}} = \dfrac{3}{2} \div \dfrac{2}{3} = \dfrac{3}{2} \cdot \dfrac{3}{2} = \dfrac{9}{4}$

47. $\dfrac{\frac{3}{2}+1}{\frac{2}{3}+1} = \dfrac{\left(\frac{3}{2}+1\right)6}{\left(\frac{2}{3}+1\right)6} = \dfrac{\frac{3}{2}\cdot 6 + 1 \cdot 6}{\frac{2}{3}\cdot 6 + 1 \cdot 6} = \dfrac{9+6}{4+6} = \dfrac{15}{10} = \dfrac{3}{2}$

48. $\dfrac{\frac{1}{x}+1}{\frac{1}{x}-1} = \dfrac{\left(\frac{1}{x}+1\right)x}{\left(\frac{1}{x}-1\right)x} = \dfrac{\frac{1}{x}\cdot x + 1 \cdot x}{\frac{1}{x}\cdot x - 1 \cdot x}$

$$= \dfrac{1+x}{1-x}$$

49. $\dfrac{1+\frac{3}{x}}{2-\frac{1}{x^2}} = \dfrac{\left(1+\frac{3}{x}\right)x^2}{\left(2-\frac{1}{x^2}\right)x^2} = \dfrac{x^2+3x}{2x^2-1}$

50. $\dfrac{\frac{2}{x-1}+\frac{x-1}{x+1}}{\frac{1}{x^2-1}} = \dfrac{\frac{2}{x-1}+\frac{x-1}{x+1}}{\frac{1}{(x+1)(x-1)}} = \dfrac{\left(\frac{2}{x-1}+\frac{x-1}{x+1}\right)(x+1)(x-1)}{\left(\frac{1}{(x+1)(x-1)}\right)(x+1)(x-1)} = \dfrac{2(x+1)+(x-1)(x-1)}{1}$

$$= 2x + 2 + x^2 - 2x + 1 = x^2 + 3$$

51. $\dfrac{\frac{a}{b}+c}{\frac{b}{a}+c} = \dfrac{\left(\frac{a}{b}+c\right)ab}{\left(\frac{b}{a}+c\right)ab} = \dfrac{a^2+abc}{b^2+abc} = \dfrac{a(a+bc)}{b(b+ac)}$

52. $$\dfrac{3}{x} = \dfrac{2}{x-1}$$

$$x(x-1)\left(\dfrac{3}{x}\right) = x(x-1)\left(\dfrac{2}{x-1}\right)$$

$$3(x-1) = 2x$$

$$3x - 3 = 2x$$

$$x = 3 \Rightarrow \text{The answer checks.}$$

53. $$\dfrac{5}{x+4} = \dfrac{3}{x+2}$$

$$(x+4)(x+2)\left(\dfrac{5}{x+4}\right) = (x+4)(x+2)\left(\dfrac{3}{x+2}\right)$$

$$5(x+2) = 3(x+4)$$

$$5x + 10 = 3x + 12$$

$$2x = 2$$

$$x = 1 \Rightarrow \text{The answer checks.}$$

54.
$$\frac{2}{3x} + \frac{1}{x} = \frac{5}{9}$$

$$9x\left(\frac{2}{3x} + \frac{1}{x}\right) = 9x\left(\frac{5}{9}\right)$$

$$6 + 9 = 5x$$

$$15 = 5x$$

$$3 = x \Rightarrow \text{The answer checks.}$$

55.
$$\frac{2x}{x+4} = \frac{3}{x-1}$$

$$(x+4)(x-1)\left(\frac{2x}{x+4}\right) = (x+4)(x-1)\left(\frac{3}{x-1}\right)$$

$$2x(x-1) = 3(x+4)$$

$$2x^2 - 2x = 3x + 12$$

$$2x^2 - 5x - 12 = 0$$

$$(2x+3)(x-4) = 0$$

$2x + 3 = 0 \quad$ **or** $\quad x - 4 = 0 \quad \Rightarrow \text{Both answers check.}$

$\quad 2x = -3 \qquad\qquad x = 4$

$\quad\; x = -\frac{3}{2} \qquad\qquad x = 4$

56.
$$\frac{2}{x-1} + \frac{3}{x+4} = \frac{-5}{x^2+3x-4}$$

$$\frac{2}{x-1} + \frac{3}{x+4} = \frac{-5}{(x+4)(x-1)}$$

$$(x-1)(x+4)\left(\frac{2}{x-1} + \frac{3}{x+4}\right) = (x-1)(x+4)\left(\frac{-5}{(x+4)(x-1)}\right)$$

$$2(x+4) + 3(x-1) = -5$$

$$2x + 8 + 3x - 3 = -5$$

$$5x = -10$$

$$x = -2 \Rightarrow \text{The answer checks.}$$

57.
$$\frac{4}{x+2} - \frac{3}{x+3} = \frac{6}{x^2+5x+6}$$

$$\frac{4}{x+2} - \frac{3}{x+3} = \frac{6}{(x+2)(x+3)}$$

$$(x+2)(x+3)\left(\frac{4}{x+2} - \frac{3}{x+3}\right) = (x+2)(x+3)\left(\frac{6}{(x+2)(x+3)}\right)$$

$$4(x+3) - 3(x+2) = 6$$

$$4x + 12 - 3x - 6 = 6$$

$$x = 0 \Rightarrow \text{The answer checks.}$$

58.

$$\frac{1}{r} = \frac{1}{r_1} + \frac{1}{r_2}$$

$$rr_1r_2 \cdot \frac{1}{r} = rr_1r_2\left(\frac{1}{r_1} + \frac{1}{r_2}\right)$$

$$r_1r_2 = rr_2 + rr_1$$

$$r_1r_2 - rr_1 = rr_2$$

$$r_1(r_2 - r) = rr_2$$

$$r_1 = \frac{rr_2}{r_2 - r}$$

59.

$$E = 1 - \frac{T_2}{T_1}$$

$$T_1E = T_1\left(1 - \frac{T_2}{T_1}\right)$$

$$T_1E = T_1 - T_2$$

$$T_1E - T_1 = -T_2$$

$$T_1(E - 1) = -T_2$$

$$T_1 = -\frac{T_2}{E - 1} = \frac{T_2}{1 - E}$$

60.

$$H = \frac{RB}{R + B}$$

$$(R + B)H = (R + B)\left(\frac{RB}{R + B}\right)$$

$$RH + BH = RB$$

$$BH = RB - RH$$

$$BH = R(B - H)$$

$$\frac{BH}{B - H} = R$$

61. Let x = hours for both pipes to empty.

1st pipe in 1 hour		2nd pipe in 1 hour		Total in 1 hour
	+		=	

$$\frac{1}{18} + \frac{1}{20} = \frac{1}{x}$$

$$180x\left(\frac{1}{18} + \frac{1}{20}\right) = 180x\left(\frac{1}{x}\right)$$

$$10x + 9x = 180$$

$$19x = 180$$

$$x = \frac{180}{19}$$

It can be emptied in $9\frac{9}{19}$ hours.

62. Let x = days for both working together.

Painter in 1 day		Owner in 1 day		Total in 1 day
	+		=	

$$\frac{1}{10} + \frac{1}{14} = \frac{1}{x}$$

$$70x\left(\frac{1}{10} + \frac{1}{14}\right) = 70x\left(\frac{1}{x}\right)$$

$$7x + 5x = 70$$

$$12x = 70$$

$$x = \frac{70}{12} = \frac{35}{6}$$

It can be painted in $5\frac{5}{6}$ days.

63. Let $r =$ the rate at which he jogs.

	d	r	t
Jogs	10	r	$\dfrac{10}{r}$
Rides	30	$r + 10$	$\dfrac{30}{r + 10}$

$$\boxed{\begin{array}{c}\text{Time he}\\\text{jogs}\end{array}} = \boxed{\begin{array}{c}\text{Time he}\\\text{rides}\end{array}}$$

$$\frac{10}{r} = \frac{30}{r + 10}$$

$$r(r+10)\left(\frac{10}{r}\right) = r(r+10)\left(\frac{30}{r+10}\right)$$

$$10(r + 10) = 30r$$

$$10r + 100 = 30r$$

$$100 = 20r$$

$$5 = r$$

He jogs 5 miles per hour.

64. Let $r =$ the speed of the wind.

	d	r	t
Downwind	400	$360 + r$	$\dfrac{400}{360 + r}$
Upwind	320	$360 - r$	$\dfrac{320}{360 - r}$

$$\boxed{\begin{array}{c}\text{Time}\\\text{downwind}\end{array}} = \boxed{\begin{array}{c}\text{Time}\\\text{upwind}\end{array}}$$

$$\frac{400}{360 + r} = \frac{320}{360 - r}$$

$$(360 + r)(360 - r)\left(\frac{400}{360 + r}\right) = (360 + r)(360 - r)\left(\frac{320}{360 - r}\right)$$

$$400(360 - r) = 320(360 + r)$$

$$144000 - 400r = 115200 + 320r$$

$$28800 = 720r$$

$$40 = r \Rightarrow \text{The wind has a speed of 40 miles per hour.}$$

Chapter 6 Test   (page 424)

1.
$$\frac{6 \text{ ft}}{3 \text{ yd}} = \frac{6 \text{ ft}}{9 \text{ ft}} = \frac{2}{3}$$

2.
$$\frac{3xy}{5xy} \stackrel{?}{=} \frac{3xt}{5xt}$$

$$3xy \cdot 5xt \stackrel{?}{=} 5xy \cdot 3xt$$

$$15x^2yt = 15x^2yt \Rightarrow \text{proportion}$$

3.
$$\frac{y}{y - 1} = \frac{y - 2}{y}$$

$$y(y) = (y - 1)(y - 2)$$

$$y^2 = y^2 - 3y + 2$$

$$3y = 2$$

$$y = \frac{2}{3}$$

4. Let $h =$ the height of the tree.

$$\frac{\text{height of tree}}{\text{shadow of tree}} = \frac{\text{height of man}}{\text{shadow of man}}$$

$$\frac{h}{30} = \frac{6}{4}$$

$$4h = 30(6)$$

$$4h = 180$$

$$h = 45 \text{ ft}$$

5. $\dfrac{48x^2y}{54xy^2} = \dfrac{6 \cdot 8x^2y}{6 \cdot 9xy^2} = \dfrac{8x}{9y}$

6. $\dfrac{2x^2 - x - 3}{4x^2 - 9} = \dfrac{(2x-3)(x+1)}{(2x+3)(2x-3)} = \dfrac{x+1}{2x+3}$

7. $\dfrac{3(x+2) - 3}{2x - 4 - (x-5)} = \dfrac{3x + 6 - 3}{2x - 4 - x + 5}$

$= \dfrac{3x+3}{x+1} = \dfrac{3(x+1)}{x+1} = 3$

8. $\dfrac{12x^2y}{15xyz} \cdot \dfrac{25y^2z}{16xt} = \dfrac{3 \cdot 4 \cdot 5 \cdot 5x^2y^3z}{3 \cdot 5 \cdot 4 \cdot 4x^2ytz} = \dfrac{5y^2}{4t}$

9. $\dfrac{x^2 + 3x + 2}{3x + 9} \cdot \dfrac{x+3}{x^2 - 4} = \dfrac{(x+2)(x+1)}{3(x+3)} \cdot \dfrac{x+3}{(x+2)(x-2)} = \dfrac{(x+2)(x+1)(x+3)}{3(x+3)(x+2)(x-2)} = \dfrac{x+1}{3(x-2)}$

10. $\dfrac{8x^2y}{25xt} \div \dfrac{16x^2y^3}{30xyt^3} = \dfrac{8x^2y}{25xt} \cdot \dfrac{30xyt^3}{16x^2y^3} = \dfrac{4 \cdot 2 \cdot 5 \cdot 2 \cdot 3x^3y^2t^3}{5 \cdot 5 \cdot 4 \cdot 4x^3y^3t} = \dfrac{3t^2}{5y}$

11. $\dfrac{x^2 - x}{3x^2 + 6x} \div \dfrac{3x - 3}{3x^3 + 6x^2} = \dfrac{x^2 - x}{3x^2 + 6x} \cdot \dfrac{3x^3 + 6x^2}{3x - 3} = \dfrac{x(x-1)}{3x(x+2)} \cdot \dfrac{3x^2(x+2)}{3(x-1)}$

$= \dfrac{3x^3(x-1)(x+2)}{9x(x-1)(x+2)} = \dfrac{x^2}{3}$

12. $\dfrac{x^2 + xy}{x - y} \cdot \dfrac{x^2 - y^2}{x^2 - 2x} \div \dfrac{x^2 + 2xy + y^2}{x^2 - 4} = \dfrac{x^2 + xy}{x - y} \cdot \dfrac{x^2 - y^2}{x^2 - 2x} \cdot \dfrac{x^2 - 4}{x^2 + 2xy + y^2}$

$= \dfrac{x(x+y)}{x - y} \cdot \dfrac{(x+y)(x-y)}{x(x-2)} \cdot \dfrac{(x+2)(x-2)}{(x+y)(x+y)}$

$= \dfrac{x(x+y)(x+y)(x-y)(x+2)(x-2)}{x(x-y)(x-2)(x+y)(x+y)} = x + 2$

13. $\dfrac{5x - 4}{x - 1} + \dfrac{5x + 3}{x - 1} = \dfrac{5x - 4 + 5x + 3}{x - 1} = \dfrac{10x - 1}{x - 1}$

14. $\dfrac{3y + 7}{2y + 3} - \dfrac{3(y - 2)}{2y + 3} = \dfrac{3y + 7 - 3(y - 2)}{2y + 3} = \dfrac{3y + 7 - 3y + 6}{2y + 3} = \dfrac{13}{2y + 3}$

15. $\dfrac{x + 1}{x} + \dfrac{x - 1}{x + 1} = \dfrac{(x+1)(x+1)}{x(x+1)} + \dfrac{x(x-1)}{x(x+1)} = \dfrac{(x+1)(x+1) + x(x-1)}{x(x+1)}$

$= \dfrac{x^2 + 2x + 1 + x^2 - x}{x(x+1)} = \dfrac{2x^2 + x + 1}{x(x+1)}$

16. $\dfrac{5x}{x - 2} - 3 = \dfrac{5x}{x - 2} - \dfrac{3}{1} = \dfrac{5x}{x - 2} - \dfrac{3(x - 2)}{x - 2} = \dfrac{5x - 3(x - 2)}{x - 2} = \dfrac{5x - 3x + 6}{x - 2} = \dfrac{2x + 6}{x - 2}$

17. $\dfrac{\frac{8x^2}{xy^3}}{\frac{4y^3}{x^2y^3}} = \dfrac{8x^2}{xy^3} \div \dfrac{4y^3}{x^2y^3} = \dfrac{8x^2}{xy^3} \cdot \dfrac{x^2y^3}{4y^3} = \dfrac{8x^4y^3}{4xy^6} = \dfrac{2x^3}{y^3}$

18. $\dfrac{1+\frac{y}{x}}{\frac{y}{x}-1}=\dfrac{\left(1+\frac{y}{x}\right)x}{\left(\frac{y}{x}-1\right)x}=\dfrac{1\cdot x+\frac{y}{x}\cdot x}{\frac{y}{x}\cdot x-1\cdot x}$

$$=\dfrac{x+y}{y-x}$$

19. $\dfrac{x}{10}-\dfrac{1}{2}=\dfrac{x}{5}$

$$10\left(\dfrac{x}{10}-\dfrac{1}{2}\right)=10\left(\dfrac{x}{5}\right)$$

$$x-5=2x$$

$$-5=x$$

20. $3x-\dfrac{2(x+3)}{3}=16-\dfrac{x+2}{2}$

$$6\left(3x-\dfrac{2x+6}{3}\right)=6\left(16-\dfrac{x+2}{2}\right)$$

$$18x-2(2x+6)=96-3(x+2)$$

$$18x-4x-12=96-3x-6$$

$$17x=102$$

$$x=6$$

21. $\dfrac{7}{x+4}-\dfrac{1}{2}=\dfrac{3}{x+4}$

$$2(x+4)\left(\dfrac{7}{x+4}-\dfrac{1}{2}\right)=2(x+4)\left(\dfrac{3}{x+4}\right)$$

$$2(7)-(x+4)=2(3)$$

$$14-x-4=6$$

$$-x=-4$$

$$x=4$$

The answer checks.

22. $H=\dfrac{RB}{R+B}$

$$(R+B)H=(R+B)\left(\dfrac{RB}{R+B}\right)$$

$$RH+BH=RB$$

$$RH=RB-BH$$

$$RH=B(R-H)$$

$$\dfrac{RH}{R-H}=B$$

23. Let $x=$ hours working together.

1st worker in 1 hour		2nd worker in 1 hour		Total in 1 hour
	+		=	

$$\dfrac{1}{7}+\dfrac{1}{9}=\dfrac{1}{x}$$

$$63x\left(\dfrac{1}{7}+\dfrac{1}{9}\right)=63x\left(\dfrac{1}{x}\right)$$

$$9x+7x=63$$

$$16x=63$$

$$x=\dfrac{63}{16}$$

They can finish in $3\frac{15}{16}$ hours.

24. Let $r=$ the speed of the current.

$$\dfrac{\text{Time downstream}}{}=\dfrac{\text{Time upstream}}{}$$

$$\dfrac{28}{23+r}=\dfrac{18}{23-r}$$

$$(23+r)(23-r)\left(\dfrac{28}{23+r}\right)=(23+r)(23-r)\left(\dfrac{18}{23-r}\right)$$

$$28(23-r)=18(23+r)$$

$$644-28r=414+18r$$

$$230=46r$$

$$5=r$$

	d	r	t
Downstream	28	$23+r$	$\dfrac{28}{23+r}$
Upstream	18	$23-r$	$\dfrac{18}{23-r}$

The current has a speed of 5 miles per hour.

25.

$$\frac{575}{\frac{1}{2}} = \frac{x}{7}$$

$$7(575) = \frac{1}{2}x$$

$$4025 = \frac{1}{2}x$$

$$2(4025) = 2\left(\frac{1}{2}x\right)$$

$$8050 = x \Rightarrow \text{The plane will lose 8,050 feet of altitude.}$$

Cumulative Review Exercises (page 425)

1. $x^2 x^5 = x^{2+5} = x^7$

2. $\left(x^2\right)^5 = x^{2\cdot5} = x^{10}$

3. $\frac{x^5}{x^2} = x^{5-2} = x^3$

4. $\left(3x^5\right)^0 = 1$

5. $(3x^2 - 2x) + (6x^3 - 3x^2 - 1) = 3x^2 - 2x + 6x^3 - 3x^2 - 1 = 6x^3 - 2x - 1$

6. $(4x^3 - 2x) - (2x^3 - 2x^2 - 3x + 1) = 4x^3 - 2x - 2x^3 + 2x^2 + 3x - 1 = 2x^3 + 2x^2 + x - 1$

7. $3(5x^2 - 4x + 3) + 2(-x^2 + 2x - 4) = 15x^2 - 12x + 9 - 2x^2 + 4x - 8 = 13x^2 - 8x + 1$

8. $4(3x^2 - 4x - 1) - 2(-2x^2 + 4x - 3) = 12x^2 - 16x - 4 + 4x^2 - 8x + 6 = 16x^2 - 24x + 2$

9. $(3x^3y^2)(-4x^2y^3) = 3(-4)x^3x^2y^2y^3 = -12x^5y^5$

10. $-5x^2(7x^3 - 2x^2 - 2) = -35x^5 + 10x^4 + 10x^2$

11. $(3x + 1)(2x + 4) = 6x^2 + 12x + 2x + 4 = 6x^2 + 14x + 4$

12. $(5x - 4y)(3x + 2y) = 15x^2 + 10xy - 12xy - 8y^2 = 15x^2 - 2xy - 8y^2$

13.
```
           x +  4
x + 3 | x² + 7x + 12
        x² + 3x
        ────────
             4x + 12
             4x + 12
             ───────
                   0
```

14.
```
              x² +  x + 1
2x - 3 | 2x³ -  x² -  x - 3
         2x³ - 3x²
         ────────
               2x² -  x
               2x² - 3x
               ───────
                     2x - 3
                     2x - 3
                     ──────
                          0
```

15. $3x^2y - 6xy^2 = 3xy(x - 2y)$

16. $3(a + b) + x(a + b) = (a + b)(3 + x)$

17. $2a + 2b + ab + b^2 = 2(a + b) + b(a + b)$
$= (a + b)(2 + b)$

18. $25p^4 - 16q^2 = \left(5p^2\right)^2 - (4q)^2$
$= \left(5p^2 + 4q\right)\left(5p^2 - 4q\right)$

19. $x^2 - 11x - 12 = (x - 12)(x + 1)$

20. $x^2 - xy - 6y^2 = (x - 3y)(x + 2y)$

21. $6a^2 - 7a - 20 = (2a - 5)(3a + 4)$

22. $8m^2 - 10mn - 3n^2 = (4m + n)(2m - 3n)$

23. $p^3 - 27q^3 = p^3 - (3q)^3 = (p - 3q)\left(p^2 + p(3q) + (3q)^2\right) = (p - 3q)(p^2 + 3pq + 9q^2)$

24. $8r^3 + 64s^3 = 8\left(r^3 + 8s^3\right) = 8\left[r^3 + (2s)^3\right] = 8(r + 2s)\left(r^2 - r \cdot 2s + (2s)^2\right)$
$$= 8(r + 2s)\left(r^2 - 2rs + 4s^2\right)$$

25. $\dfrac{4}{5}x + 6 = 18$

$\dfrac{4}{5}x = 12$

$5 \cdot \dfrac{4}{5}x = 5(12)$

$4x = 60$

$x = 15$

26. $5 - \dfrac{x + 2}{3} = 7 - x$

$3\left(5 - \dfrac{x + 2}{3}\right) = 3(7 - x)$

$15 - (x + 2) = 21 - 3x$

$15 - x - 2 = 21 - 3x$

$2x = 8$

$x = 4$

27. $6x^2 - x - 2 = 0$

$(2x + 1)(3x - 2) = 0$

$2x + 1 = 0 \quad$ **or** $\quad 3x - 2 = 0$

$2x = -1 \qquad\qquad 3x = 2$

$x = -\frac{1}{2} \qquad\qquad x = \frac{2}{3}$

28. $5x^2 = 10x$

$5x^2 - 10x = 0$

$5x(x - 2) = 0$

$5x = 0 \quad$ **or** $\quad x - 2 = 0$

$x = 0 \qquad\qquad x = 2$

29. $x^2 + 3x + 2 = 0$

$(x + 1)(x + 2) = 0$

$x + 1 = 0 \quad$ **or** $\quad x + 2 = 0$

$x = -1 \qquad\qquad x = -2$

30. $2y^2 + 5y - 12 = 0$

$(2y - 3)(y + 4) = 0$

$2y - 3 = 0 \quad$ **or** $\quad y + 4 = 0$

$2y = 3 \qquad\qquad y = -4$

$y = \frac{3}{2} \qquad\qquad y = -4$

31. $5x - 3 > 7$

$5x > 10$

$x > 2$

32. $7x - 9 < 5$

$7x < 14$

$x < 2$

33.
$$-2 < -x + 3 < 5$$
$$-5 < \quad -x \quad < 2$$
$$\frac{-5}{-1} > \quad \frac{-x}{-1} \quad > \frac{2}{-1}$$
$$5 > \quad x \quad > -2$$
$$-2 < \quad x \quad < 5$$

34.
$$0 \le \frac{4-x}{3} \le 2$$
$$3(0) \le 3 \cdot \frac{4-x}{3} \le 3(2)$$
$$0 \le \quad 4-x \quad \le 6$$
$$-4 \le \quad -x \quad \le 2$$
$$\frac{-4}{-1} \ge \quad \frac{-x}{-1} \quad \ge \frac{2}{-1}$$
$$4 \ge \quad x \quad \ge -2$$
$$-2 \le \quad x \quad \le 4$$

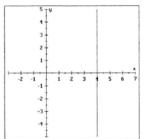

35. $4x - 3y = 12$

x	y
0	-4
3	0

36. $3x + 4y = 4y + 12$
$$3x = 12$$
$$x = 4$$

x	y
4	0
4	2

37. $f(0) = 2(0)^2 - 3 = 0 - 3 = -3$

38. $f(3) = 2(3)^2 - 3 = 2(9) - 3 = 18 - 3 = 15$

39. $f(-2) = 2(-2)^2 - 3 = 2(4) - 3$
$$= 8 - 3 = 5$$

40. $f(2x) = 2(2x)^2 - 3 = 2(4x^2) - 3$
$$= 8x^2 - 3$$

41. $\dfrac{x^2 + 2x + 1}{x^2 - 1} = \dfrac{(x+1)(x+1)}{(x+1)(x-1)} = \dfrac{x+1}{x-1}$

42. $\dfrac{x^2 + 2x - 15}{x^2 + 3x - 10} = \dfrac{(x+5)(x-3)}{(x+5)(x-2)} = \dfrac{x-3}{x-2}$

43. $\dfrac{x^2 + x - 6}{5x - 5} \cdot \dfrac{5x - 10}{x + 3} = \dfrac{(x+3)(x-2) \cdot 5(x-2)}{5(x-1) \cdot (x+3)} = \dfrac{(x-2)^2}{x-1}$

44. $\dfrac{p^2 - p - 6}{3p - 9} \div \dfrac{p^2 + 6p + 9}{p^2 - 9} = \dfrac{p^2 - p - 6}{3p - 9} \cdot \dfrac{p^2 - 9}{p^2 + 6p + 9} = \dfrac{(p-3)(p+2) \cdot (p+3)(p-3)}{3(p-3) \cdot (p+3)(p+3)}$
$$= \dfrac{(p+2)(p-3)}{3(p+3)}$$

45. $\dfrac{3x}{x+2} + \dfrac{5x}{x+2} - \dfrac{7x-2}{x+2} = \dfrac{3x + 5x - (7x-2)}{x+2} = \dfrac{3x + 5x - 7x + 2}{x+2} = \dfrac{x+2}{x+2} = 1$

46. $\dfrac{x-1}{x+1} + \dfrac{x+1}{x-1} = \dfrac{(x-1)(x-1)}{(x+1)(x-1)} + \dfrac{(x+1)(x+1)}{(x-1)(x+1)} = \dfrac{x^2-2x+1}{(x+1)(x-1)} + \dfrac{x^2+2x+1}{(x+1)(x-1)}$

$$= \dfrac{2x^2+2}{(x+1)(x-1)}$$

47. $\dfrac{a+1}{2a+4} - \dfrac{a^2}{2a^2-8} = \dfrac{a+1}{2(a+2)} - \dfrac{a^2}{2(a+2)(a-2)} = \dfrac{(a+1)(a-2)}{2(a+2)(a-2)} - \dfrac{a^2}{2(a+2)(a-2)}$

$$= \dfrac{a^2-2a+a-2-a^2}{2(a+2)(a-2)}$$

$$= \dfrac{-a-2}{2(a+2)(a-2)}$$

$$= \dfrac{-(a+2)}{2(a+2)(a-2)} = -\dfrac{1}{2(a-2)}$$

48. $\dfrac{\frac{1}{x} + \frac{1}{y}}{\frac{1}{x} - \frac{1}{y}} = \dfrac{\left(\frac{1}{x} + \frac{1}{y}\right)xy}{\left(\frac{1}{x} - \frac{1}{y}\right)xy} = \dfrac{\frac{1}{x} \cdot xy + \frac{1}{y} \cdot xy}{\frac{1}{x} \cdot xy - \frac{1}{y} \cdot xy} = \dfrac{y+x}{y-x}$

Exercise 7.1 (page 438)

1. $\left(\dfrac{t^3 t^5 t^{-6}}{t^2 t^{-4}}\right)^{-3} = \left(\dfrac{t^2 t^{-4}}{t^3 t^5 t^{-6}}\right)^3 = \left(\dfrac{t^{-2}}{t^2}\right)^3$

$$= \left(\dfrac{1}{t^4}\right)^3$$

$$= \dfrac{1}{t^{12}}$$

3. Let x = the number of pies made.

$$\boxed{\text{Expenses}} = \boxed{\text{Income}}$$

$$1200 + 3.40x = 5.95x$$
$$1200 = 2.55x$$
$$470.59 = x$$

He must sell at least 471 pies.

5. equation

7. multiplied; divided

9. contradiction

11. is greater than

13. half-open

15. positive

17.
$$2x + 1 = 13$$
$$2x + 1 - 1 = 13 - 1$$
$$2x = 12$$
$$\dfrac{2x}{2} = \dfrac{12}{2}$$
$$x = 6$$

19.
$$3(x+1) = 15$$
$$3x + 3 = 15$$
$$3x + 3 - 3 = 15 - 3$$
$$3x = 12$$
$$\dfrac{3x}{3} = \dfrac{12}{3}$$
$$x = 4$$

21.
$$2r - 5 = 1 - r$$
$$2r + r - 5 = 1 - r + r$$
$$3r - 5 + 5 = 1 + 5$$
$$3r = 6$$
$$\dfrac{3r}{3} = \dfrac{6}{3}$$
$$r = 2$$

23.
$$3(2y - 4) - 6 = 3y$$
$$6y - 12 - 6 = 3y$$
$$6y - 18 = 3y$$
$$6y - 3y - 18 = 3y - 3y$$
$$3y - 18 + 18 = 0 + 18$$
$$3y = 18$$
$$\frac{3y}{3} = \frac{18}{3}$$
$$y = 6$$

25.
$$5(5 - a) = 37 - 2a$$
$$25 - 5a = 37 - 2a$$
$$25 - 5a + 2a = 37 - 2a + 2a$$
$$25 - 25 - 3a = 37 - 25$$
$$-3a = 12$$
$$\frac{-3a}{-3} = \frac{12}{-3}$$
$$a = -4$$

27.
$$4(y + 1) = -2(4 - y)$$
$$4y + 4 = -8 + 2y$$
$$4y - 2y + 4 = -8 + 2y - 2y$$
$$2y + 4 - 4 = -8 - 4$$
$$2y = -12$$
$$\frac{2y}{2} = \frac{-12}{2}$$
$$y = -6$$

29.
$$2(a - 5) - (3a + 1) = 0$$
$$2a - 10 - 3a - 1 = 0$$
$$-a - 11 + 11 = 0 + 11$$
$$-a = 11$$
$$a = -11$$

31.
$$\frac{x}{2} - \frac{x}{3} = 4$$
$$6\left(\frac{x}{2} - \frac{x}{3}\right) = 6(4)$$
$$6\left(\frac{x}{2}\right) - 6\left(\frac{x}{3}\right) = 24$$
$$3x - 2x = 24$$
$$x = 24$$

33.
$$\frac{x}{6} + 1 = \frac{x}{3}$$
$$6\left(\frac{x}{6} + 1\right) = 6\left(\frac{x}{3}\right)$$
$$6\left(\frac{x}{6}\right) + 6(1) = 2x$$
$$x + 6 = 2x$$
$$x - 2x + 6 = 2x - 2x$$
$$-x + 6 - 6 = 0 - 6$$
$$-x = -6$$
$$x = 6$$

35.
$$\frac{a + 1}{3} + \frac{a - 1}{5} = \frac{2}{15}$$
$$15\left(\frac{a + 1}{3} + \frac{a - 1}{5}\right) = 15\left(\frac{2}{15}\right)$$
$$15\left(\frac{a + 1}{3}\right) + 15\left(\frac{a - 1}{5}\right) = 2$$
$$5(a + 1) + 3(a - 1) = 2$$
$$5a + 5 + 3a - 3 = 2$$
$$8a + 2 - 2 = 2 - 2$$
$$8a = 0$$
$$\frac{8a}{8} = \frac{0}{8}$$
$$a = 0$$

37.
$$4(2 - 3t) + 6t = -6t + 8$$
$$8 - 12t + 6t = -6t + 8$$
$$8 - 6t = -6t + 8$$
$$-6t + 8 = -6t + 8$$
Identity

39.
$$\frac{a+1}{4} + \frac{2a-3}{4} = \frac{a}{2} - 2$$
$$4\left(\frac{a+1}{4} + \frac{2a-3}{4}\right) = 4\left(\frac{a}{2} - 2\right)$$
$$4\left(\frac{a+1}{4}\right) + 4\left(\frac{2a-3}{4}\right) = 4\left(\frac{a}{2}\right) + 4(-2)$$
$$a + 1 + 2a - 3 = 2a - 8$$
$$3a - 2 = 2a - 8$$
$$3a - 2a - 2 = 2a - 2a - 8$$
$$a - 2 + 2 = -8 + 2$$
$$a = -6$$

41.
$$3(x-4) + 6 = -2(x+4) + 5x$$
$$3x - 12 + 6 = -2x - 8 + 5x$$
$$3x - 6 = 3x - 8$$
$$3x - 3x - 6 = 3x - 3x - 8$$
$$-6 \neq -8$$
Contradiction

43.
$$y(y+2) = (y+1)^2 - 1$$
$$y^2 + 2y = (y+1)(y+1) - 1$$
$$y^2 + 2y = y^2 + 2y + 1 - 1$$
$$y^2 + 2y = y^2 + 2y$$
Identity

45.
$$V = \frac{1}{3}Bh$$
$$3V = 3 \cdot \frac{1}{3}Bh$$
$$3V = Bh$$
$$\frac{3V}{h} = \frac{Bh}{h}$$
$$\frac{3V}{h} = B, \text{ or } B = \frac{3V}{h}$$

47.
$$P = 2l + 2w$$
$$P - 2l = 2l - 2l + 2w$$
$$P - 2l = 2w$$
$$\frac{P-2l}{2} = \frac{2w}{2}$$
$$\frac{P-2l}{2} = w, \text{ or } w = \frac{P-2l}{2}$$

49.
$$z = \frac{x-\mu}{\sigma}$$
$$\sigma z = \sigma \cdot \frac{x-\mu}{\sigma}$$
$$\sigma z = x - \mu$$
$$\sigma z + \mu = x, \text{ or } x = z\sigma + \mu$$

51.
$$y = mx + b$$
$$y - b = mx + b - b$$
$$y - b = mx$$
$$\frac{y-b}{m} = \frac{mx}{m}$$
$$\frac{y-b}{m} = x, \text{ or } x = \frac{y-b}{m}$$

53.
$$P = L + \frac{s}{f}i$$
$$P - L = \frac{s}{f}i$$
$$f(P-L) = f \cdot \frac{s}{f}i$$
$$f(P-L) = si$$
$$\frac{f(P-L)}{i} = \frac{si}{i}$$
$$\frac{f(P-L)}{i} = s, \text{ or } s = \frac{f(P-L)}{i}$$

55.
$$5x - 3 > 7$$
$$5x - 3 + 3 > 7 + 3$$
$$5x > 10$$
$$\frac{5x}{5} > \frac{10}{5}$$
$$x > 2$$
solution set: $(2, \infty)$

57.
$$-3x - 1 \le 5$$
$$-3x - 1 + 1 \le 5 + 1$$
$$-3x \le 6$$
$$\frac{-3x}{-3} \ge \frac{6}{-3}$$
$$x \ge -2$$
solution set: $[-2, \infty)$

59.
$$8 - 9y \ge -y$$
$$8 - 9y + y \ge -y + y$$
$$8 - 8 - 8y \ge 0 - 8$$
$$-8y \ge -8$$
$$\frac{-8y}{-8} \le \frac{-8}{-8}$$
$$y \le 1$$
solution set: $(-\infty, 1]$

61.
$$-3(a + 2) > 2(a + 1)$$
$$-3a - 6 > 2a + 2$$
$$-3a - 2a - 6 > 2a - 2a + 2$$
$$-5a - 6 + 6 > 2 + 6$$
$$-5a > 8$$
$$\frac{-5a}{-5} < \frac{8}{-5}$$
$$a < -\frac{8}{5}$$
solution set: $\left(-\infty, -\frac{8}{5}\right)$

63.
$$\frac{1}{2}y + 2 \ge \frac{1}{3}y - 4$$
$$6\left(\frac{1}{2}y + 2\right) \ge 6\left(\frac{1}{3}y - 4\right)$$
$$3y + 12 \ge 2y - 24$$
$$3y - 2y - 12 \ge 2y - 2y - 24 - 12$$
$$y \ge -36$$
solution set: $[-36, \infty)$

65.
$$-2 < -b + 3 < 5$$
$$-2 - 3 < -b + 3 - 3 < 5 - 3$$
$$-5 < -b < 2$$
$$\frac{-5}{-1} > \frac{-b}{-1} > \frac{2}{-1}$$
$$5 > b > -2, \text{ or } -2 < b < 5$$
solution set: $(-2, 5)$

67.
$$15 > 2x - 7 > 9$$
$$15 + 7 > 2x - 7 + 7 > 9 + 7$$
$$22 > 2x > 16$$
$$\frac{22}{2} > \frac{2x}{2} > \frac{16}{2}$$
$$11 > x > 8, \text{ or } 8 < x < 11$$
solution set: $(8, 11)$

69.
$$-6 < -3(x - 4) \le 24$$
$$\frac{-6}{-3} > \frac{-3(x - 4)}{-3} \ge \frac{24}{-3}$$
$$2 > x - 4 \ge -8$$
$$2 + 4 > x - 4 + 4 \ge -8 + 4$$
$$6 > x \ge -4, \text{ or } -4 \le x < 6$$
solution set: $[-4, 6)$

71. $0 \ge \frac{1}{2}x - 4 > 6$

This inequality indicates that $0 \ge 6$ (by the transitive property). Since this is not possible, there is no solution to the inequality.

73.
$$0 \le \frac{4-x}{3} \le 2$$

$$3(0) \le 3\left(\frac{4-x}{3}\right) \le 3(2)$$

$$0 \le 4 - x \le 6$$

$$0 - 4 \le 4 - 4 - x \le 6 - 4$$

$$-4 \le -x \le 2$$

$$\frac{-4}{-1} \ge \frac{-x}{-1} \ge \frac{2}{-1}$$

$$4 \ge x \ge -2, \text{ or } -2 \le x \le 4$$

solution set: $[-2, 4]$

75.
$$\begin{array}{lll} 3x + 2 < 8 & \text{or} & 2x - 3 > 11 \\ 3x + 2 - 2 < 8 - 2 & & 2x - 3 + 3 > 11 + 3 \\ 3x < 6 & & 2x > 14 \\ x < 2 & & x > 7 \end{array}$$

If $x < 2$ or $x > 7$, then the solution set is $(-\infty, 2) \cup (7, \infty)$.

77.
$$\begin{array}{lll} -4(x + 2) \ge 12 & \text{or} & 3x + 8 < 11 \\ -4x - 8 + 8 \ge 12 + 8 & & 3x + 8 - 8 < 11 - 8 \\ -4x \ge 20 & & 3x < 3 \\ \frac{-4x}{-4} \le \frac{20}{-4} & & x < 1 \\ x \le -5 \end{array}$$

If $x \le -5$ or $x < 1$, then $x < 1$. The solution set is $(-\infty, 1)$.

79.
$$x < -3 \quad \text{and} \quad x > 3$$

It is impossible for x to be both less than -3 and greater than 3. There is no solution.

81. Let $x =$ the length of the shorter piece.
Then $2x + 1 =$ the length of the other piece.

$$\boxed{\begin{array}{c}\text{shorter}\\\text{length}\end{array}} + \boxed{\begin{array}{c}\text{other}\\\text{length}\end{array}} = \boxed{\text{total length}}$$

$$x + 2x + 1 = 22$$

$$3x + 1 = 22$$

$$3x = 21$$

$$x = 7$$

The pieces have lengths of 7 feet and 15 feet.

83. Let $w =$ the width. Then $2w =$ the length.

$$\boxed{\text{Perimeter of garden}} = 72$$

$$w + 2w + w + 2w = 72$$

$$6w = 72$$

$$w = 12$$

The dimension are 12 m by 24 m.

85.

$$\boxed{\text{Total amount of fence}} = 150$$
$$x + x + x + x + x + 5 + x + 5 + x = 150$$
$$7x + 10 = 150$$
$$7x = 140$$
$$x = 20$$

The dimensions are 20 feet by 45 feet.

87.

$$\text{Price} < 42$$
$$\text{Price} - \text{Cost} < 42 - \text{Cost}$$
$$\text{Profit} < 42 - 27$$
$$p < 15$$

89. Let $x =$ the # of compact discs bought.
$$175 + 8.50x \le 330$$
$$8.50x \le 155$$
$$x \le 18.24$$
He can buy at most 18 discs.

91. **Answers may vary.**

93. The 4 was not distributed correctly in the 2nd line. It should be $4x + 12 = 16$.

Exercise 7.2 (page 445)

1.
$$3(2a - 1) = 2a$$
$$6a - 3 = 2a$$
$$4a = 3$$
$$a = \frac{3}{4}$$

3.
$$\frac{5x}{2} - 1 = \frac{x}{3} + 12$$
$$6\left(\frac{5x}{2} - 1\right) = 6\left(\frac{x}{3} + 12\right)$$
$$15x - 6 = 2x + 72$$
$$13x = 78$$
$$x = 6$$

5. x

7. 0

9. $a = -b$

11. $|8| = 8$

13. $|-12| = 12$

15. $-|2| = -2$

17. $-|-30| = -30$

19. $-(-|50|) = -(-50) = 50$

21. $|\pi - 4| = -(\pi - 4) = 4 - \pi$
($\pi - 4$ is less than zero.)

23. $|2| = 2$; $|5| = 5$;
$|2|$ is smaller.

25. $|5| = 5$; $|-8| = 8$;
$|5|$ is smaller.

27. $|-2| = 2$; $|10| = 10$;
$|-2|$ is smaller.

29. $|-3| = 3$; $-|-4| = -4$;
$-|-4|$ is smaller.

31. $-|-5| = -5$;
$-|-7| = -7$;
$-|-7|$ is smaller.

33. $-x < 0$ if $x > 0$;
$|x + 1| > 0$
$-x$ is smaller.

35.
$$|x| = 8$$
$$x = 8 \quad \text{or} \quad x = -8$$

37.
$$|x - 3| = 6$$
$$x - 3 = 6 \quad \text{or} \quad x - 3 = -6$$
$$x = 9 \qquad\qquad x = -3$$

39.
$$|2x - 3| = 5$$
$$2x - 3 = 5 \quad \text{or} \quad 2x - 3 = -5$$
$$2x = 8 \qquad\qquad 2x = -2$$
$$x = 4 \qquad\qquad x = -1$$

41.
$$|3x + 2| = 16$$
$$3x + 2 = 16 \quad \text{or} \quad 3x + 2 = -16$$
$$3x = 14 \qquad\qquad 3x = -18$$
$$x = \tfrac{14}{3} \qquad\qquad x = -6$$

43.
$$\left|\tfrac{7}{2}x + 3\right| = -5$$
Since an absolute value cannot be negative, there is no solution.

45.
$$\left|\tfrac{x}{2} - 1\right| = 3$$
$$\tfrac{x}{2} - 1 = 3 \quad \text{or} \quad \tfrac{x}{2} - 1 = -3$$
$$\tfrac{x}{2} = 4 \qquad\qquad \tfrac{x}{2} = -2$$
$$x = 8 \qquad\qquad x = -4$$

47.
$$|3 - 4x| = 5$$
$$3 - 4x = 5 \quad \text{or} \quad 3 - 4x = -5$$
$$-4x = 2 \qquad\qquad -4x = -8$$
$$x = -\tfrac{1}{2} \qquad\qquad x = 2$$

49.
$$|3x + 24| = 0$$
$$3x + 24 = 0 \quad \text{or} \quad 3x + 24 = -0$$
$$3x = -24 \qquad\qquad 3x = -24$$
$$x = -8 \qquad\qquad x = -8$$

51.
$$\left|\dfrac{3x + 48}{3}\right| = 12$$
$$\dfrac{3x + 48}{3} = 12 \quad \text{or} \quad \dfrac{3x + 48}{3} = -12$$
$$3x + 48 = 36 \qquad\qquad 3x + 48 = -36$$
$$3x = -12 \qquad\qquad 3x = -84$$
$$x = -4 \qquad\qquad x = -28$$

53.
$$|x + 3| + 7 = 10$$
$$|x + 3| = 3$$
$$x + 3 = 3 \quad \text{or} \quad x + 3 = -3$$
$$x = 0 \qquad\qquad x = -6$$

55.
$$\left|\tfrac{3}{5}x - 4\right| - 2 = -2$$
$$\left|\tfrac{3}{5}x - 4\right| = 0$$
$$\tfrac{3}{5}x - 4 = 0 \quad \text{or} \quad \tfrac{3}{5}x - 4 = -0$$
$$\tfrac{3}{5}x = 4 \qquad\qquad \tfrac{3}{5}x = 4$$
$$x = \tfrac{20}{3} \qquad\qquad x = \tfrac{20}{3}$$

57.
$$|2x + 1| = |3x + 3|$$
$$2x + 1 = 3x + 3 \quad \text{or} \quad 2x + 1 = -(3x + 3)$$
$$-x = 2 \qquad\qquad 2x + 1 = -3x - 3$$
$$x = -2 \qquad\qquad 5x = -4$$
$$x = -\tfrac{4}{5}$$

59.
$$|3x - 1| = |x + 5|$$
$$3x - 1 = x + 5 \quad \text{or} \quad 3x - 1 = -(x + 5)$$
$$2x = 6 \qquad\qquad 3x - 1 = -x - 5$$
$$x = 3 \qquad\qquad 4x = -4$$
$$x = -1$$

61.
$$|2 - x| = |3x + 2|$$
$$2 - x = 3x + 2 \quad \text{or} \quad 2 - x = -(3x + 2)$$
$$-4x = 0 \qquad\qquad 2 - x = -3x - 2$$
$$x = 0 \qquad\qquad 2x = -4$$
$$x = -2$$

63.
$$\left|\frac{x}{2}+2\right|=\left|\frac{x}{2}-2\right|$$

$\frac{x}{2}+2=\frac{x}{2}-2 \qquad$ **or** $\qquad \frac{x}{2}+2=-\left(\frac{x}{2}-2\right)$

$\qquad 0=-4 \qquad\qquad\qquad\qquad \frac{x}{2}+2=-\frac{x}{2}+2$

(no solution from this part) $\qquad\qquad\qquad x=0$

65.
$$\left|x+\tfrac{1}{3}\right|=|x-3|$$

$x+\tfrac{1}{3}=x-3 \qquad$ **or** $\qquad x+\tfrac{1}{3}=-(x-3)$

$3x+1=3x-9 \qquad\qquad\qquad x+\tfrac{1}{3}=-x+3$

$\qquad 0=-10 \qquad\qquad\qquad 3x+1=-3x+9$

(no solution from this part) $\qquad\qquad 6x=8$

$\qquad\qquad\qquad\qquad\qquad\qquad\qquad x=\tfrac{4}{3}$

67. $\qquad |3x+7|=-|8x-2|$
Since an absolute value cannot be
negative, there is no solution.

69. **Answers may vary.**

71. $|x|+k=0$
$\qquad |x|=-k$
If this equation has exactly two solutions,
then $-k>0$, or $k<0$.

73. **Answers may vary.**

75. **Answers may vary.**

Exercise 7.3 (page 450)

1. $\qquad A=p+prt$
$A-p=prt$
$\dfrac{A-p}{pr}=t$, or $t=\dfrac{A-p}{pr}$

3. $\qquad P=2w+2l$
$P-2w=2l$
$\dfrac{P-2w}{2}=l$, or $l=\dfrac{P-2w}{2}$

5. $\quad -k<x<k$

7. $\quad x<-k$ or $x>k$

9. $\quad |2x|<8$
$-8<2x<8$
$\frac{-8}{2}<\frac{2x}{2}<\frac{8}{2}$
$-4<x<4$
solution set: $(-4,4)$

11. $\quad |x+9|\le 12$
$-12\le x+9\le 12$
$-12-9\le x+9-9\le 12-9$
$-21\le x\le 3$
solution set: $[-21,3]$

13. $|3x+2|\le -3$; Since an absolute value can never be negative, this inequality has no solution.

15. $|4x - 1| \le 7$

$$-7 \le 4x - 1 \le 7$$
$$-7 + 1 \le 4x - 1 + 1 \le 7 + 1$$
$$-6 \le 4x \le 8$$
$$\frac{-6}{4} \le \frac{4x}{4} \le \frac{8}{4}$$
$$-\frac{3}{2} \le x \le 2$$

solution set: $\left[-\frac{3}{2}, 2\right]$

17. $|3 - 2x| < 7$

$$-7 < 3 - 2x < 7$$
$$-7 - 3 < 3 - 3 - 2x < 7 - 3$$
$$-10 < -2x < 4$$
$$\frac{-10}{-2} > \frac{-2x}{-2} > \frac{4}{-2}$$
$$5 > x > -2, \text{ or } -2 < x < 5$$

solution set: $(-2, 5)$

19. $|5x| > 5$

$$5x < -5 \quad \textbf{or} \quad 5x > 5$$
$$x < -1 \qquad\qquad x > 1$$

solution set: $(-\infty, -1) \cup (1, \infty)$

21. $|x - 12| > 24$

$$x - 12 < -24 \quad \textbf{or} \quad x - 12 > 24$$
$$x < -12 \qquad\qquad x > 36$$

solution set: $(-\infty, -12) \cup (36, \infty)$

23. $|3x + 2| > 14$

$$3x + 2 < -14 \quad \textbf{or} \quad 3x + 2 > 14$$
$$3x < -16 \qquad\qquad 3x > 12$$
$$x < -\frac{16}{3} \qquad\qquad x > 4$$

solution set: $\left(-\infty, -\frac{16}{3}\right) \cup (4, \infty)$

25. $|4x + 3| > -5$

Since an absolute value is always at least 0,
this inequality is true for all real numbers.

solution set: $(-\infty, \infty)$

27. $|2 - 3x| \ge 8$

$$2 - 3x \le -8 \quad \textbf{or} \quad 2 - 3x \ge 8$$
$$-3x \le -10 \qquad\qquad -3x \ge 6$$
$$x \ge \frac{10}{3} \qquad\qquad x \le -2$$

solution set: $(-\infty, -2] \cup \left[\frac{10}{3}, \infty\right)$

29. $-|2x - 3| < -7$

$$|2x - 3| > 7$$
$$2x - 3 < -7 \quad \textbf{or} \quad 2x - 3 > 7$$
$$2x < -4 \qquad\qquad 2x > 10$$
$$x < -2 \qquad\qquad x > 5$$

solution set: $(-\infty, -2) \cup (5, \infty)$

31.
$$|8x - 3| > 0$$
$$8x - 3 < -0 \quad \text{or} \quad 8x - 3 > 0$$
$$8x < 3 \qquad\qquad 8x > 3$$
$$x < \tfrac{3}{8} \qquad\qquad x > \tfrac{3}{8}$$
solution set: $\left(-\infty, \tfrac{3}{8}\right) \cup \left(\tfrac{3}{8}, \infty\right)$

$\tfrac{3}{8}$

33.
$$\left|\frac{x - 2}{3}\right| \leq 4$$
$$-4 \leq \frac{x - 2}{3} \leq 4$$
$$-12 \leq x - 2 \leq 12$$
$$-10 \leq x \leq 14$$
solution set: $[-10, 14]$

$-10 \qquad 14$

35.
$$|3x + 1| + 2 < 6$$
$$|3x + 1| < 4$$
$$-4 < 3x + 1 < 4$$
$$-5 < 3x < 3$$
$$-\tfrac{5}{3} < x < 1$$
solution set: $\left(-\tfrac{5}{3}, 1\right)$

$-\tfrac{5}{3} \qquad 1$

37.
$$3|2x + 5| \geq 9$$
$$|2x + 5| \geq 3$$
$$2x + 5 \leq -3 \quad \text{or} \quad 2x + 5 \geq 3$$
$$2x \leq -8 \qquad\qquad 2x \geq -2$$
$$x \leq -4 \qquad\qquad x \geq -1$$
solution set: $(-\infty, -4] \cup [-1, \infty)$

$-4 \qquad -1$

39.
$$|5x - 1| + 4 \leq 0$$
$$|5x - 1| \leq -4$$
Since an absolute value can never be negative, this inequality has no solution.

41.
$$\left|\tfrac{1}{3}x + 7\right| + 5 > 6$$
$$\left|\tfrac{1}{3}x + 7\right| > 1$$
$$\tfrac{1}{3}x + 7 < -1 \quad \text{or} \quad \tfrac{1}{3}x + 7 > 1$$
$$x + 21 < -3 \qquad\qquad x + 21 > 3$$
$$x < -24 \qquad\qquad x > -18$$
solution set: $(-\infty, -24) \cup (-18, \infty)$

$-24 \qquad -18$

43.
$$\left|\tfrac{1}{5}x - 5\right| + 4 > 4$$
$$\left|\tfrac{1}{5}x - 5\right| > 0$$
$$\tfrac{1}{5}x - 5 < -0 \quad \text{or} \quad \tfrac{1}{5}x - 5 > 0$$
$$\tfrac{1}{5}x < 5 \qquad\qquad \tfrac{1}{5}x > 5$$
$$x < 25 \qquad\qquad x > 25$$
solution set: $(-\infty, 25) \cup (25, \infty)$

25

45.
$$\left|\tfrac{3}{5}x + \tfrac{7}{3}\right| < 2$$
$$-2 < \tfrac{3}{5}x + \tfrac{7}{3} < 2$$
$$15(-2) < 15\left(\tfrac{3}{5}x + \tfrac{7}{3}\right) < 15(2)$$
$$-30 < 9x + 35 < 30$$
$$-65 < 9x < -5$$
$$-\tfrac{65}{9} < x < -\tfrac{5}{9}$$
solution set: $\left(-\tfrac{65}{9}, -\tfrac{5}{9}\right)$

$-\tfrac{65}{9} \qquad -\tfrac{5}{9}$

47.

$$\left|3\left(\frac{x+4}{4}\right)\right| > 0$$

$$3\left(\tfrac{x+4}{4}\right) < -0 \quad \textbf{or} \quad 3\left(\tfrac{x+4}{4}\right) > 0$$

$$\tfrac{4}{3}\cdot 3\left(\tfrac{x+4}{4}\right) < \tfrac{4}{3}(-0) \qquad \tfrac{4}{3}\cdot 3\left(\tfrac{x+4}{4}\right) > \tfrac{4}{3}\cdot 0$$

$$x+4 < 0 \qquad\qquad x+4 > 0$$

$$x < -4 \qquad\qquad x > -4$$

solution set: $(-\infty, -4) \cup (-4, \infty) \Rightarrow$

$$-4$$

49.

$$\left|\tfrac{1}{7}x + 1\right| \le 0$$

Since an absolute value can never be less than zero, the only solution for this inequality is when the absolute value is equal to 0.

$$\left|\tfrac{1}{7}x + 1\right| = 0$$

$$\tfrac{1}{7}x + 1 = 0 \quad \textbf{or} \quad \tfrac{1}{7}x + 1 = -0$$

$$\tfrac{1}{7}x = -1 \qquad\qquad \tfrac{1}{7}x = -1$$

$$x = -7 \qquad\qquad x = -7$$

solution set: $[-7, -7]$

$$-7$$

51.

$$\left|\frac{x-5}{10}\right| \le 0$$

Since an absolute value can never be less than zero, the only solution for this inequality is when the absolute value is equal to 0.

$$\left|\frac{x-5}{10}\right| = 0$$

$$\frac{x-5}{10} = 0 \quad \textbf{or} \quad \frac{x-5}{10} = -0$$

$$x - 5 = 0 \qquad\qquad x - 5 = 0$$

$$x = 5 \qquad\qquad x = 5$$

solution set: $[5, 5]$

$$5$$

53. Answers may vary.

55. $|x| + |y| > |x + y|$ when x and y have different signs.

Exercise 7.4 (page 461)

1. 1.1×10^3 ft/sec $= 1100$ ft/sec

3.

$$\frac{2}{3}(5t - 3) = 38$$

$$3 \cdot \frac{2}{3}(5t - 3) = 3(38)$$

$$2(5t - 3) = 114$$

$$10t - 6 = 114$$

$$10t = 120$$

$$t = 12$$

5. $ab + ac$

7. perfect

9. $(x^2 - xy + y^2)$

11. $2x + 8 = 2 \cdot x + 2 \cdot 4 = 2(x + 4)$

13. $2x^2 - 6x = \mathbf{2x} \cdot x - \mathbf{2x} \cdot 3 = \mathbf{2x}(x - 3)$

15. $15x^2y - 10x^2y^2 = \mathbf{5x^2y} \cdot 3 - \mathbf{5x^2y} \cdot 2y = \mathbf{5x^2y}(3 - 2y)$

17. $13ab^2c^3 - 26a^3b^2c = \mathbf{13ab^2c} \cdot c^2 - \mathbf{13ab^2c} \cdot 2a^2 = \mathbf{13ab^2c}(c^2 - 2a^2)$

19. $27z^3 + 12z^2 + 3z = \mathbf{3z} \cdot 9z^2 + \mathbf{3z} \cdot 4z + \mathbf{3z} \cdot 1 = \mathbf{3z}(9z^2 + 4z + 1)$

21. $24s^3 - 12s^2t + 6st^2 = \mathbf{6s}(4s^2) - \mathbf{6s}(2st) + \mathbf{6s}(t^2) = \mathbf{6s}(4s^2 - 2st + t^2)$

23. $\begin{aligned} -3a - 6 &= (\mathbf{-3})(a) + (\mathbf{-3})(2) \\ &= \mathbf{-3}(a + 2) \end{aligned}$ **25.** $\begin{aligned} -6x^2 - 3xy &= (\mathbf{-3x})(2x) + (\mathbf{-3x})(y) \\ &= \mathbf{-3x}(2x + y) \end{aligned}$

27. $-63u^3v^6z^9 + 28u^2v^7z^2 - 21u^3v^3z^4 = -7u^2v^3z^2(9uv^3z^7 - 4v^4 + 3uz^2)$

29. $\begin{aligned} x^{n+2} + x^{n+3} &= x^2(x^{n+2-2} + x^{n+3-2}) \\ &= x^2(x^n + x^{n+1}) \end{aligned}$ **31.** $\begin{aligned} 2y^{n+2} - 3y^{n+3} &= y^n(2y^{n+2-n} - 3y^{n+3-n}) \\ &= y^n(2y^2 - 3y^3) \end{aligned}$

33. $\begin{aligned} ax + bx + ay + by &= x(a + b) + y(a + b) \\ &= (a + b)(x + y) \end{aligned}$ **35.** $\begin{aligned} x^2 + yx + 2x + 2y &= x(x + y) + 2(x + y) \\ &= (x + y)(x + 2) \end{aligned}$

37. $3c - cd + 3d - c^2 = 3c + 3d - c^2 - cd = 3(c + d) - c(c + d) = (c + d)(3 - c)$

39. $4y + 8 + 2xy + 4x = 2(2y + 4 + xy + 2x) = 2[2(y + 2) + x(y + 2)] = 2(y + 2)(2 + x)$

41. $\begin{aligned} r_1r_2 &= rr_2 + rr_1 \\ r_1r_2 - rr_1 &= rr_2 \\ r_1(r_2 - r) &= rr_2 \\ r_1 &= \frac{rr_2}{r_2 - r} \end{aligned}$ **43.** $\begin{aligned} S(1 - r) &= a - lr \\ S - Sr &= a - lr \\ S - a &= Sr - lr \\ S - a &= r(S - l) \\ \frac{S - a}{S - l} &= r, \text{ or } r = \frac{S - a}{S - l} \end{aligned}$

45. $x^2 - 4 = x^2 - 2^2 = (x + 2)(x - 2)$ **47.** $9y^2 - 64 = (3y)^2 - 8^2 = (3y + 8)(3y - 8)$

49. $\begin{aligned} 81a^4 - 49b^2 &= (9a^2)^2 - (7b)^2 \\ &= (9a^2 + 7b)(9a^2 - 7b) \end{aligned}$ **51.** $\begin{aligned} (x + y)^2 - z^2 &= [(x + y) + z][(x + y) - z] \\ &= (x + y + z)(x + y - z) \end{aligned}$

53. $x^4 - y^4 = (x^2 + y^2)(x^2 - y^2) = (x^2 + y^2)(x + y)(x - y)$

55. $\begin{aligned} 2x^2 - 288 &= 2(x^2 - 144) \\ &= 2(x + 12)(x - 12) \end{aligned}$ **57.** $\begin{aligned} 2x^3 - 32x &= 2x(x^2 - 16) \\ &= 2x(x + 4)(x - 4) \end{aligned}$

59. $x^{2m} - y^{4n} = (x^m)^2 - (y^{2n})^2 = (x^m + y^{2n})(x^m - y^{2n})$

61. $\begin{aligned} a^2 - b^2 + a + b &= (a^2 - b^2) + (a + b) = (a + b)(a - b) + 1(a + b) \\ &= (a + b)(a - b + 1) \end{aligned}$

63. $\begin{aligned} 2x + y + 4x^2 - y^2 &= (2x + y) + (4x^2 - y^2) = (2x + y)(1) + (2x + y)(2x - y) \\ &= (2x + y)(1 + 2x - y) \end{aligned}$

65. $x^2 + 5x + 6 = (x + 2)(x + 3)$

67. $a^2 + 5a - 52 \Rightarrow$ prime

69. $2x^2 - 11x + 5 = (2x - 1)(x - 5)$

71. $6y^2 + 7y + 2 = (2y + 1)(3y + 2)$

73. $8a^2 + 6a - 9 = (4a - 3)(2a + 3)$

75. $a^2 - 3ab - 4b^2 = (a - 4b)(a + b)$

77. $2y^2 + yt - 6t^2 = (2y - 3t)(y + 2t)$

79. $5x^2 + 4x + 1 \Rightarrow$ prime

81. $8x^2 - 10x + 3 = (4x - 3)(2x - 1)$

83. $3x^2 + 12x - 63 = 3(x^2 + 4x - 21)$
$$= 3(x + 7)(x - 3)$$

85. $3x^3 - 10x^2 + 3x = x(3x^2 - 10x + 3)$
$$= x(3x - 1)(x - 3)$$

87. $a^2b^2 - 13ab^2 + 22b^2 = b^2(a^2 - 13a + 22)$
$$= b^2(a - 11)(a - 2)$$

89. $-a^2 + 4a + 32 = -(a^2 - 4a - 32)$
$$= -(a - 8)(a + 4)$$

91. $-3x^2 + 15x - 18 = -3(x^2 - 5x + 6)$
$$= -3(x - 2)(x - 3)$$

93. $-3a^2 + ab + 2b^2 = -(3a^2 - ab - 2b^2) = -(3a + 2b)(a - b)$

95. $-4x^3 - 9x + 12x^2 = -x(4x^2 - 12x + 9) = -x(2x - 3)(2x - 3) = -x(2x - 3)^2$

97. $x^4 + 8x^2 + 15 = (x^2 + 5)(x^2 + 3)$

99. $y^4 - 13y^2 + 30 = (y^2 - 10)(y^2 - 3)$

101. $a^4 - 13a^2 + 36 = (a^2 - 4)(a^2 - 9) = (a + 2)(a - 2)(a + 3)(a - 3)$

103. $x^{2n} + 2x^n + 1 = (x^n + 1)(x^n + 1)$
$$= (x^n + 1)^2$$

105. $2a^{6n} - 3a^{3n} - 2 = (2a^{3n} + 1)(a^{3n} - 2)$

107. $x^{4n} + 2x^{2n}y^{2n} + y^{4n} = (x^{2n} + y^{2n})(x^{2n} + y^{2n}) = (x^{2n} + y^{2n})^2$

109. $6x^{2n} + 7x^n - 3 = (3x^n - 1)(2x^n + 3)$

111. $x^2 + 4x + 4 - y^2 = (x^2 + 4x + 4) - y^2 = (x + 2)(x + 2) - y^2 = (x + 2)^2 - y^2$
$$= [(x + 2) + y][(x + 2) - y]$$
$$= (x + 2 + y)(x + 2 - y)$$

113. $x^2 + 2x + 1 - 9z^2 = (x^2 + 2x + 1) - 9z^2 = (x + 1)(x + 1) - 9z^2$
$$= (x + 1)^2 - (3z)^2$$
$$= [(x + 1) + 3z][(x + 1) - 3z]$$
$$= (x + 1 + 3z)(x + 1 - 3z)$$

115. $c^2 - 4a^2 + 4ab - b^2 = c^2 - (4a^2 - 4ab + b^2) = c^2 - (2a - b)(2a - b)$
$$= c^2 - (2a - b)^2$$
$$= [c + (2a - b)][c - (2a - b)]$$
$$= (c + 2a - b)(c - 2a + b)$$

117. $y^3 + 1 = y^3 + 1^3$
$$= (y + 1)(y^2 - 1y + 1^2)$$
$$= (y + 1)(y^2 - y + 1)$$

119. $a^3 - 27 = a^3 - 3^3$
$$= (a - 3)(a^2 + 3a + 3^2)$$
$$= (a - 3)(a^2 + 3a + 9)$$

121. $8 + x^3 = 2^3 + x^3$
$$= (2 + x)(2^2 - 2x + x^2)$$
$$= (2 + x)(4 - 2x + x^2)$$

123. $s^3 - t^3 = (s - t)(s^2 + st + t^2)$

125. $27x^3 + y^3 = (3x)^3 + y^3$
$$= (3x + y)((3x)^2 - 3xy + y^2)$$
$$= (3x + y)(9x^2 - 3xy + y^2)$$

127. $a^3 + 8b^3 = a^3 + (2b)^3$
$$= (a + 2b)(a^2 - 2ab + (2b)^2)$$
$$= (a + 2b)(a^2 - 2ab + 4b^2)$$

129. $27x^3 - 125y^3 = (3x)^3 - (5y)^3 = (3x - 5y)((3x)^2 + (3x)(5y) + (5y)^2)$
$$= (3x - 5y)(9x^2 + 15xy + 25y^2)$$

131. $a^6 - b^3 = (a^2)^3 - b^3 = (a^2 - b)((a^2)^2 + a^2b + b^2) = (a^2 - b)(a^4 + a^2b + b^2)$

133. $x^9 + y^6 = (x^3)^3 + (y^2)^3 = (x^3 + y^2)((x^3)^2 - x^3y^2 + (y^2)^2)$
$$= (x^3 + y^2)(x^6 - x^3y^2 + y^4)$$

135. $2x^3 + 54 = 2(x^3 + 27)$
$$= 2(x^3 + 3^3)$$
$$= 2(x + 3)(x^2 - 3x + 3^2)$$
$$= 2(x + 3)(x^2 - 3x + 9)$$

137. $-x^3 + 216 = -(x^3 - 216)$
$$= -(x^3 - 6^3)$$
$$= -(x - 6)(x^2 + 6x + 6^2)$$
$$= -(x - 6)(x^2 + 6x + 36)$$

139. $64m^3x - 8n^3x = 8x(8m^3 - n^3) = 8x((2m)^3 - n^3) = 8x(2m - n)((2m)^2 + 2mn + n^2)$
$$= 8x(2m - n)(4m^2 + 2mn + n^2)$$

141. $x^4y + 216xy^4 = xy(x^3 + 216y^3) = xy(x^3 + (6y)^3) = xy(x + 6y)(x^2 - 6xy + (6y)^2)$
$$= xy(x + 6y)(x^2 - 6xy + 36y^2)$$

143. $81r^4s^2 - 24rs^5 = 3rs^2(27r^3 - 8s^3) = 3rs^2((3r)^3 - (2s)^3)$
$$= 3rs^2(3r - 2s)((3r)^2 + (3r)(2s) + (2s)^2)$$
$$= 3rs^2(3r - 2s)(9r^2 + 6rs + 4s^2)$$

145. $125a^6b^2 + 64a^3b^5 = a^3b^2(125a^3 + 64b^3) = a^3b^2((5a)^3 + (4b)^3)$
$$= a^3b^2(5a + 4b)((5a)^2 - (5a)(4b) + (4b)^2)$$
$$= a^3b^2(5a + 4b)(25a^2 - 20ab + 16b^2)$$

147. $2mp^4 + 16mpq^3 = 2mp(p^3 + 8q^3) = 2mp(p^3 + (2q)^3) = 2mp(p + 2q)(p^2 - 2pq + (2q)^2)$
$$= 2mp(p + 2q)(p^2 - 2pq + 4q^2)$$

149. $3(x^3 + y^3) - z(x^3 + y^3) = (x^3 + y^3)(3 - z) = (x + y)(x^2 - xy + y^2)(3 - z)$

151. $(m^3 + 8n^3) + (m^3x + 8n^3x) = 1(m^3 + 8n^3) + x(m^3 + 8n^3)$
$$= (m^3 + 8n^3)(1 + x)$$
$$= (m + 2n)(m^2 - 2mn + (2n)^2)(1 + x)$$
$$= (m + 2n)(m^2 - 2mn + 4n^2)(1 + x)$$

153. $(a^4 + 27a) - (a^3b + 27b) = a(a^3 + 27) - b(a^3 + 27) = (a^3 + 27)(a - b)$
$$= (a^3 + 3^3)(a - b)$$
$$= (a + 3)(a^2 - 3a + 3^2)(a - b)$$
$$= (a + 3)(a^2 - 3a + 9)(a - b)$$

155. $y^3(y^2 - 1) - 27(y^2 - 1) = (y^2 - 1)(y^3 - 27) = (y^2 - 1^2)(y^3 - 3^3)$
$$= (y + 1)(y - 1)(y - 3)(y^2 + 3y + 9)$$

157. $x^6 - 1 = (x^3)^2 - 1^2 = (x^3 + 1)(x^3 - 1) = (x^3 + 1^3)(x^3 - 1^3)$
$$= (x + 1)(x^2 - 1x + 1^2)(x - 1)(x^2 + 1x + 1^2)$$
$$= (x + 1)(x^2 - x + 1)(x - 1)(x^2 + x + 1)$$

159. $x^{12} - y^6 = (x^6)^2 - (y^3)^2 = (x^6 + y^3)(x^6 - y^3)$
$$= ((x^2)^3 + y^3)((x^2)^3 - y^3)$$
$$= (x^2 + y)((x^2)^2 - x^2y + y^2)(x^2 - y)((x^2)^2 + x^2y + y^2)$$
$$= (x^2 + y)(x^4 - x^2y + y^2)(x^2 - y)(x^4 + x^2y + y^2)$$

161. Answers may vary.   **163. Answers may vary.**

165. $x^2 - q^2 = x^2 + 0x - q^2$: $a = 1, b = 0, c = -q^2 \Rightarrow b^2 - 4ac = 0^2 - 4(1)(-q^2) = 4q^2 = (2q)^2$.
Since $4q^2$ is a perfect square, the binomial is factorable, and the test works.

167. In the 5th line, both sides of the equation are divided by $x - y$. This is allowed as long as $x - y \neq 0$, or as long as $x \neq y$. However, the 1st line states that $x = y$, so when both sides are divided by $x - y$ in the 5th line, both sides are being divided by 0, which is not allowed.

Exercise 7.5 (page 475)

1.
$$\longleftarrow \!\!)\!\!-\!\!-\!\![\!\!\longrightarrow$$
$$\;-4 \qquad 5$$

3.
$$P = 2l + 2w$$
$$P - 2l = 2w$$
$$\frac{P - 2l}{2} = \frac{2w}{2}$$
$$\frac{P - 2l}{2} = w, \text{ or } w = \frac{P - 2l}{2}$$

5.
$$a^4 - 13a^2 + 36 = 0$$
$$\left(a^2 - 4\right)\left(a^2 - 9\right) = 36$$
$$(a + 2)(a - 2)(a + 3)(a - 3) = 36$$

$$a + 2 = 0 \quad \textbf{or} \quad a - 2 = 0 \quad \textbf{or} \quad a + 3 = 0 \quad \textbf{or} \quad a - 3 = 0$$
$$a = -2 \qquad\qquad a = 2 \qquad\qquad a = -3 \qquad\qquad a = 3$$

7. $\dfrac{a}{b}$

9. $\dfrac{ad}{bc}$

11. $\dfrac{12x^3}{3x} = \dfrac{12}{3} \cdot \dfrac{x^3}{x} = 4x^2$

13. $\dfrac{-24x^3y^4}{18x^4y^3} = -\dfrac{24}{18} \cdot \dfrac{x^3y^4}{x^4y^3} = -\dfrac{4y}{3x}$

15. $\dfrac{9y^2(y - z)}{21y(y - z)^2} = \dfrac{3y}{7(y - z)}$

17. $\dfrac{(a - b)(b - c)(c - d)}{(c - d)(b - c)(a - b)} = 1$

19. $\dfrac{x + y}{x^2 - y^2} = \dfrac{x + y}{(x + y)(x - y)} = \dfrac{1}{x - y}$

21. $\dfrac{12 - 3x^2}{x^2 - x - 2} = \dfrac{3(4 - x^2)}{(x - 2)(x + 1)} = \dfrac{3(2 + x)(2 - x)}{(x - 2)(x + 1)} = \dfrac{-3(x + 2)(x - 2)}{(x - 2)(x + 1)} = \dfrac{-3(x + 2)}{x + 1}$

23. $\dfrac{x^3 + 8}{x^2 - 2x + 4} = \dfrac{(x + 2)(x^2 - 2x + 4)}{x^2 - 2x + 4} = x + 2$

25. $\dfrac{x^2 + 2x + 1}{x^2 + 4x + 3} = \dfrac{(x + 1)(x + 1)}{(x + 1)(x + 3)} = \dfrac{x + 1}{x + 3}$

27. $\dfrac{3m - 6n}{3n - 6m} = \dfrac{3(m - 2n)}{3(n - 2m)} = \dfrac{m - 2n}{n - 2m}$

29. $\dfrac{4x^2 + 24x + 32}{16x^2 + 8x - 48} = \dfrac{4(x^2 + 6x + 8)}{8(2x^2 + x - 6)} = \dfrac{4(x + 4)(x + 2)}{8(2x - 3)(x + 2)} = \dfrac{x + 4}{2(2x - 3)}$

31. $\dfrac{3x^2 - 3y^2}{x^2 + 2y + 2x + yx} = \dfrac{3(x^2 - y^2)}{x^2 + 2x + yx + 2y} = \dfrac{3(x^2 - y^2)}{x(x + 2) + y(x + 2)} = \dfrac{3(x + y)(x - y)}{(x + 2)(x + y)}$
$$= \dfrac{3(x - y)}{x + 2}$$

33. $\dfrac{x - y}{x^3 - y^3 - x + y} = \dfrac{x - y}{(x^3 - y^3) - (x - y)} = \dfrac{x - y}{(x - y)(x^2 + xy + y^2) - 1(x - y)}$
$$= \dfrac{x - y}{(x - y)(x^2 + xy + y^2 - 1)}$$
$$= \dfrac{1}{x^2 + xy + y^2 - 1}$$

35. $\dfrac{x^2y^2}{cd} \cdot \dfrac{c^{-2}d^2}{x} = x^{2-1}y^2c^{-2-1}d^{2-1} = xy^2c^{-3}d = \dfrac{xy^2d}{c^3}$

37. $\dfrac{-x^2y^{-2}}{x^{-1}y^{-3}} \div \dfrac{x^{-3}y^2}{x^4y^{-1}} = \dfrac{-x^2y^{-2}}{x^{-1}y^{-3}} \cdot \dfrac{x^4y^{-1}}{x^{-3}y^2} = \dfrac{-x^6y^{-3}}{x^{-4}y^{-1}} = -x^{6-(-4)}y^{-3-(-1)} = -x^{10}y^{-2} = -\dfrac{x^{10}}{y^2}$

39. $\dfrac{x^2+2x+1}{x} \cdot \dfrac{x^2-x}{x^2-1} = \dfrac{(x+1)(x+1)}{x} \cdot \dfrac{x(x-1)}{(x+1)(x-1)} = x+1$

41. $\dfrac{2x^2-x-3}{x^2-1} \cdot \dfrac{x^2+x-2}{2x^2+x-6} = \dfrac{(2x-3)(x+1)}{(x+1)(x-1)} \cdot \dfrac{(x+2)(x-1)}{(2x-3)(x+2)} = 1$

43. $\dfrac{x^2-16}{x^2-25} \div \dfrac{x+4}{x-5} = \dfrac{x^2-16}{x^2-25} \cdot \dfrac{x-5}{x+4} = \dfrac{(x+4)(x-4)}{(x+5)(x-5)} \cdot \dfrac{x-5}{x+4} = \dfrac{x-4}{x+5}$

45. $\dfrac{a^2+2a-35}{12x} \div \dfrac{ax-3x}{a^2+4a-21} = \dfrac{a^2+2a-35}{12x} \cdot \dfrac{a^2+4a-21}{ax-3x}$

$\qquad = \dfrac{(a+7)(a-5)}{12x} \cdot \dfrac{(a+7)(a-3)}{x(a-3)} = \dfrac{(a+7)^2(a-5)}{12x^2}$

47. $\dfrac{3t^2-t-2}{6t^2-5t-6} \cdot \dfrac{4t^2-9}{2t^2+5t+3} = \dfrac{(3t+2)(t-1)}{(3t+2)(2t-3)} \cdot \dfrac{(2t+3)(2t-3)}{(2t+3)(t+1)} = \dfrac{t-1}{t+1}$

49. $\dfrac{3n^2+5n-2}{12n^2-13n+3} \div \dfrac{n^2+3n+2}{4n^2+5n-6} = \dfrac{3n^2+5n-2}{12n^2-13n+3} \cdot \dfrac{4n^2+5n-6}{n^2+3n+2}$

$\qquad = \dfrac{(3n-1)(n+2)}{(3n-1)(4n-3)} \cdot \dfrac{(4n-3)(n+2)}{(n+2)(n+1)} = \dfrac{n+2}{n+1}$

51. $(2x^2-15x+25) \div \dfrac{2x^2-3x-5}{x+1} = \dfrac{2x^2-15x+25}{1} \cdot \dfrac{x+1}{2x^2-3x-5}$

$\qquad = \dfrac{(2x-5)(x-5)}{1} \cdot \dfrac{x+1}{(2x-5)(x+1)} = x-5$

53. $\dfrac{x^3+y^3}{x^3-y^3} \div \dfrac{x^2-xy+y^2}{x^2+xy+y^2} = \dfrac{x^3+y^3}{x^3-y^3} \cdot \dfrac{x^2+xy+y^2}{x^2-xy+y^2} = \dfrac{(x+y)(x^2-xy+y^2)}{(x-y)(x^2+xy+y^2)} \cdot \dfrac{x^2+xy+y^2}{x^2-xy+y^2}$

$\qquad = \dfrac{x+y}{x-y}$

55. $\dfrac{m^2-n^2}{2x^2+3x-2} \cdot \dfrac{2x^2+5x-3}{n^2-m^2} = \dfrac{(m+n)(m-n)}{(2x-1)(x+2)} \cdot \dfrac{(2x-1)(x+3)}{(n+m)(n-m)} = -\dfrac{x+3}{x+2}$

57. $\dfrac{ax+ay+bx+by}{x^3-27} \cdot \dfrac{x^2+3x+9}{xc+xd+yc+yd} = \dfrac{a(x+y)+b(x+y)}{(x-3)(x^2+3x+9)} \cdot \dfrac{x^2+3x+9}{x(c+d)+y(c+d)}$

$$= \dfrac{(x+y)(a+b)}{(x-3)(x^2+3x+9)} \cdot \dfrac{x^2+3x+9}{(c+d)(x+y)}$$

$$= \dfrac{a+b}{(x-3)(c+d)}$$

59. $\dfrac{x^2-x-6}{x^2-4} \cdot \dfrac{x^2-x-2}{9-x^2} = \dfrac{(x-3)(x+2)}{(x+2)(x-2)} \cdot \dfrac{(x-2)(x+1)}{(3+x)(3-x)} = -\dfrac{x+1}{x+3}$

61. $\dfrac{2x^2+3xy+y^2}{y^2-x^2} \div \dfrac{6x^2+5xy+y^2}{2x^2-xy-y^2} = \dfrac{2x^2+3xy+y^2}{y^2-x^2} \cdot \dfrac{2x^2-xy-y^2}{6x^2+5xy+y^2}$

$$= \dfrac{(2x+y)(x+y)}{(y+x)(y-x)} \cdot \dfrac{(2x+y)(x-y)}{(2x+y)(3x+y)} = -\dfrac{2x+y}{3x+y}$$

63. $\dfrac{3x^2y^2}{6x^3y} \cdot \dfrac{-4x^7y^{-2}}{18x^{-2}y} \div \dfrac{36x}{18y^{-2}} = \dfrac{3x^2y^2}{6x^3y} \cdot \dfrac{-4x^7y^{-2}}{18x^{-2}y} \cdot \dfrac{18y^{-2}}{36x} = \dfrac{-2^3 \cdot 3^3 x^9 y^{-2}}{2^4 \cdot 3^5 x^2 y^2} = -\dfrac{x^7}{18y^4}$

65. $(4x+12) \cdot \dfrac{x^2}{2x-6} \div \dfrac{2}{x-3} = \dfrac{4x+12}{1} \cdot \dfrac{x^2}{2x-6} \cdot \dfrac{x-3}{2} = \dfrac{4(x+3)}{1} \cdot \dfrac{x^2}{2(x-3)} \cdot \dfrac{x-3}{2}$

$$= x^2(x+3)$$

67. $\dfrac{2x^2-2x-4}{x^2+2x-8} \cdot \dfrac{3x^2+15x}{x+1} \div \dfrac{4x^2-100}{x^2-x-20} = \dfrac{2x^2-2x-4}{x^2+2x-8} \cdot \dfrac{3x^2+15x}{x+1} \cdot \dfrac{x^2-x-20}{4x^2-100}$

$$= \dfrac{2(x-2)(x+1)}{(x+4)(x-2)} \cdot \dfrac{3x(x+5)}{x+1} \cdot \dfrac{(x-5)(x+4)}{4(x+5)(x-5)}$$

$$= \dfrac{3x}{2}$$

69. $\dfrac{2x^2+5x-3}{x^2+2x-3} \div \left(\dfrac{x^2+2x-35}{x^2-6x+5} \div \dfrac{x^2-9x+14}{2x^2-5x+2} \right) =$

$$= \dfrac{2x^2+5x-3}{x^2+2x-3} \div \left(\dfrac{x^2+2x-35}{x^2-6x+5} \cdot \dfrac{2x^2-5x+2}{x^2-9x+14} \right)$$

$$= \dfrac{2x^2+5x-3}{x^2+2x-3} \div \left(\dfrac{(x+7)(x-5)}{(x-5)(x-1)} \cdot \dfrac{(2x-1)(x-2)}{(x-7)(x-2)} \right)$$

$$= \dfrac{2x^2+5x-3}{x^2+2x-3} \div \dfrac{(x+7)(2x-1)}{(x-1)(x-7)} = \dfrac{(2x-1)(x+3)}{(x+3)(x-1)} \cdot \dfrac{(x-1)(x-7)}{(x+7)(2x-1)} = \dfrac{x-7}{x+7}$$

71. $\dfrac{x^2-x-12}{x^2+x-2} \div \dfrac{x^2-6x+8}{x^2-3x-10} \cdot \dfrac{x^2-3x+2}{x^2-2x-15} = \dfrac{x^2-x-12}{x^2+x-2} \cdot \dfrac{x^2-3x-10}{x^2-6x+8} \cdot \dfrac{x^2-3x+2}{x^2-2x-15}$

$$= \dfrac{(x-4)(x+3)}{(x+2)(x-1)} \cdot \dfrac{(x-5)(x+2)}{(x-4)(x-2)} \cdot \dfrac{(x-2)(x-1)}{(x-5)(x+3)} = 1$$

73. $\dfrac{3}{a+b} - \dfrac{a}{a+b} = \dfrac{3-a}{a+b}$

75. $\dfrac{3x}{2x+2} + \dfrac{x+4}{2x+2} = \dfrac{4x+4}{2x+2} = \dfrac{4(x+1)}{2(x+1)}$
$= 2$

77. $\dfrac{5x}{x+1} + \dfrac{3}{x+1} - \dfrac{2x}{x+1} = \dfrac{3x+3}{x+1} = \dfrac{3(x+1)}{x+1} = 3$

79. $\dfrac{3(x^2+x)}{x^2-5x+6} + \dfrac{-3(x^2-x)}{x^2-5x+6} = \dfrac{3x^2+3x}{x^2-5x+6} + \dfrac{-3x^2+3x}{x^2-5x+6} = \dfrac{6x}{(x-3)(x-2)}$

81. $\dfrac{a}{2} + \dfrac{2a}{5} = \dfrac{a\cdot 5}{2\cdot 5} + \dfrac{2a\cdot 2}{5\cdot 2} = \dfrac{5a}{10} + \dfrac{4a}{10}$
$= \dfrac{9a}{10}$

83. $\dfrac{3}{4x} + \dfrac{2}{3x} = \dfrac{3\cdot 3}{4x\cdot 3} + \dfrac{2\cdot 4}{3x\cdot 4} = \dfrac{9}{12x} + \dfrac{8}{12x}$
$= \dfrac{17}{12x}$

85. $\dfrac{a+b}{3} + \dfrac{a-b}{7} = \dfrac{(a+b)7}{3(7)} + \dfrac{(a-b)3}{7(3)} = \dfrac{7a+7b}{21} + \dfrac{3a-3b}{21} = \dfrac{10a+4b}{21}$

87. $\dfrac{3}{x+2} + \dfrac{5}{x-4} = \dfrac{3(x-4)}{(x+2)(x-4)} + \dfrac{5(x+2)}{(x-4)(x+2)} = \dfrac{3x-12}{(x+2)(x-4)} + \dfrac{5x+10}{(x+2)(x-4)}$
$= \dfrac{8x-2}{(x+2)(x-4)} = \dfrac{2(4x-1)}{(x+2)(x-4)}$

89. $\dfrac{x+2}{x+5} - \dfrac{x-3}{x+7} = \dfrac{(x+2)(x+7)}{(x+5)(x+7)} - \dfrac{(x-3)(x+5)}{(x+7)(x+5)} = \dfrac{x^2+9x+14}{(x+5)(x+7)} - \dfrac{x^2+2x-15}{(x+5)(x+7)}$
$= \dfrac{7x+29}{(x+5)(x+7)}$

91. $x + \dfrac{1}{x} = \dfrac{x}{1} + \dfrac{1}{x} = \dfrac{x(x)}{1(x)} + \dfrac{1}{x} = \dfrac{x^2}{x} + \dfrac{1}{x} = \dfrac{x^2+1}{x}$

93. $\dfrac{x+8}{x-3} - \dfrac{x-14}{3-x} = \dfrac{x+8}{x-3} - \dfrac{-x+14}{x-3} = \dfrac{2x-6}{x-3} = \dfrac{2(x-3)}{x-3} = 2$

95. $\dfrac{x}{x^2+5x+6} + \dfrac{x}{x^2-4} = \dfrac{x}{(x+2)(x+3)} + \dfrac{x}{(x+2)(x-2)}$
$= \dfrac{x(x-2)}{(x+2)(x+3)(x-2)} + \dfrac{x(x+3)}{(x+2)(x-2)(x+3)}$
$= \dfrac{x^2-2x}{(x+2)(x+3)(x-2)} + \dfrac{x^2+3x}{(x+2)(x+3)(x-2)}$
$= \dfrac{2x^2+x}{(x+2)(x+3)(x-2)}$

97. $\dfrac{8}{x^2-9} + \dfrac{2}{x-3} - \dfrac{6}{x} = \dfrac{8}{(x+3)(x-3)} + \dfrac{2}{x-3} - \dfrac{6}{x}$

$$= \dfrac{8x}{x(x+3)(x-3)} + \dfrac{2x(x+3)}{x(x+3)(x-3)} - \dfrac{6(x+3)(x-3)}{x(x+3)(x-3)}$$

$$= \dfrac{8x}{x(x+3)(x-3)} + \dfrac{2x^2+6x}{x(x+3)(x-3)} - \dfrac{6(x^2-9)}{x(x+3)(x-3)}$$

$$= \dfrac{8x}{x(x+3)(x-3)} + \dfrac{2x^2+6x}{x(x+3)(x-3)} - \dfrac{6x^2-54}{x(x+3)(x-3)}$$

$$= \dfrac{-4x^2+14x+54}{x(x+3)(x-3)}$$

99. $1 + x - \dfrac{x}{x-5} = \dfrac{x+1}{1} - \dfrac{x}{x-5} = \dfrac{(x+1)(x-5)}{1(x-5)} - \dfrac{x}{x-5}$

$$= \dfrac{x^2-4x-5}{x-5} - \dfrac{x}{x-5} = \dfrac{x^2-5x-5}{x-5}$$

101. $\dfrac{3}{x+1} - \dfrac{2}{x-1} + \dfrac{x+3}{x^2-1} = \dfrac{3}{x+1} - \dfrac{2}{x-1} + \dfrac{x+3}{(x+1)(x-1)}$

$$= \dfrac{3(x-1)}{(x+1)(x-1)} - \dfrac{2(x+1)}{(x+1)(x-1)} + \dfrac{x+3}{(x+1)(x-1)}$$

$$= \dfrac{3x-3}{(x+1)(x-1)} - \dfrac{2x+2}{(x+1)(x-1)} + \dfrac{x+3}{(x+1)(x-1)}$$

$$= \dfrac{2x-2}{(x+1)(x-1)} = \dfrac{2(x-1)}{(x+1)(x-1)} = \dfrac{2}{x+1}$$

103. $\dfrac{x-2}{x^2-3x} + \dfrac{2x-1}{x^2+3x} - \dfrac{2}{x^2-9} = \dfrac{x-2}{x(x-3)} + \dfrac{2x-1}{x(x+3)} - \dfrac{2}{(x+3)(x-3)}$

$$= \dfrac{(x-2)(x+3)}{x(x-3)(x+3)} + \dfrac{(2x-1)(x-3)}{x(x-3)(x+3)} - \dfrac{2x}{x(x-3)(x+3)}$$

$$= \dfrac{x^2+x-6}{x(x-3)(x+3)} + \dfrac{2x^2-7x+3}{x(x-3)(x+3)} - \dfrac{2x}{x(x-3)(x+3)}$$

$$= \dfrac{3x^2-8x-3}{x(x-3)(x+3)} = \dfrac{(3x+1)(x-3)}{x(x-3)(x+3)} = \dfrac{3x+1}{x(x+3)}$$

105. $\dfrac{\frac{4x}{y}}{\frac{6xz}{y^2}} = \dfrac{4x}{y} \div \dfrac{6xz}{y^2} = \dfrac{4x}{y} \cdot \dfrac{y^2}{6xz} = \dfrac{2y}{3z}$

107. $\dfrac{\frac{x-y}{xy}}{\frac{y-x}{x}} = \dfrac{x-y}{xy} \div \dfrac{y-x}{x} = \dfrac{x-y}{xy} \cdot \dfrac{x}{y-x} = \dfrac{x-y}{xy} \cdot \dfrac{x}{-(x-y)} = -\dfrac{1}{y}$

109. $\dfrac{\frac{1}{a} + \frac{1}{b}}{\frac{1}{a}} = \dfrac{\left(\frac{1}{a} + \frac{1}{b}\right) \cdot ab}{\frac{1}{a} \cdot ab} = \dfrac{\frac{1}{a} \cdot ab + \frac{1}{b} \cdot ab}{b} = \dfrac{b+a}{b}$

111. $\dfrac{\frac{y}{x} - \frac{x}{y}}{\frac{1}{x} + \frac{1}{y}} = \dfrac{\left(\frac{y}{x} - \frac{x}{y}\right) \cdot xy}{\left(\frac{1}{x} + \frac{1}{y}\right) \cdot xy} = \dfrac{\frac{y}{x} \cdot xy - \frac{x}{y} \cdot xy}{\frac{1}{x} \cdot xy + \frac{1}{y} \cdot xy} = \dfrac{y^2 - x^2}{y + x} = \dfrac{(y+x)(y-x)}{y+x} = y - x$

113. $\dfrac{\frac{1}{a} - \frac{1}{b}}{\frac{a}{b} - \frac{b}{a}} = \dfrac{\left(\frac{1}{a} - \frac{1}{b}\right) \cdot ab}{\left(\frac{a}{b} - \frac{b}{a}\right) \cdot ab} = \dfrac{\frac{1}{a} \cdot ab - \frac{1}{b} \cdot ab}{\frac{a}{b} \cdot ab - \frac{b}{a} \cdot ab} = \dfrac{b - a}{a^2 - b^2} = \dfrac{b - a}{(a+b)(a-b)} = -\dfrac{1}{a+b}$

115. $\dfrac{1 + \frac{6}{x} + \frac{8}{x^2}}{1 + \frac{1}{x} - \frac{12}{x^2}} = \dfrac{\left(1 + \frac{6}{x} + \frac{8}{x^2}\right) \cdot x^2}{\left(1 + \frac{1}{x} - \frac{12}{x^2}\right) \cdot x^2} = \dfrac{x^2 + 6x + 8}{x^2 + x - 12} = \dfrac{(x+4)(x+2)}{(x+4)(x-3)} = \dfrac{x+2}{x-3}$

117. $\dfrac{\frac{1}{a+1} + 1}{\frac{3}{a-1} + 1} = \dfrac{\left(\frac{1}{a+1} + 1\right)(a+1)(a-1)}{\left(\frac{3}{a-1} + 1\right)(a+1)(a-1)} = \dfrac{1(a-1) + 1(a+1)(a-1)}{3(a+1) + 1(a+1)(a-1)} = \dfrac{a - 1 + a^2 - 1}{3a + 3 + a^2 - 1}$

$= \dfrac{a^2 + a - 2}{a^2 + 3a + 2}$

$= \dfrac{(a+2)(a-1)}{(a+2)(a+1)}$

$= \dfrac{a-1}{a+1}$

119. $\dfrac{x^{-1} + y^{-1}}{x^{-1} - y^{-1}} = \dfrac{\frac{1}{x} + \frac{1}{y}}{\frac{1}{x} - \frac{1}{y}} = \dfrac{\left(\frac{1}{x} + \frac{1}{y}\right)(xy)}{\left(\frac{1}{x} - \frac{1}{y}\right)(xy)} = \dfrac{y+x}{y-x}$

121. $\dfrac{x+y}{x^{-1} + y^{-1}} = \dfrac{x+y}{\frac{1}{x} + \frac{1}{y}} = \dfrac{(x+y)(xy)}{\left(\frac{1}{x} + \frac{1}{y}\right)(xy)} = \dfrac{(x+y)xy}{y+x} = xy$

123. $\dfrac{x - y^{-2}}{y - x^{-2}} = \dfrac{x - \frac{1}{y^2}}{y - \frac{1}{x^2}} = \dfrac{\left(x - \frac{1}{y^2}\right)(x^2 y^2)}{\left(y - \frac{1}{x^2}\right)(x^2 y^2)} = \dfrac{x^3 y^2 - x^2}{x^2 y^3 - y^2} = \dfrac{x^2(xy^2 - 1)}{y^2(x^2 y - 1)}$

125. $\dfrac{1 + \frac{a}{b}}{1 - \frac{a}{1 - \frac{a}{b}}} = \dfrac{1 + \frac{a}{b}}{1 - \frac{a(b)}{(1 - \frac{a}{b})(b)}} = \dfrac{1 + \frac{a}{b}}{1 - \frac{ab}{b - a}} = \dfrac{\left(1 + \frac{a}{b}\right)(b)(b - a)}{\left(1 - \frac{ab}{b-a}\right)(b)(b-a)} = \dfrac{b(b-a) + a(b-a)}{b(b-a) - ab(b)}$

$= \dfrac{b^2 - ab + ab - a^2}{b^2 - ab - ab^2}$

$= \dfrac{b^2 - a^2}{b(b - a - ab)}$

$= \dfrac{(b+a)(b-a)}{b(b-a-ab)}$

127. $a + \dfrac{a}{1 + \frac{a}{a+1}} = a + \dfrac{a(a+1)}{\left(1 + \frac{a}{a+1}\right)(a+1)} = a + \dfrac{a(a+1)}{a+1+a} = a + \dfrac{a^2+a}{2a+1} = \dfrac{a(2a+1)}{2a+1} + \dfrac{a^2+a}{2a+1}$

$$= \dfrac{2a^2 + a + a^2 + a}{2a+1}$$

$$= \dfrac{3a^2 + 2a}{2a+1}$$

129. Answers may vary.

131. Answers may vary.

133. $\dfrac{a-3b}{2b-a} = \dfrac{-(3b-a)}{-(a-2b)} = \dfrac{3b-a}{a-2b}.$ The two answers are the same.

135. You can divide out the 4's in parts a and d.

Exercise 7.6 (page 483)

1. $f(1) = 3(1)^2 + 2(1) - 1 = 4$

3. $f(2a) = 3(2a)^2 + 2(2a) - 1$
$\qquad\quad = 12a^2 + 4a - 1$

5. $2(x^2 + 4x - 1) + 3(2x^2 - 2x + 2) = 2x^2 + 8x - 2 + 6x^2 - 6x + 6 = 8x^2 + 2x + 4$

7. $P(r)$

9.
$$\begin{array}{r|rrr} 1 & 1 & 1 & -2 \\ & & 1 & 2 \\ \hline & 1 & 2 & 0 \end{array} \Rightarrow \boxed{x+2}$$

11.
$$\begin{array}{r|rrr} 4 & 1 & -7 & 12 \\ & & 4 & -12 \\ \hline & 1 & -3 & 0 \end{array} \Rightarrow \boxed{x-3}$$

13.
$$\begin{array}{r|rrr} -4 & 1 & 6 & 8 \\ & & -4 & -8 \\ \hline & 1 & 2 & 0 \end{array} \Rightarrow \boxed{x+2}$$

15.
$$\begin{array}{r|rrr} -2 & 1 & -5 & 14 \\ & & -2 & 14 \\ \hline & 1 & -7 & 28 \end{array} \Rightarrow \boxed{x - 7 + \dfrac{28}{x+2}}$$

17.
$$\begin{array}{r|rrrr} 3 & 3 & -10 & 5 & -6 \\ & & 9 & -3 & 6 \\ \hline & 3 & -1 & 2 & 0 \end{array}$$
$$\Rightarrow \boxed{3x^2 - x + 2}$$

19.
$$\begin{array}{r|rrrr} 2 & 2 & 0 & -5 & -6 \\ & & 4 & 8 & 6 \\ \hline & 2 & 4 & 3 & 0 \end{array}$$
$$\Rightarrow \boxed{2x^2 + 4x + 3}$$

21.
$$\begin{array}{r|rrrr} -1 & 6 & 5 & 0 & 4 \\ & & -6 & 1 & -1 \\ \hline & 6 & -1 & 1 & 3 \end{array}$$
$$\Rightarrow \boxed{6x^2 - x + 1 + \dfrac{3}{x+1}}$$

23.
$$\begin{array}{r|rrrr} 0.2 & 7.2 & -2.1 & 0.5 \\ & & 1.44 & -0.132 \\ \hline & 7.2 & -0.66 & 0.368 \end{array}$$
$$\Rightarrow \boxed{7.2x - 0.66 + \dfrac{0.368}{x-0.2}}$$

25.

$$
\begin{array}{r|rrr}
-1.7 & 2.7 & 1.0 & -5.2 \\
 & & -4.59 & 6.103 \\
\hline
 & 2.7 & -3.59 & 0.903
\end{array}
$$

$\Rightarrow \boxed{2.7x - 3.59 + \frac{0.903}{x+1.7}}$

27.

$$
\begin{array}{r|rrrr}
-57 & 9 & 0 & 0 & -25 \\
 & & -513 & 29241 & -1666737 \\
\hline
 & 9 & -513 & 29241 & -1666762
\end{array}
$$

$\Rightarrow \boxed{9x^2 - 513x + 29{,}241 - \frac{1{,}666{,}762}{x+57}}$

29. $P(1) = 2(1)^3 - 4(1)^2 + 2(1) - 1 = \boxed{-1}$

$$
\begin{array}{r|rrrr}
1 & 2 & -4 & 2 & -1 \\
 & & 2 & -2 & 0 \\
\hline
 & 2 & -2 & 0 & \boxed{-1}
\end{array}
$$

31. $P(-2) = 2(-2)^3 - 4(-2)^2 + 2(-2) - 1$

$= \boxed{-37}$

$$
\begin{array}{r|rrrr}
-2 & 2 & -4 & 2 & -1 \\
 & & -4 & 16 & -36 \\
\hline
 & 2 & -8 & 18 & \boxed{-37}
\end{array}
$$

33. $P(3) = 2(3)^3 - 4(3)^2 + 2(3) - 1 = \boxed{23}$

$$
\begin{array}{r|rrrr}
3 & 2 & -4 & 2 & -1 \\
 & & 6 & 6 & 24 \\
\hline
 & 2 & 2 & 8 & \boxed{23}
\end{array}
$$

35. $P(0) = 2(0)^3 - 4(0)^2 + 2(0) - 1 = \boxed{-1}$

$$
\begin{array}{r|rrrr}
0 & 2 & -4 & 2 & -1 \\
 & & 0 & 0 & 0 \\
\hline
 & 2 & -4 & 2 & \boxed{-1}
\end{array}
$$

37. $Q(-1) = (-1)^4 - 3(-1)^3 + 2(-1)^2 + (-1) - 3 = \boxed{2}$

$$
\begin{array}{r|rrrrr}
-1 & 1 & -3 & 2 & 1 & -3 \\
 & & -1 & 4 & -6 & 5 \\
\hline
 & 1 & -4 & 6 & -5 & \boxed{2}
\end{array}
$$

39. $Q(2) = (2)^4 - 3(2)^3 + 2(2)^2 + (2) - 3$

$= \boxed{-1}$

$$
\begin{array}{r|rrrrr}
2 & 1 & -3 & 2 & 1 & -3 \\
 & & 2 & -2 & 0 & 2 \\
\hline
 & 1 & -1 & 0 & 1 & \boxed{-1}
\end{array}
$$

41. $Q(3) = (3)^4 - 3(3)^3 + 2(3)^2 + (3) - 3$

$= \boxed{18}$

$$
\begin{array}{r|rrrrr}
3 & 1 & -3 & 2 & 1 & -3 \\
 & & 3 & 0 & 6 & 21 \\
\hline
 & 1 & 0 & 2 & 7 & \boxed{18}
\end{array}
$$

43. $Q(-3) = (-3)^4 - 3(-3)^3 + 2(-3)^2 + (-3) - 3 = \boxed{174}$

$$
\begin{array}{r|rrrrr}
-3 & 1 & -3 & 2 & 1 & -3 \\
 & & -3 & 18 & -60 & 177 \\
\hline
 & 1 & -6 & 20 & -59 & \boxed{174}
\end{array}
$$

45.

$$
\begin{array}{r|rrrr}
2 & 1 & -4 & 1 & -2 \\
 & & 2 & -4 & -6 \\
\hline
 & 1 & -2 & -3 & \boxed{-8}
\end{array}
$$

47.

$$
\begin{array}{r|rrrr}
3 & 2 & 0 & 1 & 2 \\
 & & 6 & 18 & 57 \\
\hline
 & 2 & 6 & 19 & \boxed{59}
\end{array}
$$

49.

$$
\begin{array}{r|rrrrr}
-2 & 1 & -2 & 1 & -3 & 2 \\
 & & -2 & 8 & -18 & 42 \\
\hline
 & 1 & -4 & 9 & -21 & \boxed{44}
\end{array}
$$

51.

$$
\begin{array}{r|rrrrrr}
-\frac{1}{2} & 3 & 0 & 0 & 0 & 0 & 1 \\
 & & -\frac{3}{2} & \frac{3}{4} & -\frac{3}{8} & \frac{3}{16} & -\frac{3}{32} \\
\hline
 & 3 & -\frac{3}{2} & \frac{3}{4} & -\frac{3}{8} & \frac{3}{16} & \boxed{\frac{29}{32}}
\end{array}
$$

53. $\underline{3}\,\big|\,\begin{array}{cccc}1 & -3 & 5 & -15\end{array}$
$\phantom{\underline{3}\,\big|\,\,1\,\,}\begin{array}{ccc}3 & 0 & 15\end{array}$
$\phantom{\underline{3}\,\big|\,}\begin{array}{cccc}1 & 0 & 5 & \boxed{0}\end{array} \quad \Rightarrow \text{factor}$

55. $\underline{-2}\,\big|\,\begin{array}{ccc}3 & -7 & 4\end{array}$
$\phantom{\underline{-2}\,\big|\,\,3\,\,}\begin{array}{cc}-6 & 26\end{array}$
$\phantom{\underline{-2}\,\big|\,}\begin{array}{ccc}3 & -13 & \boxed{30}\end{array} \quad \Rightarrow \text{not a factor}$

57. $\underline{2}\,\big|\,\begin{array}{ccccccc}1 & 0 & 0 & 0 & 0 & 0 & 0\end{array}$
$\phantom{\underline{2}\,\big|\,\,1\,\,}\begin{array}{cccccc}2 & 4 & 8 & 16 & 32 & 64\end{array}$
$\phantom{\underline{2}\,\big|\,}\begin{array}{ccccccc}1 & 2 & 4 & 8 & 16 & 32 & \boxed{64}\end{array}$

59. **Answers may vary.**

61. $\text{remainder} = P(1) = 1^{100} - 1^{99} + 1^{98} - 1^{97} + \cdots + 1^{2} - 1 + 1 = 1$

Chapter 7 Summary (page 486)

1. $\quad 4(y - 1) = 28$
$\qquad 4y - 4 = 28$
$\qquad\quad 4y = 32$
$\qquad\quad\ \ y = 8$

2. $\quad 3(x + 7) = 42$
$\qquad 3x + 21 = 42$
$\qquad\quad 3x = 21$
$\qquad\quad\ x = 7$

3. $\quad 13(x - 9) - 2 = 7x - 5$
$\qquad 13x - 117 - 2 = 7x - 5$
$\qquad\quad 13x - 119 = 7x - 5$
$\quad 13x - 7x - 119 = 7x - 7x - 5$
$\quad\ 6x - 119 + 119 = -5 + 119$
$\qquad\qquad\quad 6x = 114$
$\qquad\qquad\ \ \dfrac{6x}{6} = \dfrac{114}{6}$
$\qquad\qquad\quad\ x = 19$

4. $\qquad \dfrac{8(x - 5)}{3} = 2(x - 4)$
$\qquad 3 \cdot \dfrac{8(x - 5)}{3} = 3 \cdot 2(x - 4)$
$\qquad\quad 8(x - 5) = 6(x - 4)$
$\qquad\quad 8x - 40 = 6x - 24$
$\quad 8x - 6x - 40 = 6x - 6x - 24$
$\quad 2x - 40 + 40 = -24 + 40$
$\qquad\qquad 2x = 16$
$\qquad\qquad \dfrac{2x}{2} = \dfrac{16}{2}$
$\qquad\qquad\ x = 8$

5. $\quad 2x + 4 = 2(x + 3) - 2$
$\quad 2x + 4 = 2x + 6 - 2$
$\quad 2x + 4 = 2x + 4$
$\qquad\quad \text{identity}$

6. $\quad (3x - 2) - x = 2(x - 4)$
$\qquad 3x - 2 - x = 2x - 8$
$\qquad\quad 2x - 2 = 2x - 8$
$\quad 2x - 2x - 2 = 2x - 2x - 8$
$\qquad\qquad -2 \neq -8$
$\qquad\quad \text{contradiction}$

7.
$$V = \frac{1}{3}\pi r^2 h$$
$$3V = 3\left(\frac{1}{3}\pi r^2 h\right)$$
$$3V = \pi r^2 h$$
$$\frac{3V}{\pi r^2} = \frac{\pi r^2 h}{\pi r^2}$$
$$\frac{3V}{\pi r^2} = h, \text{ or } h = \frac{3V}{\pi r^2}$$

8.
$$V = \frac{1}{6}ab(x+y)$$
$$6V = 6 \cdot \frac{1}{6}ab(x+y)$$
$$6V = ab(x+y)$$
$$\frac{6V}{ab} = \frac{ab(x+y)}{ab}$$
$$\frac{6V}{ab} = x + y$$
$$\frac{6V}{ab} - y = x + y - y$$
$$\frac{6V}{ab} - y = x, \text{ or } x = \frac{6V}{ab} - y$$

9. Let $x =$ the length of the first piece.
Then $3x =$ the length of the other piece.

$$\boxed{\text{Sum of lengths}} = \boxed{\text{Total length}}$$
$$x + 3x = 20$$
$$4x = 20$$
$$x = 5$$

He should cut the board 5 feet from one end.

10. Let $w =$ the width of the rectangle. Then $w + 4 =$ the length of the rectangle.

$$\boxed{\text{Perimeter}} = 28$$
$$w + w + w + 4 + w + 4 = 28$$
$$4w + 8 = 28$$
$$4w = 20$$
$$w = 5$$

The dimensions are 5 meters by 9 meters, for an area of $45\,\text{m}^2$.

11.
$$\frac{1}{3}y - 2 \geq \frac{1}{2}y + 2$$
$$6\left(\frac{1}{3}y - 2\right) \geq 6\left(\frac{1}{2}y + 2\right)$$
$$2y - 12 + 12 \geq 3y + 12 + 12$$
$$2y - 3y \geq 3y - 3y + 24$$
$$-y \geq 24$$
$$y \leq -24$$
solution set: $(-\infty, -24]$

-24

12.
$$\frac{7}{4}(x+3) < \frac{3}{8}(x-3)$$
$$8 \cdot \frac{7}{4}(x+3) < 8 \cdot \frac{3}{8}(x-3)$$
$$14(x+3) < 3(x-3)$$
$$14x + 42 < 3x - 9$$
$$14x < 3x - 51$$
$$11x < -51$$
$$x < -\frac{51}{11}$$
solution set: $\left(-\infty, -\frac{51}{11}\right)$

$-\frac{51}{11}$

13.
$$3 < 3x + 4 < 10$$
$$3 - 4 < 3x + 4 - 4 < 10 - 4$$
$$-1 < 3x < 6$$
$$-\frac{1}{3} < \frac{3x}{3} < \frac{6}{3}$$

$-\frac{1}{3} < x < 2 \Rightarrow$ solution set: $\left(-\frac{1}{3}, 2\right) \Rightarrow$

$-\frac{1}{3}$ 2

14.
$$4x > 3x + 2 > x - 3$$

$4x > 3x + 2$ and $3x + 2 > x - 3$

$4x - 3x > 3x - 3x + 2$ $\qquad$ $3x + 2 - 2 > x - 3 - 2$

$x > 2$ $\qquad\qquad$ $3x - x > x - x - 5$

$x > 2$ $\qquad\qquad$ $2x > -5$

$\qquad\qquad\qquad\qquad$ $x > -\frac{5}{2}$

If $x > 2$ **and** $x > -\frac{5}{2}$, $x > 2$. The solution set is $(2, \infty)$.

15.
$$|3x + 1| = 10$$

$3x + 1 = 10$ **or** $3x + 1 = -10$

$3x = 9$ $\qquad$ $3x = -11$

$x = 3$ $\qquad$ $x = -\frac{11}{3}$

16.
$$\left|\tfrac{3}{2}x - 4\right| = 9$$

$\frac{3}{2}x - 4 = 9$ **or** $\frac{3}{2}x - 4 = -9$

$\frac{3}{2}x = 13$ $\qquad$ $\frac{3}{2}x = -5$

$x = \frac{26}{3}$ $\qquad$ $x = -\frac{10}{3}$

17.
$$|3x + 2| = |2x - 3|$$

$3x + 2 = 2x - 3$ **or** $3x + 2 = -(2x - 3)$

$x = -5$ $\qquad$ $3x + 2 = -2x + 3$

$\qquad\qquad$ $5x = 1$

$\qquad\qquad$ $x = \frac{1}{5}$

18.
$$|5x - 4| = |4x - 5|$$

$5x - 4 = 4x - 5$ **or** $5x - 4 = -(4x - 5)$

$x = -1$ $\qquad$ $5x - 4 = -4x + 5$

$\qquad\qquad$ $9x = 9$

$\qquad\qquad$ $x = 1$

19.
$$|2x + 7| < 3$$

$-3 < 2x + 7 < 3$

$-3 - 7 < 2x + 7 - 7 < 3 - 7$

$-10 < 2x < -4$

$-5 < x < -2$

solution set: $(-5, -2)$

20.
$$|3x - 8| \geq 4$$

$3x - 8 \leq -4$ **or** $3x - 8 \geq 4$

$3x \leq 4$ $\qquad$ $3x \geq 12$

$x \leq \frac{4}{3}$ $\qquad$ $x \geq 4$

solution set: $\left(-\infty, \frac{4}{3}\right] \cup [4, \infty)$

21.
$$\left|\tfrac{3}{2}x - 14\right| \geq 0$$

Since an absolute value is always at least 0, this inequality is true for all real #s. Solution set: $(-\infty, \infty)$

22.
$$\left|\tfrac{2}{3}x + 14\right| < 0$$

Since an absolute value can never be negative, there is no solution.

23. $4x + 8 = 4(x + 2)$

24. $5x^2 y^3 - 10xy^2 = 5xy^2(xy - 2)$

25. $-8x^2 y^3 z^4 - 12x^4 y^3 z^2 = -4x^2 y^3 z^2 (2z^2 + 3x^2)$

26. $12a^6 b^4 c^2 + 15a^2 b^4 c^6 = 3a^2 b^4 c^2 (4a^4 + 5c^4)$

27. $xy + 2y + 4x + 8 = y(x + 2) + 4(x + 2)$
$\qquad = (x + 2)(y + 4)$

28. $ac + bc + 3a + 3b = c(a + b) + 3(a + b)$
$\qquad = (a + b)(c + 3)$

29. $x^{2n} + x^n = x^n(x^{2n-n} + 1) = x^n(x^n + 1)$

30. $y^{2n} - y^{3n} = y^{2n}(1 - y^{4n-3n}) = y^{2n}(1 - y^n)$

31. $x^4 + 4y + 4x^2 + x^2y = x^4 + x^2y + 4x^2 + 4y = x^2(x^2 + y) + 4(x^2 + y) = (x^2 + y)(x^2 + 4)$

32. $a^5 + b^2c + a^2c + a^3b^2 = a^5 + a^3b^2 + a^2c + b^2c = a^3(a^2 + b^2) + c(a^2 + b^2) = (a^2 + b^2)(a^3 + c)$

33. $z^2 - 16 = z^2 - 4^2 = (z + 4)(z - 4)$

34. $y^2 - 121 = y^2 - 11^2 = (y + 11)(y - 11)$

35. $2x^4 - 98 = 2(x^4 - 49) = 2(x^2 + 7)(x^2 - 7)$

36. $3x^6 - 300x^2 = 3x^2(x^4 - 100) = 3x^2(x^2 + 10)(x^2 - 10)$

37. $y^2 + 21y + 20 = (y + 20)(y + 1)$

38. $z^2 - 11z + 30 = (z - 5)(z - 6)$

39. $-x^2 - 3x + 28 = -(x^2 + 3x - 28)$
$\qquad = -(x + 7)(x - 4)$

40. $-y^2 + 5y + 24 = -\left(y^2 - 5y - 24\right)$
$\qquad = -(y - 8)(y + 3)$

41. $y^3 + y^2 - 2y = y(y^2 + y - 2)$
$\qquad = y(y + 2)(y - 1)$

42. $2a^4 + 4a^3 - 6a^2 = 2a^2(a^2 + 2a - 3)$
$\qquad = 2a^2(a + 3)(a - 1)$

43. $15x^2 - 57xy - 12y^2 = 3(5x^2 - 19xy - 4y^2) = 3(5x + y)(x - 4y)$

44. $30x^2 + 65xy + 10y^2 = 5(6x^2 + 13xy + 2y^2) = 5(6x + y)(x + 2y)$

45. $x^2 + 4x + 4 - 4p^4 = (x^2 + 4x + 4) - 4p^4 = (x + 2)^2 - (2p^2)^2 = (x + 2 + 2p^2)(x + 2 - 2p^2)$

46. $y^2 + 3y + 2 + 2x + xy = (y^2 + 3y + 2) + 2x + xy = (y + 1)(y + 2) + x(2 + y)$
$\qquad = (y + 2)(y + 1 + x)$

47. $x^3 + 343 = x^3 + 7^3$
$\qquad = (x + 7)(x^2 - 7x + 49)$

48. $a^3 - 125 = a^3 - 5^3$
$\qquad = (a - 5)(a^2 + 5a + 25)$

49. $8y^3 - 512 = 8(y^3 - 64) = 8(y - 4)(y^2 + 4y + 16)$

50. $4x^3y + 108yz^3 = 4y(x^3 + 27z^3) = 4y(x + 3z)(x^2 - 3xz + 9z^2)$

51. $\dfrac{248x^2y}{576xy^2} = \dfrac{8 \cdot 31x^2y}{8 \cdot 72xy^2} = \dfrac{31x}{72y}$

52. $\dfrac{x^2 - 49}{x^2 + 14x + 49} = \dfrac{(x + 7)(x - 7)}{(x + 7)(x + 7)} = \dfrac{x - 7}{x + 7}$

53. $\dfrac{x^2 + 4x + 4}{x^2 - x - 6} \cdot \dfrac{x^2 - 9}{x^2 + 5x + 6} = \dfrac{(x + 2)(x + 2)}{(x + 2)(x - 3)} \cdot \dfrac{(x + 3)(x - 3)}{(x + 2)(x + 3)} = 1$

54. $\dfrac{x^3 - 64}{x^2 + 4x + 16} \div \dfrac{x^2 - 16}{x + 4} = \dfrac{x^3 - 64}{x^2 + 4x + 16} \cdot \dfrac{x + 4}{x^2 - 16}$

$$= \dfrac{(x-4)(x^2+4x+16)}{x^2+4x+16} \cdot \dfrac{x+4}{(x+4)(x-4)} = 1$$

55. $\dfrac{5y}{x-y} - \dfrac{3}{x-y} = \dfrac{5y-3}{x-y}$

56. $\dfrac{3x-1}{x^2+2} + \dfrac{3(x-2)}{x^2+2} = \dfrac{3x-1+3x-6}{x^2+2}$

$$= \dfrac{6x-7}{x^2+2}$$

57. $\dfrac{3}{x+2} + \dfrac{2}{x+3} = \dfrac{3(x+3)}{(x+2)(x+3)} + \dfrac{2(x+2)}{(x+2)(x+3)} = \dfrac{3x+9+2x+4}{(x+2)(x+3)} = \dfrac{5x+13}{(x+2)(x+3)}$

58. $\dfrac{4x}{x-4} - \dfrac{3}{x+3} = \dfrac{4x(x+3)}{(x-4)(x+3)} - \dfrac{3(x-4)}{(x-4)(x+3)} = \dfrac{4x^2+12x-3x+12}{(x-4)(x+3)} = \dfrac{4x^2+9x+12}{(x-4)(x+3)}$

59. $\dfrac{x^2+3x+2}{x^2-x-6} \cdot \dfrac{3x^2-3x}{x^2-3x-4} \div \dfrac{x^2+3x+2}{x^2-2x-8} = \dfrac{x^2+3x+2}{x^2-x-6} \cdot \dfrac{3x^2-3x}{x^2-3x-4} \cdot \dfrac{x^2-2x-8}{x^2+3x+2}$

$$= \dfrac{(x+2)(x+1)}{(x-3)(x+2)} \cdot \dfrac{3x(x-1)}{(x-4)(x+1)} \cdot \dfrac{(x-4)(x+2)}{(x+2)(x+1)}$$

$$= \dfrac{3x(x-1)}{(x-3)(x+1)}$$

60. $\dfrac{x^2-x-6}{x^2-3x-10} \div \dfrac{x^2-x}{x^2-5x} \cdot \dfrac{x^2-4x+3}{x^2-6x+9} = \dfrac{x^2-x-6}{x^2-3x-10} \cdot \dfrac{x^2-5x}{x^2-x} \cdot \dfrac{x^2-4x+3}{x^2-6x+9}$

$$= \dfrac{(x-3)(x+2)}{(x-5)(x+2)} \cdot \dfrac{x(x-5)}{x(x-1)} \cdot \dfrac{(x-3)(x-1)}{(x-3)(x-3)} = 1$$

61. $\dfrac{2x}{x+1} + \dfrac{3x}{x+2} + \dfrac{4x}{x^2+3x+2} = \dfrac{2x}{x+1} + \dfrac{3x}{x+2} + \dfrac{4x}{(x+2)(x+1)}$

$$= \dfrac{2x(x+2)}{(x+1)(x+2)} + \dfrac{3x(x+1)}{(x+1)(x+2)} + \dfrac{4x}{(x+2)(x+1)}$$

$$= \dfrac{2x^2+4x+3x^2+3x+4x}{(x+1)(x+2)} = \dfrac{5x^2+11x}{(x+1)(x+2)}$$

62. $\dfrac{5x}{x-3} + \dfrac{5}{x^2-5x+6} + \dfrac{x+3}{x-2} = \dfrac{5x}{x-3} + \dfrac{5}{(x-3)(x-2)} + \dfrac{x+3}{x-2}$

$$= \dfrac{5x(x-2)}{(x-3)(x-2)} + \dfrac{5}{(x-3)(x-2)} + \dfrac{(x+3)(x-3)}{(x-3)(x-2)}$$

$$= \dfrac{5x^2-10x+5+x^2-9}{(x-3)(x-2)}$$

$$= \dfrac{6x^2-10x-4}{(x-3)(x-2)} = \dfrac{2(3x+1)(x-2)}{(x-3)(x-2)} = \dfrac{2(3x+1)}{x-3}$$

63. $\dfrac{3(x+2)}{x^2-1} - \dfrac{2}{x+1} + \dfrac{4(x+3)}{x^2-2x+1} = \dfrac{3(x+2)}{(x+1)(x-1)} - \dfrac{2}{(x+1)} + \dfrac{4(x+3)}{(x-1)(x-1)}$

$$= \dfrac{3(x+2)(x-1)}{(x+1)(x-1)(x-1)} - \dfrac{2(x-1)(x-1)}{(x+1)(x-1)(x-1)} + \dfrac{4(x+3)(x+1)}{(x+1)(x-1)(x-1)}$$

$$= \dfrac{3(x^2+x-2) - 2(x^2-2x+1) + 4(x^2+4x+3)}{(x+1)(x-1)(x-1)}$$

$$= \dfrac{3x^2+3x-6 - 2x^2+4x-2 + 4x^2+16x+12}{(x+1)(x-1)(x-1)} = \dfrac{5x^2+23x+4}{(x+1)(x-1)(x-1)}$$

64. $\dfrac{x}{x^2+4x+4} + \dfrac{2x}{x^2-4} - \dfrac{x^2-4}{x-2}$

$$= \dfrac{x}{(x+2)(x+2)} + \dfrac{2x}{(x+2)(x-2)} - \dfrac{x^2-4}{x-2}$$

$$= \dfrac{x(x-2)}{(x+2)(x+2)(x-2)} + \dfrac{2x(x+2)}{(x+2)(x-2)(x+2)} - \dfrac{(x^2-4)(x+2)(x+2)}{(x-2)(x+2)(x+2)}$$

$$= \dfrac{x^2-2x + 2x^2+4x - (x^2-4)(x^2+4x+4)}{(x+2)(x+2)(x-2)}$$

$$= \dfrac{3x^2+2x - (x^4+4x^3-16x-16)}{(x+2)(x+2)(x-2)}$$

$$= \dfrac{3x^2+2x - x^4-4x^3+16x+16}{(x+2)(x+2)(x-2)} = \dfrac{-x^4-4x^3+3x^2+18x+16}{(x+2)(x+2)(x-2)}$$

65. $\dfrac{\frac{3}{x}-\frac{2}{y}}{xy} = \dfrac{\left(\frac{3}{x}-\frac{2}{y}\right)xy}{xy(xy)} = \dfrac{3y-2x}{x^2y^2}$

66. $\dfrac{\frac{1}{x}+\frac{2}{y}}{\frac{2}{x}-\frac{1}{y}} = \dfrac{\left(\frac{1}{x}+\frac{2}{y}\right)xy}{\left(\frac{2}{x}-\frac{1}{y}\right)xy} = \dfrac{y+2x}{2y-x}$

67. $\dfrac{2x+3+\frac{1}{x}}{x+2+\frac{1}{x}} = \dfrac{\left(2x+3+\frac{1}{x}\right)x}{\left(x+2+\frac{1}{x}\right)x} = \dfrac{2x^2+3x+1}{x^2+2x+1} = \dfrac{(2x+1)(x+1)}{(x+1)(x+1)} = \dfrac{2x+1}{x+1}$

68. $\dfrac{x^{-1}-y^{-1}}{x^{-1}+y^{-1}} = \dfrac{\frac{1}{x}-\frac{1}{y}}{\frac{1}{x}+\frac{1}{y}} = \dfrac{\left(\frac{1}{x}-\frac{1}{y}\right)xy}{\left(\frac{1}{x}+\frac{1}{y}\right)xy} = \dfrac{y-x}{y+x}$

69.

2	3	2	−7	2
		6	16	18
	3	8	9	20

70.

−2	2	−4	−14	3
		−4	16	−4
	2	−8	2	−1

71.

$$\begin{array}{r|rrrr} 5 & 1 & -3 & -8 & -10 \\ & & 5 & 10 & 10 \\ \hline & 1 & 2 & 2 & \boxed{0} \end{array} \Rightarrow \text{factor}$$

72.

$$\begin{array}{r|rrrr} -5 & 1 & 4 & -5 & 5 \\ & & -5 & 5 & 0 \\ \hline & 1 & -1 & 0 & \boxed{5} \end{array} \Rightarrow \text{not a factor}$$

Chapter 7 Test (page 490)

1.
$$9(x+4)+4 = 4(x-5)$$
$$9x+36+4 = 4x-20$$
$$9x+40 = 4x-20$$
$$9x-4x+40-40 = 4x-4x-20-40$$
$$5x = -60$$
$$\frac{5x}{5} = \frac{-60}{5}$$
$$x = -12$$

2.
$$\frac{y-1}{5}+2 = \frac{2y-3}{3}$$
$$15\left(\frac{y-1}{5}+2\right) = 15\left(\frac{2y-3}{3}\right)$$
$$3(y-1)+30 = 10y-15$$
$$3y-3+30 = 10y-15$$
$$3y+27 = 10y-15$$
$$3y-10y+27-27 = 10y-10y-15-27$$
$$-7y = -42$$
$$y = 6$$

3.
$$P = L + \frac{s}{f}i$$
$$P-L = \frac{s}{f}i$$
$$f(P-L) = f\cdot\frac{s}{f}i$$
$$f(P-L) = si$$
$$\frac{f(P-L)}{s} = \frac{si}{s}$$
$$\frac{f(P-L)}{s} = i, \text{ or } i = \frac{f(P-L)}{s}$$

4.
$$n = \frac{360}{180-a}$$
$$(180-a)n = (180-a)\cdot\frac{360}{180-a}$$
$$180n - an = 360$$
$$180n - 360 = an$$
$$\frac{180n-360}{n} = \frac{an}{n}$$
$$\frac{180n-360}{n} = a, \text{ or } a = \frac{180n-360}{n}$$

5. Let $x =$ the length of the shortest piece. Then $2x$ and $6x$ are the other lengths.

$$\boxed{\text{Total length}} = 20$$
$$x+2x+6x = 20$$
$$9x = 20$$
$$x = \frac{20}{9} \text{ feet}$$
$$\text{Longest} = 6\left(\frac{20}{9}\right) = \frac{\cancel{3}\cdot 2}{1}\cdot\frac{20}{\cancel{3}\cdot 3}$$
$$= \frac{40}{3} = 13\frac{1}{3} \text{ ft}$$

The longest piece is $13\frac{1}{3}$ feet long.

6. Let $w =$ the width of the rectangle. Then $w+5 =$ the length of the rectangle.

$$\boxed{\text{Perimeter}} = 26$$
$$w+w+w+5+w+5 = 26$$
$$4w+10 = 26$$
$$4w = 16$$
$$w = 4$$

The dimensions are 4 cm by 9 cm, for an area of 36 cm^2.

7.
$$-2(2x + 3) \geq 14$$
$$-4x - 6 + 6 \geq 14 + 6$$
$$-4x \geq 20$$
$$\frac{-4x}{-4} \leq \frac{20}{-4}$$
$$x \leq -5$$
solution set: $(-\infty, -5]$

8.
$$-2 < \frac{x - 4}{3} < 4$$
$$-6 < x - 4 < 12$$
$$-6 + 4 < x - 4 + 4 \leq 12 + 4$$
$$-2 < x < 16$$
solution set: $(-2, 16)$

9.
$$|2x + 3| = 11$$
$$2x + 3 = 11 \quad \textbf{or} \quad 2x + 3 = -11$$
$$2x = 8 \qquad\qquad 2x = -14$$
$$x = 4 \qquad\qquad x = -7$$

10.
$$|3x + 4| = |x + 12|$$
$$3x + 4 = x + 12 \quad \text{or} \quad 3x + 4 = -(x + 12)$$
$$2x = 8 \qquad\qquad 3x + 4 = -x - 12$$
$$x = 4 \qquad\qquad 4x = -16$$
$$x = -4$$

11.
$$|x + 3| \leq 4$$
$$-4 \leq x + 3 \leq 4$$
$$-4 - 3 \leq x + 3 - 3 \leq 4 - 3$$
$$-7 \leq x \leq 1$$
solution set: $[-7, 1]$

12.
$$|2x - 4| > 22$$
$$2x - 4 < -22 \quad \textbf{or} \quad 2x - 4 > 22$$
$$2x < -18 \qquad\qquad 2x > 26$$
$$x < -9 \qquad\qquad x > 13$$
solution set: $(-\infty, -9) \cup (13, \infty)$

13. $3xy^2 + 6x^2y = 3xy(y + 2x)$

14. $12a^3b^2c - 3a^2b^2c^2 + 6abc^3 = 3abc(4a^2b - abc + 2c^2)$

15. $ax - xy + ay - y^2 = x(a - y) + y(a - y) = (a - y)(x + y)$

16. $ax + ay + bx + by - cx - cy = a(x + y) + b(x + y) - c(x + y) = (x + y)(a + b - c)$

17. $x^2 - 49 = x^2 - 7^2 = (x + 7)(x - 7)$

18. $2x^2 - 32 = 2(x^2 - 16) = 2(x + 4)(x - 4)$

19. $4y^4 - 64 = 4(y^4 - 16) = 4(y^2 + 4)(y^2 - 4) = 4(y^2 + 4)(y + 2)(y - 2)$

20. $b^3 + 125 = (b + 5)(b^2 - 5b + 25)$

21. $b^3 - 27 = (b - 3)(b^2 + 3b + 9)$

22.
$$3u^3 - 24 = 3(u^3 - 8)$$
$$= 3(u - 2)(u^2 + 2u + 4)$$

23. $x^2 + 8x + 15 = (x + 3)(x + 5)$

24. $6b^2 + b - 2 = (3b + 2)(2b - 1)$

25.
$$6u^2 + 9u - 6 = 3(2u^2 + 3u - 2)$$
$$= 3(2u - 1)(u + 2)$$

26. $x^2 + 6x + 9 - y^2 = (x^2 + 6x + 9) - y^2 = (x+3)^2 - y^2 = (x+3+y)(x+3-y)$

27. $\dfrac{-12x^2y^3z^2}{18x^3y^4z^2} = -\dfrac{2}{3xy}$

28. $\dfrac{2x^2 + 7x + 3}{4x + 12} = \dfrac{(2x+1)(x+3)}{4(x+3)} = \dfrac{2x+1}{4}$

29. $\dfrac{x^2y^{-2}}{x^3z^2} \cdot \dfrac{x^2z^4}{y^2z} = \dfrac{x^4y^{-2}z^4}{x^3y^2z^3} = \dfrac{xz}{y^4}$

30. $\dfrac{u^2 + 5u + 6}{u^2 - 4} \cdot \dfrac{u^2 - 5u + 6}{u^2 - 9} = \dfrac{(u+2)(u+3)}{(u+2)(u-2)} \cdot \dfrac{(u-2)(u-3)}{(u+3)(u-3)} = 1$

31. $\dfrac{x^3 + y^3}{4} \div \dfrac{x^2 - xy + y^2}{2x + 2y} = \dfrac{x^3 + y^3}{4} \cdot \dfrac{2x + 2y}{x^2 - xy + y^2} = \dfrac{(x+y)(x^2 - xy + y^2)}{4} \cdot \dfrac{2(x+y)}{x^2 - xy + y^2}$

$$= \dfrac{(x+y)^2}{2}$$

32. $\dfrac{x+2}{x+1} - \dfrac{x+1}{x+2} = \dfrac{(x+2)(x+2)}{(x+1)(x+2)} - \dfrac{(x+1)(x+1)}{(x+1)(x+2)} = \dfrac{x^2 + 4x + 4 - (x^2 + 2x + 1)}{(x+1)(x+2)}$

$$= \dfrac{x^2 + 4x + 4 - x^2 - 2x - 1}{(x+1)(x+2)}$$

$$= \dfrac{2x + 3}{(x+1)(x+2)}$$

33. $\dfrac{\dfrac{2u^2w^3}{v^2}}{\dfrac{4uw^4}{uv}} = \dfrac{2u^2w^3}{v^2} \div \dfrac{4uw^4}{uv} = \dfrac{2u^2w^3}{v^2} \cdot \dfrac{uv}{4uw^4} = \dfrac{2u^3w^3v}{4uv^2w^4} = \dfrac{u^2}{2vw}$

34. $\dfrac{\dfrac{x}{y} + \dfrac{1}{2}}{\dfrac{x}{2} - \dfrac{1}{y}} = \dfrac{\left(\dfrac{x}{y} + \dfrac{1}{2}\right)2y}{\left(\dfrac{x}{2} - \dfrac{1}{y}\right)2y} = \dfrac{2x + y}{xy - 2}$

35.

$$
\begin{array}{r}
x^2 - 5x + 10 \\
x + 1 \overline{\smash{\big)}\, x^3 - 4x^2 + 5x + 3} \\
\underline{x^3 + x^2} \\
-5x^2 + 5x \\
\underline{-5x^2 - 5x} \\
10x + 3 \\
\underline{10x + 10} \\
-7
\end{array}
$$

36.

$$
\begin{array}{r|rrrr}
2 & 4 & 3 & 2 & -1 \\
  &   & 8 & 22 & 48 \\
\hline
  & 4 & 11 & 24 & \boxed{47}
\end{array}
$$

Exercise 8.1 (page 499)

1.

3. $x^2 - x = x(x-1)$

5. $x^3 - 1 = (x-1)(x^2 + x + 1)$

7. origin

9. y-coordinate

11. x-axis

13. horizontal

15-21.

23. $(2, 4)$

25. $(-2, -1)$

27. $(4, 0)$

29. $(0, 0)$

31. $x + y = 4$

x	y
-1	5
0	4
2	2

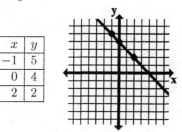

33. $2x - y = 3$

x	y
-1	-5
0	-3
3	3

35.

$$3x + 4y = 12$$
$$3x + 4(0) = 12$$
$$3x = 12$$
$$x = 4$$
x-intercept: $(4, 0)$

$$3x + 4y = 12$$
$$3(0) + 4y = 12$$
$$4y = 12$$
$$y = 3$$
y-intercept: $(0, 3)$

$3x + 4y = 12$

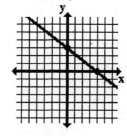

37.
$$y = -3x + 2$$
$$0 = -3x + 2$$
$$3x = 2$$
$$x = \tfrac{2}{3}$$
x-intercept: $\left(\tfrac{2}{3}, 0\right)$

$$y = -3x + 2$$
$$y = -3(0) + 2$$
$$y = 2$$
y-intercept: $(0, 2)$

$$y = -3x + 2$$

39.
$$3y = 6x - 9$$
$$3(0) = 6x - 9$$
$$0 = 6x - 9$$
$$-6x = -9$$
$$x = \tfrac{-9}{-6} = \tfrac{3}{2}$$
x-intercept: $\left(\tfrac{3}{2}, 0\right)$

$$3y = 6x - 9$$
$$3y = 6(0) - 9$$
$$3y = 0 - 9$$
$$3y = -9$$
$$y = \tfrac{-9}{3} = -3$$
y-intercept: $(0, -3)$

$$3y = 6x - 9$$

41.
$$3x + 4y - 8 = 0$$
$$3x + 4(0) - 8 = 0$$
$$3x - 8 = 0$$
$$3x = 8$$
$$x = \tfrac{8}{3}$$
x-intercept: $\left(\tfrac{8}{3}, 0\right)$

$$3x + 4y - 8 = 0$$
$$3(0) + 4y - 8 = 0$$
$$4y - 8 = 0$$
$$4y = 8$$
$$y = 2$$
y-intercept: $(0, 2)$

$$3x + 4y - 8 = 0$$

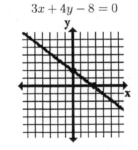

43. $x = 3$
vertical line with
x-coordinate of 3

$x = 3$

45.
$$-3y + 2 = 5$$
$$-3y = 3$$
$$y = -1$$
horizontal line
with y-coordinate
of -1

$y = -1$

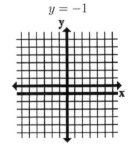

47. $x = \tfrac{x_1 + x_2}{2} = \tfrac{0+6}{2} = \tfrac{6}{2} = 3$
$y = \tfrac{y_1 + y_2}{2} = \tfrac{0+8}{2} = \tfrac{8}{2} = 4$
midpoint: $(3, 4)$

49. $x = \tfrac{x_1 + x_2}{2} = \tfrac{6+12}{2} = \tfrac{18}{2} = 9$
$y = \tfrac{y_1 + y_2}{2} = \tfrac{8+16}{2} = \tfrac{24}{2} = 12$
midpoint: $(9, 12)$

51. $x = \frac{x_1+x_2}{2} = \frac{2+5}{2} = \frac{7}{2}$

$y = \frac{y_1+y_2}{2} = \frac{4+8}{2} = \frac{12}{2} = 6$

midpoint: $\left(\frac{7}{2}, 6\right)$

53. $x = \frac{x_1+x_2}{2} = \frac{-2+3}{2} = \frac{1}{2}$

$y = \frac{y_1+y_2}{2} = \frac{-8+4}{2} = \frac{-4}{2} = -2$

midpoint: $\left(\frac{1}{2}, -2\right)$

55. $x = \frac{x_1+x_2}{2} = \frac{-3+(-5)}{2} = \frac{-8}{2} = -4$

$y = \frac{y_1+y_2}{2} = \frac{5+(-5)}{2} = \frac{0}{2} = 0$

midpoint: $(-4, 0)$

57. $x = \frac{x_1+x_2}{2} = \frac{a+4a}{2} = \frac{5a}{2}$

$y = \frac{y_1+y_2}{2} = \frac{b+3b}{2} = \frac{4b}{2} = 2b$

midpoint: $\left(\frac{5a}{2}, 2b\right)$

59. $x = \frac{x_1+x_2}{2} = \frac{(a-b)+(a+b)}{2} = \frac{2a}{2} = a$

$y = \frac{y_1+y_2}{2} = \frac{b+3b}{2} = \frac{4b}{2} = 2b$

midpoint: $(a, 2b)$

61. $x = \frac{x_1+x_2}{2}$ $y = \frac{y_1+y_2}{2}$

$-2 = \frac{-8+x_2}{2}$ $3 = \frac{5+y_2}{2}$

$-4 = -8 + x_2$ $6 = 5 + y_2$

$4 = x_2$ $1 = y_2$

Q has coordinates $(4, 1)$.

63. To find the value of the house after 5 years, let $x = 5$:

$y = 7500x + 125{,}000$

$\quad = 7500(5) + 125{,}000$

$\quad = 37{,}500 + 125{,}000$

$\quad = 162{,}500$

It will be worth \$162,500 after 5 years.

To find the value of the house after 10 years, let $x = 10$:

$y = 7500x + 125{,}000$

$\quad = 7500(10) + 125{,}000$

$\quad = 75{,}000 + 125{,}000$

$\quad = 200{,}000$

It will be worth \$200,000 after 10 years.

65. To find the number of TVs sold, let $p = 150$:

$p = -\frac{1}{10}q + 170$

$150 = -\frac{1}{10}q + 170$

$\frac{1}{10}q = 20$

$q = 200$

200 TVs will be sold at a price of \$150.

67. Let $V = 60$ and find v.

$V = \frac{nv}{N}$

$60 = \frac{12v}{20}$

$1200 = 12v$

$100 = v$

The smaller gear has a speed of 100 revolutions per minute.

69. Answers may vary.

71. If the line $y = ax + b$ passes through only quadrants I and II, then $a = 0$ and $b > 0$.

Exercise 8.2   (page 509)

1. $(x^3y^2)^3 = x^9y^6$

3. $(x^{-3}y^2)^{-4} = x^{12}y^{-8} = \frac{x^{12}}{y^8}$

5. $\left(\frac{3x^2y^3}{8}\right)^0 = 1$

SECTION 8.2

7. $y; x$

9. $\dfrac{y_2 - y_1}{x_2 - x_1}$

11. run

13. vertical

15. Parallel

17. $m = \dfrac{\Delta y}{\Delta x} = \dfrac{5 - (-3)}{2 - (-2)} = \dfrac{8}{4} = 2$

19. $m = \dfrac{\Delta y}{\Delta x} = \dfrac{9 - 0}{3 - 0} = \dfrac{9}{3} = 3$

21. $m = \dfrac{\Delta y}{\Delta x} = \dfrac{1 - 8}{6 - (-1)} = \dfrac{-7}{7} = -1$

23. $m = \dfrac{\Delta y}{\Delta x} = \dfrac{2 - (-1)}{-6 - 3} = \dfrac{3}{-9} = -\dfrac{1}{3}$

25. $m = \dfrac{\Delta y}{\Delta x} = \dfrac{5 - 5}{-9 - 7} = \dfrac{0}{-16} = 0$

27. $m = \dfrac{\Delta y}{\Delta x} = \dfrac{-2 - (-5)}{-7 - (-7)} = \dfrac{3}{0}$: undefined

29. $m = \dfrac{\Delta y}{\Delta x} = \dfrac{2.5 - 3.7}{3.7 - 2.5} = \dfrac{-1.2}{1.2} = -1$

31. Find two points on the line:

$$
\begin{array}{ll}
\text{Let } x = 0: & \text{Let } y = 0: \\
3x + 2y = 12 & 3x + 2y = 12 \\
3(0) + 2y = 12 & 3x + 2(0) = 12 \\
2y = 12 & 3x = 12 \\
y = 6 & x = 4 \\
(0, 6) & (4, 0)
\end{array}
$$

$$m = \dfrac{\Delta y}{\Delta x} = \dfrac{6 - 0}{0 - 4} = \dfrac{6}{-4} = -\dfrac{3}{2}$$

33. Find two points on the line:

$$
\begin{array}{ll}
\text{Let } x = 0: & \text{Let } y = 0: \\
3x = 4y - 2 & 3x = 4y - 2 \\
3(0) = 4y - 2 & 3x = 4(0) - 2 \\
0 = 4y - 2 & 3x = -2 \\
2 = 4y & x = -\dfrac{2}{3} \\
\dfrac{1}{2} = y & \left(-\dfrac{2}{3}, 0\right) \\
\left(0, \dfrac{1}{2}\right) &
\end{array}
$$

$$m = \dfrac{\Delta y}{\Delta x} = \dfrac{\frac{1}{2} - 0}{0 - \left(-\frac{2}{3}\right)} = \dfrac{\frac{1}{2}}{\frac{2}{3}} = \dfrac{3}{4}$$

35. Find two points on the line:

$$
\begin{array}{ll}
\text{Let } x = 0: & \text{Let } y = 0: \\
y = \dfrac{x - 4}{2} & y = \dfrac{x - 4}{2} \\
y = \dfrac{0 - 4}{2} & 0 = \dfrac{x - 4}{2} \\
y = \dfrac{-4}{2} & 0 = x - 4 \\
y = -2 & 4 = x \\
(0, -2) & (4, 0)
\end{array}
$$

$$m = \dfrac{\Delta y}{\Delta x} = \dfrac{-2 - 0}{0 - 4} = \dfrac{-2}{-4} = \dfrac{1}{2}$$

37. $4y = 3(y + 2)$
$4y = 3y + 6$
$y = 6$; horizontal line $\Rightarrow m = 0$

39. negative

41. positive

43. undefined

45. $m_1 \neq m_2 \Rightarrow$ not parallel

$m_1 \cdot m_2 = 3\left(-\dfrac{1}{3}\right) = -1 \Rightarrow$ perpendicular

47. $m_1 \neq m_2 \Rightarrow$ not parallel
$m_1 \cdot m_2 = 4(0.25) = 1 \Rightarrow$ not perpendicular

49. $m_1 = m_2 \Rightarrow$ parallel

51. $m_{\overline{PQ}} = \dfrac{\Delta y}{\Delta x} = \dfrac{2-4}{4-3} = \dfrac{-2}{1} = -2$
same slope $\Rightarrow$ parallel

53. $m_{\overline{PQ}} = \dfrac{\Delta y}{\Delta x} = \dfrac{5-1}{6-(-2)} = \dfrac{4}{8} = \dfrac{1}{2}$
opposite reciprocal slope $\Rightarrow$ perpendicular

55. $m_{\overline{PQ}} = \dfrac{\Delta y}{\Delta x} = \dfrac{6-4}{6-5} = \dfrac{2}{1} = 2$
neither parallel nor perpendicular

57. $m_{\overline{PQ}} = \dfrac{\Delta y}{\Delta x} = \dfrac{8-4}{4-(-2)} = \dfrac{4}{6} = \dfrac{2}{3}$

$m_{\overline{PR}} = \dfrac{\Delta y}{\Delta x} = \dfrac{12-4}{8-(-2)} = \dfrac{8}{10} = \dfrac{4}{5}$

Since they do not have the same slope,
the three points are not on the same line.

59. $m_{\overline{PQ}} = \dfrac{\Delta y}{\Delta x} = \dfrac{0-10}{-6-(-4)} = \dfrac{-10}{-2} = 5$

$m_{\overline{PR}} = \dfrac{\Delta y}{\Delta x} = \dfrac{5-10}{-1-(-4)} = \dfrac{-5}{3} = -\dfrac{5}{3}$

Since they do not have the same slope,
the three points are not on the same line.

61. $m_{\overline{PQ}} = \dfrac{\Delta y}{\Delta x} = \dfrac{8-4}{0-(-2)} = \dfrac{4}{2} = 2$

$m_{\overline{PR}} = \dfrac{\Delta y}{\Delta x} = \dfrac{12-4}{2-(-2)} = \dfrac{8}{4} = 2$

Since they have the same slope and a common
point, the three points are on the same line.

63. On the x-axis, all y-coordinates are 0.
The equation is $y = 0$, and the slope is 0.

65. Call the points $A(-3, 4)$, $B(4, 1)$ and $C(-1, -1)$. Compute these slopes:

$m_{\overline{AB}} = \dfrac{\Delta y}{\Delta x} = \dfrac{1-4}{4-(-3)} = \dfrac{-3}{7} = -\dfrac{3}{7}$

$m_{\overline{AC}} = \dfrac{\Delta y}{\Delta x} = \dfrac{-1-4}{-1-(-3)} = \dfrac{-5}{2} = -\dfrac{5}{2}$

$m_{\overline{BC}} = \dfrac{\Delta y}{\Delta x} = \dfrac{-1-1}{-1-4} = \dfrac{-2}{-5} = \dfrac{2}{5}$

Since $\overline{AC}$ and $\overline{BC}$ are perpendicular, it is a right triangle.

67. Call the points $A(a, 0)$, $B(0, a)$, $C(-a, 0)$ and $D(0, -a)$. Compute these slopes:

$m_{\overline{AB}} = \dfrac{\Delta y}{\Delta x} = \dfrac{a-0}{0-a} = \dfrac{a}{-a} = -1$

$m_{\overline{BC}} = \dfrac{\Delta y}{\Delta x} = \dfrac{0-a}{-a-0} = \dfrac{-a}{-a} = 1$

$m_{\overline{CD}} = \dfrac{\Delta y}{\Delta x} = \dfrac{-a-0}{0-(-a)} = \dfrac{-a}{a} = -1$

$m_{\overline{DA}} = \dfrac{\Delta y}{\Delta x} = \dfrac{0-(-a)}{a-0} = \dfrac{a}{a} = 1$

Thus, $\overline{AB} \perp \overline{BC}$, $\overline{BC} \perp \overline{CD}$, $\overline{CD} \perp \overline{DA}$, and $\overline{DA} \perp \overline{AB}$.

69. Call the points $A(0,0)$, $B(0,a)$, $C(b,c)$ and $D(b, a+c)$. Compute these slopes:

$$m_{\overline{AB}} = \frac{\Delta y}{\Delta x} = \frac{0-a}{0-0} = \frac{-a}{0} \Rightarrow \text{undefined}$$

$$m_{\overline{BD}} = \frac{\Delta y}{\Delta x} = \frac{a-(a+c)}{0-b} = \frac{a-a-c}{-b} = \frac{-c}{-b} = \frac{c}{b}$$

$$m_{\overline{DC}} = \frac{\Delta y}{\Delta x} = \frac{c-(a+c)}{b-b} = \frac{c-a-c}{0} = \frac{-a}{0} \Rightarrow \text{undefined}$$

$$m_{\overline{CA}} = \frac{\Delta y}{\Delta x} = \frac{c-0}{b-0} = \frac{c}{b}$$

Thus, $\overline{AB} \parallel \overline{DC}$ and $\overline{BD} \parallel \overline{CA}$ and the figure is a parallelogram.

71. $\text{slope} = \dfrac{\text{rise}}{\text{run}} = \dfrac{32 \text{ ft}}{1 \text{ mi}} = \dfrac{32 \text{ ft}}{5280 \text{ ft}} = \dfrac{1}{165}$

73. $m = \dfrac{\Delta y}{\Delta x} = \dfrac{2}{50} = 0.04$; $m = \dfrac{\Delta y}{\Delta x} = \dfrac{5}{50} = 0.1$; $m = \dfrac{\Delta y}{\Delta x} = \dfrac{8}{50} = 0.16$

75. $\text{slope} = \dfrac{\text{rise}}{\text{run}} = \dfrac{18 \text{ ft}}{5 \text{ ft}} = \dfrac{18}{5}$

77. Let x represent the number of years, and let y represent the sales. Then we know two points on the line: $(1, 85000)$ and $(3, 125000)$. Find the slope:

$$m = \frac{\Delta y}{\Delta x} = \frac{125000 - 85000}{3-1} = \frac{40000}{2} = \$20{,}000 \text{ per year.}$$

79. Answers may vary.

81. Find two points on the line:

Let $x = 0$:   Let $y = 0$:

$$\begin{array}{ll} Ax + By = C & Ax + By = C \\ A(0) + By = C & Ax + B(0) = C \\ By = C & Ax = C \\ y = \frac{C}{B} & x = \frac{C}{A} \end{array}$$

$$m = \frac{\Delta y}{\Delta x} = \frac{\frac{C}{B} - 0}{0 - \frac{C}{A}} = \frac{\frac{C}{B}}{-\frac{C}{A}} = -\frac{A}{B}$$

$\left(0, \frac{C}{B}\right)$ is on the line. $\left(\frac{C}{A}, 0\right)$ is on the line.

83. If the three points are on a line, then all slopes must be equal:

$$m = \frac{\Delta y}{\Delta x} = \frac{7-10}{5-7} = \frac{-3}{-2} = \frac{3}{2} \qquad m = \frac{\Delta y}{\Delta x} = \frac{a-7}{3-5} = \frac{a-7}{-2}$$

$$\frac{a-7}{-2} = \frac{3}{2}$$

$$-2 \cdot \frac{a-7}{-2} = -2 \cdot \frac{3}{2}$$

$$a - 7 = -3$$

$$a = 4$$

Exercise 8.3 (page 522)

1.
$$3(x+2) + x = 5x$$
$$3x + 6 + x = 5x$$
$$4x + 6 = 5x$$
$$6 = x$$

3.
$$\frac{5(2-x)}{3} - 1 = x + 5$$
$$\frac{10 - 5x}{3} = x + 6$$
$$3 \cdot \frac{10 - 5x}{3} = 3(x + 6)$$
$$10 - 5x = 3x + 18$$
$$10 = 8x + 18$$
$$-8 = 8x$$
$$-1 = x$$

5. We want the alloy to contain 25% gold. Let x = the amount of copper added to the alloy.

$$\boxed{\begin{array}{c}\text{Gold at}\\\text{start}\end{array}} + \boxed{\begin{array}{c}\text{Gold}\\\text{added}\end{array}} = \boxed{\begin{array}{c}\text{Gold at}\\\text{end}\end{array}}$$
$$20 + 0 = 0.25(60 + x)$$
$$20 = 15 + 0.25x$$
$$5 = 0.25x$$
$$20 = x$$

20 oz of copper should be added to the alloy.

7. $y - y_1 = m(x - x_1)$

9. $Ax + By = C$

11. perpendicular

13.
$$y - y_1 = m(x - x_1)$$
$$y - 7 = 5(x - 0)$$
$$y - 7 = 5x$$
$$-5x + y = 7$$
$$5x - y = -7$$

15.
$$y - y_1 = m(x - x_1)$$
$$y - 0 = -3(x - 2)$$
$$y = -3x + 6$$
$$3x + y = 6$$

17. Notice that the line goes through $P(2, 5)$ and the point $(-1, 3)$. Find the slope:
$$m = \frac{\Delta y}{\Delta x} = \frac{5 - 3}{2 - (-1)} = \frac{2}{3}$$
Use point-slope form to find the equation:
$$y - y_1 = m(x - x_1)$$
$$y - 5 = \frac{2}{3}(x - 2)$$
$$3(y - 5) = 3 \cdot \frac{2}{3}(x - 2)$$
$$3y - 15 = 2x - 4$$
$$-2x + 3y = 11$$
$$2x - 3y = -11$$

19. Find the slope of the line:
$$m = \frac{\Delta y}{\Delta x} = \frac{4 - 0}{4 - 0} = \frac{4}{4} = 1$$
Use point-slope form to find the equation:
$$y - y_1 = m(x - x_1)$$
$$y - 0 = 1(x - 0)$$
$$y = x$$

21. Find the slope of the line:

$$m = \frac{\Delta y}{\Delta x} = \frac{4 - (-3)}{3 - 0} = \frac{7}{3}$$

Use point-slope form to find the equation:

$$y - y_1 = m(x - x_1)$$
$$y - 4 = \tfrac{7}{3}(x - 3)$$
$$y - 4 = \tfrac{7}{3}x - 7$$
$$y = \tfrac{7}{3}x - 3$$

23. Find the slope of the line:

$$m = \frac{\Delta y}{\Delta x} = \frac{4 - (-3)}{-2 - 2} = \frac{7}{-4} = -\frac{7}{4}$$

Use point-slope form to find the equation:

$$y - y_1 = m(x - x_1)$$
$$y - 4 = -\tfrac{7}{4}(x - (-2))$$
$$y - 4 = -\tfrac{7}{4}(x + 2)$$
$$y - 4 = -\tfrac{7}{4}x - \tfrac{7}{2}$$
$$y = -\tfrac{7}{4}x - \tfrac{7}{2} + 4$$
$$y = -\tfrac{7}{4}x + \tfrac{1}{2}$$

25.
$$y = mx + b$$
$$y = 3x + 17$$

27.
$$y = mx + b$$
$$5 = -7(7) + b$$
$$5 = -49 + b$$
$$54 = b$$
$$y = -7x + 54$$

29.
$$y = mx + b$$
$$-4 = 0(2) + b$$
$$-4 = b$$
$$y = 0x + (-4)$$
$$y = -4$$

31. Find the slope of the line:

$$m = \frac{\Delta y}{\Delta x} = \frac{8 - 10}{6 - 2} = \frac{-2}{4} = -\frac{1}{2}$$
$$y = mx + b$$
$$8 = -\frac{1}{2}(6) + b$$
$$8 = -3 + b$$
$$11 = b$$
$$y = -\frac{1}{2}x + 11$$

33.
$$y + 1 = x$$
$$y = x - 1$$
$$m = 1; b = -1 \Rightarrow (0, -1)$$

35.
$$x = \tfrac{3}{2}y - 3$$
$$2x = 3y - 6$$
$$-3y = -2x + 6$$
$$y = \tfrac{2}{3}x + 2$$
$$m = \tfrac{2}{3}; b = 2 \Rightarrow (0, 2)$$

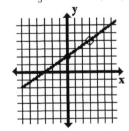

37.
$$3(y - 4) = -2(x - 3)$$
$$3y - 12 = -2x + 6$$
$$3y = -2x + 18$$
$$y = -\frac{2}{3}x + 6$$
$$m = -\tfrac{2}{3}; b = 6 \Rightarrow (0, 6)$$

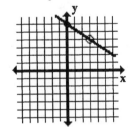

39. $3x - 2y = 8$
$-2y = -3x + 8$
$y = \frac{3}{2}x - 4$
$m = \frac{3}{2}; b = -4 \Rightarrow (0, -4)$

41. $-2(x + 3y) = 5$
$-2x - 6y = 5$
$-6y = 2x + 5$
$y = -\frac{1}{3}x - \frac{5}{6}$
$m = -\frac{1}{3}; b = -\frac{5}{6} \Rightarrow \left(0, -\frac{5}{6}\right)$

43. $x = \dfrac{2y - 4}{7}$
$7x = 2y - 4$
$-2y = -7x - 4$
$y = \frac{7}{2}x + 2$
$m = \frac{7}{2}; b = 2 \Rightarrow (0, 2)$

45. $y = 3x + 4 \qquad y = 3x - 7$
$m = 3 \qquad\qquad m = 3$
parallel

47. $x + y = 2 \qquad y = x + 5$
$y = -x + 2 \qquad m = 1$
$m = -1$
perpendicular

49. $y = 3x + 7 \qquad 2y = 6x - 9$
$m = 3 \qquad\qquad y = 3x - \dfrac{9}{2}$
$m = 3$
parallel

51. $x = 3y + 4 \qquad y = -3x + 7$
$-3y = -x + 4 \qquad m = -3$
$y = \frac{1}{3}x - \frac{4}{3}$
$m = \frac{1}{3}$
perpendicular

53. $y = 3 \qquad x = 4$
horizontal line   vertical line
perpendicular

55. $x = \dfrac{y - 2}{3} \qquad 3(y - 3) + x = 0$
$3x = y - 2 \qquad\quad 3y - 9 + x = 0$
$-y = -3x - 2 \qquad\quad 3y = -x + 9$
$y = 3x + 2 \qquad\qquad y = -\frac{1}{3}x + 3$
$m = 3 \qquad\qquad\qquad m = -\frac{1}{3}$
perpendicular

57. Find the slope of the given line:
$y = 4x - 7 \Rightarrow m = 4$
Use the parallel slope.
$y - y_1 = m(x - x_1)$
$y - 0 = 4(x - 0)$
$y = 4x$

59. Find the slope of the given line:
$4x - y = 7$
$-y = -4x + 7$
$y = 4x - 7 \Rightarrow m = 4$
Use the parallel slope.
$y - y_1 = m(x - x_1)$
$y - 5 = 4(x - 2)$
$y - 5 = 4x - 8$
$y = 4x - 3$

61. Find the slope of the given line:

$$x = \frac{5}{4}y - 2$$
$$4x = 5y - 8$$
$$-5y = -4x - 8$$
$$y = \frac{4}{5}x + \frac{8}{5} \Rightarrow m = \frac{4}{5}$$

Use the parallel slope.

$$y - y_1 = m(x - x_1)$$
$$y - (-2) = \frac{4}{5}(x - 4)$$
$$y + 2 = \frac{4}{5}x - \frac{16}{5}$$
$$y = \frac{4}{5}x - \frac{16}{5} - 2$$
$$y = \frac{4}{5}x - \frac{26}{5}$$

63. Find the slope of the given line:

$$y = 4x - 7 \Rightarrow m = 4$$

Use the perpendicular slope.

$$y - y_1 = m(x - x_1)$$
$$y - 0 = -\frac{1}{4}(x - 0)$$
$$y = -\frac{1}{4}x$$

65. Find the slope of the given line:

$$4x - y = 7$$
$$-y = -4x + 7$$
$$y = 4x - 7 \Rightarrow m = 4$$

Use the perpendicular slope.

$$y - y_1 = m(x - x_1)$$
$$y - 5 = -\frac{1}{4}(x - 2)$$
$$y - 5 = -\frac{1}{4}x + \frac{1}{2}$$
$$y = -\frac{1}{4}x + \frac{1}{2} + 5$$
$$y = -\frac{1}{4}x + \frac{11}{2}$$

67. Find the slope of the given line:

$$x = \frac{5}{4}y - 2$$
$$4x = 5y - 8$$
$$-5y = -4x - 8$$
$$y = \frac{4}{5}x + \frac{8}{5} \Rightarrow m = \frac{4}{5}$$

Use the perpendicular slope.

$$y - y_1 = m(x - x_1)$$
$$y - (-2) = -\frac{5}{4}(x - 4)$$
$$y + 2 = -\frac{5}{4}x + 5$$
$$y = -\frac{5}{4}x + 3$$

69. Find the slope of the 1st line.   Find the slope of the 2nd line.   perpendicular

$$m = -\frac{A}{B} = -\frac{4}{5} \qquad m = -\frac{A}{B} = -\frac{5}{-4} = \frac{5}{4}$$

71. Find the slope of the 1st line.   Find the slope of the 2nd line.   parallel

$$m = -\frac{A}{B} = -\frac{2}{3} \qquad m = -\frac{A}{B} = -\frac{6}{9} = -\frac{2}{3}$$

73. The line $y = 3$ is horizontal, so any perpendicular line will be vertical.
Find the midpoint of the described segment:

$$x = \frac{x_1 + x_2}{2} = \frac{2 + (-6)}{2} = \frac{-4}{2} = -2; \quad y = \frac{y_1 + y_2}{2} = \frac{4 + 10}{2} = \frac{14}{2} = 7$$

The vertical line through the point $(-2, 7)$ is $x = -2$.

75. The line $x = 3$ is vertical, so any parallel line will be vertical.
Find the midpoint of the described segment:

$$x = \frac{x_1 + x_2}{2} = \frac{2 + 8}{2} = \frac{10}{2} = 5; \quad y = \frac{y_1 + y_2}{2} = \frac{-4 + 12}{2} = \frac{8}{2} = 4$$

The vertical line through the point $(5, 4)$ is $x = 5$.

77. $Ax + By = C$

$$By = -Ax + C$$

$$\frac{By}{B} = \frac{-Ax + C}{B}$$

$$y = -\frac{A}{B}x + \frac{C}{B} \Rightarrow m = -\frac{A}{B}; \ b = \frac{C}{B}$$

79. Let x represent the number of years since the truck was purchased, and let y represent the value of the truck. Then we know two points on the depreciation line: $(0, 19984)$ and $(8, 1600)$. Find the slope and then use point-slope form to find the depreciation equation.

$$m = \frac{\Delta y}{\Delta x} = \frac{19984 - 1600}{0 - 8} = \frac{18384}{-8} = -2298 \qquad y - y_1 = m(x - x_1)$$

$$y - 19984 = -2298(x - 0)$$

$$y = -2298x + 19984$$

81. Let x represent the number of years since the painting was purchased, and let y represent the value of the painting. Then we know two points on the appreciation line: $(0, 250000)$ and $(5, 500000)$. Find the slope and then use point-slope form to find the appreciation equation.

$$m = \frac{\Delta y}{\Delta x} = \frac{500000 - 250000}{5 - 0} = \frac{250000}{5} = 50000 \qquad y - y_1 = m(x - x_1)$$

$$y - 250000 = 50000(x - 0)$$

$$y = 50{,}000x + 250{,}000$$

83. Let x represent the number of years since the TV was purchased, and let y represent the value of the TV. Then we know two points on the depreciation line: $(0, 1750)$ and $(3, 800)$. Find the slope and then use point-slope form to find the depreciation equation.

$$m = \frac{\Delta y}{\Delta x} = \frac{1750 - 800}{0 - 3} = \frac{950}{-3} = -\frac{950}{3} \qquad y - y_1 = m(x - x_1)$$

$$y - 1750 = -\frac{950}{3}(x - 0)$$

$$y = -\frac{950}{3}x + 1750$$

85. Let x represent the number of years since the copy machine was purchased, and let y represent its value. Since the value decreases \$180 per year, the slope of the depreciation line is -180. Since the copier is worth \$1750 new, the y-intercept is 1750. Thus the depreciation equation can be found:

$y = mx + b$ \qquad Now let $x = 7$: \quad $y = -180x + 1750 = -180(7) + 1750$

$y = -180x + 1750$ \hspace{4cm} $= -1260 + 1750 = \$490$

87. Let x represent the number of years since the house was purchased, and let y represent its value. Since the value increases \$4000 per year, the slope of the depreciation line is 4000. We also know a point on the line: $(2, 122000)$. Use point-slope form to find the appreciation equation.

$y - y_1 = m(x - x_1)$ \hspace{2cm} Substitute $x = 10$ and find its value:

$y - 122000 = 4000(x - 2)$ \hspace{1.5cm} $y = 4000x + 114000 = 4000(10) + 114000$

$y - 122000 = 4000x - 8000$ \hspace{2cm} $= 40000 + 114000 = \$154{,}000$

$y = 4000x + 114000$

89. The relationship between number of copies and the total cost is linear. Let x represent the number of copies (in hundreds), and let y represent the total cost. The slope of the line will be the charge per one hundred copies, or 15. We also know a point on the line: $(3, 75)$. Use point-slope form:

$$y - y_1 = m(x - x_1) \quad \text{Let } x = 10 \text{ and find the cost:}$$
$$y - 75 = 15(x - 3) \quad y = 15x + 30 = 15(10) + 30 = 150 + 30 = \$180$$
$$y - 75 = 15x - 45$$
$$y = 15x + 30$$

91-97. Answers may vary.

99. To pass through II and IV, the slope must be negative. To pass through II and not III, the y-intercept must be positive. Thus $a < 0$ and $b > 0$.

Exercise 8.4 (page 532)

1. $\dfrac{y+2}{2} = 4(y+2)$

    $2 \cdot \dfrac{y+2}{2} = 2 \cdot 4(y+2)$

    $y + 2 = 8(y+2)$

    $y + 2 = 8y + 16$

    $-7y = 14$

    $y = -2$

3. $\dfrac{2a}{3} + \dfrac{1}{2} = \dfrac{6a-1}{6}$

    $6\left(\dfrac{2a}{3} + \dfrac{1}{2}\right) = 6 \cdot \dfrac{6a-1}{6}$

    $2(2a) + 3 = 6a - 1$

    $4a + 3 = 6a - 1$

    $-2a = -4$

    $a = 2$

5. input

7. x

9. function; input; output

11. range

13. 0

15. $mx + b$

17. $y = 2x + 3$ **is a function**, since each value of x corresponds to exactly one value of y.

19. $y = 2x^2$ **is a function**, since each value of x corresponds to exactly one value of y.

21. $y = 3 + 7x^2$ **is a function**, since each value of x corresponds to exactly one value of y.

23. $x = |y|$ **is not a function**, since $x = 2$ corresponds to both $y = 2$ and $y = -2$.

25. $f(x) = 3x$
    $f(3) = 3(3) = 9$
    $f(-1) = 3(-1) = -3$

27. $f(x) = 2x - 3$
    $f(3) = 2(3) - 3 = 3$
    $f(-1) = 2(-1) - 3 = -5$

29. $f(x) = 7 + 5x$
    $f(3) = 7 + 5(3) = 22$
    $f(-1) = 7 + 5(-1) = 2$

31. $f(x) = 9 - 2x$
    $f(3) = 9 - 2(3) = 3$
    $f(-1) = 9 - 2(-1) = 11$

33. $f(x) = x^2$
    $f(2) = 2^2 = 4$
    $f(3) = 3^2 = 9$

35. $f(x) = x^3 - 1$
    $f(2) = 2^3 - 1 = 8 - 1$
    $= 7$
    $f(3) = 3^3 - 1 = 27 - 1$
    $= 26$

37. $f(x) = (x+1)^2$
$f(2) = (2+1)^2 = 3^2 = 9$
$f(3) = (3+1)^2 = 4^2 = 16$

39. $f(x) = 2x^2 - x$
$f(2) = 2(2)^2 - 2$
$\quad = 2(4) - 2$
$\quad = 8 - 2 = 6$
$f(3) = 2(3)^2 - 3$
$\quad = 2(9) - 3$
$\quad = 18 - 3 = 15$

41. $f(x) = |x| + 2$
$f(2) = |2| + 2$
$\quad = 2 + 2 = 4$
$f(-2) = |-2| + 2$
$\quad = 2 + 2 = 4$

43. $f(x) = x^2 - 2$
$f(2) = 2^2 - 2$
$\quad = 4 - 2 = 2$
$f(-2) = (-2)^2 - 2$
$\quad = 4 - 2 = 2$

45. $f(x) = \dfrac{1}{x+3}$
$f(2) = \dfrac{1}{2+3} = \dfrac{1}{5}$
$f(-2) = \dfrac{1}{-2+3} = \dfrac{1}{1} = 1$

47. $f(x) = \dfrac{x}{x-3}$
$f(2) = \dfrac{2}{2-3} = \dfrac{2}{-1}$
$\quad = -2$
$f(-2) = \dfrac{-2}{-2-3} = \dfrac{-2}{-5}$
$\quad = \dfrac{2}{5}$

49. $g(x) = 2x$
$g(w) = 2w$
$g(w+1) = 2(w+1) = 2w + 2$

51. $g(x) = 3x - 5$
$g(w) = 3w - 5$
$g(w+1) = 3(w+1) - 5 = 3w + 3 - 5$
$\quad = 3w - 2$

53. $f(3) + f(2) = [2(3) + 1] + [2(2) + 1]$
$\quad = 6 + 1 + 4 + 1 = 12$

55. $f(b) - f(a) = [2b + 1] - [2a + 1]$
$\quad = 2b + 1 - 2a - 1 = 2b - 2a$

57. $f(b) - 1 = 2b + 1 - 1 = 2b$

59. $f(0) + f\left(-\tfrac{1}{2}\right) = [2(0) + 1] + \left[2\left(-\tfrac{1}{2}\right) + 1\right]$
$\quad = 0 + 1 - 1 + 1 = 1$

61. domain $= \{-2, 4, 6\}$; range $= \{3, 5, 7\}$

63. To ensure that the denominator is not 0, x cannot equal 4. The domain is $(-\infty, 4) \cup (4, \infty)$.
Because the numerator will never be 0, the range is $(-\infty, 0) \cup (0, \infty)$.

65. not a function

67. The domain is the set of all x-coordinates on the graph. The domain is $(-\infty, \infty)$.
The range is the set of all y-coordinates on the graph. The range is $(-\infty, \infty)$.
Since each vertical line passes through the graph at most once, it is the graph of a function.

69. $f(x) = 2x - 1$
$D = (-\infty, \infty)$
$R = (-\infty, \infty)$

71. $2x - 3y = 6$
$D = (-\infty, \infty)$
$R = (-\infty, \infty)$

73. $y = 3x^2 + 2 \Rightarrow$ This is not a linear function because the exponent on x is 2.

75. $x = 3y - 4 \Rightarrow y = \dfrac{1}{3}x + \dfrac{4}{3} \Rightarrow$ This is a linear function $\left(m = \dfrac{1}{3}, b = \dfrac{4}{3} \right)$.

77. Let $t = 3$ and find s.
$$s = f(t) = -16t^2 + 256t$$
$$= -16(3)^2 + 256(3)$$
$$= -16(9) + 768$$
$$= -144 + 768$$
$$= 624$$
The bullet will have a height of 624 ft.

79. Let $h = 16$ and find t.
$$h = -16t^2 + 32t$$
$$16 = -16t^2 + 32t$$
$$16t^2 - 32t + 16 = 0$$
$$16(t^2 - 2t + 1) = 0$$
$$16(t - 1)(t - 1) = 0$$
$$t - 1 = 0 \quad \textbf{OR} \quad t - 1 = 0$$
$$t = 1 \qquad\qquad t = 1$$
It will take 1 second to reach the hand.

81. Let $C = 25$ and find F.
$$F(C) = \frac{9}{5}C + 32$$
$$= \frac{9}{5}(25) + 32$$
$$= 45 + 32$$
$$= 77$$
The temperature is $77° F$.

83. Set $R(x)$ equal to $C(x)$ and solve for x.
$$R(x) = C(x)$$
$$120x = 57.50x + 12000$$
$$62.50x = 12000$$
$$x = \frac{12000}{62.50}$$
$$x = 192$$
The company must sell 192 players.

85. **Answers may vary.**

87. $f(x) + g(x) = 2x + 1 + x^2$
$\qquad\qquad\qquad = x^2 + 2x + 1 = g(x) + f(x)$

Exercise 8.5 (page 546)

1. 41, 43, 47

3. $a \cdot b = b \cdot a$

5. 1

7. squaring

9. absolute value

11. horizontal

13. 2; down

15. 4; to the left

17. rational

19. $f(x) = x^2 - 3$
Shift $y = x^2$ down 3.

21. $f(x) = (x-1)^3$
Shift $y = x^3$ right 1.

23. $f(x) = |x| - 2$
Shift $y = |x|$ down 2.

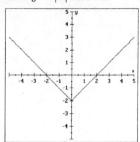

25. $f(x) = |x-1|$
Shift $y = |x|$ right 1.

27. $f(x) = x^2 + 8$
x: $[-4, 4]$, y: $[-4, 4]$

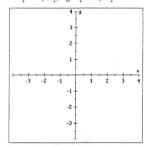

x: $[-7, 7]$, y: $[-2, 12]$

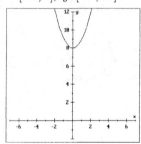

29. $f(x) = |x + 5|$
x: $[-4, 4]$, y: $[-4, 4]$

x: $[-12, 2]$, y: $[-4, 10]$

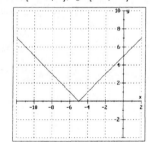

31. $f(x) = (x - 6)^2$
$x: [-4, 4], \; y: [-4, 4]$

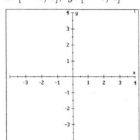

33. $f(x) = x^3 + 8$
$x: [-4, 4], \; y: [-4, 4]$

35. $f(x) = x^2 - 5$
Shift $f(x) = x^2$ D 5.

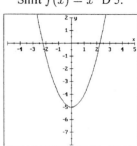

$x: [-2, 12], \; y: [-2, 12]$

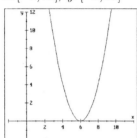

$x: [-10, 10], \; y: [-4, 16]$

37. $f(x) = (x - 1)^3$
Shift $f(x) = x^3$ R 1.

39. $f(x) = |x - 2| - 1$
Shift $f(x) = |x|$ D 1, R 2.

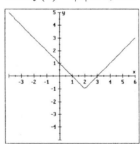

41. $f(x) = (x + 1)^3 - 2$
Shift $f(x) = x^3$ D 2, L 1.

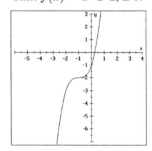

43. $f(x) = -x^2$

Reflect $f(x) = x^2$ about x.

45. $f(x) = -(x-2)^2 - 3$

Shift $f(x) = -x^2$ R2, D3.

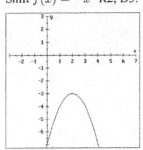

47. $t = f(r) = \dfrac{600}{r}$

$f(30) = \dfrac{600}{30}$

$\quad\;\; = 20$

$t = 20$ hours

49. $t = f(r) = \dfrac{600}{r}$

$f(50) = \dfrac{600}{50}$

$\quad\;\; = 12$

$t = 12$ hours

51. $C = f(p) = \dfrac{50{,}000p}{100 - p}$

$f(10) = \dfrac{50{,}000(10)}{100 - 10}$

$\quad\;\; = \dfrac{500{,}000}{90}$

$\quad\;\; = 5555.56$

$C = \$5555.56$

53. $C = f(p) = \dfrac{50{,}000p}{100 - p}$

$f(50) = \dfrac{50{,}000(50)}{100 - 50}$

$\quad\;\; = \dfrac{2{,}500{,}000}{50}$

$\quad\;\; = 50{,}000.00$

$C = \$50{,}000$

55. $f(x) = \dfrac{x}{x - 2}$

domain: $(-\infty, 2) \cup (2, \infty)$
range: $(-\infty, 1) \cup (1, \infty)$

57. $f(x) = \dfrac{x + 1}{x^2 - 4}$

domain: $(-\infty, -2) \cup (-2, 2) \cup (2, \infty)$
range: $(-\infty, \infty)$

59. $c = f(x) = 1.25x + 700$

61.   Refer to #59.

$c = f(500) = 1.25(500) + 700 = \$1325$

63.   Refer to #59. The mean cost $= \bar{c} = f(x) = \dfrac{1.25x + 700}{x}$.

$f(1000) = \dfrac{1.25(1000) + 700}{1000} = \dfrac{1950}{1000} = \$1.95$

65. $c = f(n) = 0.09n + 7.50$

67. Refer to **#65**. $c = f(775) = 0.09(775) + 7.50 = 69.75 + 7.50 = \$77.25$

69. Refer to **#65**. The mean cost $= \bar{c} = f(x) = \dfrac{0.09n + 7.50}{n}$.

$f(1000) = \dfrac{0.09(1000) + 7.50}{1000} = \dfrac{97.50}{1000} = \$0.0975 = 9.75\cent$

71. $c = f(x) = 350x + 5000$

73. Refer to **#71**.
$c = f(120) = 350(120) + 5000 = \$47,000$

75. **Answers may vary.**

Exercise 8.6 (page 554)

1. $(x^2 x^3)^2 = (x^2)^2 (x^3)^2 = x^4 x^6 = x^{10}$

3. $\dfrac{b^0 - 2b^0}{b^0} = \dfrac{1 - 2}{1} = \dfrac{-1}{1} = -1$

5. $35,000 = 3.5 \times 10^4$

7. $2.5 \times 10^{-3} = 0.0025$

9. proportion

11. direct

13. rational

15. joint

17. direct

19. neither

21.
$\dfrac{x}{5} = \dfrac{15}{25}$
$25x = 75$
$x = 3$

23.
$\dfrac{r - 2}{3} = \dfrac{r}{5}$
$5(r - 2) = 3r$
$5r - 10 = 3r$
$2r = 10$
$r = 5$

25.
$\dfrac{3}{n} = \dfrac{2}{n + 1}$
$3(n + 1) = 2n$
$3n + 3 = 2n$
$n = -3$

27.
$\dfrac{5}{5z + 3} = \dfrac{2z}{2z^2 + 6}$
$5(2z^2 + 6) = 2z(5z + 3)$
$10z^2 + 30 = 10z^2 + 6z$
$30 = 6z$
$5 = z$

29.
$\dfrac{2}{c} = \dfrac{c - 3}{2}$
$4 = c^2 - 3c$
$0 = c^2 - 3c - 4$
$0 = (c + 1)(c - 4)$
$c + 1 = 0 \quad \text{or} \quad c - 4 = 0$
$c = -1 \qquad\qquad c = 4$

31.
$\dfrac{2}{3x} = \dfrac{6x}{36}$
$72 = 18x^2$
$0 = 18x^2 - 72$
$0 = 18(x + 2)(x - 2)$
$x + 2 = 0 \quad \text{or} \quad x - 2 = 0$
$x = -2 \qquad\qquad x = 2$

33.
$\dfrac{2(x + 3)}{3} = \dfrac{4(x - 4)}{5}$
$10(x + 3) = 12(x - 4)$
$10x + 30 = 12x - 48$
$-2x = -78$
$x = 39$

35.
$$\frac{1}{x+3} = \frac{-2x}{x+5}$$
$$x + 5 = -2x^2 - 6x$$
$$2x^2 + 7x + 5 = 0$$
$$(2x + 5)(x + 1)$$
$$2x + 5 = 0 \quad \text{or} \quad x + 1 = 0$$
$$x = -\tfrac{5}{2} \qquad\qquad x = -1$$

37. $A = kp^2$ **39.** $v = \dfrac{k}{r^3}$ **41.** $B = kmn$ **43.** $P = \dfrac{ka^2}{j^3}$

45. L varies jointly with m and n.

47. E varies jointly with a and the square of b.

49. X varies directly with x^2 and inversely with y^2.

51. R varies directly with L and inversely with d^2.

53. $A = kr^2$
$A = \pi r^2$
$A = \pi(6 \text{ in.})^2$
$\boxed{A = 36\pi \text{ in.}^2}$

55. $d = kg$ $d = 24g$
$288 = k(12)$ $d = 24(18)$
$24 = k$ $\boxed{d = 432 \text{ mi}}$

57. $t = \dfrac{k}{n}$ $t = \dfrac{250}{n}$
$10 = \dfrac{k}{25}$ $t = \dfrac{250}{10}$
$250 = k$ $\boxed{t = 25 \text{ days}}$

59. $V = \dfrac{k}{P}$ $V = \dfrac{120}{P}$
$20 = \dfrac{k}{6}$ $V = \dfrac{120}{10}$
$120 = k$ $\boxed{V = 12 \text{ in.}^3}$

61. $f = \dfrac{k}{l}$ $f = \dfrac{512}{l}$
$256 = \dfrac{k}{2}$ $f = \dfrac{512}{6}$
$512 = k$ $\boxed{f = 85\tfrac{1}{3}}$

63. $V_1 = klwh$
$V_2 = k(2l)(3w)(2h) = 12klwh = 12V_1$
The volume is multiplied by 12.

65. $g = khr^2$
$g = 23.5hr^2$
$g = 23.5(20)(7.5)^2$
$\boxed{g = 26{,}437.5 \text{ gallons}}$

67. $V = kC$
$6 = k(2)$
$3 = k$
The resistance is 3 ohms.

69. $D = \dfrac{k}{wd^3}$ $\qquad$ $D = \dfrac{281.6}{wd^3}$

$\qquad$ $1.1 = \dfrac{k}{4(4)^3}$ $\qquad$ $D = \dfrac{281.6}{2(8)^3}$

$\qquad$ $1.1 = \dfrac{k}{256}$ $\qquad$ $D = \dfrac{281.6}{1024}$

$\qquad$ $281.6 = k$ $\qquad$ $\boxed{D = 0.275 \text{ in.}}$

71. $P = \dfrac{kT}{V}$ $\qquad$ $P = \dfrac{\frac{1}{273}T}{V}$

$\qquad$ $1 = \dfrac{k(273)}{1}$ $\qquad$ $1 = \dfrac{\frac{1}{273}T}{2}$

$\qquad$ $\dfrac{1}{273} = k$ $\qquad$ $2 = \dfrac{1}{273}T$

$\qquad$ $\qquad$ $\boxed{546 \text{ K} = T}$

73. Answers may vary.

75. Answers may vary.

77. This is not direct variation. For this to be direct variation, one temperature would have to be a constant multiple of the other.

79. No. Answers may vary.

Chapter 8 Summary (page 559)

1. $\quad x + y = 4 \quad\quad x + y = 4$
$\quad\ 0 + y = 4 \quad\quad x + 0 = 4$
$\quad\quad\quad\ y = 4 \quad\quad\quad\ x = 4$
$\quad\quad\ (0, 4) \quad\quad\quad\ (4, 0)$

2. $\quad 2x - y = 8 \quad\quad 2x - y = 8$
$\quad 2(0) - y = 8 \quad\quad 2x - 0 = 8$
$\quad\quad\ -y = 8 \quad\quad\quad 2x = 8$
$\quad\quad\quad\ y = -8 \quad\quad\quad\ x = 4$
$\quad\quad\ (0, -8) \quad\quad\quad\ (4, 0)$

3. $\quad y = 3x + 4 \quad\quad y = 3x + 4$
$\quad y = 3(0) + 4 \quad\quad y = 3(1) + 4$
$\quad y = 0 + 4 \quad\quad\quad y = 3 + 4$
$\quad y = 4 \quad\quad\quad\quad\ y = 7$
$\quad\quad (0, 4) \quad\quad\quad\quad (1, 7)$

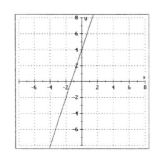

4.
$$x = 4 - 2y \quad\quad x = 4 - 2y$$
$$0 = 4 - 2y \quad\quad x = 4 - 2(0)$$
$$2y = 4 \quad\quad\quad x = 4$$
$$y = 2 \quad\quad\quad\quad (4, 0)$$
$$(0, 2)$$

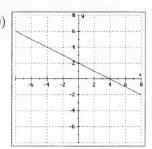

5. $\quad y = 4$ (horizontal)

6. $\quad x = -2$ (vertical)

7.
$$2(x + 3) = x + 2$$
$$2x + 6 = x + 2$$
$$x = -4 \text{ (vertical)}$$

8.
$$3y = 2(y - 1)$$
$$3y = 2y - 2$$
$$y = -2 \text{ (horiozntal)}$$

9. $\quad x = \dfrac{x_1 + x_2}{2} = \dfrac{-3 + 6}{2} = \dfrac{3}{2}; \; y = \dfrac{y_1 + y_2}{2} = \dfrac{5 + 11}{2} = \dfrac{16}{2} = 8;$ midpoint: $\left(\dfrac{3}{2}, 8\right)$

10. $\quad m = \dfrac{\Delta y}{\Delta x} = \dfrac{8 - 5}{5 - 2} = \dfrac{3}{3} = 1$

11. $\quad m = \dfrac{\Delta y}{\Delta x} = \dfrac{12 - (-2)}{6 - (-3)} = \dfrac{14}{9}$

12. $\quad m = \dfrac{\Delta y}{\Delta x} = \dfrac{-6 - 4}{-5 - (-3)} = \dfrac{-10}{-2} = 5$

13. $\quad m = \dfrac{\Delta y}{\Delta x} = \dfrac{-9 - (-4)}{-6 - 5} = \dfrac{-5}{-11} = \dfrac{5}{11}$

14. $m = \dfrac{\Delta y}{\Delta x} = \dfrac{4-4}{8-(-2)} = \dfrac{0}{10} = 0$

15. $m = \dfrac{\Delta y}{\Delta x} = \dfrac{8-(-4)}{-5-(-5)} = \dfrac{12}{0}$
undefined slope

16. $2x - 3y = 18$
$-3y = -2x + 18$
$\dfrac{-3y}{-3} = \dfrac{-2x}{-3} + \dfrac{18}{-3}$
$y = \frac{2}{3}x - 6 \Rightarrow m = \frac{2}{3}$

17. $2x + y = 8$
$y = -2x + 8 \Rightarrow m = -2$

18. $-2(x-3) = 10$
$-2x + 6 = 10$
$-2x = 4$
$x = -2$
The slope is undefined (vertical).

19. $3y + 1 = 7$
$3y = 6$
$y = 2$
$m = 0$ (horizontal)

20. $m_1 \neq m_2 \Rightarrow$ not parallel
$m_1 \cdot m_2 = 4\left(-\frac{1}{4}\right) = -1 \Rightarrow$ perpendicular

21. $m_1 = m_2 \Rightarrow$ parallel

22. $m_1 \neq m_2 \Rightarrow$ not parallel
$m_1 \cdot m_2 = 0.5\left(-\frac{1}{2}\right) = -0.25$
$\Rightarrow$ not perpendicular

23. $m_1 \neq m_2 \Rightarrow$ not parallel
$m_1 \cdot m_2 = 5(-0.2) = -1$
$\Rightarrow$ perpendicular

24. Let $y =$ sales and let $x =$ year of business.
$m = \dfrac{\Delta y}{\Delta x} = \dfrac{y_2 - y_1}{x_1 - x_1} = \dfrac{130,000 - 65,000}{4 - 1} = \dfrac{65,000}{3} \approx \$21,666.67$ per year

25. $y - y_1 = m(x - x_1)$
$y - 5 = 3(x + 8)$
$y - 5 = 3x + 24$
$-3x + y = 29$
$3x - y = -29$

26. $m = \dfrac{\Delta y}{\Delta x} = \dfrac{4-(-9)}{-2-6} = \dfrac{13}{-8} = -\dfrac{13}{8}$
$y - y_1 = m(x - x_1)$
$y - 4 = -\frac{13}{8}(x + 2)$
$8(y - 4) = 8\left(-\frac{13}{8}\right)(x + 2)$
$8y - 32 = -13(x + 2)$
$8y - 32 = -13x - 26$
$13x + 8y = 6$

27. Find the slope of the given line:

$3x - 2y = 7$

$-2y = -3x + 7$

$y = \frac{3}{2}x - \frac{7}{2} \Rightarrow m = \frac{3}{2}$

Use the parallel slope:

$y - y_1 = m(x - x_1)$

$y + 5 = \frac{3}{2}(x + 3)$

$2(y + 5) = 2 \cdot \frac{3}{2}(x + 3)$

$2y + 10 = 3(x + 3)$

$2y + 10 = 3x + 9$

$-3x + 2y = -1 \Rightarrow 3x - 2y = 1$

28. Find the slope of the given line:

$3x - 2y = 7$

$-2y = -3x + 7$

$y = \frac{3}{2}x - \frac{7}{2} \Rightarrow m = \frac{3}{2}$

Use the perpendicular slope:

$y - y_1 = m(x - x_1)$

$y + 5 = -\frac{2}{3}(x + 3)$

$3(y + 5) = 3\left(-\frac{2}{3}\right)(x + 3)$

$3y + 15 = -2(x + 3)$

$3y + 15 = -2x - 6$

$2x + 3y = -21$

29. $y = \frac{2}{3}x + 1;\ m = \frac{2}{3};\ b = 1$

30. $2x + 3y = 8 \Rightarrow m = -\dfrac{A}{B} = -\dfrac{2}{3}$

$3x - 2y = 10 \Rightarrow m = -\dfrac{A}{B} = -\dfrac{3}{-2} = \dfrac{3}{2}$

perpendicular

31. Let x represent the number of years since the copy machine was purchased, and let y represent its value. We know two points on the line: $(0, 8700)$ and $(5, 100)$.

Find the slope:

$m = \dfrac{\Delta y}{\Delta x} = \dfrac{8700 - 100}{0 - 5} = \dfrac{8600}{-5} = -1720$

Find the equation of the line:

$y - y_1 = m(x - x_1)$

$y - 8700 = -1720(x - 0)$

$y = -1720x + 8700$

32. $y = 6x - 4$ **is a function**, since each value of x corresponds to exactly one value of y.

33. $y = 4 - x$ **is a function**, since each value of x corresponds to exactly one value of y.

34. $y^2 = x$ **is not a function**, since $x = 9$ corresponds to both $y = 3$ and $y = -3$.

35. $|y| = x^2$ **is not a function**, since $x = 2$ corresponds to both $y = 4$ and $y = -4$.

36. $f(x) = 3x + 2$

$f(-3) = 3(-3) + 2 = -9 + 2 = -7$

37. $g(x) = x^2 - 4$

$g(8) = 8^2 - 4 = 64 - 4 = 60$

38. $g(x) = x^2 - 4$

$g(-2) = (-2)^2 - 4 = 4 - 4 = 0$

39. $f(x) = 3x + 2$

$f(5) = 3(5) + 2 = 15 + 2 = 17$

40. Since any value can be substituted for x, the domain is $(-\infty, \infty)$. The range is also $(-\infty, \infty)$.

41. Since any value can be substituted for x, the domain is $(-\infty, \infty)$. The range is also $(-\infty, \infty)$.

42. Since any value can be substituted for x, the domain is $(-\infty, \infty)$. Since $x^2 \geq 0$, $x^2 + 1$ must be greater than or equal to 1. Thus the range is $[1, \infty)$.

43. Since $x = 2$ makes the denominator equal to 0, the domain is $(-\infty, 2) \cup (2, \infty)$. Since the numerator will never equal 0, the fraction will never equal 0. The range is $(-\infty, 0) \cup (0, \infty)$.

44. Since $x = 3$ makes the denominator equal to 0, the domain is $(-\infty, 3) \cup (3, \infty)$. Since the numerator will never equal 0, the fraction will never equal 0. The range is $(-\infty, 0) \cup (0, \infty)$.

45. x can be any real number, so the domain is $(-\infty, \infty)$. y always has a value of 7, so the range is $\{7\}$.

46. Since each vertical line passes through the graph at most once, it is a function.

47. Since a vertical line can pass through the graph more than once, it is not a function.

48. Since a vertical line can pass through the graph more than once, it is not a function.

49. Since each vertical line passes through the graph at most once, it is a function.

50.
$$f(x) = x^2 - 3$$
Shift $y = x^2$ down 3.

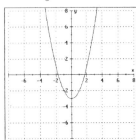

51.
$$f(x) = |x| - 4$$
Shift $y = |x|$ down 4.

52. $f(x) = (x-2)^3$

Shift $y = x^3$ right 2.

53. $f(x) = (x+4)^2 - 3$

Shift $y = x^2$ down 3, left 4.

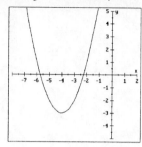

54. $f(x) = -x^3 - 2$

Shift $y = -x^3$ down 2.

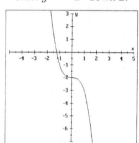

55. $f(x) = -|x-1| + 2$

Shift $y = -|x|$ up 2, right 1.

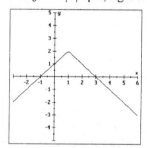

56-61. Compare your graphs to the graphs in numbers **50-55.**

62. $f(x) = \frac{2}{x-2}$

domain $= (-\infty, 2) \cup (2, \infty)$

range $= (-\infty, 0) \cup (0, \infty)$

63. $f(x) = \frac{x}{x+3}$

domain $= (-\infty, -3) \cup (-3, \infty)$

range $= (-\infty, 1) \cup (1, \infty)$

64.
$$\frac{x+1}{8} = \frac{4x-2}{23}$$
$$23(x+1) = 8(4x-2)$$
$$23x + 23 = 32x - 16$$
$$-9x = -39$$
$$x = \frac{-39}{-9} = \frac{13}{3}$$

65.
$$\frac{1}{x+6} = \frac{x+10}{12}$$
$$12 = (x+6)(x+10)$$
$$12 = x^2 + 16x + 60$$
$$0 = x^2 + 16x + 48$$
$$0 = (x+12)(x+4)$$
$$x + 12 = 0 \qquad \textbf{or} \quad x + 4 = 0$$
$$x = -12 \qquad\qquad x = -4$$

66.
$$y = kx \qquad y = \tfrac{1}{6}x$$
$$2 = k(12) \qquad y = \tfrac{1}{6}(12)$$
$$\tfrac{1}{6} = k \qquad \boxed{y = 2}$$

67.
$$y = \frac{k}{x} \qquad y = \frac{72}{x}$$
$$3 = \frac{k}{24} \qquad y = \frac{72}{12}$$
$$72 = k \qquad \boxed{y = 6}$$

68.
$$y = kxz$$
$$3 = k(24)(4)$$
$$3 = 96k$$
$$\boxed{\tfrac{3}{96} = \tfrac{1}{32} = k}$$

69.
$$y = \frac{kt}{x}$$
$$64 = \frac{k(8)}{2}$$
$$128 = 8k$$
$$\boxed{16 = k}$$

Chapter 8 Test (page 564)

1.
$$2x - 5y = 10 \qquad 2x - 5y = 10$$
$$2(0) - 5y = 10 \qquad 2x - 5(0) = 10$$
$$-5y = 10 \qquad\qquad 2x = 10$$
$$y = -2 \qquad\qquad x = 5$$
$$(0, -2) \qquad\qquad (5, 0)$$

2.
$$x = \frac{x_1 + x_2}{2} = \frac{-3+4}{2} = \frac{1}{2}$$
$$y = \frac{y_1 + y_2}{2} = \frac{3 + (-2)}{2} = \frac{1}{2}$$
$$\text{midpoint: } \left(\tfrac{1}{2}, \tfrac{1}{2}\right)$$

3.
$$y = \frac{x-3}{5} \qquad y = \frac{x-3}{5}$$
$$0 = \frac{x-3}{5} \qquad y = \frac{0-3}{5}$$
$$5(0) = 5 \cdot \frac{x-3}{5} \qquad y = -\frac{3}{5}$$
$$0 = x - 3 \qquad y\text{-intercept: } \left(0, -\frac{3}{5}\right)$$
$$3 = x$$
x-intercept: $(3, 0)$

4. $x - 7 = 0$
$x = 7$
vertical line

5. $m = \dfrac{\Delta y}{\Delta x} = \dfrac{4-8}{-2-6} = \dfrac{-4}{-8} = \dfrac{1}{2}$

6. $2x - 3y = 8$
$$-3y = -2x + 8$$
$$y = \frac{2}{3}x - \frac{8}{3} \Rightarrow m = \frac{2}{3}$$

7. The graph of $x = 12$ is a vertical line, so the slope is undefined.

8. The graph of $y = 12$ is a horizontal line, so the slope is 0.

9. $y - y_1 = m(x - x_1)$
$$y + 5 = \frac{2}{3}(x - 4)$$
$$y + 5 = \frac{2}{3}x - \frac{8}{3}$$
$$y = \frac{2}{3}x - \frac{8}{3} - 5$$
$$y = \frac{2}{3}x - \frac{23}{3}$$

10. $m = \dfrac{\Delta y}{\Delta x} = \dfrac{6 - (-10)}{-2 - (-4)} = \dfrac{16}{2} = 8$
$$y - y_1 = m(x - x_1)$$
$$y - 6 = 8(x + 2)$$
$$y - 6 = 8x + 16$$
$$-8x + y = 22$$
$$8x - y = -22$$

11. $-2(x - 3) = 3(2y + 5)$
$$-2x + 6 = 6y + 15$$
$$6y + 15 = -2x + 6$$
$$6y = -2x - 9$$
$$y = \frac{-2}{6}x - \frac{9}{6}$$
$$y = -\frac{1}{3}x - \frac{3}{2}$$
$$m = -\frac{1}{3}, b = -\frac{3}{2} \Rightarrow \left(0, -\frac{3}{2}\right)$$

12. $4x - y = 12$
$$-y = -4x + 12$$
$$y = 4x - 12 \Rightarrow m = 4$$
$$y = \frac{1}{4}x + 3 \Rightarrow m = \frac{1}{4}$$
neither parallel nor perpendicular

13. $y = -\frac{2}{3}x + 4 \Rightarrow m = -\frac{2}{3}$
$$2y = 3x - 3$$
$$y = \frac{3}{2}x - \frac{3}{2} \Rightarrow m = \frac{3}{2}$$
perpendicular

14. $y = \frac{3}{2}x - 7 \Rightarrow m = \frac{3}{2}$
Use the parallel slope:
$$y - y_1 = m(x - x_1)$$
$$y - 0 = \frac{3}{2}(x - 0)$$
$$y = \frac{3}{2}x$$

15. $y = -\frac{2}{3}x - 7 \Rightarrow m = -\frac{2}{3}$
Use the perpendicular slope:
$y - y_1 = m(x - x_1)$
$y - 6 = \frac{3}{2}(x + 3)$
$y - 6 = \frac{3}{2}x + \frac{9}{2}$
$y = \frac{3}{2}x + \frac{21}{2}$

16. $|y| = x$ **is not a function**, since $x = 2$ corresponds to both $y = 2$ and $y = -2$.

17. Domain: $(-\infty, \infty)$; Range: $[0, \infty)$

18. Domain: $(-\infty, \infty)$; Range: $(-\infty, \infty)$

19. $f(x) = 3x + 1$
$f(3) = 3(3) + 1 = 9 + 1 = 10$

20. $g(x) = x^2 - 2$
$g(0) = 0^2 - 2 = 0 - 2 = -2$

21. $f(x) = 3x + 1$
$f(a) = 3a + 1$

22. $g(x) = x^2 - 2$
$g(-x) = (-x)^2 - 2 = x^2 - 2$

23. function

24. not a function

25. $f(x) = x^2 - 1$
Shift $y = x^2$ down 1.

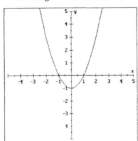

26. $f(x) = -|x + 2|$
Shift $y = -|x|$ left 2.

27. $\dfrac{3}{x - 2} = \dfrac{x + 3}{2x}$
$6x = (x + 3)(x - 2)$
$6x = x^2 + x - 6$
$0 = x^2 - 5x - 6$
$0 = (x - 6)(x + 1)$
$x - 6 = 0$ **or** $x + 1 = 0$
$x = 6$ $\qquad x = -1$

28. $y = kx \qquad y = \frac{2}{15}x$
$4 = k(30) \qquad y = \frac{2}{15}(9)$
$\frac{4}{30} = k \qquad y = \frac{18}{15}$
$\frac{2}{15} = k \qquad \boxed{y = \frac{6}{5}}$

29. $V = \dfrac{k}{t} \qquad V = \dfrac{1100}{t}$
$55 = \dfrac{k}{20} \qquad 75 = \dfrac{1100}{t}$
$1100 = k \qquad 75t = 1100$
$\boxed{t = \frac{44}{3}}$

30. function

Cumulative Review Exercises (page 566)

1. natural numbers: $1, 2, 6, 7$

2. whole numbers: $0, 1, 2, 6, 7$

3. rational numbers: $-2, 0, 1, 2, \frac{13}{12}, 6, 7$

4. irrational numbers: $\sqrt{5}, \pi$

5. negative numbers: -2

6. real numbers: $-2, 0, 1, 2, \frac{13}{12}, 6, 7, \sqrt{5}, \pi$

7. prime numbers: $2, 7$

8. composite numbers: 6

9. even numbers: $-2, 0, 2, 6$

10. odd numbers: $1, 7$

11. $-2 < x \le 5 \Rightarrow$

12. $[-5, 0) \cup [3, 6] \Rightarrow$

13. $-|5| + |-3| = -5 + 3 = -2$

14. $\dfrac{|-5| + |-3|}{-|4|} = \dfrac{5+3}{-4} = \dfrac{8}{-4} = -2$

15. $2 + 4 \cdot 5 = 2 + 20 = 22$

16. $\dfrac{8-4}{2-4} = \dfrac{4}{-2} = -2$

17. $20 \div (-10 \div 2) = 20 \div (-5) = -4$

18. $\dfrac{6 + 3(6+4)}{2(3-9)} = \dfrac{6 + 3(10)}{2(-6)} = \dfrac{6+30}{-12}$
$$= \dfrac{36}{-12} = -3$$

19. $-x - 2y = -2 - 2(-3) = -2 + 6 = 4$

20. $\dfrac{x^2 - y^2}{2x + y} = \dfrac{2^2 - (-3)^2}{2(2) + (-3)} = \dfrac{4-9}{4-3} = \dfrac{-5}{1}$
$$= -5$$

21. associative property of addition

22. distributive property

23. commutative property of addition

24. associative property of multiplication

25. $\left(x^2 y^3\right)^4 = \left(x^2\right)^4 \left(y^3\right)^4 = x^8 y^{12}$

26. $\dfrac{c^4 c^8}{\left(c^5\right)^2} = \dfrac{c^{12}}{c^{10}} = c^2$

27. $\left(-\dfrac{a^3 b^{-2}}{ab}\right)^{-1} = \left(-\dfrac{a^2}{b^3}\right)^{-1} = -\dfrac{b^3}{a^2}$

28. $\left(\dfrac{-3a^3 b^{-2}}{6a^{-2} b^3}\right)^0 = 1$

29. $0.00000497 = 4.97 \times 10^{-6}$

30. $9.32 \times 10^8 = 932{,}000{,}000$

Exercise 9.1 (page 577)

1. $\dfrac{x^2 + 7x + 12}{x^2 - 16} = \dfrac{(x+4)(x+3)}{(x+4)(x-4)} = \dfrac{x+3}{x-4}$

3. $\dfrac{x^2 - x - 6}{x^2 - 2x - 3} \cdot \dfrac{x^2 - 1}{x^2 + x - 2} = \dfrac{(x-3)(x+2)}{(x-3)(x+1)} \cdot \dfrac{(x+1)(x-1)}{(x+2)(x-1)} = 1$

5. $\dfrac{3}{m+1} + \dfrac{3m}{m-1} = \dfrac{3(m-1)}{(m+1)(m-1)} + \dfrac{3m(m+1)}{(m+1)(m-1)} = \dfrac{3m^2 + 6m - 3}{(m+1)(m-1)} = \dfrac{3(m^2 + 2m - 1)}{(m+1)(m-1)}$

7. $\left(5x^2\right)^2$

9. positive

11. 3; up

13. x

15. odd

17. 0

19. $3x^2$

21. $a^2 + b^3$

23. $\sqrt{121} = \sqrt{11^2} = 11$

25. $-\sqrt{64} = -\sqrt{8^2} = -8$

27. $\sqrt{\frac{1}{9}} = \sqrt{\left(\frac{1}{3}\right)^2} = \frac{1}{3}$

29. $-\sqrt{\frac{25}{49}} = -\sqrt{\left(\frac{5}{7}\right)^2} = -\frac{5}{7}$

31. $\sqrt{-25}$: not a real number

33. $\sqrt{0.16} = 0.4$

35. $\sqrt{(-4)^2} = \sqrt{16} = 4$

37. $\sqrt{-36}$: not a real number

39. $\sqrt{12} \approx 3.4641$

41. $\sqrt{679.25} \approx 26.0624$

43. $\sqrt{4x^2} = \sqrt{(2x)^2} = |2x| = 2|x|$

45. $\sqrt{9a^4} = \sqrt{(3a^2)^2} = |3a^2| = 3a^2$

47. $\sqrt{(t+5)^2} = |t+5|$

49. $\sqrt{(-5b)^2} = |-5b| = 5|b|$

51. $\sqrt{a^2 + 6a + 9} = \sqrt{(a+3)^2} = |a+3|$

53. $\sqrt{t^2 + 24t + 144} = \sqrt{(t+12)^2} = |t+12|$

55. $\sqrt[3]{1} = \sqrt[3]{1^3} = 1$

57. $\sqrt[3]{-125} = \sqrt[3]{(-5)^3} = -5$

59. $\sqrt[3]{-\frac{8}{27}} = \sqrt[3]{\left(-\frac{2}{3}\right)^3} = -\frac{2}{3}$

61. $\sqrt[3]{0.064} = 0.4$

63. $\sqrt[3]{8a^3} = \sqrt[3]{(2a)^3} = 2a$

65. $\sqrt[3]{-1000p^3q^3} = \sqrt[3]{(-10pq)^3} = -10pq$

67. $\sqrt[3]{-\frac{1}{8}m^6n^3} = \sqrt[3]{\left(-\frac{1}{2}m^2n\right)^3} = -\frac{1}{2}m^2n$

69. $\sqrt[3]{0.008z^9} = \sqrt[3]{(0.2z^3)^3} = 0.2z^3$

71. $\sqrt[4]{81} = \sqrt[4]{3^4} = 3$

73. $-\sqrt[5]{243} = -\sqrt[5]{3^5} = -3$

75. $\sqrt[5]{-32} = \sqrt[5]{(-2)^5} = -2$

77. $\sqrt[4]{\frac{16}{625}} = \sqrt[4]{\left(\frac{2}{5}\right)^4} = \frac{2}{5}$

79. $-\sqrt[5]{-\frac{1}{32}} = -\sqrt[5]{\left(-\frac{1}{2}\right)^5} = -\left(-\frac{1}{2}\right) = \frac{1}{2}$

81. $\sqrt[4]{-256} \Rightarrow$ not a real number

83. $\sqrt[4]{16x^4} = \sqrt[4]{(2x)^4} = |2x| = 2|x|$

85. $\sqrt[3]{8a^3} = \sqrt[3]{(2a)^3} = 2a$

87. $\sqrt[4]{\frac{1}{16}x^4} = \sqrt[4]{\left(\frac{1}{2}x\right)^4} = \left|\frac{1}{2}x\right| = \frac{1}{2}|x|$

89. $\sqrt[4]{x^{12}} = \sqrt[4]{(x^3)^4} = |x^3|$

91. $\sqrt[5]{-x^5} = \sqrt[5]{(-x)^5} = -x$

93. $\sqrt[3]{-27a^6} = \sqrt[3]{(-3a^2)^3} = -3a^2$

95. $\sqrt[25]{(x+2)^{25}} = x + 2$

97. $\sqrt[8]{0.00000001x^{16}y^8} = \sqrt[8]{(0.1x^2y)^8} = |0.1x^2y| = 0.1x^2|y|$

99. $f(4) = \sqrt{4-4} = \sqrt{0} = 0$

101. $f(20) = \sqrt{20-4} = \sqrt{16} = 4$

103. $g(9) = \sqrt{9-8} = \sqrt{1} = 1$

105. $g(8.25) = \sqrt{8.25 - 8} = \sqrt{0.25} = 0.5$

107. $f(4) = \sqrt{4^2 + 1} = \sqrt{17} \approx 4.1231$

109. $f(2.35) = \sqrt{(2.35)^2 + 1} = \sqrt{6.5225}$
≈ 2.5539

111. $f(x) = \sqrt{x+4}$; Shift $y = \sqrt{x}$ left 4.

113. $f(x) = -\sqrt{x} - 3$; Reflect $y = \sqrt{x}$ about the x-axis and shift down 3.

$D = [-4, \infty); R = [0, \infty)$

$D = [0, \infty); R = (-\infty, -3]$

115. mean $= \dfrac{2 + 5 + 5 + 6 + 7}{5} = \dfrac{25}{5} = 5$

117. $s_{\bar{x}} = \dfrac{s}{\sqrt{N}} = \dfrac{65}{\sqrt{30}} \approx 11.8673$

Original term	Mean	Difference (term−mean)	Square of difference
2	5	−3	9
5	5	0	0
5	5	0	0
6	5	1	1
7	5	2	4

st. dev. $= \sqrt{\dfrac{9 + 0 + 0 + 1 + 4}{5}} \approx 1.67$

119. $r = \sqrt{\dfrac{A}{\pi}} = \sqrt{\dfrac{9\pi}{\pi}} = \sqrt{9} = 3$ units

121. $t = \dfrac{\sqrt{s}}{4} = \dfrac{\sqrt{256}}{4} = \dfrac{16}{4} = 4$ seconds

123. $I = \sqrt{\dfrac{P}{18}} = \sqrt{\dfrac{980}{18}} \approx \sqrt{54.44} \approx 7.4$ amps   **125. Answers may vary.**

127. $\sqrt{x^2 - 4x + 4} = \sqrt{(x-2)^2} = |x - 2|.\ |x - 2| = x - 2$ when $x - 2 \geq 0$, or $x \geq 2$.

Exercise 9.2 (page 584)

1. $(4x + 2)(3x - 5) = 12x^2 - 14x - 10$ **3.** $(5t + 4s)(3t - 2s) = 15t^2 + 2ts - 8s^2$

5.   hypotenuse   **7.** $a^2 + b^2 = c^2$ **9.**   distance

11.
$$c^2 = a^2 + b^2$$
$$c^2 = 6^2 + 8^2$$
$$c^2 = 36 + 64$$
$$c^2 = 100$$
$$c = \sqrt{100}$$
$$c = 10 \text{ ft}$$

13.
$$c^2 = a^2 + b^2$$
$$82^2 = a^2 + 18^2$$
$$6724 = a^2 + 324$$
$$6400 = a^2$$
$$\sqrt{6400} = a$$
$$80 \text{ m} = a$$

15.
$$c^2 = a^2 + b^2$$
$$50^2 = 14^2 + b^2$$
$$2500 = 196 + b^2$$
$$2304 = b^2$$
$$\sqrt{2304} = b$$
$$48 \text{ in.} = b$$

17.   Let $x =$ the length of the diagonal.
$$7^2 + 7^2 = x^2$$
$$49 + 49 = x^2$$
$$98 = x^2$$
$$\sqrt{98} = x$$
$$9.9 \text{ cm} \approx x$$

19.
$$\begin{aligned} d &= \sqrt{(x_2 - x_1)^2 + (y_2 - y_1)^2} \\ &= \sqrt{(0 - 3)^2 + [0 - (-4)]^2} \\ &= \sqrt{(-3)^2 + (4)^2} \\ &= \sqrt{9 + 16} \\ &= \sqrt{25} = 5 \end{aligned}$$

21. $d = \sqrt{(x_2 - x_1)^2 + (y_2 - y_1)^2} = \sqrt{(2 - 5)^2 + (4 - 8)^2} = \sqrt{(-3)^2 + (-4)^2} = \sqrt{9 + 16}$
$$= \sqrt{25} = 5$$

23. $d = \sqrt{(x_2 - x_1)^2 + (y_2 - y_1)^2} = \sqrt{(-2 - 3)^2 + (-8 - 4)^2} = \sqrt{(-5)^2 + (-12)^2} = \sqrt{25 + 144}$
$$= \sqrt{169} = 13$$

25. $d = \sqrt{(x_2 - x_1)^2 + (y_2 - y_1)^2} = \sqrt{(6 - 12)^2 + (8 - 16)^2} = \sqrt{(-6)^2 + (-8)^2} = \sqrt{36 + 64}$
$$= \sqrt{100} = 10$$

27. $d = \sqrt{(x_2 - x_1)^2 + (y_2 - y_1)^2} = \sqrt{[-3 - (-5)]^2 + [5 - (-5)]^2} = \sqrt{2^2 + 10^2} = \sqrt{4 + 100}$
$$= \sqrt{104} \approx 10.2$$

29.   Let the points be represented by $A(5, 1)$, $B(7, 0)$ and $C(3, 0)$. Find the length of $\overline{AB}$ and $\overline{AC}$:
$$\overline{AB}: \ \sqrt{(5 - 7)^2 + (1 - 0)^2} = \sqrt{(-2)^2 + 1^2} = \sqrt{5}$$
$$\overline{AC}: \ \sqrt{(5 - 3)^2 + (1 - 0)^2} = \sqrt{(2)^2 + 1^2} = \sqrt{5}$$
Since $\overline{AB}$ and $\overline{AC}$ have the same length, $(5, 1)$ is equidistant from $(7, 0)$ and $(3, 0)$.

31. Let the points be represented by $A(-2, 4)$, $B(2, 8)$ and $C(6, 4)$. Find the length of each side:
$\overline{AB}:\ \sqrt{(-2-2)^2 + (4-8)^2} = \sqrt{(-4)^2 + (-4)^2} = \sqrt{32}$
$\overline{AC}:\ \sqrt{(-2-6)^2 + (4-4)^2} = \sqrt{(-8)^2 + 0^2} = \sqrt{64}$
$\overline{BC}:\ \sqrt{(2-6)^2 + (8-4)^2} = \sqrt{(-4)^2 + 4^2} = \sqrt{32}$
Since $\overline{AB}$ and $\overline{BC}$ have the same length, the triangle is isosceles.

33. $d^2 = 5^2 + 12^2$
$d^2 = 25 + 144$
$d^2 = 169$
$d = 13$ ft

35. Let $x = $ distance to 2nd
$x^2 = 90^2 + 90^2$
$x^2 = 8100 + 8100$
$x^2 = 16200$
$x \approx 127$ ft

37. Refer to the diagram provided. The 3rd baseman is at B, so $\overline{BC}$ has a length of 10 ft. Let $\overline{AC}$ and $\overline{AB}$ both have a length x.
$x^2 + x^2 = 10^2$
$2x^2 = 100$
$x^2 = 50$
$x = \sqrt{50} \approx 7.1$ ft

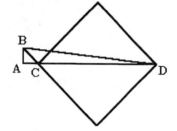

From #35, the length of $\overline{CD} \approx 127.3$ ft, so $\overline{AD}$ has a length of about $127.3 + 7.1 = 134.4$ ft. Let $y = $ the length of $\overline{BD}$.
$y^2 = (7.1)^2 + (134.4)^2$
$y^2 = 18113.77$
$y = \sqrt{18113.77} \approx 135$ ft

39. $d = \sqrt{a^2 + b^2 + c^2} = \sqrt{12^2 + 24^2 + 17^2} = \sqrt{1009} = 31.76 \Rightarrow$ The racket will not fit.

41. $d = \sqrt{a^2 + b^2 + c^2} = \sqrt{21^2 + 21^2 + 21^2} = \sqrt{1323} = 36.4 \Rightarrow$ The femur will fit.

43. Let $x = $ direct distance from A to D.
$x^2 = 52^2 + (105 + 60)^2$
$x^2 = 52^2 + 165^2$
$x^2 = 29929$
$x = 173$ yd

45. Let $x = $ half the length of the stretched wire.
$x^2 = 20^2 + 1^2$
$x^2 = 401$
$x = 20.025$ ft
The stretched wire has a length of 40.05 ft. It has been stretched by 0.05 ft.

47. $A = 6\sqrt[3]{V^2} = 6\sqrt[3]{8^2} = 6\sqrt[3]{64} = 6(4) = 24$ cm^2

49. Answers may vary.

51. $I = \dfrac{703w}{h^2} = \dfrac{703(104)}{(54.1)^2} = \dfrac{73112}{2926.81} \approx 25$

SECTION 9.3

Exercise 9.3 (page 593)

1. $5x - 4 < 11$

$5x < 15$

$x < 3$

3. $\dfrac{4}{5}(r - 3) > \dfrac{2}{3}(r + 2)$

$15 \cdot \dfrac{4}{5}(r - 3) > 15 \cdot \dfrac{2}{3}(r + 2)$

$12(r - 3) > 10(r + 2)$

$2r > 56$

$r > 28$

5. Let x = pints of water added (0% alcohol).

$\boxed{\text{Alcohol at start}} + \boxed{\text{Alcohol added}} = \boxed{\text{Alcohol at end}}$

$0.20(5) + 0(x) = 0.15(5 + x)$

$1 + 0 = 0.75 + 0.15x$

$0.25 = 0.15x$

$x = \dfrac{0.25}{0.15} = \dfrac{5}{3} \Rightarrow 1\frac{2}{3}$ pints of water should be added.

7. $a \cdot a \cdot a \cdot a$

9. a^{mn}

11. $\dfrac{a^n}{b^n}$

13. $\dfrac{1}{a^n}; 0$

15. $\left(\dfrac{b}{a}\right)^n$

17. $|x|$

19. $7^{1/3} = \sqrt[3]{7}$

21. $8^{1/5} = \sqrt[5]{8}$

23. $(3x)^{1/4} = \sqrt[4]{3x}$

25. $\left(\frac{1}{2}x^3y\right)^{1/4} = \sqrt[4]{\frac{1}{2}x^3y}$

27. $(4a^2b^3)^{1/5} = \sqrt[5]{4a^2b^3}$

29. $(x^2 + y^2)^{1/2} = \sqrt{x^2 + y^2}$

31. $\sqrt{11} = 11^{1/2}$

33. $\sqrt[4]{3a} = (3a)^{1/4}$

35. $3\sqrt[5]{a} = 3a^{1/5}$

37. $\sqrt[6]{\frac{1}{7}abc} = \left(\frac{1}{7}abc\right)^{1/6}$

39. $\sqrt[5]{\frac{1}{2}mn} = \left(\frac{1}{2}mn\right)^{1/5}$

41. $\sqrt[3]{a^2 - b^2} = (a^2 - b^2)^{1/3}$

43. $4^{1/2} = 2$

45. $27^{1/3} = 3$

47. $16^{1/4} = 2$

49. $32^{1/5} = 2$

51. $\left(\frac{1}{4}\right)^{1/2} = \frac{1}{2}$

53. $\left(\frac{1}{8}\right)^{1/3} = \frac{1}{2}$

55. $-16^{1/4} = -2$

57. $(-27)^{1/3} = -3$

59. $(-64)^{1/2}$ not a real number

61. $0^{1/3} = 0$

63. $(25y^2)^{1/2} = \left[(5y)^2\right]^{1/2} = |5y| = 5|y|$

65. $(16x^4)^{1/4} = \left[(2x)^4\right]^{1/4} = |2x| = 2|x|$

67. $(243x^5)^{1/5} = \left[(3x)^5\right]^{1/5} = 3x$

69. $(-64x^8)^{1/4} \Rightarrow$ not a real number

71. $36^{3/2} = \left(36^{1/2}\right)^3 = 6^3 = 216$

73. $81^{3/4} = \left(81^{1/4}\right)^3 = 3^3 = 27$

75. $144^{3/2} = \left(144^{1/2}\right)^3 = 12^3 = 1728$

77. $\left(\frac{1}{8}\right)^{2/3} = \left[\left(\frac{1}{8}\right)^{1/3}\right]^2 = \left(\frac{1}{2}\right)^2 = \frac{1}{4}$

79. $\left(25x^4\right)^{3/2} = \left[\left(25x^4\right)^{1/2}\right]^3 = \left(5x^2\right)^3$
$$= 125x^6$$

81. $\left(\frac{8x^3}{27}\right)^{2/3} = \left[\left(\frac{8x^3}{27}\right)^{1/3}\right]^2 = \left(\frac{2x}{3}\right)^2$
$$= \frac{4x^2}{9}$$

83. $4^{-1/2} = \frac{1}{4^{1/2}} = \frac{1}{2}$

85. $4^{-3/2} = \frac{1}{4^{3/2}} = \frac{1}{\left(4^{1/2}\right)^3} = \frac{1}{2^3} = \frac{1}{8}$

87. $\left(16x^2\right)^{-3/2} = \frac{1}{\left(16x^2\right)^{3/2}} = \frac{1}{\left[\left(16x^2\right)^{1/2}\right]^3} = \frac{1}{\left(4x\right)^3} = \frac{1}{64x^3}$

89. $\left(-27y^3\right)^{-2/3} = \frac{1}{\left(-27y^3\right)^{2/3}} = \frac{1}{\left[\left(-27y^3\right)^{1/3}\right]^2} = \frac{1}{\left(-3y\right)^2} = \frac{1}{9y^2}$

91. $\left(-32p^5\right)^{-2/5} = \frac{1}{\left(-32p^5\right)^{2/5}} = \frac{1}{\left[\left(-32p^5\right)^{1/5}\right]^2} = \frac{1}{\left(-2p\right)^2} = \frac{1}{4p^2}$

93. $\left(\frac{1}{4}\right)^{-3/2} = \left(\frac{4}{1}\right)^{3/2} = 4^{3/2} = \left(4^{1/2}\right)^3 = 2^3 = 8$

95. $\left(\frac{27}{8}\right)^{-4/3} = \left(\frac{8}{27}\right)^{4/3} = \left[\left(\frac{8}{27}\right)^{1/3}\right]^4 = \left(\frac{2}{3}\right)^4 = \frac{16}{81}$

97. $\left(-\frac{8x^3}{27}\right)^{-1/3} = \left(-\frac{27}{8x^3}\right)^{1/3} = -\frac{3}{2x}$

99. $5^{4/9}5^{4/9} = 5^{4/9+4/9} = 5^{8/9}$

101. $\left(4^{1/5}\right)^3 = 4^{(1/5)\cdot 3} = 4^{3/5}$

103. $\frac{9^{4/5}}{9^{3/5}} = 9^{4/5-3/5} = 9^{1/5}$

105. $\frac{7^{1/2}}{7^0} = 7^{1/2-0} = 7^{1/2}$

107. $6^{-2/3}6^{-4/3} = 6^{-6/3} = 6^{-2} = \frac{1}{36}$

109. $\frac{2^{5/6}2^{1/3}}{2^{1/2}} = \frac{2^{7/6}}{2^{1/2}} = 2^{4/6} = 2^{2/3}$

111. $a^{2/3}a^{1/3} = a^{3/3} = a^1 = a$

113. $\left(a^{2/3}\right)^{1/3} = a^{(2/3)(1/3)} = a^{2/9}$

115. $\left(a^{1/2}b^{1/3}\right)^{3/2} = a^{3/4}b^{1/2}$

117. $\left(mn^{-2/3}\right)^{-3/5} = m^{-3/5}n^{2/5} = \frac{n^{2/5}}{m^{3/5}}$

119. $\frac{\left(4x^3y\right)^{1/2}}{\left(9xy\right)^{1/2}} = \frac{2x^{3/2}y^{1/2}}{3x^{1/2}y^{1/2}} = \frac{2x}{3}$

121. $(27x^{-3})^{-1/3} = (27)^{-1/3}x = \dfrac{1}{3}x$

123. $y^{1/3}(y^{2/3} + y^{5/3}) = y^{3/3} + y^{6/3} = y + y^2$

125. $x^{3/5}(x^{7/5} - x^{2/5} + 1) = x^{10/5} - x^{5/5} + x^{3/5} = x^2 - x + x^{3/5}$

127. $(x^{1/2} + 2)(x^{1/2} - 2) = x^{2/2} - 2x^{1/2} + 2x^{1/2} - 4 = x - 4$

129. $(x^{2/3} - x)(x^{2/3} + x) = x^{4/3} + x^{5/3} - x^{5/3} - x^2 = x^{4/3} - x^2$

131. $(x^{2/3} + y^{2/3})^2 = (x^{2/3} + y^{2/3})(x^{2/3} + y^{2/3}) = x^{4/3} + x^{2/3}y^{2/3} + x^{2/3}y^{2/3} + y^{4/3}$
$$= x^{4/3} + 2x^{2/3}y^{2/3} + y^{4/3}$$

133. $(a^{3/2} - b^{3/2})^2 = (a^{3/2} - b^{3/2})(a^{3/2} - b^{3/2}) = a^{6/2} - a^{3/2}b^{3/2} - a^{3/2}b^{3/2} + b^{6/2}$
$$= a^3 - 2a^{3/2}b^{3/2} + b^3$$

135. $\sqrt[6]{p^3} = (p^3)^{1/6} = p^{3/6} = p^{1/2} = \sqrt{p}$

137. $\sqrt[4]{25b^2} = (5^2b^2)^{1/4} = 5^{1/2}b^{1/2} = \sqrt{5b}$

139. Answers may vary.

141. $16^{2/4} = 2^2 = 4; \; 16^{1/2} = 4$

Exercise 9.4 (page 603)

1. $3x^2y^3(-5x^3y^{-4}) = -15x^5y^{-1} = \dfrac{-15x^5}{y}$

3. $(3t + 2)^2 = (3t + 2)(3t + 2)$
$$= 9t^2 + 12t + 4$$

5.
$$\begin{array}{r} 3p + \quad 4 + \frac{-5}{2p-5} \\ 2p - 5 \,\overline{\big)\, 6p^2 - 7p - 25} \\ \underline{6p^2 - 15p} \\ 8p - 25 \\ \underline{8p - 20} \\ -5 \end{array}$$

7. $\sqrt[r]{a}\sqrt[n]{b}$

9. $\sqrt{6}\sqrt{6} = \sqrt{36} = 6$

11. $\sqrt{t}\sqrt{t} = \sqrt{t^2} = t$

13. $\sqrt[3]{5x^2}\sqrt[3]{25x} = \sqrt[3]{125x^3} = 5x$

15. $\dfrac{\sqrt{500}}{\sqrt{5}} = \sqrt{\dfrac{500}{5}} = \sqrt{100} = 10$

17. $\dfrac{\sqrt{98x^3}}{\sqrt{2x}} = \sqrt{\dfrac{98x^3}{2x}} = \sqrt{49x^2} = 7x$

19. $\dfrac{\sqrt{180ab^4}}{\sqrt{5ab^2}} = \sqrt{\dfrac{180ab^4}{5ab^2}} = \sqrt{36b^2} = 6b$

21. $\dfrac{\sqrt[3]{48}}{\sqrt[3]{6}} = \sqrt[3]{\dfrac{48}{6}} = \sqrt[3]{8} = 2$

23. $\dfrac{\sqrt[3]{189a^4}}{\sqrt[3]{7a}} = \sqrt[3]{\dfrac{189a^4}{7a}} = \sqrt[3]{27a^3} = 3a$

25. $\sqrt{20} = \sqrt{4 \cdot 5} = \sqrt{4}\sqrt{5} = 2\sqrt{5}$

27. $-\sqrt{200} = -\sqrt{100 \cdot 2} = -\sqrt{100}\sqrt{2}$
$$= -10\sqrt{2}$$

29. $\sqrt[3]{80} = \sqrt[3]{8 \cdot 10} = \sqrt[3]{8}\sqrt[3]{10} = 2\sqrt[3]{10}$

31. $\sqrt[3]{-81} = \sqrt[3]{-27 \cdot 3} = \sqrt[3]{-27}\sqrt[3]{3} = -3\sqrt[3]{3}$

33. $\sqrt[4]{32} = \sqrt[4]{16 \cdot 2} = \sqrt[4]{16}\sqrt[4]{2} = 2\sqrt[4]{2}$

35. $\sqrt[5]{96} = \sqrt[5]{32 \cdot 3} = \sqrt[5]{32}\sqrt[5]{3} = 2\sqrt[5]{3}$

37. $\sqrt{\dfrac{7}{9}} = \dfrac{\sqrt{7}}{\sqrt{9}} = \dfrac{\sqrt{7}}{3}$

39. $\sqrt[3]{\dfrac{7}{64}} = \dfrac{\sqrt[3]{7}}{\sqrt[3]{64}} = \dfrac{\sqrt[3]{7}}{4}$

41. $\sqrt[4]{\dfrac{3}{10,000}} = \dfrac{\sqrt[4]{3}}{\sqrt[4]{10,000}} = \dfrac{\sqrt[4]{3}}{10}$

43. $\sqrt[5]{\dfrac{3}{32}} = \dfrac{\sqrt[5]{3}}{\sqrt[5]{32}} = \dfrac{\sqrt[5]{3}}{2}$

45. $\sqrt{50x^2} = \sqrt{25x^2 \cdot 2} = \sqrt{25x^2}\sqrt{2} = 5x\sqrt{2}$

47. $\sqrt{32b} = \sqrt{16 \cdot 2b} = \sqrt{16}\sqrt{2b} = 4\sqrt{2b}$

49. $-\sqrt{112a^3} = -\sqrt{16a^2 \cdot 7a} = -\sqrt{16a^2}\sqrt{7a}$
$$= -4a\sqrt{7a}$$

51. $\sqrt{175a^2b^3} = \sqrt{25a^2b^2 \cdot 7b} = \sqrt{25a^2b^2}\sqrt{7b}$
$$= 5ab\sqrt{7b}$$

53. $-\sqrt{300xy} = -\sqrt{100 \cdot 3xy} = -\sqrt{100}\sqrt{3xy}$
$$= -10\sqrt{3xy}$$

55. $\sqrt[3]{-54x^6} = \sqrt[3]{-27x^6 \cdot 2} = \sqrt[3]{-27x^6}\sqrt[3]{2}$
$$= -3x^2\sqrt[3]{2}$$

57. $\sqrt[3]{16x^{12}y^3} = \sqrt[3]{8x^{12}y^3 \cdot 2} = \sqrt[3]{8x^{12}y^3}\sqrt[3]{2}$
$$= 2x^4y\sqrt[3]{2}$$

59. $\sqrt[4]{32x^{12}y^4} = \sqrt[4]{16x^{12}y^4 \cdot 2} = \sqrt[4]{16x^{12}y^4}\sqrt[4]{2}$
$$= 2x^3y\sqrt[4]{2}$$

61. $\sqrt{\dfrac{z^2}{16x^2}} = \dfrac{\sqrt{z^2}}{\sqrt{16x^2}} = \dfrac{z}{4x}$

63. $\sqrt[4]{\dfrac{5x}{16z^4}} = \dfrac{\sqrt[4]{5x}}{\sqrt[4]{16z^4}} = \dfrac{\sqrt[4]{5x}}{2z}$

65. $4\sqrt{2x} + 6\sqrt{2x} = 10\sqrt{2x}$

67. $8\sqrt[5]{7a^2} - 7\sqrt[5]{7a^2} = \sqrt[5]{7a^2}$

69. $\sqrt{3} + \sqrt{27} = \sqrt{3} + \sqrt{9}\sqrt{3}$
$$= \sqrt{3} + 3\sqrt{3} = 4\sqrt{3}$$

71. $\sqrt{2} - \sqrt{8} = \sqrt{2} - \sqrt{4}\sqrt{2}$
$$= \sqrt{2} - 2\sqrt{2} = -\sqrt{2}$$

73. $\sqrt{98} - \sqrt{50} = \sqrt{49}\sqrt{2} - \sqrt{25}\sqrt{2} = 7\sqrt{2} - 5\sqrt{2} = 2\sqrt{2}$

75. $3\sqrt{24} + \sqrt{54} = 3\sqrt{4}\sqrt{6} + \sqrt{9}\sqrt{6} = 3(2)\sqrt{6} + 3\sqrt{6} = 6\sqrt{6} + 3\sqrt{6} = 9\sqrt{6}$

77. $\sqrt[3]{24} + \sqrt[3]{3} = \sqrt[3]{8}\sqrt[3]{3} + \sqrt[3]{3} = 2\sqrt[3]{3} + \sqrt[3]{3} = 3\sqrt[3]{3}$

79. $\sqrt[3]{32} - \sqrt[3]{108} = \sqrt[3]{8}\sqrt[3]{4} - \sqrt[3]{27}\sqrt[3]{4} = 2\sqrt[3]{4} - 3\sqrt[3]{4} = -\sqrt[3]{4}$

81. $2\sqrt[3]{125} - 5\sqrt[3]{64} = 2(5) - 5(4) = 10 - 20 = -10$

83. $14\sqrt[4]{32} - 15\sqrt[4]{162} = 14\sqrt[4]{16}\sqrt[4]{2} - 15\sqrt[4]{81}\sqrt[4]{2} = 14(2)\sqrt[4]{2} - 15(3)\sqrt[4]{2} = 28\sqrt[4]{2} - 45\sqrt[4]{2}$
$$= -17\sqrt[4]{2}$$

85. $3\sqrt[4]{512} + 2\sqrt[4]{32} = 3\sqrt[4]{256}\sqrt[4]{2} + 2\sqrt[4]{16}\sqrt[4]{2} = 3(4)\sqrt[4]{2} + 2(2)\sqrt[4]{2} = 12\sqrt[4]{2} + 4\sqrt[4]{2} = 16\sqrt[4]{2}$

87. $\sqrt{98} - \sqrt{50} - \sqrt{72} = \sqrt{49}\sqrt{2} - \sqrt{25}\sqrt{2} - \sqrt{36}\sqrt{2} = 7\sqrt{2} - 5\sqrt{2} - 6\sqrt{2} = -4\sqrt{2}$

89. $\sqrt{18} + \sqrt{300} - \sqrt{243} = \sqrt{9}\sqrt{2} + \sqrt{100}\sqrt{3} - \sqrt{81}\sqrt{3} = 3\sqrt{2} + 10\sqrt{3} - 9\sqrt{3} = 3\sqrt{2} + \sqrt{3}$

91. $2\sqrt[3]{16} - \sqrt[3]{54} - 3\sqrt[3]{128} = 2\sqrt[3]{8}\sqrt[3]{2} - \sqrt[3]{27}\sqrt[3]{2} - 3\sqrt[3]{64}\sqrt[3]{2} = 2(2)\sqrt[3]{2} - 3\sqrt[3]{2} - 3(4)\sqrt[3]{2}$
$$= 4\sqrt[3]{2} - 3\sqrt[3]{2} - 12\sqrt[3]{2} = -11\sqrt[3]{2}$$

93. $\sqrt{25y^2z} - \sqrt{16y^2z} = \sqrt{25y^2}\sqrt{z} - \sqrt{16y^2}\sqrt{z} = 5y\sqrt{z} - 4y\sqrt{z} = y\sqrt{z}$

95. $\sqrt{36xy^2} + \sqrt{49xy^2} = \sqrt{36y^2}\sqrt{x} + \sqrt{49y^2}\sqrt{x} = 6y\sqrt{x} + 7y\sqrt{x} = 13y\sqrt{x}$

97. $2\sqrt[3]{64a} + 2\sqrt[3]{8a} = 2\sqrt[3]{64}\sqrt[3]{a} + 2\sqrt[3]{8}\sqrt[3]{a} = 2(4)\sqrt[3]{a} + 2(2)\sqrt[3]{a} = 8\sqrt[3]{a} + 4\sqrt[3]{a} = 12\sqrt[3]{a}$

99. $\sqrt{y^5} - \sqrt{9y^5} - \sqrt{25y^5} = \sqrt{y^4}\sqrt{y} - \sqrt{9y^4}\sqrt{y} - \sqrt{25y^4}\sqrt{y} = y^2\sqrt{y} - 3y^2\sqrt{y} - 5y^2\sqrt{y}$
$$= -7y^2\sqrt{y}$$

101. $\sqrt[5]{x^6y^2} + \sqrt[5]{32x^6y^2} + \sqrt[5]{x^6y^2} = \sqrt[5]{x^5}\sqrt[5]{xy^2} + \sqrt[5]{32x^5}\sqrt[5]{xy^2} + \sqrt[5]{x^5}\sqrt[5]{xy^2}$
$$= x\sqrt[5]{xy^2} + 2x\sqrt[5]{xy^2} + x\sqrt[5]{xy^2} = 4x\sqrt[5]{xy^2}$$

103. $\sqrt{x^2 + 2x + 1} + \sqrt{x^2 + 2x + 1} = \sqrt{(x+1)^2} + \sqrt{(x+1)^2} = x + 1 + x + 1 = 2x + 2$

105. $x = 2.00;$
$h = 2\sqrt{2} \approx 2.83$

107. $h = 2(5) = 10.00;$
$x = 5\sqrt{3} \approx 8.66$

109. $x = \frac{9.37}{2} \approx 4.69;$
$y \approx 4.69\sqrt{3} \approx 8.11$

111. $x = y = \frac{17.12}{\sqrt{2}} \approx 12.11$

113. Answers may vary.

115. $\sqrt{a+b} = \sqrt{a} + \sqrt{b}$ if either a or b equals 0 and the other is nonnegative.

Exercise 9.5 (page 611)

1. $\dfrac{2}{3-a} = 1$
$$2 = 3 - a$$
$$a = 1$$

3. $\dfrac{8}{b-2} + \dfrac{3}{2-b} = -\dfrac{1}{b}$

$\dfrac{8}{b-2} + \dfrac{-3}{b-2} = -\dfrac{1}{b}$

$\dfrac{5}{b-2} = \dfrac{-1}{b}$

$5b = -b + 2$

$6b = 2$

$b = \dfrac{2}{6} = \dfrac{1}{3}$

5. $2; \sqrt{7}; \sqrt{5}$

7. FOIL

9. conjugate

11. $\sqrt{2}\sqrt{8} = \sqrt{16} = 4$

13. $\sqrt{5}\sqrt{10} = \sqrt{50} = \sqrt{25}\sqrt{2} = 5\sqrt{2}$

15. $2\sqrt{3}\sqrt{6} = 2\sqrt{18} = 2\sqrt{9}\sqrt{2} = 2(3)\sqrt{2}$
$= 6\sqrt{2}$

17. $\sqrt[3]{5}\sqrt[3]{25} = \sqrt[3]{125} = 5$

19. $\left(3\sqrt[3]{9}\right)\left(2\sqrt[3]{3}\right) = 6\sqrt[3]{27} = 6(3) = 18$

21. $\sqrt[3]{2}\sqrt[3]{12} = \sqrt[3]{24} = \sqrt[3]{8}\sqrt[3]{3} = 2\sqrt[3]{3}$

23. $\sqrt{ab^3}\sqrt{ab} = \sqrt{a^2b^4} = ab^2$

25. $\sqrt{5ab}\sqrt{5a} = \sqrt{25a^2b} = \sqrt{25a^2}\sqrt{b}$
$= 5a\sqrt{b}$

27. $\sqrt[3]{5r^2s}\sqrt[3]{2r} = \sqrt[3]{10r^3s} = \sqrt[3]{r^3}\sqrt[3]{10s}$
$= r\sqrt[3]{10s}$

29. $\sqrt[3]{a^5b}\sqrt[3]{16ab^5} = \sqrt[3]{16a^6b^6} = \sqrt[3]{8a^6b^6}\sqrt[3]{2}$
$= 2a^2b^2\sqrt[3]{2}$

31. $\sqrt{x(x+3)}\sqrt{x^3(x+3)} = \sqrt{x^4(x+3)^2} = x^2(x+3)$

33. $\sqrt[3]{6x^2(y+z)^2}\sqrt[3]{18x(y+z)} = \sqrt[3]{108x^3(y+z)^3} = \sqrt[3]{27x^3(y+z)^3}\sqrt[3]{4} = 3x(y+z)\sqrt[3]{4}$

35. $3\sqrt{5}\left(4 - \sqrt{5}\right) = 12\sqrt{5} - 3\sqrt{25} = 12\sqrt{5} - 3(5) = 12\sqrt{5} - 15$

37. $3\sqrt{2}\left(4\sqrt{3} + 2\sqrt{7}\right) = 12\sqrt{6} + 6\sqrt{14}$

39. $-2\sqrt{5x}\left(4\sqrt{2x} - 3\sqrt{3}\right) = -8\sqrt{10x^2} + 6\sqrt{15x} = -8x\sqrt{10} + 6\sqrt{15x}$

41. $\left(\sqrt{2} + 1\right)\left(\sqrt{2} - 3\right) = \sqrt{4} - 3\sqrt{2} + \sqrt{2} - 3 = 2 - 2\sqrt{2} - 3 = -1 - 2\sqrt{2}$

43. $\left(4\sqrt{x} + 3\right)\left(2\sqrt{x} - 5\right) = 8\sqrt{x^2} - 20\sqrt{x} + 6\sqrt{x} - 15 = 8x - 14\sqrt{x} - 15$

45. $\left(\sqrt{5z} + \sqrt{3}\right)\left(\sqrt{5z} + \sqrt{3}\right) = \sqrt{25z^2} + \sqrt{15z} + \sqrt{15z} + \sqrt{9} = 5z + 2\sqrt{15z} + 3$

47. $\left(\sqrt{3x}-\sqrt{2y}\right)\left(\sqrt{3x}+\sqrt{2y}\right) = \sqrt{9x^2}+\sqrt{6xy}-\sqrt{6xy}-\sqrt{4y^2} = 3x-2y$

49. $\left(2\sqrt{3a}-\sqrt{b}\right)\left(\sqrt{3a}+3\sqrt{b}\right) = 2\sqrt{9a^2}+6\sqrt{3ab}-\sqrt{3ab}-3\sqrt{b^2} = 6a+5\sqrt{3ab}-3b$

51. $\left(3\sqrt{2r}-2\right)^2 = \left(3\sqrt{2r}-2\right)\left(3\sqrt{2r}-2\right) = 9\sqrt{4r^2}-6\sqrt{2r}-6\sqrt{2r}+4 = 18r-12\sqrt{2r}+4$

53. $-2\left(\sqrt{3x}+\sqrt{3}\right)^2 = -2\left(\sqrt{3x}+\sqrt{3}\right)\left(\sqrt{3x}+\sqrt{3}\right)$

$$= -2\left(\sqrt{9x^2}+\sqrt{9x}+\sqrt{9x}+\sqrt{9}\right)$$

$$= -2\left(3x+3\sqrt{x}+3\sqrt{x}+3\right) = -2\left(3x+6\sqrt{x}+3\right) = -6x-12\sqrt{x}-6$$

55. $\sqrt{\dfrac{1}{7}} = \dfrac{\sqrt{1}}{\sqrt{7}} = \dfrac{1\sqrt{7}}{\sqrt{7}\sqrt{7}} = \dfrac{\sqrt{7}}{7}$

57. $\sqrt{\dfrac{2}{3}} = \dfrac{\sqrt{2}}{\sqrt{3}} = \dfrac{\sqrt{2}\sqrt{3}}{\sqrt{3}\sqrt{3}} = \dfrac{\sqrt{6}}{3}$

59. $\dfrac{\sqrt{5}}{\sqrt{8}} = \dfrac{\sqrt{5}\sqrt{2}}{\sqrt{8}\sqrt{2}} = \dfrac{\sqrt{10}}{\sqrt{16}} = \dfrac{\sqrt{10}}{4}$

61. $\dfrac{\sqrt{8}}{\sqrt{2}} = \sqrt{\dfrac{8}{2}} = \sqrt{4} = 2$

63. $\dfrac{1}{\sqrt[3]{2}} = \dfrac{1\sqrt[3]{4}}{\sqrt[3]{2}\sqrt[3]{4}} = \dfrac{\sqrt[3]{4}}{\sqrt[3]{8}} = \dfrac{\sqrt[3]{4}}{2}$

65. $\dfrac{3}{\sqrt[3]{9}} = \dfrac{3\sqrt[3]{3}}{\sqrt[3]{9}\sqrt[3]{3}} = \dfrac{3\sqrt[3]{3}}{\sqrt[3]{27}} = \dfrac{3\sqrt[3]{3}}{3} = \sqrt[3]{3}$

67. $\dfrac{\sqrt[3]{2}}{\sqrt[3]{9}} = \dfrac{\sqrt[3]{2}\sqrt[3]{3}}{\sqrt[3]{9}\sqrt[3]{3}} = \dfrac{\sqrt[3]{6}}{\sqrt[3]{27}} = \dfrac{\sqrt[3]{6}}{3}$

69. $\dfrac{\sqrt{8x^2y}}{\sqrt{xy}} = \sqrt{\dfrac{8x^2y}{xy}} = \sqrt{8x} = 2\sqrt{2x}$

71. $\dfrac{\sqrt{10xy^2}}{\sqrt{2xy^3}} = \sqrt{\dfrac{10xy^2}{2xy^3}} = \sqrt{\dfrac{5}{y}} = \dfrac{\sqrt{5}\sqrt{y}}{\sqrt{y}\sqrt{y}}$

$$= \dfrac{\sqrt{5y}}{y}$$

73. $\dfrac{\sqrt[3]{4a^2}}{\sqrt[3]{2ab}} = \sqrt[3]{\dfrac{4a^2}{2ab}} = \sqrt[3]{\dfrac{2a}{b}} = \dfrac{\sqrt[3]{2a}\sqrt[3]{b^2}}{\sqrt[3]{b}\sqrt[3]{b^2}}$

$$= \dfrac{\sqrt[3]{2ab^2}}{b}$$

75. $\dfrac{1}{\sqrt[4]{4}} = \dfrac{1\sqrt[4]{4}}{\sqrt[4]{4}\sqrt[4]{4}} = \dfrac{\sqrt[4]{4}}{\sqrt[4]{16}} = \dfrac{\sqrt[4]{4}}{2}$

77. $\dfrac{1}{\sqrt[5]{16}} = \dfrac{1\sqrt[5]{2}}{\sqrt[5]{16}\sqrt[5]{2}} = \dfrac{\sqrt[5]{2}}{\sqrt[5]{32}} = \dfrac{\sqrt[5]{2}}{2}$

79. $\dfrac{1}{\sqrt{2}-1} = \dfrac{1\left(\sqrt{2}+1\right)}{\left(\sqrt{2}-1\right)\left(\sqrt{2}+1\right)} = \dfrac{\sqrt{2}+1}{\sqrt{4}-1} = \dfrac{\sqrt{2}+1}{2-1} = \dfrac{\sqrt{2}+1}{1} = \sqrt{2}+1$

81. $\dfrac{\sqrt{2}}{\sqrt{5}+3} = \dfrac{\sqrt{2}\left(\sqrt{5}-3\right)}{\left(\sqrt{5}+3\right)\left(\sqrt{5}-3\right)} = \dfrac{\sqrt{2}\left(\sqrt{5}-3\right)}{\sqrt{25}-9} = \dfrac{\sqrt{2}\left(\sqrt{5}-3\right)}{5-9} = \dfrac{\sqrt{2}\left(\sqrt{5}-3\right)}{-4}$

$$= \dfrac{\sqrt{10}-3\sqrt{2}}{-4}$$

$$= \dfrac{3\sqrt{2}-\sqrt{10}}{4}$$

83. $\dfrac{\sqrt{3}+1}{\sqrt{3}-1} = \dfrac{\left(\sqrt{3}+1\right)\left(\sqrt{3}+1\right)}{\left(\sqrt{3}-1\right)\left(\sqrt{3}+1\right)} = \dfrac{\sqrt{9}+\sqrt{3}+\sqrt{3}+1}{\sqrt{9}-1} = \dfrac{4+2\sqrt{3}}{2} = \dfrac{2\left(2+\sqrt{3}\right)}{2}$

$$= 2+\sqrt{3}$$

85. $\dfrac{\sqrt{7}-\sqrt{2}}{\sqrt{2}+\sqrt{7}} = \dfrac{\left(\sqrt{7}-\sqrt{2}\right)\left(\sqrt{2}-\sqrt{7}\right)}{\left(\sqrt{2}+\sqrt{7}\right)\left(\sqrt{2}-\sqrt{7}\right)} = \dfrac{\sqrt{14}-7-2+\sqrt{14}}{\sqrt{4}-\sqrt{49}} = \dfrac{2\sqrt{14}-9}{-5} = \dfrac{9-2\sqrt{14}}{5}$

87. $\dfrac{2}{\sqrt{x}+1} = \dfrac{2\left(\sqrt{x}-1\right)}{\left(\sqrt{x}+1\right)\left(\sqrt{x}-1\right)} = \dfrac{2\left(\sqrt{x}-1\right)}{\sqrt{x^2}-1} = \dfrac{2\left(\sqrt{x}-1\right)}{x-1} = \dfrac{2\sqrt{x}-2}{x-1}$

89. $\dfrac{x}{\sqrt{x}-4} = \dfrac{x\left(\sqrt{x}+4\right)}{\left(\sqrt{x}-4\right)\left(\sqrt{x}+4\right)} = \dfrac{x\left(\sqrt{x}+4\right)}{\sqrt{x^2}-16} = \dfrac{x\left(\sqrt{x}+4\right)}{x-16} = \dfrac{x\sqrt{x}+4x}{x-16}$

91. $\dfrac{2z-1}{\sqrt{2z}-1} = \dfrac{(2z-1)\left(\sqrt{2z}+1\right)}{\left(\sqrt{2z}-1\right)\left(\sqrt{2z}+1\right)} = \dfrac{(2z-1)\left(\sqrt{2z}+1\right)}{\sqrt{4z^2}-1} = \dfrac{(2z-1)\left(\sqrt{2z}+1\right)}{2z-1}$

$$= \sqrt{2z}+1$$

93. $\dfrac{\sqrt{x}-\sqrt{y}}{\sqrt{x}+\sqrt{y}} = \dfrac{\left(\sqrt{x}-\sqrt{y}\right)\left(\sqrt{x}-\sqrt{y}\right)}{\left(\sqrt{x}+\sqrt{y}\right)\left(\sqrt{x}-\sqrt{y}\right)} = \dfrac{\sqrt{x^2}-\sqrt{xy}-\sqrt{xy}+\sqrt{y^2}}{\sqrt{x^2}-\sqrt{y^2}} = \dfrac{x-2\sqrt{xy}+y}{x-y}$

95. $\dfrac{\sqrt{3}+1}{2} = \dfrac{\left(\sqrt{3}+1\right)\left(\sqrt{3}-1\right)}{2\left(\sqrt{3}-1\right)} = \dfrac{\sqrt{9}-1}{2\left(\sqrt{3}-1\right)} = \dfrac{2}{2\left(\sqrt{3}-1\right)} = \dfrac{1}{\sqrt{3}-1}$

97. $\dfrac{\sqrt{x}+3}{x} = \dfrac{\left(\sqrt{x}+3\right)\left(\sqrt{x}-3\right)}{x\left(\sqrt{x}-3\right)} = \dfrac{\sqrt{x^2}-9}{x\left(\sqrt{x}-3\right)} = \dfrac{x-9}{x\sqrt{x}-3x}$

99. $\dfrac{\sqrt{x}+\sqrt{y}}{\sqrt{x}} = \dfrac{\left(\sqrt{x}+\sqrt{y}\right)\left(\sqrt{x}-\sqrt{y}\right)}{\sqrt{x}\left(\sqrt{x}-\sqrt{y}\right)} = \dfrac{\sqrt{x^2}-\sqrt{y^2}}{\sqrt{x^2}-\sqrt{xy}} = \dfrac{x-y}{x-\sqrt{xy}}$

101. If the area of the aperture is again cut in half, the area will equal $9\pi/4$ cm^2.

$$A = \pi r^2 \qquad \qquad f\text{-number} = \frac{f}{d}$$

$$\frac{9\pi}{4} = \pi \left(\frac{d}{2}\right)^2 \qquad \qquad = \frac{12}{3}$$

$$\frac{9\pi}{4} = \frac{\pi d^2}{4} \qquad \qquad = 4$$

$$36\pi = 4\pi d^2$$

$$\frac{36\pi}{4\pi} = d^2$$

$$9 = d^2$$

$$3 = d$$

103. Answers may vary.

105. $\dfrac{\sqrt{x}-3}{4} = \dfrac{(\sqrt{x}-3)(\sqrt{x}+3)}{4(\sqrt{x}+3)} = \dfrac{x-9}{4(\sqrt{x}+3)}$

Exercise 9.6 (page 619)

1. $f(0) = 3(0)^2 - 4(0) + 2 = 2$

3. $f(2) = 3(2)^2 - 4(2) + 2 = 6$

5. $x^n = y^n$

7. square

9. extraneous

11.
$$\sqrt{5x-6} = 2$$
$$\left(\sqrt{5x-6}\right)^2 = 2^2$$
$$5x - 6 = 4$$
$$5x = 10$$
$$x = 2$$
The answer checks.

13.
$$\sqrt{6x+1} + 2 = 7$$
$$\sqrt{6x+1} = 5$$
$$\left(\sqrt{6x+1}\right)^2 = 5^2$$
$$6x + 1 = 25$$
$$6x = 24$$
$$x = 4$$
The answer checks.

15.
$$2\sqrt{4x+1} = \sqrt{x+4}$$
$$\left(2\sqrt{4x+1}\right)^2 = \left(\sqrt{x+4}\right)^2$$
$$4(4x+1) = x + 4$$
$$16x + 4 = x + 4$$
$$15x = 0$$
$$x = 0$$
The answer checks.

17.
$$\sqrt[3]{7n-1} = 3$$
$$\left(\sqrt[3]{7n-1}\right)^3 = 3^3$$
$$7n - 1 = 27$$
$$7n = 28$$
$$n = 4$$
The answer checks.

19.
$$\sqrt[4]{10p+1} = \sqrt[4]{11p-7}$$
$$\left(\sqrt[4]{10p+1}\right)^4 = \left(\sqrt[4]{11p-7}\right)^4$$
$$10p+1 = 11p-7$$
$$-p = -8$$
$$p = 8$$
The answer checks.

21.
$$x = \frac{\sqrt{12x-5}}{2}$$
$$2x = \sqrt{12x-5}$$
$$(2x)^2 = \left(\sqrt{12x-5}\right)^2$$
$$4x^2 = 12x-5$$
$$4x^2 - 12x + 5 = 0$$
$$(2x-1)(2x-5)$$
$$2x-1=0 \quad \text{or} \quad 2x-5=0$$
$$2x = 1 \qquad\qquad 2x = 5$$
$$x = \tfrac{1}{2} \qquad\qquad x = \tfrac{5}{2}$$
Both answers check.

23.
$$\sqrt{x+2} = \sqrt{4-x}$$
$$\left(\sqrt{x+2}\right)^2 = \left(\sqrt{4-x}\right)^2$$
$$x+2 = 4-x$$
$$2x = 2$$
$$x = 1$$
The answer checks.

25.
$$2\sqrt{x} = \sqrt{5x-16}$$
$$\left(2\sqrt{x}\right)^2 = \left(\sqrt{5x-16}\right)^2$$
$$4x = 5x-16$$
$$-x = -16$$
$$x = 16$$
The answer checks.

27.
$$r-9 = \sqrt{2r-3}$$
$$(r-9)^2 = \left(\sqrt{2r-3}\right)^2$$
$$r^2 - 18r + 81 = 2r-3$$
$$r^2 - 20r + 84 = 0$$
$$(r-14)(r-6) = 0$$
$$r-14=0 \quad \text{or} \quad r-6=0$$
$$r = 14 \qquad\qquad r = 6$$
$$\text{solution} \qquad\qquad \text{not a solution}$$

29.
$$\sqrt{-5x+24} = 6-x$$
$$\left(\sqrt{-5x+24}\right)^2 = (6-x)^2$$
$$-5x+24 = 36 - 12x + x^2$$
$$0 = x^2 - 7x + 12$$
$$0 = (x-3)(x-4)$$
$$x-3=0 \quad \text{or} \quad x-4=0$$
$$x = 3 \qquad\qquad x = 4$$
$$\text{solution} \qquad\qquad \text{solution}$$

31.
$$\sqrt{y+2} = 4-y$$
$$\left(\sqrt{y+2}\right) = (4-y)^2$$
$$y+2 = 16 - 8y + y^2$$
$$0 = y^2 - 9y + 14$$
$$0 = (y-2)(y-7)$$
$$y-2=0 \quad \text{or} \quad y-7=0$$
$$y = 2 \qquad\qquad y = 7$$
$$\text{solution} \qquad\qquad \text{not a solution}$$

33.
$$\sqrt{x}\sqrt{x+16} = 15$$
$$\left(\sqrt{x}\sqrt{x+16}\right)^2 = 15^2$$
$$x(x+16) = 225$$
$$x^2 + 16x - 225 = 0$$
$$(x-9)(x+25) = 0$$
$$x-9=0 \quad \text{or} \quad x+25=0$$
$$x = 9 \qquad\qquad x = -25$$
$$\text{solution} \qquad\qquad \text{not a solution}$$

35.
$$\sqrt[3]{x^3 - 7} = x - 1$$
$$\left(\sqrt[3]{x^3 - 7}\right)^3 = (x-1)^3$$
$$x^3 - 7 = x^3 - 3x^2 + 3x - 1$$
$$3x^2 - 3x - 6 = 0$$
$$3(x-2)(x+1) = 0$$
$$x - 2 = 0 \quad \textbf{or} \quad x + 1 = 0$$
$$x = 2 \qquad\qquad x = -1$$
$$\text{solution} \qquad\qquad \text{solution}$$

37.
$$\sqrt[4]{x^4 + 4x^2 - 4} = -x$$
$$\left(\sqrt[4]{x^4 + 4x^2 - 4}\right)^4 = (-x)^4$$
$$x^4 + 4x^2 - 4 = x^4$$
$$4x^2 - 4 = 0$$
$$4(x+1)(x-1) = 0$$
$$x + 1 = 0 \quad \textbf{or} \quad x - 1 = 0$$
$$x = -1 \qquad\qquad x = 1$$
$$\text{solution} \qquad\qquad \text{not a solution}$$

39.
$$\sqrt[4]{12t + 4} + 2 = 0$$
$$\sqrt[4]{12t + 4} = -2$$
$$\left(\sqrt[4]{12t + 4}\right)^4 = (-2)^4$$
$$12t + 4 = 16$$
$$12t = 12$$
$$t = 1$$
The answer does not check. $\Rightarrow$ no solution

41.
$$\sqrt{2y + 1} = 1 - 2\sqrt{y}$$
$$\left(\sqrt{2y + 1}\right)^2 = \left(1 - 2\sqrt{y}\right)^2$$
$$2y + 1 = 1 - 4\sqrt{y} + 4y$$
$$4\sqrt{y} = 2y$$
$$\left(4\sqrt{y}\right)^2 = (2y)^2$$
$$16y = 4y^2$$
$$0 = 4y^2 - 16y$$
$$0 = 4y(y - 4)$$
$$4y = 0 \quad \textbf{or} \quad y - 4 = 0$$
$$y = 0 \qquad\qquad y = 4$$
$$\text{solution} \qquad\qquad \text{not a solution}$$

43.
$$\sqrt{y + 7} + 3 = \sqrt{y + 4}$$
$$\left(\sqrt{y+7} + 3\right)^2 = \left(\sqrt{y+4}\right)^2$$
$$y + 7 + 6\sqrt{y+7} + 9 = y + 4$$
$$6\sqrt{y+7} = -12$$
$$\left(6\sqrt{y+7}\right)^2 = (-12)^2$$
$$36(y+7) = 144$$
$$y + 7 = 4$$
$$y = -3$$
The answer does not check. $\Rightarrow$ no solution

45.
$$\sqrt{v} + \sqrt{3} = \sqrt{v + 3}$$
$$\left(\sqrt{v} + \sqrt{3}\right)^2 = \left(\sqrt{v+3}\right)^2$$
$$v + 2\sqrt{3v} + 3 = v + 3$$
$$2\sqrt{3v} = 0$$
$$\left(2\sqrt{3v}\right)^2 = 0^2$$
$$12v = 0$$
$$v = 0$$
The answer checks.

47.
$$2 + \sqrt{u} = \sqrt{2u + 7}$$
$$\left(2 + \sqrt{u}\right)^2 = \left(\sqrt{2u + 7}\right)^2$$
$$4 + 4\sqrt{u} + u = 2u + 7$$
$$4\sqrt{u} = u + 3$$
$$\left(4\sqrt{u}\right)^2 = (u + 3)^2$$
$$16u = u^2 + 6u + 9$$
$$0 = u^2 - 10u + 9$$
$$0 = (u - 9)(u - 1)$$
$$u - 9 = 0 \quad \textbf{or} \quad u - 1 = 0$$
$$u = 9 \qquad\qquad u = 1$$
solution $\qquad$ solution

49.
$$\sqrt{6t + 1} - 3\sqrt{t} = -1$$
$$\sqrt{6t + 1} = 3\sqrt{t} - 1$$
$$\left(\sqrt{6t + 1}\right)^2 = \left(3\sqrt{t} - 1\right)^2$$
$$6t + 1 = 9t - 6\sqrt{t} + 1$$
$$6\sqrt{t} = 3t$$
$$\left(6\sqrt{t}\right)^2 = (3t)^2$$
$$36t = 9t^2$$
$$0 = 9t^2 - 36t$$
$$0 = 9t(t - 4)$$
$$9t = 0 \quad \textbf{or} \quad t - 4 = 0$$
$$t = 0 \qquad\qquad t = 4$$
not a solution $\qquad$ solution

51.
$$\sqrt{2x + 5} + \sqrt{x + 2} = 5$$
$$\sqrt{2x + 5} = 5 - \sqrt{x + 2}$$
$$\left(\sqrt{2x + 5}\right)^2 = \left(5 - \sqrt{x + 2}\right)^2$$
$$2x + 5 = 25 - 10\sqrt{x + 2} + x + 2$$
$$10\sqrt{x + 2} = -x + 22$$
$$\left(10\sqrt{x + 2}\right)^2 = (-x + 22)^2$$
$$100(x + 2) = x^2 - 44x + 484$$
$$0 = x^2 - 144x + 284$$
$$0 = (x - 142)(x - 2)$$
$$x - 142 = 0 \quad \textbf{or} \quad x - 2 = 0$$
$$x = 142 \qquad\qquad x = 2$$
not a solution $\qquad$ solution

53.
$$\sqrt{z - 1} + \sqrt{z + 2} = 3$$
$$\sqrt{z - 1} = 3 - \sqrt{z + 2}$$
$$\left(\sqrt{z - 1}\right)^2 = \left(3 - \sqrt{z + 2}\right)^2$$
$$z - 1 = 9 - 6\sqrt{z + 2} + z + 2$$
$$6\sqrt{z + 2} = 12$$
$$\sqrt{z + 2} = 2$$
$$\left(\sqrt{z + 2}\right)^2 = 2^2$$
$$z + 2 = 4$$
$$z = 2: \text{ The answer checks.}$$

55.
$$\sqrt{x - 5} - \sqrt{x + 3} = 4$$
$$\sqrt{x - 5} = \sqrt{x + 3} + 4$$
$$\left(\sqrt{x - 5}\right)^2 = \left(\sqrt{x + 3} + 4\right)^2$$
$$x - 5 = x + 3 + 8\sqrt{x + 3} + 16$$
$$-24 = 8\sqrt{x + 3}$$
$$-3 = \sqrt{x + 3}$$
$$(-3)^2 = \left(\sqrt{x + 3}\right)^2$$
$$9 = x + 3$$
$$6 = x$$
The answer does not check. $\Rightarrow$ no solution

57.
$$\sqrt{x+1} + \sqrt{3x} = \sqrt{5x+1}$$
$$\left(\sqrt{x+1} + \sqrt{3x}\right)^2 = \left(\sqrt{5x+1}\right)^2$$
$$x + 1 + 2\sqrt{3x(x+1)} + 3x = 5x + 1$$
$$2\sqrt{3x^2 + 3x} = x$$
$$\left(2\sqrt{3x^2 + 3x}\right)^2 = x^2$$
$$12x^2 + 12x = x^2$$
$$11x^2 + 12x = 0$$
$$x(11x + 12) = 0$$
$$x = 0 \quad \textbf{or} \quad 11x + 12 = 0$$
$$x = -\tfrac{12}{11}$$
solution $\qquad$ not a solution

59.
$$\sqrt{\sqrt{a} + \sqrt{a+8}} = 2$$
$$\left(\sqrt{\sqrt{a} + \sqrt{a+8}}\right)^2 = 2^2$$
$$\sqrt{a} + \sqrt{a+8} = 4$$
$$\left(\sqrt{a} + \sqrt{a+8}\right)^2 = 4^2$$
$$a + 2\sqrt{a(a+8)} + a + 8 = 16$$
$$2\sqrt{a^2 + 8a} = -2a + 8$$
$$\left(2\sqrt{a^2 + 8a}\right)^2 = (-2a + 8)^2$$
$$4a^2 + 32a = 4a^2 - 32a + 64$$
$$64a = 64$$
$$a = 1: \text{ The answer checks.}$$

61.
$$\frac{6}{\sqrt{x+5}} = \sqrt{x}$$
$$\left(\frac{6}{\sqrt{x+5}}\right)^2 = \left(\sqrt{x}\right)^2$$
$$\frac{36}{x+5} = x$$
$$36 = x^2 + 5x$$
$$0 = x^2 + 5x - 36$$
$$0 = (x+9)(x-4)$$
$$x + 9 = 0 \quad \textbf{or} \quad x - 4 = 0$$
$$x = -9 \qquad\qquad x = 4$$
not a solution $\qquad$ solution

63.

$$\sqrt{x+2} + \sqrt{2x-3} = \sqrt{11-x}$$

$$\left(\sqrt{x+2} + \sqrt{2x-3}\right)^2 = \left(\sqrt{11-x}\right)^2$$

$$x+2+2\sqrt{(x+2)(2x-3)} + 2x-3 = 11-x$$

$$2\sqrt{2x^2+x-6} = -4x+12$$

$$\left(2\sqrt{2x^2+x-6}\right)^2 = (-4x+12)^2$$

$$8x^2 + 4x - 24 = 16x^2 - 96x + 144$$

$$0 = 8x^2 - 100x + 168$$

$$0 = 4(2x-21)(x-2)$$

$$2x - 21 = 0 \quad \textbf{or} \quad x - 2 = 0$$

$$x = \tfrac{21}{2} \qquad\qquad x = 2$$

not a solution   solution

65.

$$s = 1.45\sqrt{r}$$

$$65 = 1.45\sqrt{r}$$

$$65^2 = \left(1.45\sqrt{r}\right)^2$$

$$4225 = 2.1025r$$

$$2010 \text{ ft} \approx r$$

67.

$$v = \sqrt[3]{\dfrac{P}{0.02}}$$

$$v = \sqrt[3]{\dfrac{500}{0.02}}$$

$$v = \sqrt[3]{25000}$$

$$v \approx 29 \text{ mph}$$

69. $r = 1 - \sqrt[n]{\dfrac{T}{C}} = 1 - \sqrt[5]{\dfrac{9000}{22000}} \approx 1 - \sqrt[5]{0.40909} \approx 1 - 0.836 \approx 0.164 \approx 16\%$

71. Graph $y = \sqrt{5x}$ and $y = \sqrt{100 - 3x^2}$:

equilibrium price: $x = \$5$

73.

$$r = \sqrt[4]{\dfrac{8kl}{\pi R}}$$

$$r^4 = \left(\sqrt[4]{\dfrac{8kl}{\pi R}}\right)^4$$

$$r^4 = \dfrac{8kl}{\pi R}$$

$$\pi R r^4 = 8kl$$

$$R = \dfrac{8kl}{\pi r^4}$$

75. Answers may vary.

77.
$$\sqrt[3]{2x} = \sqrt{x}$$
$$\left(\sqrt[3]{2x}\right)^2 = \left(\sqrt{x}\right)^2$$
$$\left[\left(\sqrt[3]{2x}\right)^2\right]^3 = x^3$$
$$4x^2 = x^3$$
$$0 = x^3 - 4x^2$$
$$0 = x^2(x - 4)$$
$$x = 0 \text{ or } x = 4$$

Exercise 9.7 (page 630)

1. $\dfrac{x^2 - x - 6}{9 - x^2} \cdot \dfrac{x^2 + x - 6}{x^2 - 4} = \dfrac{(x-3)(x+2)}{(3+x)(3-x)} \cdot \dfrac{(x+3)(x-2)}{(x+2)(x-2)} = -1$

3. Let $w =$ the speed of the wind.

	Rate	Time	Dist.
Downwind	$200 + w$	$\frac{330}{200+w}$	330
Upwind	$200 - w$	$\frac{330}{200-w}$	330

$$\frac{330}{200 + w} + \frac{330}{200 - w} = \frac{10}{3}$$
$$\left(\frac{330}{200 + w} + \frac{330}{200 - w}\right)3(200 + w)(200 - w) = \frac{10}{3} \cdot 3(200 + w)(200 - w)$$
$$330(3)(200 - w) + 330(3)(200 + w) = 10(40,000 - w^2)$$
$$198,000 - 990w + 198,000 + 990w = 400,000 - 10w^2$$
$$10w^2 = 4000$$
$$w^2 = 400$$
$$w = 20 \text{ mph}$$

5. imaginary

7. -1

9. 1

11. $\dfrac{\sqrt{a}}{\sqrt{b}}$

13. $5; 7$

15. conjugates

17. $\sqrt{-9} = \sqrt{-1 \cdot 9} = \sqrt{i^2 \cdot 3^2} = 3i$

19. $\sqrt{-36} = \sqrt{-1 \cdot 36} = \sqrt{i^2 \cdot 6^2} = 6i$

21. $\sqrt{-7} = \sqrt{-1 \cdot 7} = \sqrt{i^2 \cdot 7} = i\sqrt{7}$

23. $3 + 7i \stackrel{?}{=} \sqrt{9} + (5 + 2)i$
$3 + 7i \stackrel{?}{=} 3 + 7i$
They are equal.

25. $8 + 5i \stackrel{?}{=} 2^3 + \sqrt{25}i^3$

$8 + 5i \stackrel{?}{=} 8 + 5(-i)$

$8 + 5i \stackrel{?}{=} 8 - 5i$

They are not equal.

27. $\sqrt{4} + \sqrt{-4} \stackrel{?}{=} 2 - 2i$

$2 + \sqrt{-1 \cdot 4} \stackrel{?}{=} 2 - 2i$

$2 + 2i = 2 - 2i$

They are not equal.

29. $(3 + 4i) + (5 - 6i) = 3 + 4i + 5 - 6i$
$$= 8 - 2i$$

31. $(7 - 3i) - (4 + 2i) = 7 - 3i - 4 - 2i$
$$= 3 - 5i$$

33. $(8 + 5i) + (7 + 2i) = 8 + 5i + 7 + 2i$
$$= 15 + 7i$$

35. $(1 + i) - 2i + (5 - 7i) = 1 + i - 2i + 5 - 7i$
$$= 6 - 8i$$

37. $(5 + 3i) - (3 - 5i) + \sqrt{-1} = 5 + 3i - 3 + 5i + i = 2 + 9i$

39. $(-8 - \sqrt{3}i) - (7 - 3\sqrt{3}i) = -8 - \sqrt{3}i - 7 + 3\sqrt{3}i = -15 + 2\sqrt{3}i$

41. $3i(2 - i) = 6i - 3i^2 = 6i - 3(-1) = 6i + 3 = 3 + 6i$

43. $-5i(5 - 5i) = -25i + 25i^2 = -25i + 25(-1) = -25i - 25 = -25 - 25i$

45. $(2 + i)(3 - i) = 6 - 2i + 3i - i^2 = 6 + i - (-1) = 6 + i + 1 = 7 + i$

47. $(2 - 4i)(3 + 2i) = 6 + 4i - 12i - 8i^2 = 6 - 8i - 8(-1) = 6 - 8i + 8 = 14 - 8i$

49. $(2 + \sqrt{2}i)(3 - \sqrt{2}i) = 6 - 2\sqrt{2}i + 3\sqrt{2}i - 2i^2 = 6 + \sqrt{2}i - 2(-1) = 8 + \sqrt{2}i$

51. $(8 - \sqrt{-1})(-2 - \sqrt{-16}) = (8 - i)(-2 - 4i) = -16 - 32i + 2i + 4i^2 = -16 - 30i - 4$
$$= -20 - 30i$$

53. $(2 + i)^2 = (2 + i)(2 + i) = 4 + 2i + 2i + i^2 = 4 + 4i - 1 = 3 + 4i$

55. $(2 + 3i)^2 = (2 + 3i)(2 + 3i) = 4 + 6i + 6i + 9i^2 = 4 + 12i - 9 = -5 + 12i$

57. $i(5 + i)(3 - 2i) = i(15 - 10i + 3i - 2i^2) = i(15 - 7i + 2) = i(17 - 7i) = 17i - 7i^2 = 7 + 17i$

59. $(2 + i)(2 - i)(1 + i) = (4 - 2i + 2i - i^2)(1 + i) = 5(1 + i) = 5 + 5i$

61. $(3 + i)[(3 - 2i) + (2 + i)] = (3 + i)(5 - i) = 15 - 3i + 5i - i^2 = 16 + 2i$

63. $\dfrac{1}{i} = \dfrac{1i^3}{ii^3} = \dfrac{i^3}{i^4} = \dfrac{i^3}{1} = i^3 = -i = 0 - i$

65. $\dfrac{4}{5i^3} = \dfrac{4i}{5i^3i} = \dfrac{4i}{5i^4} = \dfrac{4i}{5(1)} = \dfrac{4}{5}i = 0 + \dfrac{4}{5}i$

67. $\dfrac{3i}{8\sqrt{-9}} = \dfrac{3i}{8(3i)} = \dfrac{1}{8} = \dfrac{1}{8} + 0i$

69. $\dfrac{-3}{5i^5} = \dfrac{-3i^3}{5i^5i^3} = \dfrac{-3(-i)}{5i^8} = \dfrac{3i}{5} = 0 + \dfrac{3}{5}i$

71. $\dfrac{5}{2-i} = \dfrac{5(2+i)}{(2-i)(2+i)} = \dfrac{5(2+i)}{4-i^2} = \dfrac{5(2+i)}{5} = 2+i$

73. $\dfrac{13i}{5+i} = \dfrac{13i(5-i)}{(5+i)(5-i)} = \dfrac{65i-13i^2}{25-i^2} = \dfrac{13+65i}{26} = \dfrac{13}{26} + \dfrac{65}{26}i = \dfrac{1}{2} + \dfrac{5}{2}i$

75. $\dfrac{-12}{7-\sqrt{-1}} = \dfrac{-12}{7-i} = \dfrac{-12(7+i)}{(7-i)(7+i)} = \dfrac{-84-12i}{49-i^2} = \dfrac{-84-12i}{50} = \dfrac{-84}{50} - \dfrac{12}{50}i = -\dfrac{42}{25} - \dfrac{6}{25}i$

77. $\dfrac{5i}{6+2i} = \dfrac{5i(6-2i)}{(6+2i)(6-2i)} = \dfrac{30i-10i^2}{36-4i^2} = \dfrac{10+30i}{40} = \dfrac{10}{40} + \dfrac{30}{40}i = \dfrac{1}{4} + \dfrac{3}{4}i$

79. $\dfrac{3-2i}{3+2i} = \dfrac{(3-2i)(3-2i)}{(3+2i)(3-2i)} = \dfrac{9-6i-6i+4i^2}{9-4i^2} = \dfrac{5-12i}{13} = \dfrac{5}{13} - \dfrac{12}{13}i$

81. $\dfrac{3+2i}{3+i} = \dfrac{(3+2i)(3-i)}{(3+i)(3-i)} = \dfrac{9-3i+6i-2i^2}{9-i^2} = \dfrac{11+3i}{10} = \dfrac{11}{10} + \dfrac{3}{10}i$

83. $\dfrac{\sqrt{5}-\sqrt{3}i}{\sqrt{5}+\sqrt{3}i} = \dfrac{\left(\sqrt{5}-\sqrt{3}i\right)\left(\sqrt{5}-\sqrt{3}i\right)}{\left(\sqrt{5}+\sqrt{3}i\right)\left(\sqrt{5}-\sqrt{3}i\right)} = \dfrac{5-\sqrt{15}i-\sqrt{15}i+3i^2}{5-3i^2} = \dfrac{2-2\sqrt{15}i}{8}$

$$= \dfrac{2}{8} - \dfrac{2\sqrt{15}}{8}i$$

$$= \dfrac{1}{4} - \dfrac{\sqrt{15}}{4}i$$

85. $\left(\dfrac{i}{3+2i}\right)^2 = \dfrac{i^2}{(3+2i)^2} = \dfrac{-1}{(3+2i)(3+2i)} = \dfrac{-1}{9+12i+4i^2} = \dfrac{-1}{5+12i} = \dfrac{-1(5-12i)}{(5+12i)(5-12i)}$

$$= \dfrac{-5+12i}{25-144i^2}$$

$$= -\dfrac{5}{169} + \dfrac{12}{169}i$$

87. $\dfrac{i(3-i)}{3+i} = \dfrac{(3i-i^2)(3-i)}{(3+i)(3-i)} = \dfrac{(1+3i)(3-i)}{9-i^2} = \dfrac{3-i+9i-3i^2}{10} = \dfrac{6+8i}{10} = \dfrac{3}{5} + \dfrac{4}{5}i$

89. $\dfrac{(2-5i)-(5-2i)}{5-i} = \dfrac{2-5i-5+2i}{5-i} = \dfrac{-3-3i}{5-i} = \dfrac{(-3-3i)(5+i)}{(5-i)(5+i)} = \dfrac{-15-3i-15i-3i^2}{25-i^2}$

$$= \dfrac{-12-18i}{26}$$

$$= \dfrac{-12}{26} - \dfrac{18}{26}i$$

$$= -\dfrac{6}{13} - \dfrac{9}{13}i$$

91. $i^{21} = i^{20}i^1 = (i^4)^5 i = 1^5 i = i$

93. $i^{27} = i^{24}i^3 = (i^4)^6 i^3 = 1^6 i^3 = i^3 = -i$

95. $i^{100} = (i^4)^{25} = 1^{25} = 1$

97. $i^{97} = i^{96}i^1 = (i^4)^{24} i = 1^{24} i = i$

99. $|6 + 8i| = \sqrt{6^2 + 8^2} = \sqrt{36 + 64}$
$$= \sqrt{100} = 10$$

101. $|12 - 5i| = \sqrt{12^2 + (-5)^2} = \sqrt{144 + 25}$
$$= \sqrt{169} = 13$$

103. $|5 + 7i| = \sqrt{5^2 + 7^2} = \sqrt{25 + 49} = \sqrt{74}$

105. $\left| \dfrac{3}{5} - \dfrac{4}{5}i \right| = \sqrt{\left(\dfrac{3}{5}\right)^2 + \left(\dfrac{4}{5}\right)^2} = \sqrt{\dfrac{9}{25} + \dfrac{16}{25}} = \sqrt{\dfrac{25}{25}} = \sqrt{1} = 1$

107.
$$x^2 - 2x + 26 = 0$$
$$(1 - 5i)^2 - 2(1 - 5i) + 26 = 0$$
$$(1 - 5i)(1 - 5i) - 2 + 10i + 26 = 0$$
$$1 - 10i + 25i^2 + 24 + 10i = 0$$
$$1 - 10i - 25 + 24 + 10i = 0$$
$$0 = 0$$

109.
$$x^4 - 3x^2 - 4 = 0$$
$$i^4 - 3i^2 - 4 = 0$$
$$1 - 3(-1) - 4 = 0$$
$$1 + 3 - 4 = 0$$
$$0 = 0$$

111. $V = IR = (2 - 3i)(2 + i) = 4 + 2i - 6i - 3i^2 = 4 - 4i - 3(-1) = 4 - 4i + 3 = 7 - 4i$ volts

113. $Z = \dfrac{V}{I} = \dfrac{1.7 + 0.5i}{0.5i} = \dfrac{(1.7 + 0.5i)i}{(0.5i)i} = \dfrac{1.7i + 0.5i^2}{0.5i^2} = \dfrac{-0.5 + 1.7i}{-0.5} = 1 - 3.4i$

115. Answers may vary.

117. $\dfrac{3 - i}{2} = \dfrac{(3 - i)(3 + i)}{2(3 + i)} = \dfrac{9 - i^2}{2(3 + i)} = \dfrac{10}{2(3 + i)} = \dfrac{5}{3 + i}$

Chapter 9 Summary (page 633)

1. $\sqrt{49} = \sqrt{7^2} = 7$

2. $-\sqrt{121} = -\sqrt{11^2} = -11$

3. $-\sqrt{36} = -\sqrt{6^2} = -6$

4. $\sqrt{225} = \sqrt{15^2} = 15$

5. $\sqrt[3]{-27} = \sqrt[3]{(-3)^3} = -3$

6. $-\sqrt[3]{216} = -\sqrt[3]{6^3} = -6$

7. $\sqrt[4]{625} = \sqrt[4]{5^4} = 5$

8. $\sqrt[5]{-32} = \sqrt[5]{(-2)^5} = -2$

9. $\sqrt{25x^2} = \sqrt{5^2 x^2} = |5x| = 5|x|$

10. $\sqrt{x^2 + 4x + 4} = \sqrt{(x + 2)^2} = |x + 2|$

11. $\sqrt[3]{27a^6 b^3} = 3a^2 b$

12. $\sqrt[4]{256x^8 y^4} = |4x^2 y| = 4x^2 |y|$

13. $y = f(x) = \sqrt{x+2}$; Shift $y = \sqrt{x}$ left 2.

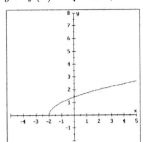

14. $y = f(x) = -\sqrt{x-1}$; Reflect $y = \sqrt{x}$ about the x-axis and shift right 1.

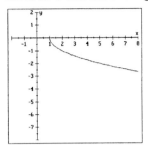

15. $y = f(x) = -\sqrt{x} + 2$; Reflect $y = \sqrt{x}$ about the x-axis and shift up 2.

16. $y = f(x) = -\sqrt[3]{x} + 3$; Reflect $y = \sqrt[3]{x}$ about the x-axis and shift up 3.

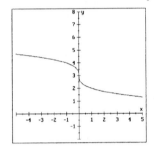

17. mean $= \dfrac{4+8+12+16+20}{5} = \dfrac{60}{5} = 12$

18.

Original term	Mean	Difference (term−mean)	Square of difference
4	12	−8	64
8	12	−4	16
12	12	0	0
16	12	4	16
20	12	8	64

$$s = \sqrt{\frac{64 + 16 + 0 + 16 + 64}{5}}$$
$$= \sqrt{\frac{160}{5}}$$
$$= \sqrt{32}$$
$$\approx 5.7$$

19. $d = 1.4\sqrt{h}$
$d = 1.4\sqrt{4.7} \approx 3.0$ miles

20. $d = 1.4\sqrt{h}$
$4 = 1.4\sqrt{h}$
$16 = 1.96h$
$8.2 \text{ ft} = h$

21. Let x = one-half of d.
$$125^2 = x^2 + 117^2$$
$$15625 = x^2 + 13689$$
$$1936 = x^2$$
$$44 = x$$
$$d = 2x = 2(44) = 88 \text{ yd}$$

22. Let x = one-half of d.
$$8900^2 = x^2 + 3900^2$$
$$79{,}210{,}000 = x^2 + 15{,}210{,}000$$
$$64{,}000{,}000 = x^2$$
$$8000 = x$$
$$d = 2x = 2(8000) = 16{,}000 \text{ yd}$$

23. $d = \sqrt{(x_2 - x_1)^2 + (y_2 - y_1)^2} = \sqrt{(0 - 5)^2 + [0 - (-12)]^2} = \sqrt{(-5)^2 + 12^2} = \sqrt{169} = 13$

24. $d = \sqrt{(x_2 - x_1)^2 + (y_2 - y_1)^2} = \sqrt{(-4 - (-2))^2 + (6 - 8)^2} = \sqrt{(-2)^2 + (-2)^2} = \sqrt{8} \approx 2.83$

25. $25^{1/2} = 5$

26. $-36^{1/2} = -6$

27. $9^{3/2} = \left(9^{1/2}\right)^3 = 3^3 = 27$

28. $16^{3/2} = \left(16^{1/2}\right)^3 = 4^3$
$= 64$

29. $(-8)^{1/3} = -2$

30. $-8^{2/3} = -(8^{1/3})^2 = -2^2$
$= -4$

31. $8^{-2/3} = \dfrac{1}{8^{2/3}} = \dfrac{1}{(8^{1/3})^2} = \dfrac{1}{2^2} = \dfrac{1}{4}$

32. $8^{-1/3} = \dfrac{1}{8^{1/3}} = \dfrac{1}{2}$

33. $-49^{5/2} = -(49^{1/2})^5 = -7^5 = -16{,}807$

34. $\dfrac{1}{25^{5/2}} = \dfrac{1}{(25^{1/2})^5} = \dfrac{1}{5^5} = \dfrac{1}{3125}$

35. $\left(\dfrac{1}{4}\right)^{-3/2} = 4^{3/2} = (4^{1/2})^3 = 2^3 = 8$

36. $\left(\dfrac{4}{9}\right)^{-3/2} = \left(\dfrac{9}{4}\right)^{3/2} = \left[\left(\dfrac{9}{4}\right)^{1/2}\right]^3$
$= \left(\dfrac{3}{2}\right)^3 = \dfrac{27}{8}$

37. $(27x^3 y)^{1/3} = 3xy^{1/3}$

38. $(81x^4 y^2)^{1/4} = 3xy^{1/2}$

39. $(25x^3 y^4)^{3/2} = 125x^{9/2} y^6$

40. $(8u^2 v^3)^{-2/3} = \dfrac{1}{4u^{4/3} v^2}$

41. $5^{1/4} 5^{1/2} = 5^{1/4+1/2} = 5^{3/4}$

42. $a^{3/7} a^{2/7} = a^{3/7+2/7} = a^{5/7}$

43. $u^{1/2}\left(u^{1/2} - u^{-1/2}\right) = u^{2/2} - u^0 = u - 1$

44. $v^{2/3}\left(v^{1/3} + v^{4/3}\right) = v^{3/3} + v^{6/3} = v + v^2$

45. $\left(x^{1/2} + y^{1/2}\right)^2 = \left(x^{1/2} + y^{1/2}\right)\left(x^{1/2} + y^{1/2}\right) = x^{2/2} + x^{1/2} y^{1/2} + x^{1/2} y^{1/2} + y^{2/2}$
$= x + 2x^{1/2} y^{1/2} + y$

46. $\left(a^{2/3} + b^{2/3}\right)\left(a^{2/3} - b^{2/3}\right) = a^{4/3} - a^{2/3} b^{2/3} + a^{2/3} b^{2/3} - b^{4/3} = a^{4/3} - b^{4/3}$

47. $\sqrt[6]{5^2} = 5^{2/6} = 5^{1/3} = \sqrt[3]{5}$

48. $\sqrt[8]{x^4} = x^{4/8} = x^{1/2} = \sqrt{x}$

49. $\sqrt[9]{27a^3 b^6} = (3^3 a^3 b^6)^{1/9} = 3^{3/9} a^{3/9} b^{6/9} = 3^{1/3} a^{1/3} b^{2/3} = \sqrt[3]{3ab^2}$

50. $\sqrt[4]{25a^2b^2} = 5^{2/4}a^{2/4}b^{2/4} = 5^{1/2}a^{1/2}b^{1/2} = \sqrt{5ab}$

51. $\sqrt{240} = \sqrt{16 \cdot 15} = \sqrt{16}\sqrt{15} = 4\sqrt{15}$ **52.** $\sqrt[3]{54} = \sqrt[3]{27 \cdot 2} = \sqrt[3]{27}\sqrt[3]{2} = 3\sqrt[3]{2}$

53. $\sqrt[4]{32} = \sqrt[4]{16 \cdot 2} = \sqrt[4]{16}\sqrt[4]{2} = 2\sqrt[4]{2}$ **54.** $\sqrt[5]{96} = \sqrt[5]{32 \cdot 3} = \sqrt[5]{32}\sqrt[5]{3} = 2\sqrt[5]{3}$

55. $\sqrt{8x^3} = \sqrt{4x^2}\sqrt{2x} = 2x\sqrt{2x}$ **56.** $\sqrt{18x^4y^3} = \sqrt{9x^4y^2}\sqrt{2y} = 3x^2y\sqrt{2y}$

57. $\sqrt[3]{16x^5y^4} = \sqrt[3]{8x^3y^3}\sqrt[3]{2x^2y} = 2xy\sqrt[3]{2x^2y}$ **58.** $\sqrt[3]{54x^7y^3} = \sqrt[3]{27x^6y^3}\sqrt[3]{2x} = 3x^2y\sqrt[3]{2x}$

59. $\dfrac{\sqrt{32x^3}}{\sqrt{2x}} = \sqrt{\dfrac{32x^3}{2x}} = \sqrt{16x^2} = 4x$ **60.** $\dfrac{\sqrt[3]{16x^5}}{\sqrt[3]{2x^2}} = \sqrt[3]{\dfrac{16x^5}{2x^2}} = \sqrt[3]{8x^3} = 2x$

61. $\sqrt[3]{\dfrac{2a^2b}{27x^3}} = \dfrac{\sqrt[3]{2a^2b}}{3x}$ **62.** $\sqrt{\dfrac{17xy}{64a^4}} = \dfrac{\sqrt{17xy}}{8a^2}$

63. $\sqrt{2} + \sqrt{8} = \sqrt{2} + \sqrt{4}\sqrt{2} = \sqrt{2} + 2\sqrt{2}$
$= 3\sqrt{2}$ **64.** $\sqrt{20} - \sqrt{5} = \sqrt{4}\sqrt{5} - \sqrt{5} = 2\sqrt{5} - \sqrt{5}$
$= \sqrt{5}$

65. $2\sqrt[3]{3} - \sqrt[3]{24} = 2\sqrt[3]{3} - \sqrt[3]{8}\sqrt[3]{3} = 2\sqrt[3]{3} - 2\sqrt[3]{3} = 0$

66. $\sqrt[4]{32} + 2\sqrt[4]{162} = \sqrt[4]{16}\sqrt[4]{2} + 2\sqrt[4]{81}\sqrt[4]{2} = 2\sqrt[4]{2} + 2(3)\sqrt[4]{2} = 2\sqrt[4]{2} + 6\sqrt[4]{2} = 8\sqrt[4]{2}$

67. $2x\sqrt{8} + 2\sqrt{200x^2} + \sqrt{50x^2} = 2x\sqrt{4}\sqrt{2} + 2\sqrt{100x^2}\sqrt{2} + \sqrt{25x^2}\sqrt{2}$
$= 2x(2)\sqrt{2} + 2(10x)\sqrt{2} + 5x\sqrt{2}$
$= 4x\sqrt{2} + 20x\sqrt{2} + 5x\sqrt{2} = 29x\sqrt{2}$

68. $3\sqrt{27a^3} - 2a\sqrt{3a} + 5\sqrt{75a^3} = 3\sqrt{9a^2}\sqrt{3a} - 2a\sqrt{3a} + 5\sqrt{25a^2}\sqrt{3a}$
$= 3(3a)\sqrt{3a} - 2a\sqrt{3a} + 5(5a)\sqrt{3a}$
$= 9a\sqrt{3a} - 2a\sqrt{3a} + 25a\sqrt{3a} = 32a\sqrt{3a}$

69. $\sqrt[3]{54} - 3\sqrt[3]{16} + 4\sqrt[3]{128} = \sqrt[3]{27}\sqrt[3]{2} - 3\sqrt[3]{8}\sqrt[3]{2} + 4\sqrt[3]{64}\sqrt[3]{2}$
$= 3\sqrt[3]{2} - 3(2)\sqrt[3]{2} + 4(4)\sqrt[3]{2} = 3\sqrt[3]{2} - 6\sqrt[3]{2} + 16\sqrt[3]{2} = 13\sqrt[3]{2}$

70. $2\sqrt[4]{32x^5} + 4\sqrt[4]{162x^5} - 5x\sqrt[4]{512x} = 2\sqrt[4]{16x^4}\sqrt[4]{2x} + 4\sqrt[4]{81x^4}\sqrt[4]{2x} - 5x\sqrt[4]{256}\sqrt[4]{2x}$
$= 2(2x)\sqrt[4]{2x} + 4(3x)\sqrt[4]{2x} - 5x(4)\sqrt[4]{2x}$
$= 4x\sqrt[4]{2x} + 12x\sqrt[4]{2x} - 20x\sqrt[4]{2x} = -4x\sqrt[4]{2x}$

71. hypotenuse $= 7\sqrt{2}$ m   **72.** shorter leg $= \dfrac{1}{2}\left(12\sqrt{3}\right) = 6\sqrt{3}$ cm
longer leg $= \sqrt{3}\left(6\sqrt{3}\right) = 6(3) = 18$ cm

73. $x = 5\sqrt{2} \approx 7.07$ in.

74. $x = \sqrt{3}(\frac{1}{2} \cdot 10) = 5\sqrt{3} \approx 8.66$ cm

75. $\left(2\sqrt{5}\right)\left(3\sqrt{2}\right) = 6\sqrt{10}$

76. $2\sqrt{6}\sqrt{216} = 2\sqrt{6}\sqrt{36}\sqrt{6} = 2(6)(6) = 72$

77. $\sqrt{9x}\sqrt{x} = \sqrt{9x^2} = 3x$

78. $\sqrt[3]{3}\sqrt[3]{9} = \sqrt[3]{27} = 3$

79. $-\sqrt[3]{2x^2}\sqrt[3]{4x} = -\sqrt[3]{8x^3}$
$= -2x$

80. $-\sqrt[4]{256x^5y^{11}}\sqrt[4]{625x^9y^3} = -\sqrt[4]{256x^4y^8}\sqrt[4]{xy^3}\sqrt[4]{625x^8}\sqrt[4]{xy^3} = -4xy^2(5x^2)\sqrt[4]{x^2y^6}$
$= -20x^3y^2\sqrt[4]{y^4}\sqrt[4]{x^2y^2}$
$= -20x^3y^3\sqrt[4]{x^2y^2} = -20x^3y^3\sqrt{xy}$

81. $\sqrt{2}\left(\sqrt{8} - 3\right) = \sqrt{16} - 3\sqrt{2} = 4 - 3\sqrt{2}$

82. $\sqrt{2}\left(\sqrt{2} + 3\right) = \sqrt{4} + 3\sqrt{2} = 2 + 3\sqrt{2}$

83. $\sqrt{5}\left(\sqrt{2} - 1\right) = \sqrt{10} - \sqrt{5}$

84. $\sqrt{3}\left(\sqrt{3} + \sqrt{2}\right) = \sqrt{9} + \sqrt{6} = 3 + \sqrt{6}$

85. $\left(\sqrt{2} + 1\right)\left(\sqrt{2} - 1\right) = \sqrt{4} - \sqrt{2} + \sqrt{2} - 1 = 1$

86. $\left(\sqrt{3} + \sqrt{2}\right)\left(\sqrt{3} + \sqrt{2}\right) = \sqrt{9} + \sqrt{6} + \sqrt{6} + \sqrt{4} = 5 + 2\sqrt{6}$

87. $\left(\sqrt{x} + \sqrt{y}\right)\left(\sqrt{x} - \sqrt{y}\right) = \sqrt{x^2} - \sqrt{xy} + \sqrt{xy} - \sqrt{y^2} = x - y$

88. $\left(2\sqrt{u} + 3\right)\left(3\sqrt{u} - 4\right) = 6\sqrt{u^2} - 8\sqrt{u} + 9\sqrt{u} - 12 = 6u + \sqrt{u} - 12$

89. $\dfrac{1}{\sqrt{3}} = \dfrac{1\sqrt{3}}{\sqrt{3}\sqrt{3}} = \dfrac{\sqrt{3}}{3}$

90. $\dfrac{\sqrt{3}}{\sqrt{5}} = \dfrac{\sqrt{3}\sqrt{5}}{\sqrt{5}\sqrt{5}} = \dfrac{\sqrt{15}}{5}$

91. $\dfrac{x}{\sqrt{xy}} = \dfrac{x\sqrt{xy}}{\sqrt{xy}\sqrt{xy}} = \dfrac{x\sqrt{xy}}{xy} = \dfrac{\sqrt{xy}}{y}$

92. $\dfrac{\sqrt[3]{uv}}{\sqrt[3]{u^5v^7}} = \dfrac{\sqrt[3]{uv}\sqrt[3]{uv^2}}{\sqrt[3]{u^5v^7}\sqrt[3]{uv^2}} = \dfrac{\sqrt[3]{u^2v^3}}{\sqrt[3]{u^6v^9}} = \dfrac{v\sqrt[3]{u^2}}{u^2v^3} = \dfrac{\sqrt[3]{u^2}}{u^2v^2}$

93. $\dfrac{2}{\sqrt{2}-1} = \dfrac{2\left(\sqrt{2}+1\right)}{\left(\sqrt{2}-1\right)\left(\sqrt{2}+1\right)} = \dfrac{2\left(\sqrt{2}+1\right)}{\sqrt{4}-1} = \dfrac{2\left(\sqrt{2}+1\right)}{2-1} = \dfrac{2\left(\sqrt{2}+1\right)}{1} = 2\left(\sqrt{2}+1\right)$

94. $\dfrac{\sqrt{2}}{\sqrt{3}-1} = \dfrac{\sqrt{2}\left(\sqrt{3}+1\right)}{\left(\sqrt{3}-1\right)\left(\sqrt{3}+1\right)} = \dfrac{\sqrt{2}\left(\sqrt{3}+1\right)}{\sqrt{9}-1} = \dfrac{\sqrt{2}\left(\sqrt{3}+1\right)}{3-1} = \dfrac{\sqrt{2}\left(\sqrt{3}+1\right)}{2}$

$$= \dfrac{\sqrt{6}+\sqrt{2}}{2}$$

95. $\dfrac{2x-32}{\sqrt{x}+4} = \dfrac{(2x-32)\left(\sqrt{x}-4\right)}{\left(\sqrt{x}+4\right)\left(\sqrt{x}-4\right)} = \dfrac{2(x-16)\left(\sqrt{x}-4\right)}{\sqrt{x^2}-16} = \dfrac{2(x-16)\left(\sqrt{x}-4\right)}{x-16} = 2\left(\sqrt{x}-4\right)$

96. $\dfrac{\sqrt{a}+1}{\sqrt{a}-1} = \dfrac{\left(\sqrt{a}+1\right)\left(\sqrt{a}+1\right)}{\left(\sqrt{a}-1\right)\left(\sqrt{a}+1\right)} = \dfrac{\sqrt{a^2}+\sqrt{a}+\sqrt{a}+1}{\sqrt{a^2}-1} = \dfrac{a+2\sqrt{a}+1}{a-1}$

97. $\dfrac{\sqrt{3}}{5} = \dfrac{\sqrt{3}\sqrt{3}}{5\sqrt{3}} = \dfrac{3}{5\sqrt{3}}$

98. $\dfrac{\sqrt[3]{9}}{3} = \dfrac{\sqrt[3]{9}\sqrt[3]{3}}{3\sqrt[3]{3}} = \dfrac{\sqrt[3]{27}}{3\sqrt[3]{3}} = \dfrac{3}{3\sqrt[3]{3}} = \dfrac{1}{\sqrt[3]{3}}$

99. $\dfrac{3-\sqrt{x}}{2} = \dfrac{\left(3-\sqrt{x}\right)\left(3+\sqrt{x}\right)}{2\left(3+\sqrt{x}\right)} = \dfrac{9-\sqrt{x^2}}{2\left(3+\sqrt{x}\right)} = \dfrac{9-x}{2\left(3+\sqrt{x}\right)}$

100. $\dfrac{\sqrt{a}-\sqrt{b}}{\sqrt{a}} = \dfrac{\left(\sqrt{a}-\sqrt{b}\right)\left(\sqrt{a}+\sqrt{b}\right)}{\sqrt{a}\left(\sqrt{a}+\sqrt{b}\right)} = \dfrac{\sqrt{a^2}-\sqrt{b^2}}{\sqrt{a^2}+\sqrt{ab}} = \dfrac{a-b}{a+\sqrt{ab}}$

101.
$$\sqrt{y+3} = \sqrt{2y-19}$$
$$\left(\sqrt{y+3}\right)^2 = \left(\sqrt{2y-19}\right)^2$$
$$y+3 = 2y-19$$
$$-y = -22$$
$$y = 22$$
The answer checks.

102.
$$u = \sqrt{25u-144}$$
$$u^2 = \left(\sqrt{25u-144}\right)^2$$
$$u^2 = 25u-144$$
$$u^2 - 25u + 144 = 0$$
$$(u-9)(u-16) = 0$$
$$u = 9 \text{ or } u = 16; \text{ Both answers check.}$$

103.
$$r = \sqrt{12r-27}$$
$$r^2 = \left(\sqrt{12r-27}\right)^2$$
$$r^2 = 12r-27$$
$$r^2 - 12r + 27 = 0$$
$$(r-9)(r-3) = 0$$
$$r = 9 \text{ or } r = 3; \text{ Both answers check.}$$

104.
$$\sqrt{z+1} + \sqrt{z} = 2$$
$$\sqrt{z+1} = 2 - \sqrt{z}$$
$$\left(\sqrt{z+1}\right)^2 = \left(2-\sqrt{z}\right)^2$$
$$z+1 = 4 - 4\sqrt{z} + z$$
$$4\sqrt{z} = 3$$
$$\left(4\sqrt{z}\right)^2 = 3^2$$
$$16z = 9$$
$$z = \tfrac{9}{16}; \text{ The answer checks.}$$

105. $\sqrt{2x+5} - \sqrt{2x} = 1$

$\sqrt{2x+5} = 1 + \sqrt{2x}$

$\left(\sqrt{2x+5}\right)^2 = \left(1 + \sqrt{2x}\right)^2$

$2x + 5 = 1 + 2\sqrt{2x} + 2x$

$4 = 2\sqrt{2x}$

$4^2 = \left(2\sqrt{2x}\right)^2$

$16 = 8x$

$2 = x$; The answer checks.

106. $\sqrt[3]{x^3+8} = x + 2$

$\left(\sqrt[3]{x^3+8}\right)^3 = (x+2)^3$

$x^3 + 8 = x^3 + 6x^2 + 12x + 8$

$0 = 6x^2 + 12x$

$0 = 6x(x+2)$

$x = 0$ or $x = -2$; Both answers check.

107. $(5 + 4i) + (7 - 12i) = 5 + 4i + 7 - 12i = 12 - 8i$

108. $(-6 - 40i) - (-8 + 28i) = -6 - 40i + 8 - 28i = 2 - 68i$

109. $(-32 + \sqrt{-144}) - (64 + \sqrt{-81}) = -32 + \sqrt{144i^2} - 64 - \sqrt{81i^2} = -32 + 12i - 64 - 9i$
$$= -96 + 3i$$

110. $(-8 + \sqrt{-8}) + (6 - \sqrt{-32}) = -8 + \sqrt{4i^2}\sqrt{2} + 6 - \sqrt{16i^2}\sqrt{2}$
$$= -8 + 2i\sqrt{2} + 6 - 4i\sqrt{2} = -2 - 2\sqrt{2}i$$

111. $(2 - 7i)(-3 + 4i) = -6 + 8i + 21i - 28i^2 = -6 + 29i + 28 = 22 + 29i$

112. $(-5 + 6i)(2 + i) = -10 - 5i + 12i + 6i^2 = -10 + 7i - 6 = -16 + 7i$

113. $(5 - \sqrt{-27})(-6 + \sqrt{-12}) = (5 - 3i\sqrt{3})(-6 + 2i\sqrt{3}) = -30 + 10i\sqrt{3} + 18i\sqrt{3} - 6i^2(3)$
$$= -30 + 28i\sqrt{3} + 18 = -12 + 28\sqrt{3}i$$

114. $(2 + \sqrt{-128})(3 - \sqrt{-98}) = (2 + 8i\sqrt{2})(3 - 7i\sqrt{2}) = 6 - 14i\sqrt{2} + 24i\sqrt{2} - 56i^2(2)$
$$= 6 + 10i\sqrt{2} + 112 = 118 + 10\sqrt{2}i$$

115. $\dfrac{3}{4i} = \dfrac{3}{4i} \cdot \dfrac{i}{i} = \dfrac{3i}{4i^2} = \dfrac{3i}{-4} = -\dfrac{3}{4}i = 0 - \dfrac{3}{4}i$ **116.** $\dfrac{-2}{5i^3} = \dfrac{-2}{5i^3} \cdot \dfrac{i}{i} = \dfrac{-2i}{5i^4} = \dfrac{-2i}{5} = 0 - \dfrac{2}{5}i$

117. $\dfrac{6}{2+i} = \dfrac{6}{2+i} \cdot \dfrac{2-i}{2-i} = \dfrac{6(2-i)}{4-i^2} = \dfrac{6(2-i)}{5} = \dfrac{12-6i}{5} = \dfrac{12}{5} - \dfrac{6}{5}i$

118. $\dfrac{7}{3-i} = \dfrac{7}{3-i} \cdot \dfrac{3+i}{3+i} = \dfrac{7(3+i)}{9-i^2} = \dfrac{7(3+i)}{10} = \dfrac{21+7i}{10} = \dfrac{21}{10} + \dfrac{7}{10}i$

119. $\dfrac{4+i}{4-i} = \dfrac{4+i}{4-i} \cdot \dfrac{4+i}{4+i} = \dfrac{16+8i+i^2}{16-i^2} = \dfrac{15+8i}{17} = \dfrac{15}{17} + \dfrac{8}{17}i$

120. $\dfrac{3-i}{3+i} = \dfrac{3-i}{3+i} \cdot \dfrac{3-i}{3-i} = \dfrac{9-6i+i^2}{9-i^2} = \dfrac{8-6i}{10} = \dfrac{8}{10} - \dfrac{6}{10}i = \dfrac{4}{5} - \dfrac{3}{5}i$

121. $\dfrac{3}{5+\sqrt{-4}} = \dfrac{3}{5+2i} = \dfrac{3}{5+2i} \cdot \dfrac{5-2i}{5-2i} = \dfrac{3(5-2i)}{25-4i^2} = \dfrac{3(5-2i)}{29} = \dfrac{15-6i}{29} = \dfrac{15}{29} - \dfrac{6}{29}i$

122. $\dfrac{2}{3-\sqrt{-9}} = \dfrac{2}{3-3i} = \dfrac{2}{3-3i} \cdot \dfrac{3+3i}{3+3i} = \dfrac{2(3+3i)}{9-9i^2} = \dfrac{2(3+3i)}{18} = \dfrac{3+3i}{9} = \dfrac{3}{9} + \dfrac{3}{9}i = \dfrac{1}{3} + \dfrac{1}{3}i$

123. $|9+12i| = \sqrt{9^2+12^2} = \sqrt{81+144} = \sqrt{225} = 15 = 15+0i$

124. $|24-10i| = \sqrt{24^2+(-10)^2} = \sqrt{576+100} = \sqrt{676} = 26 = 26+0i$

125. $i^{12} = (i^4)^3 = 1^3 = 1$

126. $i^{583} = i^{580}i^3 = (i^4)^{145}i^3 = 1^{145}(-i) = -i$

Chapter 9 Test   (page 639)

1. $\sqrt{49} = \sqrt{7^2} = 7$

2. $\sqrt[3]{64} = \sqrt[3]{4^3} = 4$

3. $\sqrt{4x^2} = \sqrt{(2x)^2} = |2x| = 2|x|$

4. $\sqrt[3]{8x^3} = \sqrt[3]{(2x)^3} = 2x$

5. $f(x) = \sqrt{x-2}$; Shift $y = \sqrt{x}$ right 2.

domain $= [2, \infty)$, range $= [0, \infty)$

6. $f(x) = \sqrt[3]{x} + 3$; Shift $y = \sqrt[3]{x}$ up 3.

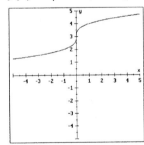

domain $= (-\infty, \infty)$, range $= (-\infty, \infty)$

7.
$$53^2 = 45^2 + h^2$$
$$2809 = 2025 + h^2$$
$$784 = h^2$$
$$28 \text{ in.} = h$$

8.
$$2^2 = \left(\dfrac{w}{2}\right)^2 + (1.9)^2$$
$$4 = \dfrac{w^2}{4} + 3.61$$
$$0.39 = \dfrac{w^2}{4}$$
$$1.56 = w^2$$
$$1.25 \text{ meters} = w$$

9. $d = \sqrt{(x_2-x_1)^2 + (y_2-y_1)^2} = \sqrt{(6-0)^2 + (8-0)^2} = \sqrt{6^2+8^2} = \sqrt{100} = 10$

10. $d = \sqrt{(x_2 - x_1)^2 + (y_2 - y_1)^2} = \sqrt{(-2 - 22)^2 + (5 - 12)^2} = \sqrt{(-24)^2 + (-7)^2} = \sqrt{625} = 25$

11. $16^{1/4} = 2$

12. $27^{2/3} = \left(27^{1/3}\right)^2 = 3^2 = 9$

13. $36^{-3/2} = \dfrac{1}{36^{3/2}} = \dfrac{1}{\left(36^{1/2}\right)^3} = \dfrac{1}{6^3} = \dfrac{1}{216}$

14. $\left(-\dfrac{8}{27}\right)^{-2/3} = \left(-\dfrac{27}{8}\right)^{2/3} = \left[\left(-\dfrac{27}{8}\right)^{1/3}\right]^2 = \left(-\dfrac{3}{2}\right)^2 = \dfrac{9}{4}$

15. $\dfrac{2^{5/3}2^{1/6}}{2^{1/2}} = \dfrac{2^{10/6}2^{1/6}}{2^{3/6}} = 2^{10/6+1/6-3/6} = 2^{8/6} = 2^{4/3}$

16. $\dfrac{(8x^3y)^{1/2}(8xy^5)^{1/2}}{(x^3y^6)^{1/3}} = \dfrac{8^{1/2}x^{3/2}y^{1/2}8^{1/2}x^{1/2}y^{5/2}}{x^{3/3}y^{6/3}} = \dfrac{8^{2/2}x^{4/2}y^{6/2}}{xy^2} = \dfrac{8x^2y^3}{xy^2} = 8xy$

17. $\sqrt{48} = \sqrt{16}\sqrt{3} = 4\sqrt{3}$

18. $\sqrt{250x^3y^5} = \sqrt{25x^2y^4}\sqrt{10xy} = 5xy^2\sqrt{10xy}$

19. $\dfrac{\sqrt[3]{24x^{15}y^4}}{\sqrt[3]{y}} = \sqrt[3]{\dfrac{24x^{15}y^4}{y}} = \sqrt[3]{24x^{15}y^3} = \sqrt[3]{8x^{15}y^3}\sqrt[3]{3} = 2x^5y\sqrt[3]{3}$

20. $\sqrt{\dfrac{3a^5}{48a^7}} = \sqrt{\dfrac{1}{16a^2}} = \dfrac{1}{4a}$

21. $\sqrt{12x^2} = \sqrt{4x^2}\sqrt{3} = 2|x|\sqrt{3}$

22. $\sqrt{8x^6} = \sqrt{4x^6}\sqrt{2} = 2|x^3|\sqrt{2}$

23. $\sqrt[3]{81x^3} = \sqrt[3]{27x^3}\sqrt[3]{3} = 3x\sqrt[3]{3}$

24. $\sqrt{18x^4y^9} = \sqrt{9x^4y^8}\sqrt{2y} = |3x^2y^4|\sqrt{2y} = 3x^2y^4\sqrt{2y}$

25. $\sqrt{12} - \sqrt{27} = \sqrt{4}\sqrt{3} - \sqrt{9}\sqrt{3} = 2\sqrt{3} - 3\sqrt{3} = -\sqrt{3}$

26. $2\sqrt[3]{40} - \sqrt[3]{5000} + 4\sqrt[3]{625} = 2\sqrt[3]{8}\sqrt[3]{5} - \sqrt[3]{1000}\sqrt[3]{5} + 4\sqrt[3]{125}\sqrt[3]{5} = 2(2)\sqrt[3]{5} - 10\sqrt[3]{5} + 4(5)\sqrt[3]{5}$
$$= 4\sqrt[3]{5} - 10\sqrt[3]{5} + 20\sqrt[3]{5}$$
$$= 14\sqrt[3]{5}$$

27. $2\sqrt{48y^5} - 3y\sqrt{12y^3} = 2\sqrt{16y^4}\sqrt{3y} - 3y\sqrt{4y^2}\sqrt{3y} = 2(4y^2)\sqrt{3y} - 3y(2y)\sqrt{3y}$
$$= 8y^2\sqrt{3y} - 6y^2\sqrt{3y} = 2y^2\sqrt{3y}$$

28. $\sqrt[4]{768z^5} + z\sqrt[4]{48z} = \sqrt[4]{256z^4}\sqrt[4]{3z} + z\sqrt[4]{16}\sqrt[4]{3z} = 4z\sqrt[4]{3z} + 2z\sqrt[4]{3z} = 6z\sqrt[4]{3z}$

29. $-2\sqrt{xy}\left(3\sqrt{x} + \sqrt{xy^3}\right) = -6\sqrt{x^2y} - 2\sqrt{x^2y^4} = -6x\sqrt{y} - 2xy^2$

30. $\left(3\sqrt{2}+\sqrt{3}\right)\left(2\sqrt{2}-3\sqrt{3}\right) = 6\sqrt{4}-9\sqrt{6}+2\sqrt{6}-3\sqrt{9} = 12-7\sqrt{6}-9 = 3-7\sqrt{6}$

31. $\dfrac{1}{\sqrt{5}} = \dfrac{1\sqrt{5}}{\sqrt{5}\sqrt{5}} = \dfrac{\sqrt{5}}{5}$

32. $\dfrac{3t-1}{\sqrt{3t}-1} = \dfrac{(3t-1)\left(\sqrt{3t}+1\right)}{\left(\sqrt{3t}-1\right)\left(\sqrt{3t}+1\right)} = \dfrac{(3t-1)\left(\sqrt{3t}+1\right)}{\sqrt{9t^2}-1} = \dfrac{(3t-1)\left(\sqrt{3t}+1\right)}{3t-1} = \sqrt{3t}+1$

33. $\dfrac{\sqrt{3}}{\sqrt{7}} = \dfrac{\sqrt{3}\sqrt{3}}{\sqrt{7}\sqrt{3}} = \dfrac{3}{\sqrt{21}}$

34. $\dfrac{\sqrt{a}+\sqrt{b}}{\sqrt{a}-\sqrt{b}} = \dfrac{\left(\sqrt{a}+\sqrt{b}\right)\left(\sqrt{a}-\sqrt{b}\right)}{\left(\sqrt{a}-\sqrt{b}\right)\left(\sqrt{a}-\sqrt{b}\right)} = \dfrac{\sqrt{a^2}-\sqrt{b^2}}{\sqrt{a^2}-\sqrt{ab}-\sqrt{ab}+\sqrt{b^2}} = \dfrac{a-b}{a-2\sqrt{ab}+b}$

35. $\sqrt[3]{6n+4}-4 = 0$

$\qquad \sqrt[3]{6n+4} = 4$

$\qquad \left(\sqrt[3]{6n+4}\right)^3 = 4^3$

$\qquad\qquad 6n+4 = 64$

$\qquad\qquad\quad 6n = 60$

$\qquad\qquad\quad\ n = 10$

The answer checks.

36. $\qquad 1-\sqrt{u} = \sqrt{u-3}$

$\qquad \left(1-\sqrt{u}\right)^2 = \left(\sqrt{u-3}\right)^2$

$\qquad 1-2\sqrt{u}+u = u-3$

$\qquad\qquad 4 = 2\sqrt{u}$

$\qquad\qquad 4^2 = \left(2\sqrt{u}\right)^2$

$\qquad\qquad 16 = 4u$

$\qquad\qquad 4 = u$

The answer does not check. $\Rightarrow$ no solution

37. $(2+4i)+(-3+7i) = 2+4i-3+7i = -1+11i$

38. $(3-\sqrt{-9})-(-1+\sqrt{-16}) = 3-3i+1-4i = 4-7i$

39. $2i(3-4i) = 6i-8i^2 = 6i+8 = 8+6i$

40. $(3+2i)(-4-i) = -12-3i-8i-2i^2 = -12-11i+2 = -10-11i$

41. $\dfrac{1}{i\sqrt{2}} = \dfrac{1}{i\sqrt{2}}\cdot\dfrac{i\sqrt{2}}{i\sqrt{2}} = \dfrac{i\sqrt{2}}{2i^2} = -\dfrac{\sqrt{2}i}{2} = 0-\dfrac{\sqrt{2}}{2}i$

42. $\dfrac{2+i}{3-i} = \dfrac{2+i}{3-i}\cdot\dfrac{3+i}{3+i} = \dfrac{6+5i+i^2}{9-i^2} = \dfrac{5+5i}{10} = \dfrac{1}{2}+\dfrac{1}{2}i$

Exercise 10.1 (page 650)

1.
$$\frac{t+9}{2}+\frac{t+2}{5}=\frac{8}{5}+4t$$
$$5(t+9)+2(t+2)=2(8)+40t$$
$$5t+45+2t+4=16+40t$$
$$33=33t$$
$$1=t$$

3.
$$3(t-3)+3t\le 2(t+1)+t+1$$
$$3t-9+3t\le 2t+2+t+1$$
$$3t\le 12$$
$$t\le 4$$

5. $x=\sqrt{c};\ x=-\sqrt{c}$

7. positive or negative

9.
$$6x^2+12x=0$$
$$6x(x+2)=0$$
$$6x=0\quad\text{or}\quad x+2=0$$
$$x=0\qquad\qquad x=-2$$

11.
$$2y^2-50=0$$
$$2(y+5)(y-5)=0$$
$$y+5=0\quad\text{or}\quad y-5=0$$
$$y=-5\qquad\qquad y=5$$

13.
$$r^2+6r+8=0$$
$$(r+2)(r+4)=0$$
$$r+2=0\quad\text{or}\quad r+4=0$$
$$r=-2\qquad\qquad r=-4$$

15.
$$7x-6=x^2$$
$$0=x^2-7x+6$$
$$0=(x-6)(x-1)$$
$$x-6=0\quad\text{or}\quad x-1=0$$
$$x=6\qquad\qquad x=1$$

17.
$$2z^2-5z+2=0$$
$$(2z-1)(z-2)=0$$
$$2z-1=0\quad\text{or}\quad z-2=0$$
$$z=\tfrac{1}{2}\qquad\qquad z=2$$

19.
$$6s^2+11s-10=0$$
$$(2s+5)(3s-2)=0$$
$$2s+5=0\quad\text{or}\quad 3s-2=0$$
$$s=-\tfrac{5}{2}\qquad\qquad s=\tfrac{2}{3}$$

21. $x^2=36$
$$x=\pm\sqrt{36}=\pm 6$$

23. $z^2=5$
$$z=\pm\sqrt{5}$$

25.
$$3x^2-16=0$$
$$3x^2=16$$
$$x^2=\frac{16}{3}$$
$$x=\pm\sqrt{\frac{16}{3}}=\pm\frac{4}{\sqrt{3}}=\pm\frac{4\sqrt{3}}{3}$$

27.
$$(y+1)^2=1$$
$$y+1=\pm\sqrt{1}$$
$$y+1=\pm 1$$
$$y=-1\pm 1$$
$$y=0\text{ or }y=-2$$

29.
$$(s-7)^2-9=0$$
$$(s-7)^2=9$$
$$s-7=\pm\sqrt{9}$$
$$s-7=\pm 3$$
$$s=7\pm 3$$
$$s=10\text{ or }s=4$$

31.
$$(x+5)^2-3=0$$
$$(x+5)^2=3$$
$$x+5=\pm\sqrt{3}$$
$$x=-5\pm\sqrt{3}$$

33.
$$(x-2)^2-5=0$$
$$(x-2)^2=5$$
$$x-2=\pm\sqrt{5}$$
$$x=2\pm\sqrt{5}$$

35. $p^2 + 16 = 0$

$$p^2 = -16$$
$$p = \pm\sqrt{-16}$$
$$p = \pm 4i$$

37. $4m^2 + 81 = 0$

$$4m^2 = -81$$
$$m^2 = -\frac{81}{4}$$
$$m = \pm\sqrt{-\frac{81}{4}}$$
$$m = \pm\frac{9}{2}i$$

39. $x^2 + 2x - 8 = 0$

$$x^2 + 2x = 8$$
$$x^2 + 2x + 1 = 8 + 1$$
$$(x+1)^2 = 9$$
$$x + 1 = \pm 3$$
$$x = -1 \pm 3$$
$$x = 2 \text{ or } x = -4$$

41. $x^2 - 6x + 8 = 0$

$$x^2 - 6x = -8$$
$$x^2 - 6x + 9 = -8 + 9$$
$$(x-3)^2 = 1$$
$$x - 3 = \pm 1$$
$$x = 3 \pm 1$$
$$x = 4 \text{ or } x = 2$$

43. $x^2 + 5x + 4 = 0$

$$x^2 + 5x = -4$$
$$x^2 + 5x + \frac{25}{4} = -4 + \frac{25}{4}$$
$$\left(x + \frac{5}{2}\right)^2 = \frac{9}{4}$$
$$x + \frac{5}{2} = \pm\frac{3}{2}$$
$$x = -\frac{5}{2} \pm \frac{3}{2}$$
$$x = -\frac{2}{2} = -1 \text{ or } x = -\frac{8}{2} = -4$$

45.

$$x + 1 = 2x^2$$
$$-2x^2 + x = -1$$
$$x^2 - \frac{1}{2}x = \frac{1}{2}$$
$$x^2 - \frac{1}{2}x + \frac{1}{16} = \frac{1}{2} + \frac{1}{16}$$
$$\left(x - \frac{1}{4}\right)^2 = \frac{9}{16}$$
$$x - \frac{1}{4} = \pm\frac{3}{4}$$
$$x = \frac{1}{4} \pm \frac{3}{4}$$
$$x = \frac{4}{4} = 1 \text{ or } x = -\frac{2}{4} = -\frac{1}{2}$$

47.

$$6x^2 + 11x + 3 = 0$$
$$6x^2 + 11x = -3$$
$$x^2 + \frac{11}{6}x = -\frac{1}{2}$$
$$x^2 + \frac{11}{6}x + \frac{121}{144} = -\frac{1}{2} + \frac{121}{144}$$
$$\left(x + \frac{11}{12}\right)^2 = \frac{49}{144}$$
$$x + \frac{11}{12} = \pm\frac{7}{12}$$
$$x = -\frac{11}{12} \pm \frac{7}{12}$$
$$x = -\frac{4}{12} = -\frac{1}{3} \text{ or } x = -\frac{18}{12} = -\frac{3}{2}$$

49.

$$9 - 6r = 8r^2$$
$$-8r^2 - 6r = -9$$
$$r^2 + \frac{3}{4}r = \frac{9}{8}$$
$$r^2 + \frac{3}{4}r + \frac{9}{64} = \frac{9}{8} + \frac{9}{64}$$
$$\left(r + \frac{3}{8}\right)^2 = \frac{81}{64}$$
$$r + \frac{3}{8} = \pm\frac{9}{8}$$
$$r = -\frac{3}{8} \pm \frac{9}{8}$$
$$r = \tfrac{6}{8} = \tfrac{3}{4} \text{ or } r = -\tfrac{12}{8} = -\tfrac{3}{2}$$

51.

$$\frac{7x + 1}{5} = -x^2$$
$$7x + 1 = -5x^2$$
$$5x^2 + 7x = -1$$
$$x^2 + \frac{7}{5}x = -\frac{1}{5}$$
$$x^2 + \frac{7}{5}x + \frac{49}{100} = -\frac{1}{5} + \frac{49}{100}$$
$$\left(x + \frac{7}{10}\right)^2 = \frac{29}{100}$$
$$x + \frac{7}{10} = \pm\frac{\sqrt{29}}{10}$$
$$x = -\frac{7}{10} \pm \frac{\sqrt{29}}{10}$$

53.

$$p^2 + 2p + 2 = 0$$
$$p^2 + 2p = -2$$
$$p^2 + 2p + 1 = -2 + 1$$
$$(p + 1)^2 = -1$$
$$p + 1 = \pm\sqrt{-1}$$
$$p + 1 = \pm i$$
$$p = -1 \pm i$$

55.

$$y^2 + 8y + 18 = 0$$
$$y^2 + 8y = -18$$
$$y^2 + 8y + 16 = -18 + 16$$
$$(y + 4)^2 = -2$$
$$y + 4 = \pm\sqrt{-2}$$
$$y = -4 \pm i\sqrt{2}$$

57.

$$3m^2 - 2m + 3 = 0$$
$$m^2 - \frac{2}{3}m = -1$$
$$m^2 - \frac{2}{3}m + \frac{1}{9} = -1 + \frac{1}{9}$$
$$\left(m - \frac{1}{3}\right)^2 = -\frac{8}{9}$$
$$m - \frac{1}{3} = \pm\sqrt{-\frac{8}{9}}$$
$$m = \frac{1}{3} \pm \frac{\sqrt{8}}{3}i$$
$$m = \frac{1}{3} \pm \frac{2\sqrt{2}}{3}i$$

59.

$$f(x) = 0$$
$$2x^2 + x - 5 = 0$$
$$x^2 + \frac{1}{2}x = \frac{5}{2}$$
$$x^2 + \frac{1}{2}x + \frac{1}{16} = \frac{5}{2} + \frac{1}{16}$$
$$\left(x + \frac{1}{4}\right)^2 = \frac{41}{16}$$
$$x + \frac{1}{4} = \pm\sqrt{\frac{41}{16}}$$
$$x + \frac{1}{4} = \pm\frac{\sqrt{41}}{4}$$
$$x = -\frac{1}{4} \pm \frac{\sqrt{41}}{4}$$

61.
$$f(x) = 0$$
$$x^2 + x - 3 = 0$$
$$x^2 + x = 3$$
$$x^2 + x + \frac{1}{4} = 3 + \frac{1}{4}$$
$$\left(x + \frac{1}{2}\right)^2 = \frac{13}{4}$$
$$x + \frac{1}{2} = \pm\sqrt{\frac{13}{4}}$$
$$x = -\frac{1}{2} \pm \frac{\sqrt{13}}{2}$$

63.
$$s = 16t^2$$
$$256 = 16t^2$$
$$16 = t^2$$
$$\pm\sqrt{16} = t$$
$$\pm 4 = t$$
$t = 4$ is the only answer that makes sense, so it will take 4 seconds.

65.
$$s^2 = 10.5l$$
$$s^2 = 10.5(500)$$
$$s^2 = 5250$$
$$s = \pm\sqrt{5250}$$
$$s \approx \pm 72.5$$
s must be positive, so the speed was about 72.5 mph.

67.
$$A = P(1 + r)^t$$
$$9193.60 = 8500(1 + r)^2$$
$$\frac{9193.60}{8500} = (1 + r)^2$$
$$1.0816 = (1 + r)^2$$
$$\pm\sqrt{1.0816} = \sqrt{(1 + r)^2}$$
$$\pm 1.04 = 1 + r$$
$$-1 \pm 1.04 = r$$
$r = 0.04$ or $r = -2.04$; r must be positive, so $r = 0.04$, or 4%.

69. Answers may vary.

71. $\left(\frac{1}{2}\sqrt{3}\right)^2 = \left(\frac{\sqrt{3}}{2}\right)^2 = \frac{3}{4}$

Exercise 10.2 (page 656)

1.
$$Ax + By = C$$
$$By = -Ax + C$$
$$B = \frac{-Ax + C}{y}$$

3. $\sqrt{24} = \sqrt{4}\sqrt{6} = 2\sqrt{6}$

5. $\frac{3}{\sqrt{3}} = \frac{3}{\sqrt{3}} \cdot \frac{\sqrt{3}}{\sqrt{3}} = \frac{3\sqrt{3}}{3} = \sqrt{3}$

7. $3; -2; 6$

9. $x^2 + 3x + 2 = 0$
$a = 1, b = 3, c = 2$

$$x = \frac{-b \pm \sqrt{b^2 - 4ac}}{2a}$$

$$= \frac{-3 \pm \sqrt{3^2 - 4(1)(2)}}{2(1)}$$

$$= \frac{-3 \pm \sqrt{9 - 8}}{2}$$

$$= \frac{-3 \pm \sqrt{1}}{2} = \frac{-3 \pm 1}{2}$$

$x = -\frac{2}{2} = -1$ or $x = -\frac{4}{2} = -2$

11. $a^2 - 2a - 15 = 0$
$a = 1, b = -2, c = -15$

$$a = \frac{-b \pm \sqrt{b^2 - 4ac}}{2a}$$

$$= \frac{2 \pm \sqrt{(-2)^2 - 4(1)(-15)}}{2(1)}$$

$$= \frac{2 \pm \sqrt{4 + 60}}{2}$$

$$= \frac{2 \pm \sqrt{64}}{2} = \frac{2 \pm 8}{2}$$

$x = \frac{10}{2} = 5$ or $x = -\frac{6}{2} = -3$

13. $x^2 + 12x = -36$
$x^2 + 12x + 36 = 0$
$a = 1, b = 12, c = 36$

$$x = \frac{-b \pm \sqrt{b^2 - 4ac}}{2a}$$

$$= \frac{-12 \pm \sqrt{12^2 - 4(1)(36)}}{2(1)}$$

$$= \frac{-12 \pm \sqrt{144 - 144}}{2}$$

$$= \frac{-12 \pm \sqrt{0}}{2} = \frac{-12 \pm 0}{2}$$

$x = -\frac{12}{2} = -6$ or $x = -\frac{12}{2} = -6$

15. $2x^2 - x - 3 = 0$
$a = 2, b = -1, c = -3$

$$x = \frac{-b \pm \sqrt{b^2 - 4ac}}{2a}$$

$$= \frac{1 \pm \sqrt{(-1)^2 - 4(2)(-3)}}{2(2)}$$

$$= \frac{1 \pm \sqrt{1 + 24}}{4}$$

$$= \frac{1 \pm \sqrt{25}}{4} = \frac{1 \pm 5}{4}$$

$x = \frac{6}{4} = \frac{3}{2}$ or $x = -\frac{4}{4} = -1$

17. $5x^2 + 5x + 1 = 0$
$a = 5, b = 5, c = 1$

$$x = \frac{-b \pm \sqrt{b^2 - 4ac}}{2a}$$

$$= \frac{-5 \pm \sqrt{5^2 - 4(5)(1)}}{2(5)}$$

$$= \frac{-5 \pm \sqrt{25 - 20}}{10}$$

$$= \frac{-5 \pm \sqrt{5}}{10}$$

$$= \frac{-5}{10} \pm \frac{\sqrt{5}}{10} = -\frac{1}{2} \pm \frac{\sqrt{5}}{10}$$

19. $8u = -4u^2 - 3$
$4u^2 + 8u + 3 = 0$
$a = 4, b = 8, c = 3$

$$u = \frac{-b \pm \sqrt{b^2 - 4ac}}{2a}$$

$$= \frac{-8 \pm \sqrt{8^2 - 4(4)(3)}}{2(4)}$$

$$= \frac{-8 \pm \sqrt{64 - 48}}{8}$$

$$= \frac{-8 \pm \sqrt{16}}{8} = \frac{-8 \pm 4}{8}$$

$u = \frac{-4}{8} = -\frac{1}{2}$ or $u = \frac{-12}{8} = -\frac{3}{2}$

21. $16y^2 + 8y - 3 = 0$

$a = 16, b = 8, c = -3$

$$y = \frac{-b \pm \sqrt{b^2 - 4ac}}{2a}$$

$$= \frac{-8 \pm \sqrt{8^2 - 4(16)(-3)}}{2(16)}$$

$$= \frac{-8 \pm \sqrt{64 + 192}}{32}$$

$$= \frac{-8 \pm \sqrt{256}}{32} = \frac{-8 \pm 16}{32}$$

$y = \frac{8}{32} = \frac{1}{4}$ or $y = \frac{-24}{32} = -\frac{3}{4}$

23. $\frac{x^2}{2} + \frac{5}{2}x = -1$

$$x^2 + 5x = -2$$

$$x^2 + 5x + 2 = 0$$

$$a = 1, b = 5, c = 2$$

$$x = \frac{-b \pm \sqrt{b^2 - 4ac}}{2a}$$

$$= \frac{-5 \pm \sqrt{5^2 - 4(1)(2)}}{2(1)}$$

$$= \frac{-5 \pm \sqrt{25 - 8}}{2}$$

$$= \frac{-5 \pm \sqrt{17}}{2} = -\frac{5}{2} \pm \frac{\sqrt{17}}{2}$$

25. $x^2 + 2x + 2 = 0$

$a = 1, b = 2, c = 2$

$$x = \frac{-b \pm \sqrt{b^2 - 4ac}}{2a}$$

$$= \frac{-2 \pm \sqrt{2^2 - 4(1)(2)}}{2(1)}$$

$$= \frac{-2 \pm \sqrt{4 - 8}}{2}$$

$$= \frac{-2 \pm \sqrt{-4}}{2}$$

$$= \frac{-2 \pm \sqrt{-1 \cdot 4}}{2}$$

$$= \frac{-2 \pm 2i}{2} = \frac{-2}{2} \pm \frac{2i}{2} = -1 \pm i$$

27. $2x^2 + x + 1 = 0$

$a = 2, b = 1, c = 1$

$$x = \frac{-b \pm \sqrt{b^2 - 4ac}}{2a}$$

$$= \frac{-1 \pm \sqrt{1^2 - 4(2)(1)}}{2(2)}$$

$$= \frac{-1 \pm \sqrt{1 - 8}}{4}$$

$$= \frac{-1 \pm \sqrt{-7}}{4}$$

$$= \frac{-1 \pm \sqrt{-1 \cdot 7}}{4}$$

$$= \frac{-1 \pm i\sqrt{7}}{4} = -\frac{1}{4} \pm \frac{\sqrt{7}}{4}i$$

29. $3x^2 - 4x = -2$

$$3x^2 - 4x + 2 = 0$$

$$a = 3, b = -4, c = 2$$

$$x = \frac{-b \pm \sqrt{b^2 - 4ac}}{2a}$$

$$= \frac{4 \pm \sqrt{(-4)^2 - 4(3)(2)}}{2(3)}$$

$$= \frac{4 \pm \sqrt{16 - 24}}{6}$$

$$x = \frac{4 \pm \sqrt{16 - 24}}{6}$$

$$= \frac{4 \pm \sqrt{-8}}{6}$$

$$= \frac{4 \pm \sqrt{-1 \cdot 4 \cdot 2}}{6}$$

$$= \frac{4 \pm 2i\sqrt{2}}{6}$$

$$= \frac{4}{6} \pm \frac{2\sqrt{2}}{6}i = \frac{2}{3} \pm \frac{\sqrt{2}}{3}i$$

31.

$$3x^2 - 2x = -3$$
$$3x^2 - 2x + 3 = 0$$
$$a = 3, b = -2, c = 3$$
$$x = \frac{-b \pm \sqrt{b^2 - 4ac}}{2a}$$
$$= \frac{2 \pm \sqrt{(-2)^2 - 4(3)(3)}}{2(3)}$$
$$= \frac{2 \pm \sqrt{4 - 36}}{6}$$

$$x = \frac{2 \pm \sqrt{4 - 36}}{6}$$
$$= \frac{2 \pm \sqrt{-32}}{6}$$
$$= \frac{2 \pm \sqrt{-1 \cdot 16 \cdot 2}}{6}$$
$$= \frac{2 \pm 4i\sqrt{2}}{6}$$
$$= \frac{2}{6} \pm \frac{4\sqrt{2}}{6}i = \frac{1}{3} \pm \frac{2\sqrt{2}}{3}i$$

33.

$$f(x) = 0$$
$$4x^2 + 4x - 19 = 0$$
$$a = 4, b = 4, c = -19$$
$$x = \frac{-b \pm \sqrt{b^2 - 4ac}}{2a}$$
$$= \frac{-4 \pm \sqrt{4^2 - 4(4)(-19)}}{2(4)}$$
$$= \frac{-4 \pm \sqrt{16 + 304}}{8}$$
$$= \frac{-4 \pm \sqrt{320}}{8}$$
$$= -\frac{4}{8} \pm \frac{8\sqrt{5}}{8} = -\frac{1}{2} \pm \sqrt{5}$$

35.

$$f(x) = 0$$
$$3x^2 + 2x + 2 = 0$$
$$a = 3, b = 2, c = 2$$
$$x = \frac{-b \pm \sqrt{b^2 - 4ac}}{2a}$$
$$= \frac{-2 \pm \sqrt{2^2 - 4(3)(2)}}{2(3)}$$
$$= \frac{-2 \pm \sqrt{4 - 24}}{6}$$
$$= \frac{-2 \pm \sqrt{-20}}{6}$$
$$= -\frac{2}{6} \pm \frac{2i\sqrt{5}}{6} = -\frac{1}{3} \pm \frac{\sqrt{5}}{3}i$$

37.

$$0.7x^2 - 3.5x - 25 = 0$$
$$a = 0.7, b = -3.5, c = -25$$
$$x = \frac{-b \pm \sqrt{b^2 - 4ac}}{2a}$$
$$= \frac{3.5 \pm \sqrt{(-3.5)^2 - 4(0.7)(-25)}}{2(0.7)}$$
$$= \frac{3.5 \pm \sqrt{12.25 + 70}}{1.4}$$
$$= \frac{3.5 \pm \sqrt{82.25}}{1.4}$$
$$= \frac{3.5 \pm 9.069}{1.4}$$
$$x = 8.98 \text{ or } x = -3.98$$

39.

$$C = \frac{N^2 - N}{2}$$
$$2C = N^2 - N$$
$$N^2 - N - 2C = 0$$
$$a = 1, b = -1, c = -2C$$
$$N = \frac{-b \pm \sqrt{b^2 - 4ac}}{2a}$$
$$= \frac{1 \pm \sqrt{(-1)^2 - 4(1)(-2C)}}{2(1)}$$
$$= \frac{1 \pm \sqrt{1 + 8C}}{2}$$

41. Let x and $x + 2$ represent the integers.
$$x(x + 2) = 288$$
$$x^2 + 2x - 288 = 0$$
$$(x + 18)(x - 16) = 0$$
$$x = -18 \text{ or } x = 16$$
Since the integers are positive, they must be 16 and 18.

43. Let x and $x + 1$ represent the integers.
$$x^2 + (x + 1)^2 = 85$$
$$x^2 + x^2 + 2x + 1 = 85$$
$$2x^2 + 2x - 84 = 0$$
$$2(x + 7)(x - 6) = 0$$
$$x = -7 \text{ or } x = 6$$
Since the integers are positive, they must be 6 and 7.

45. $(x - 3)(x - 5) = 0$
$$x^2 - 8x + 15 = 0$$

47. $(x - 2)(x - 3)(x + 4) = 0$
$$(x^2 - 5x + 6)(x + 4) = 0$$
$$x^3 - x^2 - 14x + 24 = 0$$

49. $\quad$ length $\cdot$ width $=$ Area
$$(x + 4)x = 96$$
$$x^2 + 4x - 96 = 0$$
$$(x + 12)(x - 8) = 0$$
$$x = -12 \text{ or } x = 8$$
Since the width is positive, the dimensions are 8 ft by 12 ft.

51. Let $s =$ the length of a side.
$$\text{Area} = \text{perimeter}$$
$$s^2 = 4s$$
$$s^2 - 4s = 0$$
$$s(s - 4) = 0$$
$$s = 0 \text{ or } s = 4$$
Since the length cannot be 0, the length of a side is 4 units.

53. Let b represent the base.
Then $3b + 5$ represents the height.
$$\tfrac{1}{2}\text{base} \cdot \text{height} = \text{Area}$$
$$\tfrac{1}{2}b(3b + 5) = 6$$
$$b(3b + 5) = 12$$
$$3b^2 + 5b - 12 = 0$$
$$(3b - 4)(b + 3) = 0$$
$$b = \tfrac{4}{3} \text{ or } b = -3$$
Since the base is positive, it must be $\tfrac{4}{3}$ cm.

55. Let r = the slower rate. Then $r + 20$ = the faster rate.

	Rate	Time	Dist.
Slower	r	$\frac{150}{r}$	150
Faster	$r + 20$	$\frac{150}{r+20}$	150

$$\boxed{\text{Faster time}} + 2 = \boxed{\text{Slower time}}$$

$$\frac{150}{r + 20} + 2 = \frac{150}{r}$$

$$\left(\frac{150}{r + 20} + 2\right)(r)(r + 20) = \frac{150}{r} \cdot r(r + 20)$$

$$150r + 2r(r + 20) = 150(r + 20)$$

$$2r^2 + 40r - 3000 = 0$$

$$2(r + 50)(r - 30) = 0$$

$r = -50$ or $r = 30$ $r = 30$ is the only answer that makes sense.

Her original speed was 30 mph.

57. Let x = the number of 10¢ increases. Then the ticket price will be $4 + 0.10x$, while the projected attendance will be $300 - 5x$, for total receipts of $(4 + 0.10x)(300 - 5x)$.

$$\text{Total} = 1248$$

$$(4 + 0.10x)(300 - 5x) = 1248$$

$$1200 + 10x - 0.5x^2 = 1248$$

$$-0.5x^2 + 10x - 48 = 0$$

$$x^2 - 20x + 96 = 0$$

$$(x - 12)(x - 8) = 0$$

$x = 12$ or $x = 8 \Rightarrow 4 + 0.10(12) = 5.20; 4 + 0.10(8) = 4.80$

The ticket price would be either \$5.20 or \$4.80.

59. Let x = the number of additional subscribers. Then the profit per subscriber will be $20 + 0.01x$, for a total profit of $(20 + 0.01x)(3000 + x)$.

$$\text{Total profit} = 120000$$

$$(20 + 0.01x)(3000 + x) = 120000$$

$$60{,}000 + 50x + 0.01x^2 = 120000$$

$$0.01x^2 + 50x - 60{,}000 = 0$$

$$x^2 + 5000x - 6{,}000{,}000 = 0$$

$$(x + 6000)(x - 1000) = 0$$

$x = -6000$ (impossible) or $x = 1000 \Rightarrow$ The total number of subscribers would be 4000.

61. Let w = the constant width.

$$\text{Frame area} = \text{Picture area}$$
$$(12 + 2w)(10 + 2w) - 12(10) = 12(10)$$
$$120 + 44w + 4w^2 - 240 = 0$$
$$4w^2 + 44w - 120 = 0$$
$$4(w^2 + 11w - 30) = 0$$
$$w^2 + 11w - 30 = 0 \Rightarrow a = 1, b = 11, c = -30$$

$$w = \frac{-11 \pm \sqrt{11^2 - 4(1)(-30)}}{2(1)}$$

$$= \frac{-11 \pm \sqrt{121 + 120}}{2} = \frac{-11 \pm \sqrt{241}}{2} = \frac{-11 \pm 15.52}{2} = \frac{4.52}{2} \text{ or } \frac{-26.52}{2} \text{ (impossible)}$$

$$w = 2.26 \text{ in.}$$

63. Let $[\text{H}^+]$ (and then $[\text{A}^-]$) = x and $[\text{HA}] = 0.1 - x$.

$$\frac{[\text{H}^+][\text{A}^-]}{[\text{HA}]} = 4 \times 10^{-4}$$

$$\frac{x^2}{0.1 - x} = 4 \times 10^{-4}$$

$$x^2 = 4 \times 10^{-5} - \left(4 \times 10^{-4}\right)x$$

$$x^2 + \left(4 \times 10^{-4}\right)x - 4 \times 10^{-5} = 0$$

$$x = \frac{-b \pm \sqrt{b^2 - 4ac}}{2a} = \frac{-4 \times 10^{-4} \pm \sqrt{(4 \times 10^{-4})^2 - 4(1)(-4 \times 10^{-5})}}{2(1)}$$

$$\approx \frac{-4 \times 10^{-4} \pm 0.012655}{2} \approx \frac{0.012255}{2} \text{ or } -\frac{0.013055}{2} \text{ (impossible)}$$

The concentration is about 0.00613 M = 6.13×10^{-3} M.

65. Answers may vary.

67. $x^2 + 2\sqrt{2}x - 6 = 0$

$a = 1, b = 2\sqrt{2}, c = -6$

$$x = \frac{-b \pm \sqrt{b^2 - 4ac}}{2a}$$

$$= \frac{-2\sqrt{2} \pm \sqrt{\left(2\sqrt{2}\right)^2 - 4(1)(-6)}}{2(1)}$$

$$= \frac{-2\sqrt{2} \pm \sqrt{8 + 24}}{2}$$

$$= \frac{-2\sqrt{2} \pm \sqrt{32}}{2} = \frac{-2\sqrt{2} \pm 4\sqrt{2}}{2}$$

$$x = \frac{2\sqrt{2}}{2} = \sqrt{2} \text{ or } x = \frac{-6\sqrt{2}}{2} = -3\sqrt{2}$$

69. $x^2 - 3ix - 2 = 0$

$a = 1, b = -3i, c = -2$

$$x = \frac{-b \pm \sqrt{b^2 - 4ac}}{2a}$$

$$= \frac{3i \pm \sqrt{(-3i)^2 - 4(1)(-2)}}{2(1)}$$

$$= \frac{3i \pm \sqrt{9i^2 + 8}}{2}$$

$$= \frac{3i \pm \sqrt{-1}}{2} = \frac{3i \pm i}{2}$$

$$x = \frac{4i}{2} = 2i \text{ or } x = \frac{2i}{2} = i$$

Exercise 10.3 (page 664)

1.
$$\frac{1}{4} + \frac{1}{t} = \frac{1}{2t}$$
$$\left(\frac{1}{4} + \frac{1}{t}\right)4t = \frac{1}{2t} \cdot 4t$$
$$t + 4 = 2$$
$$t = -2$$

3.
$$m = \frac{\Delta y}{\Delta x} = \frac{-4 - 5}{-2 - 3} = \frac{-9}{-5} = \frac{9}{5}$$

5. $b^2 - 4ac$

7. rational; unequal

9. $4x^2 - 4x + 1 = 0; a = 4, b = -4, c = 1$
$$b^2 - 4ac = (-4)^2 - 4(4)(1)$$
$$= 16 - 16 = 0$$
The solutions are rational and equal.

11. $5x^2 + x + 2 = 0; a = 5, b = 1, c = 2$
$$b^2 - 4ac = 1^2 - 4(5)(2)$$
$$= 1 - 40 = -39$$
The solutions are complex conjugates.

13.
$$2x^2 = 4x - 1$$
$$2x^2 - 4x + 1 = 0; a = 2, b = -4, c = 1$$
$$b^2 - 4ac = (-4)^2 - 4(2)(1)$$
$$= 16 - 8 = 8$$
The solutions are irrational and unequal.

15.
$$x(2x - 3) = 20$$
$$2x^2 - 3x - 20 = 0; a = 2, b = -3, c = -20$$
$$b^2 - 4ac = (-3)^2 - 4(2)(-20)$$
$$= 9 + 160 = 169$$
The solutions are rational and unequal.

17. $x^2 + kx + 9 = 0; a = 1, b = k, c = 9$
Set the discriminant equal to 0:
$$b^2 - 4ac = 0$$
$$k^2 - 4(1)(9) = 0$$
$$k^2 - 36 = 0$$
$$k^2 = 36$$
$$k = \pm 6$$

19.
$$9x^2 + 4 = -kx$$
$$9x^2 + kx + 4 = 0; a = 9, b = k, c = 4$$
Set the discriminant equal to 0:
$$b^2 - 4ac = 0$$
$$k^2 - 4(9)(4) = 0$$
$$k^2 - 144 = 0$$
$$k^2 = 144$$
$$k = \pm 12$$

21. $(k - 1)x^2 + (k - 1)x + 1 = 0$
$a = k - 1, b = k - 1, c = 1$
Set the discriminant equal to 0:
$$b^2 - 4ac = 0$$
$$(k - 1)^2 - 4(k - 1)(1) = 0$$
$$k^2 - 2k + 1 - 4k + 4 = 0$$
$$k^2 - 6k + 5 = 0$$
$$(k - 5)(k - 1) = 0$$
$$k - 5 = 0 \quad \textbf{or} \quad k - 1 = 0$$
$$k = 5 \qquad\qquad k = 1$$
$$\text{doesn't work}$$

23. $(k + 4)x^2 + 2kx + 9 = 0$
$a = k + 4, b = 2k, c = 9$
Set the discriminant equal to 0:
$$b^2 - 4ac = 0$$
$$(2k)^2 - 4(k + 4)(9) = 0$$
$$4k^2 - 36(k + 4) = 0$$
$$4k^2 - 36k - 144 = 0$$
$$4(k^2 - 9k - 36) = 0$$
$$4(k - 12)(k + 3) = 0$$
$$k - 12 = 0 \quad \textbf{or} \quad k + 3 = 0$$
$$k = 12 \qquad\qquad k = -3$$

25. $1492x^2 + 1776x - 1984 = 0$

$a = 1492, b = 1776, c = -1984$

$b^2 - 4ac = (1776)^2 - 4(1492)(-1984)$

$\qquad = 3{,}154{,}176 + 11{,}840{,}512$

$\qquad = 14{,}994{,}688$

The solutions are real numbers.

27. $\qquad 3x^2 + 4x = k$

$\qquad 3x^2 + 4x - k = 0$

$\qquad a = 3, b = 4, c = -k$

Set the discriminant less than 0:

$\qquad b^2 - 4ac < 0$

$\qquad 4^2 - 4(3)(-k) < 0$

$\qquad 16 + 12k < 0$

$\qquad\qquad 12k < -16$

$\qquad\qquad k < -\dfrac{16}{12}, \text{ or } k < -\dfrac{4}{3}$

29. $\qquad x^4 - 17x^2 + 16 = 0$

$\qquad (x^2 - 16)(x^2 - 1) = 0$

$\qquad x^2 - 16 = 0 \quad \textbf{or} \quad x^2 - 1 = 0$

$\qquad\qquad x^2 = 16 \qquad\qquad x^2 = 1$

$\qquad\qquad x = \pm 4 \qquad\qquad x = \pm 1$

31. $\qquad x^4 - 3x^2 = -2$

$\qquad x^4 - 3x^2 + 2 = 0$

$\qquad (x^2 - 2)(x^2 - 1) = -0$

$\qquad x^2 - 2 = 0 \quad \textbf{or} \quad x^2 - 1 = 0$

$\qquad\qquad x^2 = 2 \qquad\qquad x^2 = 1$

$\qquad\qquad x = \pm\sqrt{2} \qquad\qquad x = \pm 1$

33. $\qquad\qquad x^4 = 6x^2 - 5$

$\qquad x^4 - 6x^2 + 5 = 0$

$\qquad (x^2 - 5)(x^2 - 1) = -0$

$\qquad x^2 - 5 = 0 \quad \textbf{or} \quad x^2 - 1 = 0$

$\qquad\qquad x^2 = 5 \qquad\qquad x^2 = 1$

$\qquad\qquad x = \pm\sqrt{5} \qquad\qquad x = \pm 1$

35. $\qquad 2x^4 - 10x^2 = -8$

$\qquad 2x^4 - 10x^2 + 8 = 0$

$\qquad 2(x^2 - 4)(x^2 - 1) = -0$

$\qquad x^2 - 4 = 0 \quad \textbf{or} \quad x^2 - 1 = 0$

$\qquad\qquad x^2 = 4 \qquad\qquad x^2 = 1$

$\qquad\qquad x = \pm 2 \qquad\qquad x = \pm 1$

37. $\qquad\qquad 2x^4 + 24 = 26x^2$

$\qquad 2x^4 - 26x^2 + 24 = 0$

$\qquad 2(x^2 - 12)(x^2 - 1) = -0$

$\qquad x^2 - 12 = 0 \quad \textbf{or} \quad x^2 - 1 = 0$

$\qquad\qquad x^2 = 12 \qquad\qquad x^2 = 1$

$\qquad\qquad x = \pm\sqrt{12} \qquad\qquad x = \pm 1$

$\qquad\qquad\quad = \pm 2\sqrt{3}$

39. $\qquad 2x + x^{1/2} - 3 = 0$

$\qquad \left(2x^{1/2} + 3\right)\left(x^{1/2} - 1\right) = 0$

$\qquad 2x^{1/2} + 3 = 0 \quad \textbf{or} \quad x^{1/2} - 1 = 0$

$\qquad\qquad 2x^{1/2} = -3 \qquad\qquad x^{1/2} = 1$

$\qquad\qquad x^{1/2} = -\dfrac{3}{2} \qquad\quad \left(x^{1/2}\right)^2 = 1^2$

$\qquad \left(x^{1/2}\right)^2 = \left(-\dfrac{3}{2}\right)^2 \qquad\qquad x = 1$

$\qquad\qquad x = \dfrac{9}{4} \qquad\qquad\qquad \text{Solution}$

Does not check

41.
$$3x + 5x^{1/2} + 2 = 0$$
$$\left(3x^{1/2} + 2\right)\left(x^{1/2} + 1\right) = 0$$

$3x^{1/2} + 2 = 0 \qquad$ **or** $\ x^{1/2} + 1 = 0$

$\qquad 3x^{1/2} = -2 \qquad\qquad\qquad x^{1/2} = -1$

$\qquad x^{1/2} = -\frac{2}{3} \qquad\qquad \left(x^{1/2}\right)^2 = (-1)^2$

$\qquad \left(x^{1/2}\right)^2 = \left(-\frac{2}{3}\right)^2 \qquad\qquad x = 1$

$\qquad\qquad x = \frac{4}{9} \qquad\qquad$ Does not check

Does not check

43.
$$x^{2/3} + 5x^{1/3} + 6 = 0$$
$$\left(x^{1/3} + 2\right)\left(x^{1/3} + 3\right) = 0$$

$x^{1/3} + 2 = 0 \qquad$ **or** $\ x^{1/3} + 3 = 0$

$\qquad x^{1/3} = -2 \qquad\qquad\qquad x^{1/3} = -3$

$\qquad \left(x^{1/3}\right)^3 = (-2)^3 \qquad\qquad \left(x^{1/3}\right)^3 = (-3)^3$

$\qquad\qquad x = -8 \qquad\qquad\qquad x = -27$

$\qquad$ Solution $\qquad\qquad\qquad$ Solution

45.
$$x^{2/3} - 2x^{1/3} - 3 = 0$$
$$\left(x^{1/3} - 3\right)\left(x^{1/3} + 1\right) = 0$$

$x^{1/3} - 3 = 0 \qquad$ **or** $\ x^{1/3} + 1 = 0$

$\qquad x^{1/3} = 3 \qquad\qquad\qquad x^{1/3} = -1$

$\qquad \left(x^{1/3}\right)^3 = (3)^3 \qquad\qquad \left(x^{1/3}\right)^3 = (-1)^3$

$\qquad\qquad x = 27 \qquad\qquad\qquad x = -1$

$\qquad$ Solution $\qquad\qquad\qquad$ Solution

47.
$$x + 5 + \frac{4}{x} = 0$$
$$x\left(x + 5 + \frac{4}{x}\right) = x(0)$$
$$x^2 + 5x + 4 = 0$$
$$(x + 4)(x + 1) = 0$$

$x + 4 = 0 \qquad$ **or** $\ x + 1 = 0$

$\qquad x = -4 \qquad\qquad\qquad x = -1$

49.
$$x + 1 = \frac{20}{x}$$
$$x + 1 - \frac{20}{x} = 0$$
$$x\left(x + 1 - \frac{20}{x}\right) = x(0)$$
$$x^2 + x - 20 = 0$$
$$(x + 5)(x - 4) = 0$$

$x + 5 = 0 \qquad$ **or** $\ x - 4 = 0$

$\qquad x = -5 \qquad\qquad\qquad x = 4$

51.
$$\frac{1}{x - 1} + \frac{3}{x + 1} = 2$$
$$\left(\frac{1}{x - 1} + \frac{3}{x + 1}\right)(x - 1)(x + 1) = 2(x + 1)(x - 1)$$
$$1(x + 1) + 3(x - 1) = 2(x^2 - 1)$$
$$x + 1 + 3x - 3 = 2x^2 - 2$$
$$0 = 2x^2 - 4x$$

$0 = 2x(x - 2) \qquad 2x = 0 \quad$ **or** $\ x - 2 = 0$

$\qquad\qquad\qquad\qquad x = 0 \qquad\qquad x = 2$

53.

$$\frac{1}{x+2} + \frac{24}{x+3} = 13$$

$$\left(\frac{1}{x+2} + \frac{24}{x+3}\right)(x+2)(x+3) = 13(x+2)(x+3)$$

$$1(x+3) + 24(x+2) = 13(x^2 + 5x + 6)$$

$$x + 3 + 24x + 48 = 13x^2 + 65x + 78$$

$$0 = 13x^2 + 40x + 27$$

$$0 = (13x + 27)(x + 1) \qquad 13x + 27 = 0 \qquad \textbf{or} \quad x + 1 = 0$$

$$13x = -27 \qquad\qquad x = -1$$

$$x = -\frac{27}{13}$$

55.

$$x^{-4} - 2x^{-2} + 1 = 0$$

$$(x^{-2} - 1)(x^{-2} - 1) = 0$$

$$x^{-2} - 1 = 0 \quad \text{or} \quad x^{-2} - 1 = 0$$

$$x^{-2} = 1 \qquad\qquad x^{-2} = 1$$

$$\frac{1}{x^2} = 1 \qquad\qquad \frac{1}{x^2} = 1$$

$$1 = x^2 \qquad\qquad 1 = x^2$$

$$\pm 1 = x \qquad\qquad \pm 1 = x$$

57.

$$x + \frac{2}{x-2} = 0$$

$$\left(x + \frac{2}{x-2}\right)(x-2) = 0(x-2)$$

$$x(x-2) + 2 = 0$$

$$x^2 - 2x + 2 = 0$$

$$x^2 - 2x = -2$$

$$x^2 - 2x + 1 = -2 + 1$$

$$(x - 1)^2 = -1$$

$$x - 1 = \pm\sqrt{-1}$$

$$x = 1 \pm i$$

59.

$$x^2 + y^2 = r^2$$

$$x^2 = r^2 - y^2$$

$$x = \pm\sqrt{r^2 - y^2}$$

61.

$$I = \frac{k}{d^2}$$

$$Id^2 = k$$

$$d^2 = \frac{k}{I}$$

$$d = \pm\sqrt{\frac{k}{I}} = \pm\frac{\sqrt{kI}}{I}$$

63.

$$xy^2 + 3xy + 7 = 0; \ a = x, b = 3x, c = 7$$

$$y = \frac{-b \pm \sqrt{b^2 - 4ac}}{2a}$$

$$= \frac{-3x \pm \sqrt{(3x)^2 - 4(x)(7)}}{2x}$$

$$= \frac{-3x \pm \sqrt{9x^2 - 28x}}{2x}$$

65.

$$\sigma = \sqrt{\frac{\Sigma x^2}{N} - \mu^2}$$

$$\sigma^2 = \frac{\Sigma x^2}{N} - \mu^2$$

$$\mu^2 = \frac{\Sigma x^2}{N} - \sigma^2$$

67. $12x^2 - 5x - 2 = 0; a = 12, b = -5, c = -2$

$(4x + 1)(3x - 2) = 0$

$4x + 1 = 0$ **or** $3x - 2 = 0$ $\qquad -\dfrac{b}{a} = -\dfrac{-5}{12} = \dfrac{5}{12}$ $\qquad \dfrac{c}{a} = \dfrac{-2}{12} = -\dfrac{1}{6}$

$\qquad 4x = -1 \qquad\qquad 3x = 2 \qquad -\dfrac{1}{4} + \dfrac{2}{3} = -\dfrac{3}{12} + \dfrac{8}{12} = \dfrac{5}{12} \qquad \left(-\dfrac{1}{4}\right)\left(\dfrac{2}{3}\right) = -\dfrac{1}{6}$

$\qquad\quad x = -\tfrac{1}{4} \qquad\qquad x = \tfrac{2}{3}$

69. $2x^2 + 5x + 1 = 0; a = 2, b = 5, c = 1; -\dfrac{b}{a} = -\dfrac{5}{2}; \dfrac{c}{a} = \dfrac{1}{2}$

$$x = \frac{-b \pm \sqrt{b^2 - 4ac}}{2a} = \frac{-5 \pm \sqrt{5^2 - 4(2)(1)}}{2(2)} = \frac{-5 \pm \sqrt{17}}{4} = -\frac{5}{4} \pm \frac{\sqrt{17}}{4}$$

$$\frac{-5 + \sqrt{17}}{4} + \frac{-5 - \sqrt{17}}{4} = \frac{-10}{4} = -\frac{5}{2}$$

$$\left(\frac{-5 + \sqrt{17}}{4}\right)\left(\frac{-5 - \sqrt{17}}{4}\right) = \frac{25 + 5\sqrt{17} - 5\sqrt{17} - 17}{16} = \frac{8}{16} = \frac{1}{2}$$

71. $3x^2 - 2x + 4 = 0; a = 3, b = -2, c = 4; -\dfrac{b}{a} = -\dfrac{-2}{3} = \dfrac{2}{3}; \dfrac{c}{a} = \dfrac{4}{3}$

$$x = \frac{-b \pm \sqrt{b^2 - 4ac}}{2a} = \frac{-(-2) \pm \sqrt{(-2)^2 - 4(3)(4)}}{2(3)} = \frac{2 \pm \sqrt{-44}}{6} = \frac{1}{3} \pm \frac{i\sqrt{11}}{3}$$

$$\frac{1 + i\sqrt{11}}{3} + \frac{1 - i\sqrt{11}}{3} = \frac{2}{3}$$

$$\left(\frac{1 + i\sqrt{11}}{3}\right)\left(\frac{1 - i\sqrt{11}}{3}\right) = \frac{1 - i\sqrt{11} + i\sqrt{11} - 11i^2}{9} = \frac{1 + 11}{9} = \frac{12}{9} = \frac{4}{3}$$

73. $x^2 + 2x + 5 = 0; a = 1, b = 2, c = 5; -\dfrac{b}{a} = -\dfrac{2}{1} = -2; \dfrac{c}{a} = \dfrac{5}{1} = 5$

$$x = \frac{-b \pm \sqrt{b^2 - 4ac}}{2a} = \frac{-2 \pm \sqrt{2^2 - 4(1)(5)}}{2(1)} = \frac{-2 \pm \sqrt{-16}}{2} = \frac{-2 \pm 4i}{2} = -1 \pm 2i$$

$(-1 + 2i) + (-1 - 2i) = -2; \quad (-1 + 2i)(-1 - 2i) = 1 + 2i - 2i - 4i^2 = 1 + 4 = 5$

75. Answers may vary. $\qquad\qquad\qquad$ **77.** No

Exercise 10.4 (page 674)

1. $3x + 5 = 5x - 15$

$\qquad -2x = -20$

$\qquad\quad x = 10$

3. Let $t =$ the time of the second train.
Then $t + 3 =$ the time of the first train.

	Rate	Time	Dist.
First	30	$t+3$	$30(t+3)$
Second	55	t	$55t$

1st distance $=$ 2nd distance
$$30(t+3) = 55t$$
$$30t + 90 = 55t$$
$$-25t = -90$$
$$t = \tfrac{-90}{-25} = \tfrac{18}{5} = 3\tfrac{3}{5} \text{ hours}$$

5. $f(x) = ax^2 + bx + c; \ a \neq 0$

7. vertex

9. upward

11. to the right

13. upward

15. $f(x) = x^2$
vertex: $(0,0)$; opens U

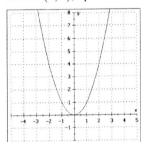

17. $f(x) = x^2 + 2$
vertex: $(0,2)$; opens U

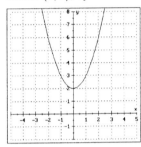

19. $f(x) = -(x-2)^2$
vertex: $(2,0)$; opens D

21. $f(x) = (x-3)^2 + 2$
vertex: $(3,2)$; opens U

23. $f(x) = x^2 + x - 6$
$f(x) = \left(x + \tfrac{1}{2}\right)^2 - \tfrac{25}{4}$
vertex: $\left(-\tfrac{1}{2}, -\tfrac{25}{4}\right)$; opens U

25. $f(x) = -2x^2 + 4x + 1$
$f(x) = -2(x-1)^2 + 3$
vertex: $(1,3)$; opens D

27. $y = (x-1)^2 + 2; V(1,2)$; axis: $x = 1$

29. $y = 2(x+3)^2 - 4;$
$V(-3,-4)$; axis: $x = -3$

31. $y = -3x^2 \Rightarrow y = -3(x-0)^2 + 0$
$V(0,0)$; axis: $x = 0$

33. $y = 2x^2 - 4x$
$y = 2(x^2 - 2x)$
$y = 2(x^2 - 2x + 1) - 2$
$y = 2(x-1)^2 - 2$
$V(1, -2)$; axis: $x = 1$

35. $y = -4x^2 + 16x + 5$
$y = -4(x^2 - 4x) + 5$
$y = -4(x^2 - 4x + 4) + 5 + 16$
$y = -4(x-2)^2 + 21$
$V(2, 21)$; axis: $x = 2$

37. $y - 7 = 6x^2 - 5x$
$y = 6\left(x^2 - \dfrac{5}{6}x\right) + 7$
$y = 6\left(x^2 - \dfrac{5}{6}x + \dfrac{25}{144}\right) + 7 - \dfrac{25}{24}$
$y = 6\left(x - \dfrac{5}{12}\right)^2 + \dfrac{143}{24}$
$V\left(\dfrac{5}{12}, \dfrac{143}{24}\right)$; axis: $x = \dfrac{5}{12}$

39. $y - 2 = (x - 5)^2$
$y = (x-5)^2 + 2 \Rightarrow V(5, 2)$

41. $y = 2x^2 - x + 1$

$V(0.25, 0.88)$

43. $y = 7 + x - x^2$

$V(0.5, 7.25)$

45. $y = x^2 + x - 6$

solution set: $\{2, -3\}$

47. $y = 0.5x^2 - 0.7x - 3$

solution set: $\{-1.85, 3.25\}$

49. Since the graph of the height equation is a parabola, the max. height occurs at the vertex.

$s = 48t - 16t^2$

$s = -16(t^2 - 3t)$

$s = -16\left(t^2 - 3t + \frac{9}{4}\right) + 36$

$s = -16\left(t - \frac{3}{2}\right)^2 + 36$

$V\left(\frac{3}{2}, 36\right) \Rightarrow$ max. height $= 36$ ft

To find the time it takes for the ball to return to earth, set $s = 0$ and solve for t.

$s = 48t - 16t^2$

$0 = 48t - 16t^2$

$0 = 16t(3 - t)$

$t = 0$ or $t = 3$

The ball returns to earth after 3 seconds.

(0 seconds is when the ball is originally thrown.)

51. Let $w =$ the width of the rectangle.
Then $100 - w =$ the length.

$A = w(100 - w)$

$A = -w^2 + 100w$

$A = -(w^2 - 100w + 2500) + 2500$

$A = -(w - 50)^2 + 2500$

dim: 50 ft by 50 ft; area $= 2500$ ft^2

53. Let $w =$ the width of the rectangle.
Then $150 - w =$ the length.

$A = w(150 - w)$

$A = -w^2 + 150w$

$A = -(w^2 - 150w + 5625) + 5625$

$A = -(w - 75)^2 + 5625$

dim: 75 ft by 75 ft; area $= 5625$ ft^2

55. Graph $y = R = -\frac{x^2}{1000} + 10x$ and find the x-coordinate of the vertex:

5000 stereos should be sold.

57. Graph $y = R = -\frac{x^2}{728} + 9x$ and find the x- and y- coordinates of the vertex:

Max. revenue $= \$14{,}742$; # radios $= 3276$

59. Let x = the number of $1 increases to the price. Then the sales will be $4000 - 100x$, and the revenue will be $(30 + x)(4000 - 100x)$. Find the vertex of the parabola $y = (30 + x)(4000 - 100x)$.

The price should increase $5, to a total of $35.

61. Graph $y = 50x(1 - x)$ and find the x-coordinate(s) when $y = 9.375$.

$p = 0.25$ or $p = 0.75$

63. **Answers may vary.**

65. Graph $y = x^2 + x + 1$ to find x-intercept(s):

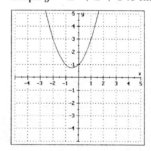

There are no x-intercepts, which means there is no solution to the equation.

Exercise 10.5 (page 684)

1. $y = kx$

3. $t = kxy$

5. $y = 3x - 4$
$m = 3$

7. greater

9. undefined

11. $x^2 - 5x + 4 < 0$
$(x - 4)(x - 1) < 0$

$x - 4 \quad --------0\ ++++$
$x - 1 \quad ---0++++++++++$

solution set: $(1, 4)$

13. $x^2 - 8x + 15 > 0$
$(x - 5)(x - 3) > 0$

$x - 5 \quad --------0+++++$
$x - 3 \quad --- 0+++++++++++$

solution set: $(-\infty, 3) \cup (5, \infty)$

15. $x^2 + x - 12 \leq 0$

$(x-3)(x+4) \leq 0$

$x - 3 \quad ---------0++++$
$x + 4 \quad --- \quad 0++++++++++$

$$\xleftarrow{\quad} \underset{-4}{[} \underline{\quad\quad} \underset{3}{]} \xrightarrow{\quad}$$

solution set: $[-4, 3]$

17. $x^2 + 2x \geq 15$

$x^2 + 2x - 15 \geq 0$

$(x-3)(x+5) \geq 0$

$x - 3 \quad --------0++++$
$x + 5 \quad --- \quad 0++++++++++$

$$\xleftarrow{\quad} \underset{-5}{]} \underline{\quad\quad} \underset{3}{[} \xrightarrow{\quad}$$

solution set: $(-\infty, -5] \cup [3, \infty)$

19. $x^2 + 8x < -16$

$x^2 + 8x + 16 < 0$

$(x+4)(x+4) < 0$

$x + 4 \quad ---- \quad 0+++++++$
$x + 4 \quad ---- \quad 0+++++++$

$$\xleftarrow{\quad} \underset{-4}{} \xrightarrow{\quad}$$

Since the product is never negative, there is no solution.

21. $x^2 \geq 9$

$x^2 - 9 \geq 0$

$(x-3)(x+3) \geq 0$

$x - 3 \quad ----------0++++$
$x + 3 \quad --- \quad 0++++++++++$

$$\xleftarrow{\quad} \underset{-3}{]} \underline{\quad\quad} \underset{3}{[} \xrightarrow{\quad}$$

solution set: $(-\infty, -3] \cup [3, \infty)$

23. $2x^2 - 50 < 0$

$2(x-5)(x+5) < 0$

$x - 5 \quad ---------0++++$
$x + 5 \quad --- \quad 0++++++ ++++$

$$\xleftarrow{\quad} \underset{-5}{(} \underline{\quad\quad} \underset{5}{)} \xrightarrow{\quad}$$

solution set: $(-5, 5)$

25. $\dfrac{1}{x} < 2$

$\dfrac{1}{x} - 2 < 0$

$\dfrac{1}{x} - \dfrac{2x}{x} < 0$

$\dfrac{1 - 2x}{x} < 0$

$1 - 2x \quad +++++++++ \ 0 ---$
$x \qquad\quad ---0++++++++++$

$$\xleftarrow{\quad} \underset{0}{)} \underline{\quad\quad} \underset{\frac{1}{2}}{(} \xrightarrow{\quad}$$

solution set: $(-\infty, 0) \cup \left(\dfrac{1}{2}, \infty\right)$

27. $\dfrac{4}{x} \geq 2$

$\dfrac{4}{x} - 2 \geq 0$

$\dfrac{4}{x} - \dfrac{2x}{x} \geq 0$

$\dfrac{4 - 2x}{x} \geq 0$

$4 - 2x \quad +++++++++0 ---$
$x \qquad\quad ---0+++++++++++$

$$\xleftarrow{\quad} \underset{0}{(} \underline{\quad\quad} \underset{2}{]} \xrightarrow{\quad}$$

solution set: $(0, 2]$

29.
$$-\frac{5}{x} < 3$$
$$-\frac{5}{x} - 3 < 0$$
$$\frac{-5}{x} - \frac{3x}{x} < 0$$
$$\frac{-5 - 3x}{x} < 0$$

$-5 - 3x$ $\;+{+}{+}\;0{-}{-}{-}{-}{-}{-}{-}{-}$
x $\quad\;\;{-}{-}{-}{-}{-}{-}{-}{-}0{+}{+}{+}{+}$

solution set: $\left(-\infty, -\dfrac{5}{3}\right) \cup (0, \infty)$

31.
$$\frac{x^2 - x - 12}{x - 1} < 0$$
$$\frac{(x - 4)(x + 3)}{x - 1} < 0$$

$x - 4$ $\;{-}{-}{-}{-}{-}{-}{-}{-}\;\;{-}{-}{-}\;0{+}{+}{+}{+}$
$x - 1$ $\;{-}{-}{-}{-}{-}{-}{-}{-}\;0{+}{+}{+}\;\;{+}{+}{+}{+}{+}$
$x + 3$ $\;{-}{-}{-}\;0{+}{+}{+}{+}{+}{+}\;\;{+}{+}{+}{+}\;\;{+}{+}{+}{+}{+}$

solution set: $(-\infty, -3) \cup (1, 4)$

33.
$$\frac{x^2 + x - 20}{x + 2} \geq 0$$
$$\frac{(x - 4)(x + 5)}{x + 2} \geq 0$$

$x - 4$ $\;{-}{-}{-}{-}{-}{-}{-}{-}\;\;{-}{-}{-}{-}\;0{+}{+}{+}{+}$
$x + 2$ $\;{-}{-}{-}{-}{-}{-}{-}{-}\;0{+}{+}{+}{+}\;{+}{+}{+}{+}{+}$
$x + 5$ $\;{-}{-}{-}\;0{+}{+}{+}{+}{+}{+}{+}{+}{+}{+}{+}{+}\;{+}{+}{+}{+}{+}$

solution set: $[-5, -2) \cup [4, \infty)$

35.
$$\frac{x^2 - 4x + 4}{x + 4} < 0$$
$$\frac{(x - 2)(x - 2)}{x + 4} < 0$$

$x - 2$ $\;{-}{-}{-}{-}{-}{-}{-}{-}\;0{+}{+}{+}{+}$
$x - 2$ $\;{-}{-}{-}{-}{-}{-}{-}{-}\;0{+}{+}{+}{+}$
$x + 4$ $\;{-}{-}{-}\;0{+}{+}{+}{+}{+}{+}\;{+}{+}{+}{+}{+}$

solution set: $(-\infty, -4)$

37.
$$\frac{6x^2 - 5x + 1}{2x + 1} > 0$$
$$\frac{(2x - 1)(3x - 1)}{2x + 1} > 0$$

$2x - 1$ $\;{-}{-}{-}{-}{-}{-}{-}{-}\;\;{-}{-}{-}{-}\;0{+}{+}{+}{+}$
$3x - 1$ $\;{-}{-}{-}{-}{-}{-}{-}{-}\;0{+}{+}{+}{+}{+}{+}{+}{+}{+}$
$2x + 1$ $\;{-}{-}{-}\;0{+}{+}{+}{+}{+}{+}{+}{+}{+}{+}{+}{+}{+}{+}{+}$

solution set: $\left(-\dfrac{1}{2}, \dfrac{1}{3}\right) \cup \left(\dfrac{1}{2}, \infty\right)$

39.
$$\frac{3}{x - 2} < \frac{4}{x}$$
$$\frac{3}{x - 2} - \frac{4}{x} < 0$$
$$\frac{3x}{x(x - 2)} - \frac{4(x - 2)}{x(x - 2)} < 0$$
$$\frac{-x + 8}{x(x - 2)} < 0$$

$-x + 8$ $\;{+}{+}{+}{+}{+}{+}{+}{+}{+}{+}{+}{+}{+}{+}0{-}{-}{-}$
$x - 2$ $\;{-}{-}{-}{-}{-}{-}{-}\;0{+}{+}{+}{+}{+}{+}{+}{+}$
x $\quad\;{-}{-}{-}\;0{+}{+}{+}{+}\;{+}{+}{+}{+}{+}{+}{+}{+}{+}$

solution set: $(0, 2) \cup (8, \infty)$

41.
$$\frac{-5}{x+2} \geq \frac{4}{2-x}$$

$$\frac{-5}{x+2} - \frac{4}{2-x} \geq 0$$

$$\frac{-5(2-x)}{(x+2)(2-x)} - \frac{4(x+2)}{(x+2)(2-x)} \geq 0$$

$$\frac{x-18}{(x+2)(2-x)} \geq 0$$

$x - 18$ $\;- - - - - - - - - \;- - - -\,0++++$
$2 - x$ $\;+++++++++++\,0- - - - - -$
$x + 2$ $\;- - -\;0+++++++++++++++++$

$$\longleftarrow \;)\!\!-\!\!-\!\!-\!\!-\!\!-\;(\!\!-\!\!-\!\!-\;]\longrightarrow$$
$$\qquad -2 \qquad\quad 2 \quad\; 18$$

solution set: $(-\infty, -2) \cup (2, 18]$

43.
$$\frac{7}{x-3} \geq \frac{2}{x+4}$$

$$\frac{7}{x-3} - \frac{2}{x+4} \geq 0$$

$$\frac{7(x+4)}{(x-3)(x+4)} - \frac{2(x-3)}{(x-3)(x+4)} \geq 0$$

$$\frac{5x+34}{(x-3)(x+4)} \geq 0$$

$x - 3$ $\;- - - - - - - - \;- - - -\,0++++$
$x + 4$ $\;- - - - - - - -\;0++++\;+++++$
$5x + 34$ $\;- - -\;0++++++\;+++++\;+++++$

$$\longleftarrow \;[\!\!-\!\!-\!\!-\!\!-\;)\;-\!\!-\!\!(\!\!-\!\!-\!\!\longrightarrow$$
$$\quad -\tfrac{34}{5} \qquad\; -4 \quad\; 3$$

solution set: $\left[-\dfrac{34}{5}, -4\right) \cup (3, \infty)$

45.
$$\frac{x}{x+4} \leq \frac{1}{x+1}$$

$$\frac{x}{x+4} - \frac{1}{x+1} \leq 0$$

$$\frac{x(x+1)}{(x+4)(x+1)} - \frac{1(x+4)}{(x+4)(x+1)} \leq 0$$

$$\frac{x^2-4}{(x+4)(x+1)} \leq 0$$

$$\frac{(x+2)(x-2)}{(x+4)(x+1)} \leq 0$$

$x - 2$ $\;- - - - - - - - - - - \;- - - -\,0+++$
$x + 1$ $\;- - - - - - - - - -\;0++++\;++++$
$x + 2$ $\;- - - - - -\,0++++\;+++++\;++++$
$x + 4$ $\;- -\;0++++++++++\;+++++\;++++$

$$\longleftarrow \;(\!\!-\!\!-\;]\;-\!\!-\;(\!\!-\!\!-\;]\longrightarrow$$
$$\quad -4 \quad\; -2 \quad -1 \quad 2$$

solution set: $(-4, -2] \cup (-1, 2]$

47.
$$\frac{x}{x+16} > \frac{1}{x+1}$$

$$\frac{x}{x+16} - \frac{1}{x+1} > 0$$

$$\frac{x(x+1)}{(x+16)(x+1)} - \frac{1(x+16)}{(x+16)(x+1)} > 0$$

$$\frac{x^2-16}{(x+16)(x+1)} > 0$$

$$\frac{(x+4)(x-4)}{(x+16)(x+1)} > 0$$

$x - 4$ $\;- - - - - - - \;- - - - \;- - -\,0+++$
$x + 1$ $\;- - - - - - - \;- - - -\;0++++\;++++$
$x + 4$ $\;- - - - - - -\;0++++\;+++++\;++++$
$x + 16$ $\;- - -\;0++++\;+++++\;+++++\;++++$

$$\longleftarrow \;)\;-\!\!-\;(\!\!-\!\!-\;)\;-\!\!(\!\!-\!\!\longrightarrow$$
$$\quad -16 \quad\; -4 \quad -1 \quad 4$$

solution set: $(-\infty, -16) \cup (-4, -1) \cup (4, \infty)$

49.
$$(x+2)^2 > 0$$
$$(x+2)(x+2) > 0$$

$x + 2$ $\;- - - - -\,0+++++++$
$x + 2$ $\;- - - - -\,0+++++++$

$$\longleftarrow \;)\;(\!\!-\!\!-\!\!-\!\!\longrightarrow$$
$$\qquad -2$$

solution set: $(-\infty, -2) \cup (-2, \infty)$

51. $x^2 - 2x - 3 < 0$

Graph $y = x^2 - 2x - 3$ and find the x-coordinates of points below the x-axis.

$(-1, 3)$

53. $\frac{x+3}{x-2} > 0$

Graph $y = \frac{x+3}{x-2}$ and find the x-coordinates of points above the x-axis.

$(-\infty, -3) \cup (2, \infty)$

55. $y < x^2 + 1$

57. $y \leq x^2 + 5x + 6$

59. $y \geq (x-1)^2$

61. $-x^2 - y + 6 > -x$

$-x^2 + x + 6 > y$

63. $y < |x + 4|$

65. $y \leq -|x| + 2$

67. **Answers may vary.**

69. It will be positive if 4, 2 or 0 factors are negative.

Exercise 10.6 (page 692)

1. $\dfrac{3x^2 + x - 14}{4 - x^2} = \dfrac{(3x+7)(x-2)}{(2+x)(2-x)} = -\dfrac{3x+7}{x+2}$

3. $\dfrac{8 + 2x - x^2}{12 + x - 3x^2} \div \dfrac{3x^2 + 5x - 2}{3x - 1} = \dfrac{x^2 - 2x - 8}{3x^2 - x - 12} \cdot \dfrac{3x - 1}{3x^2 + 5x - 2}$

   $\qquad = \dfrac{(x - 4)(x + 2)}{3x^2 - x - 12} \cdot \dfrac{3x - 1}{(3x - 1)(x + 2)} = \dfrac{x - 4}{3x^2 - x - 12}$

5. $f(x) + g(x)$ 7. $f(x)g(x)$ 9. domain   11. $f(x)$

13. $f + g = f(x) + g(x) = 3x + 4x = 7x$
    domain $= (-\infty, \infty)$

15. $f \cdot g = f(x) \cdot g(x) = 3x \cdot 4x = 12x^2$
    domain $= (-\infty, \infty)$

17. $g - f = g(x) - f(x) = 4x - 3x = x$
    domain $= (-\infty, \infty)$

19. $g/f = \dfrac{g(x)}{f(x)} = \dfrac{4x}{3x} = \dfrac{4}{3}$ (for $x \neq 0$)
    domain $= (-\infty, 0) \cup (0, \infty)$

21. $f + g = f(x) + g(x) = 2x + 1 + x - 3 = 3x - 2$; domain $= (-\infty, \infty)$

23. $f \cdot g = f(x) \cdot g(x) = (2x + 1)(x - 3) = 2x^2 - 5x - 3$; domain $= (-\infty, \infty)$

25. $g - f = g(x) - f(x) = (x - 3) - (2x + 1) = x - 3 - 2x - 1 = -x - 4$; domain $= (-\infty, \infty)$

27. $g/f = \dfrac{g(x)}{f(x)} = \dfrac{x - 3}{2x + 1}$; domain $= \left(-\infty, -\dfrac{1}{2}\right) \cup \left(-\dfrac{1}{2}, \infty\right)$

29. $f - g = f(x) - g(x) = (3x - 2) - (2x^2 + 1) = 3x - 2 - 2x^2 - 1 = -2x^2 + 3x - 3$
    domain $= (-\infty, \infty)$

31. $f/g = \dfrac{f(x)}{g(x)} = \dfrac{3x - 2}{2x^2 + 1}$; domain $= (-\infty, \infty)$

33. $f - g = f(x) - g(x) = (x^2 - 1) - (x^2 - 4) = x^2 - 1 - x^2 + 4 = 3$; domain $= (-\infty, \infty)$

35. $g/f = \dfrac{g(x)}{f(x)} = \dfrac{x^2 - 4}{x^2 - 1} = \dfrac{(x + 2)(x - 2)}{(x + 1)(x - 1)}$; domain $= (-\infty, -1) \cup (-1, 1) \cup (1, \infty)$

37. $(f \circ g)(2) = f(g(2)) = f(2^2 - 1) = f(3) = 2(3) + 1 = 7$

39. $(g \circ f)(-3) = g(f(-3)) = g(2(-3) + 1) = g(-5) = (-5)^2 - 1 = 24$

41. $(f \circ g)(0) = f(g(0)) = f(0^2 - 1) = f(-1) = 2(-1) + 1 = -1$

43. $(f \circ g)\left(\dfrac{1}{2}\right) = f\left(g\left(\dfrac{1}{2}\right)\right) = f\left(\left(\dfrac{1}{2}\right)^2 - 1\right) = f\left(-\dfrac{3}{4}\right) = 2\left(-\dfrac{3}{4}\right) + 1 = -\dfrac{3}{2} + 1 = -\dfrac{1}{2}$

45. $(f \circ g)(x) = f(g(x)) = f(x^2 - 1) = 2(x^2 - 1) + 1 = 2x^2 - 2 + 1 = 2x^2 - 1$

47. $(g \circ f)(2x) = g(f(2x)) = g(2(2x) + 1) = g(4x + 1) = (4x + 1)^2 - 1 = 16x^2 + 8x + 1 - 1$
$$= 16x^2 + 8x$$

49. $(f \circ g)(4) = f(g(4)) = f(4^2 + 4) = f(20) = 3(20) - 2 = 58$

51. $(g \circ f)(-3) = g(f(-3)) = g(3(-3) - 2) = g(-11) = (-11)^2 + (-11) = 110$

53. $(g \circ f)(0) = g(f(0)) = g(3(0) - 2) = g(-2) = (-2)^2 + (-2) = 2$

55. $(g \circ f)(x) = g(f(x)) = g(3x - 2) = (3x - 2)^2 + 3x - 2 = 9x^2 - 12x + 4 + 3x - 2$
$$= 9x^2 - 9x + 2$$

57. $\dfrac{f(x + h) - f(x)}{h} = \dfrac{2(x + h) + 3 - (2x + 3)}{h} = \dfrac{2x + 2h + 3 - 2x - 3}{h} = \dfrac{2h}{h} = 2$

59. $\dfrac{f(x + h) - f(x)}{h} = \dfrac{(x + h)^2 - x^2}{h} = \dfrac{x^2 + 2xh + h^2 - x^2}{h} = \dfrac{2xh + h^2}{h} = 2x + h$

61. $\dfrac{f(x + h) - f(x)}{h} = \dfrac{2(x + h)^2 - 1 - (2x^2 - 1)}{h} = \dfrac{2x^2 + 4xh + 2h^2 - 1 - 2x^2 + 1}{h}$
$$= \dfrac{4xh + 2h^2}{h} = 4x + 2h$$

63. $\dfrac{f(x + h) - f(x)}{h} = \dfrac{(x + h)^2 + (x + h) - (x^2 + x)}{h} = \dfrac{x^2 + 2xh + h^2 + x + h - x^2 - x}{h}$
$$= \dfrac{2xh + h^2 + h}{h} = 2x + h + 1$$

65. $\dfrac{f(x + h) - f(x)}{h} = \dfrac{(x + h)^2 + 3(x + h) - 4 - (x^2 + 3x - 4)}{h}$
$$= \dfrac{x^2 + 2xh + h^2 + 3x + 3h - 4 - x^2 - 3x + 4}{h}$$
$$= \dfrac{2xh + h^2 + 3h}{h} = 2x + h + 3$$

67. $\dfrac{f(x + h) - f(x)}{h} = \dfrac{2(x + h)^2 + 3(x + h) - 7 - (2x^2 + 3x - 7)}{h}$
$$= \dfrac{2x^2 + 4xh + 2h^2 + 3x + 3h - 7 - 2x^2 - 3x + 7}{h}$$
$$= \dfrac{4xh + 2h^2 + 3h}{h} = 4x + 2h + 3$$

69. $\dfrac{f(x) - f(a)}{x - a} = \dfrac{(2x + 3) - (2a + 3)}{x - a} = \dfrac{2x + 3 - 2a - 3}{x - a} = \dfrac{2x - 2a}{x - a} = \dfrac{2(x - a)}{x - a} = 2$

71. $\dfrac{f(x) - f(a)}{x - a} = \dfrac{x^2 - a^2}{x - a} = \dfrac{(x + a)(x - a)}{x - a} = x + a$

73. $\begin{aligned} \dfrac{f(x) - f(a)}{x - a} &= \dfrac{(2x^2 - 1) - (2a^2 - 1)}{x - a} = \dfrac{2x^2 - 1 - 2a^2 + 1}{x - a} = \dfrac{2x^2 - 2a^2}{x - a} \\ &= \dfrac{2(x + a)(x - a)}{x - a} \\ &= 2(x + a) = 2x + 2a \end{aligned}$

75. $\begin{aligned} \dfrac{f(x) - f(a)}{x - a} &= \dfrac{(x^2 + x) - (a^2 + a)}{x - a} = \dfrac{x^2 + x - a^2 - a}{x - a} = \dfrac{x^2 - a^2 + x - a}{x - a} \\ &= \dfrac{(x + a)(x - a) + 1(x - a)}{x - a} \\ &= \dfrac{(x - a)(x + a + 1)}{x - a} = x + a + 1 \end{aligned}$

77. $\begin{aligned} \dfrac{f(x) - f(a)}{x - a} &= \dfrac{(x^2 + 3x - 4) - (a^2 + 3a - 4)}{x - a} = \dfrac{x^2 + 3x - 4 - a^2 - 3a + 4}{x - a} \\ &= \dfrac{x^2 - a^2 + 3x - 3a}{x - a} \\ &= \dfrac{(x + a)(x - a) + 3(x - a)}{x - a} \\ &= \dfrac{(x - a)(x + a + 3)}{x - a} = x + a + 3 \end{aligned}$

79. $\begin{aligned} \dfrac{f(x) - f(a)}{x - a} &= \dfrac{(2x^2 + 3x - 7) - (2a^2 + 3a - 7)}{x - a} = \dfrac{2x^2 + 3x - 7 - 2a^2 - 3a + 7}{x - a} \\ &= \dfrac{2x^2 - 2a^2 + 3x - 3a}{x - a} \\ &= \dfrac{2(x + a)(x - a) + 3(x - a)}{x - a} \\ &= \dfrac{(x - a)(2(x + a) + 3)}{x - a} \\ &= 2(x + a) + 3 = 2x + 2a + 3 \end{aligned}$

81. $(f \circ g)(x) = f(g(x)) = f(2x - 5) = (2x - 5) + 1 = 2x - 4$
$(g \circ f)(x) = g(f(x)) = g(x + 1) = 2(x + 1) - 5 = 2x + 2 - 5 = 2x - 3$

83. $f(a) = a^2 + 2a - 3; \ f(h) = h^2 + 2h - 3 \Rightarrow f(a) + f(h) = a^2 + h^2 + 2a + 2h - 6$
$\begin{aligned} f(a + h) &= (a + h)^2 + 2(a + h) - 3 = a^2 + 2ah + h^2 + 2a + 2h - 3 \\ &= a^2 + h^2 + 2ah + 2a + 2h - 3 \end{aligned}$

85. $\dfrac{f(x+h)-f(x)}{h} = \dfrac{(x+h)^3 - 1 - (x^3-1)}{h} = \dfrac{x^3 + 3x^2h + 3xh^2 + h^3 - 1 - x^3 + 1}{h}$

$$= \dfrac{3x^2h + 3xh^2 + h^3}{h}$$

$$= \dfrac{h(3x^2 + 3xh + h^2)}{h} = 3x^2 + 3xh + h^2$$

87. $F(t) = 2700 - 200t;\ C(F) = \frac{5}{9}(F - 32)$

$C(F(t)) = C(2700 - 200t) = \frac{5}{9}(2700 - 200t - 32) = \frac{5}{9}(2668 - 200t)$

89. **Answers may vary.** **91.** It is associative. Examples will vary.

Exercise 10.7 (page 701)

1. $3 - \sqrt{-64} = 3 - \sqrt{64i^2} = 3 - 8i$

3. $(3 + 4i)(2 - 3i) = 6 - 9i + 8i - 12i^2 = 6 - i - 12(-1) = 6 - i + 12 = 18 - i$

5. $|6 - 8i| = \sqrt{6^2 + (-8)^2} = \sqrt{36 + 64} = \sqrt{100} = 10$

7. one-to-one **9.** 2 **11.** x

13. Each input has a different output. one-to-one **15.** The inputs $x = 2$ and $x = -2$ have the same output. not one-to-one

17. one-to-one **19.** one-to-one **21.** not one-to-one **23.** one-to-one

25. inverse $= \{(2,3),(1,2),(0,1)\}$. Since each x-coordinate is paired with only one y-coordinate, the inverse relation **is a function**.

27. inverse $= \{(2,1),(3,2),(3,1),(5,1)\}$. Since $x = 3$ is paired with more than one y-coordinate, the inverse relation **is not a function**.

29. inverse $= \{(1,1),(4,2),(9,3),(16,4)\}$. Since each x-coordinate is paired with only one y-coordinate, the inverse relation **is a function**.

31.

$f(x) = 3x + 1$

$y = 3x + 1$

$x = 3y + 1$

$x - 1 = 3y$

$\dfrac{x-1}{3} = y$

$\dfrac{1}{3}x - \dfrac{1}{3} = y$

$f^{-1}(x) = \frac{1}{3}x - \frac{1}{3}$

$f \circ f^{-1}$

$f[f^{-1}(x)] = f\left(\frac{1}{3}x - \frac{1}{3}\right)$

$= 3 \cdot \left(\frac{1}{3}x - \frac{1}{3}\right) + 1$

$= x - 1 + 1$

$= x$

$f^{-1} \circ f$

$f^{-1}[f(x)] = f^{-1}(3x + 1)$

$= \frac{1}{3}(3x + 1) - \frac{1}{3}$

$= x + \frac{1}{3} - \frac{1}{3}$

$= x$

33.

$$x + 4 = 5y$$
$$y = f(x) = \frac{x+4}{5}$$
$$x = \frac{y+4}{5}$$
$$5x = y + 4$$
$$5x - 4 = y$$
$$f^{-1}(x) = 5x - 4$$

$$\underline{f \circ f^{-1}}$$
$$f\left[f^{-1}(x)\right] = f(5x - 4)$$
$$= \frac{(5x - 4) + 4}{5}$$
$$= \frac{5x}{5}$$
$$= x$$

$$\underline{f^{-1} \circ f}$$
$$f^{-1}[f(x)] = f^{-1}\left(\frac{x+4}{5}\right)$$
$$= 5 \cdot \left(\frac{x+4}{5}\right) - 4$$
$$= x + 4 - 4$$
$$= x$$

35.

$$f(x) = \frac{x-4}{5}$$
$$y = \frac{x-4}{5}$$
$$x = \frac{y-4}{5}$$
$$5x = y - 4$$
$$5x + 4 = y$$
$$f^{-1}(x) = 5x + 4$$

$$\underline{f \circ f^{-1}}$$
$$f\left[f^{-1}(x)\right] = f(5x + 4)$$
$$= \frac{(5x + 4) - 4}{5}$$
$$= \frac{5x}{5}$$
$$= x$$

$$\underline{f^{-1} \circ f}$$
$$f^{-1}[f(x)] = f^{-1}\left(\frac{x-4}{5}\right)$$
$$= 5 \cdot \left(\frac{x-4}{5}\right) + 4$$
$$= x - 4 + 4$$
$$= x$$

37.

$$4x - 5y = 20$$
$$5y = 4x - 20$$
$$y = f(x) = \frac{4}{5}x - 4$$
$$x = \frac{4}{5}y - 4$$
$$x + 4 = \frac{4}{5}y$$
$$\frac{5}{4}(x + 4) = y$$
$$f^{-1}(x) = \frac{5}{4}x + 5$$

$$\underline{f \circ f^{-1}}$$
$$f\left[f^{-1}(x)\right] = f\left(\tfrac{5}{4}x + 5\right)$$
$$= \tfrac{4}{5}\left(\tfrac{5}{4}x + 5\right) - 4$$
$$= x + 4 - 4$$
$$= x$$

$$\underline{f^{-1} \circ f}$$
$$f^{-1}[f(x)] = f^{-1}\left(\tfrac{4}{5}x - 4\right)$$
$$= \tfrac{5}{4}\left(\tfrac{4}{5}x - 4\right) + 5$$
$$= x - 5 + 5$$
$$= x$$

39.

$$y = 4x + 3$$
$$x = 4y + 3$$
$$x - 3 = 4y$$
$$\frac{x-3}{4} = y$$

41.

$$x = \frac{y-2}{3}$$
$$y = \frac{x-2}{3}$$

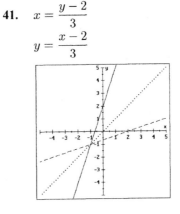

43. $3x - y = 5$
$3y - x = 5$
$\quad 3y = x + 5$
$\quad\quad y = \dfrac{x + 5}{3}$

45. $3(x + y) = 2x + 4$
$3(y + x) = 2y + 4$
$3y + 3x = 2y + 4$
$\quad\quad y = 4 - 3x$

47. $\quad\quad y = x^2 + 4$
$\quad\quad x = y^2 + 4$
$\quad x - 4 = y^2$
$\pm\sqrt{x - 4} = y$
The relation **is not** a function.

49. $\quad y = x^3$
$\quad x = y^3$
$\sqrt[3]{x} = y$
The relation **is** a function.

51. $y = |x|$
$x = |y|$
The relation **is not** a function.

53. $\quad\quad\quad y = 2x^3 - 3$
$\quad\quad\quad x = 2y^3 - 3$
$\quad\quad x + 3 = 2y^3$
$\quad\quad \dfrac{x + 3}{2} = y^3$
$\sqrt[3]{\dfrac{x + 3}{2}} = \sqrt[3]{y^3}$, or $y = f^{-1}(x) = \sqrt[3]{\dfrac{x + 3}{2}}$

55. $y = x^2 + 1$
inverse: $x = y^2 + 1$

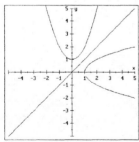

57. $y = \sqrt{x}$
inverse: $x = \sqrt{y}$

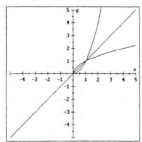

59. **Answers may vary.**

61.
$$y = \frac{x+1}{x-1}$$
$$x = \frac{y+1}{y-1}$$
$$x(y-1) = y+1$$
$$xy - x = y+1$$
$$xy - y = x+1$$
$$y(x-1) = x+1$$
$$y = \frac{x+1}{x-1}$$

Chapter 10 Summary (page 705)

1.
$$12x^2 + x - 6 = 0$$
$$(4x+3)(3x-2) = 0$$
$$4x + 3 = 0 \quad \textbf{or} \quad 3x - 2 = 0$$
$$x = -\tfrac{3}{4} \qquad\qquad x = \tfrac{2}{3}$$

2.
$$6x^2 + 17x + 5 = 0$$
$$(2x+5)(3x+1) = 0$$
$$2x + 5 = 0 \quad \textbf{or} \quad 3x + 1 = 0$$
$$x = -\tfrac{5}{2} \qquad\qquad x = -\tfrac{1}{3}$$

3.
$$15x^2 + 2x - 8 = 0$$
$$(3x-2)(5x+4) = 0$$
$$3x - 2 = 0 \quad \textbf{or} \quad 5x + 4 = 0$$
$$x = \tfrac{2}{3} \qquad\qquad x = -\tfrac{4}{5}$$

4.
$$(x+2)^2 = 36$$
$$x + 2 = \pm\sqrt{36}$$
$$x + 2 = \pm 6$$
$$x = -2 \pm 6$$
$$x = 4 \quad \text{or} \quad x = -8$$

5.
$$x^2 + 6x + 8 = 0$$
$$x^2 + 6x = -8$$
$$x^2 + 6x + 9 = -8 + 9$$
$$(x+3)^2 = 1$$
$$x + 3 = \pm 1$$
$$x = -3 \pm 1$$
$$x = -2 \quad \text{or} \quad x = -4$$

6.
$$2x^2 - 9x + 7 = 0$$
$$x^2 - \frac{9}{2}x + \frac{7}{2} = 0$$
$$x^2 - \frac{9}{2}x = -\frac{7}{2}$$
$$x^2 - \frac{9}{2}x + \frac{81}{16} = -\frac{56}{16} + \frac{81}{16}$$
$$\left(x - \frac{9}{4}\right)^2 = \frac{25}{16}$$
$$x - \frac{9}{4} = \pm\frac{5}{4}$$
$$x = \frac{9}{4} \pm \frac{5}{4}$$
$$x = \frac{7}{2} \quad \text{or} \quad x = 1$$

7.
$$2x^2 - x - 5 = 0$$
$$x^2 - \frac{1}{2}x - \frac{5}{2} = 0$$
$$x^2 - \frac{1}{2}x = \frac{5}{2}$$
$$x^2 - \frac{1}{2}x + \frac{1}{16} = \frac{5}{2} + \frac{1}{16}$$
$$\left(x - \frac{1}{4}\right)^2 = \frac{41}{16}$$
$$x - \frac{1}{4} = \pm\frac{\sqrt{41}}{4}$$
$$x = \frac{1}{4} \pm \frac{\sqrt{41}}{4}$$

8.
$$x^2 - 8x - 9 = 0$$
$$a = 1, b = -8, c = -9$$
$$x = \frac{-b \pm \sqrt{b^2 - 4ac}}{2a}$$
$$= \frac{-(-8) \pm \sqrt{(-8)^2 - 4(1)(-9)}}{2(1)}$$
$$= \frac{8 \pm \sqrt{64 + 36}}{2}$$
$$= \frac{8 \pm \sqrt{100}}{2} = \frac{8 \pm 10}{2}$$
$$x = \frac{18}{2} = 9 \text{ or } x = \frac{-2}{2} = -1$$

9.
$$x^2 - 10x = 0$$
$$a = 1, b = -10, c = 0$$
$$x = \frac{-b \pm \sqrt{b^2 - 4ac}}{2a}$$
$$= \frac{-(-10) \pm \sqrt{(-10)^2 - 4(1)(0)}}{2(1)}$$
$$= \frac{10 \pm \sqrt{100 + 0}}{2}$$
$$= \frac{10 \pm \sqrt{100}}{2} = \frac{10 \pm 10}{2}$$
$$x = \frac{20}{2} = 10 \text{ or } x = \frac{0}{2} = 0$$

10.
$$2x^2 + 13x - 7 = 0$$
$$a = 2, b = 13, c = -7$$
$$x = \frac{-b \pm \sqrt{b^2 - 4ac}}{2a}$$
$$= \frac{-(13) \pm \sqrt{13^2 - 4(2)(-7)}}{2(2)}$$
$$= \frac{-13 \pm \sqrt{169 + 56}}{4}$$
$$= \frac{-13 \pm \sqrt{225}}{4} = \frac{-13 \pm 15}{4}$$
$$x = \frac{2}{4} = \frac{1}{2} \text{ or } x = \frac{-28}{4} = -7$$

11.
$$3x^2 + 20x - 7 = 0$$
$$a = 3, b = 20, c = -7$$
$$x = \frac{-b \pm \sqrt{b^2 - 4ac}}{2a}$$
$$= \frac{-20 \pm \sqrt{(20)^2 - 4(3)(-7)}}{2(3)}$$
$$= \frac{-20 \pm \sqrt{400 + 84}}{6}$$
$$= \frac{-20 \pm \sqrt{484}}{6} = \frac{-20 \pm 22}{6}$$
$$x = \frac{2}{6} = \frac{1}{3} \text{ or } x = \frac{-42}{6} = -7$$

12.
$$2x^2 - x - 2 = 0$$
$$a = 2, b = -1, c = -2$$
$$x = \frac{-b \pm \sqrt{b^2 - 4ac}}{2a}$$
$$= \frac{-(-1) \pm \sqrt{(-1)^2 - 4(2)(-2)}}{2(2)}$$
$$= \frac{1 \pm \sqrt{1 + 16}}{4}$$
$$= \frac{1 \pm \sqrt{17}}{4} = \frac{1}{4} \pm \frac{\sqrt{17}}{4}$$

13. $x^2 + x + 2 = 0$

$a = 1, b = 1, c = 2$

$x = \dfrac{-b \pm \sqrt{b^2 - 4ac}}{2a}$

$= \dfrac{-1 \pm \sqrt{1^2 - 4(1)(2)}}{2(1)}$

$= \dfrac{-1 \pm \sqrt{1 - 8}}{2}$

$= \dfrac{-1 \pm \sqrt{-7}}{2} = -\dfrac{1}{2} \pm \dfrac{\sqrt{7}}{2}i$

14. Let w represent the original width.
Then $w + 2$ represents the original length.
The new dimensions are then $2w$ and
$2(w + 2) = 2w + 4$.
Old Area $+ 72 =$ New Area
$w(w + 2) + 72 = 2w(2w + 4)$
$w^2 + 2w + 72 = 4w^2 + 8w$
$0 = 3w^2 + 6w - 72$
$0 = 3(w + 6)(w - 4)$
$w = -6$ or $w = 4$
Since the width is positive, the
dimensions are 4 cm by 6 cm.

15. Let w represent the original width.
Then $w + 1$ represents the original length.
The new dimensions are then $2w$ and
$3(w + 1) = 3w + 3$.
Old Area $+ 30 =$ New Area
$w(w + 1) + 30 = 2w(3w + 3)$
$w^2 + w + 30 = 6w^2 + 6w$
$0 = 5w^2 + 5w - 30$
$0 = 5(w + 3)(w - 2)$
$w = -3$ or $w = 2$
Since the width is positive, the
dimensions are 2 ft by 3 ft.

16. When the rocket hits the ground, $h = 0$:
$h = 112t - 16t^2$
$0 = 112t - 16t^2$
$0 = 16t(7 - t)$
$t = 0$ or $t = 7$
It hits the ground after 7 seconds.

17. The maximum height occurs at the vertex:
$h = 112t - 16t^2$
$h = -16t^2 + 112t$
$h = -16(t^2 - 7t)$
$h = -16\left(t^2 - 7t + \dfrac{49}{4}\right) + 196$

$h = -16\left(t^2 - 7t + \dfrac{49}{4}\right) + 196$

$h = -16\left(t - \dfrac{7}{2}\right)^2 + 196$

Vertex: $\left(\dfrac{7}{2}, 196\right) \Rightarrow$ max. height $= 196$ ft

18. $3x^2 + 4x - 3 = 0$
$a = 3, b = 4, c = -3$
$b^2 - 4ac = 4^2 - 4(3)(-3)$
$= 16 + 36 = 52$
irrational unequal solutions

19. $4x^2 - 5x + 7 = 0$
$a = 4, b = -5, c = 7$
$b^2 - 4ac = (-5)^2 - 4(4)(7)$
$= 25 - 112 = -87$
complex conjugate solutions

20. $(k - 8)x^2 + (k + 16)x = -49$

$(k - 8)x^2 + (k + 16)x + 49 = 0$

$a = k - 8, b = k + 16, c = 49$

Set the discriminant equal to 0:

$$b^2 - 4ac = 0$$

$$(k + 16)^2 - 4(k - 8)(49) = 0$$

$$k^2 + 32k + 256 - 196k + 1568 = 0$$

$$k^2 - 164k + 1824 = 0$$

$$(k - 12)(k - 152) = 0$$

$k - 12 = 0 \quad$ **or** $\quad k - 152 = 0$

$\qquad k = 12 \qquad\qquad k = 152$

21. $3x^2 + 4x = k + 1$

$3x^2 + 4x - k - 1 = 0$

$a = 3, b = 4, c = -k - 1$

Set the discriminant ≥ 0:

$$b^2 - 4ac \geq 0$$

$$4^2 - 4(3)(-k - 1) \geq 0$$

$$16 + 12k + 12 \geq 0$$

$$12k \geq -28$$

$$k \geq -\frac{28}{12}$$

$$k \geq -\frac{7}{3}$$

22. $x - 13x^{1/2} + 12 = 0$

$\left(x^{1/2} - 12\right)\left(x^{1/2} - 1\right) = 0$

$x^{1/2} - 12 = 0 \quad$ **or** $\quad x^{1/2} - 1 = 0$

$\quad x^{1/2} = 12 \qquad\qquad x^{1/2} = 1$

$\left(x^{1/2}\right)^2 = (12)^2 \qquad \left(x^{1/2}\right)^2 = 1^2$

$\qquad x = 144 \qquad\qquad x = 1$

$\qquad$ Solution. $\qquad\qquad$ Solution.

23. $a^{2/3} + a^{1/3} - 6 = 0$

$\left(a^{1/3} - 2\right)\left(a^{1/3} + 3\right) = 0$

$a^{1/3} - 2 = 0 \quad$ **or** $\quad a^{1/3} + 3 = 0$

$\quad a^{1/3} = 2 \qquad\qquad a^{1/3} = -3$

$\left(a^{1/3}\right)^3 = (2)^3 \qquad \left(a^{1/3}\right)^3 = (-3)^3$

$\qquad a = 8 \qquad\qquad a = -27$

$\qquad$ Solution. $\qquad\qquad$ Solution.

24.

$$\frac{1}{x + 1} - \frac{1}{x} = -\frac{1}{x + 1}$$

$$\left(\frac{1}{x + 1} - \frac{1}{x}\right)(x)(x + 1) = -\frac{1}{x + 1}(x)(x + 1)$$

$$1(x) - 1(x + 1) = -x$$

$$x - x - 1 = -x$$

$$-1 = -x$$

$$1 = x$$

25.

$$\frac{6}{x + 2} + \frac{6}{x + 1} = 5$$

$$\left(\frac{6}{x + 2} + \frac{6}{x + 1}\right)(x + 2)(x + 1) = 5(x + 2)(x + 1)$$

$$6(x + 1) + 6(x + 2) = 5(x^2 + 3x + 2)$$

$$6x + 6 + 6x + 12 = 5x^2 + 15x + 10$$

$$0 = 5x^2 + 3x - 8$$

$$0 = (5x + 8)(x - 1) \qquad 5x + 8 = 0 \quad \textbf{or} \quad x - 1 = 0$$

$$x = -\tfrac{8}{5} \qquad\qquad x = 1$$

26. $3x^2 - 14x + 3 = 0$

$$\text{sum} = -\frac{b}{a} = -\frac{-14}{3} = \frac{14}{3}$$

27. $3x^2 - 14x + 3 = 0$

$$\text{product} = \frac{c}{a} = \frac{3}{3} = 1$$

28. $y = 2x^2 - 3$
$y = 2(x - 0)^2 - 3$

vertex: $(0, -3)$

29. $y = -2x^2 - 1$
$y = -2(x - 0)^2 - 1$

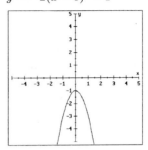

vertex: $(0, -1)$

30. $y = -4(x - 2)^2 + 1$

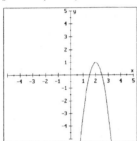

vertex: $(2, 1)$

31. $y = 5x^2 + 10x - 1$
$\quad = 5(x^2 + 2x) - 1$
$\quad = 5(x^2 + 2x + 1) - 1 - 5$
$\quad = 5(x + 1)^2 - 6$

vertex: $(-1, -6)$

32. $\quad x^2 + 2x - 35 > 0$
$\quad (x + 7)(x - 5) > 0$

$x - 5 \quad --------0\ ++++$
$x + 7 \quad ---\ 0++++++\ ++++$

$\xleftarrow{\hspace{1cm}}\ )\ \underline{\hspace{1.2cm}}\ (\xrightarrow{\hspace{1cm}}$
$\qquad\quad -7 \qquad\quad 5$

solution set: $(-\infty, -7) \cup (5, \infty)$

33. $\quad x^2 + 7x - 18 < 0$
$\quad (x - 2)(x + 9) < 0$

$x - 2 \quad ---------\ 0++++$
$x + 9 \quad ---\quad 0+++++++++++$

$\xleftarrow{\hspace{1cm}}\ (\underline{\hspace{1.2cm}})\xrightarrow{\hspace{0.8cm}}$
$\qquad\quad -9 \qquad\ 2$

solution set: $(-9, 2)$

34.
$$\frac{3}{x} \le 5$$
$$\frac{3}{x} - 5 \le 0$$
$$\frac{3}{x} - \frac{5x}{x} \le 0$$
$$\frac{3 - 5x}{x} \le 0$$

$$3 - 5x \quad ++++++++\ 0\ ---$$
$$x \qquad\quad ---0++++++++++$$

solution set: $(-\infty, 0) \cup \left[\frac{3}{5}, \infty\right)$

35.
$$\frac{2x^2 - x - 28}{x - 1} > 0$$
$$\frac{(2x + 7)(x - 4)}{x - 1} > 0$$

$$x - 4 \quad --------\ ----\ 0++++$$
$$x - 1 \quad --------\ 0++++++++++$$
$$2x + 7 \quad ---\ 0+++++++++++++++$$

solution set: $\left(-\frac{7}{2}, 1\right) \cup (4, \infty)$

36. $x^2 + 2x - 35 > 0$
Graph $y = x^2 + 2x - 35$
and find the x-coordinates
of points above the x-axis.

$(-\infty, -7) \cup (5, \infty)$

37. $x^2 + 7x - 18 < 0$
Graph $y = x^2 + 7x - 18$
and find the x-coordinates
of points below the x-axis.

$(-9, 2)$

38. $\frac{3}{x} \le 5 \Rightarrow \frac{3}{x} - 5 \le 0$
Graph $y = (3/x) - 5$
and find the x-coordinates
of points below or on the x-axis.

$(-\infty, 0) \cup \left[\frac{3}{5}, \infty\right)$

39. $\frac{2x^2 - x - 28}{x - 1} > 0$
Graph $y = (2x^2 - x - 28)/(x - 1)$
and find the x-coordinates
of points above the x-axis.

$\left(-\frac{7}{2}, 1\right) \cup (4, \infty)$

40. $y < \dfrac{1}{2}x^2 - 1$

41. $y \geq -|x|$

42. $f + g = f(x) + g(x) = 2x + x + 1$
$$= 3x + 1$$

43. $f - g = f(x) - g(x) = 2x - (x + 1)$
$$= x - 1$$

44. $f \cdot g = f(x)g(x) = 2x(x + 1) = 2x^2 + 2x$

45. $f/g = \dfrac{f(x)}{g(x)} = \dfrac{2x}{x+1}$

46. $(f \circ g)(2) = f(g(2)) = f(2 + 1)$
$$= f(3) = 2(3) = 6$$

47. $(g \circ f)(-1) = g(f(-1)) = g(2(-1))$
$$= g(-2)$$
$$= -2 + 1 = -1$$

48. $(f \circ g)(x) = f(g(x)) = f(x + 1) = 2(x + 1)$

49. $(g \circ f)(x) = g(f(x)) = g(2x) = 2x + 1$

50. $f(x) = 2(x - 3)$

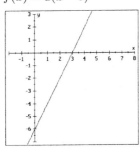

one-to-one

51. $f(x) = x(2x - 3)$

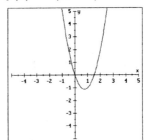

not one-to-one

52. $f(x) = -3(x-2)^2 + 5$

not one-to-one

53. $f(x) = |x|$

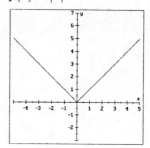

not one-to-one

54.
$$y = 6x - 3$$
$$x = 6y - 3$$
$$x + 3 = 6y$$
$$\tfrac{x+3}{6} = y, \text{ or } y = f^{-1}(x) = \tfrac{x+3}{6}$$

55.
$$y = 4x + 5$$
$$x = 4y + 5$$
$$x - 5 = 4y$$
$$\tfrac{x-5}{4} = y, \text{ or } y = f^{-1}(x) = \tfrac{x-5}{4}$$

56.
$$y = 2x^2 - 1$$
$$x = 2y^2 - 1$$
$$x + 1 = 2y^2$$
$$\tfrac{x+1}{2} = y^2$$
$$\sqrt{\tfrac{x+1}{2}} = y, \text{ or } y = f^{-1}(x) = \sqrt{\tfrac{x+1}{2}}$$

57.
$$y = |x|$$
$$x = |y|$$

Chapter 10 Test (page 709)

1.
$$x^2 + 3x - 18 = 0$$
$$(x + 6)(x - 3) = 0$$
$$x + 6 = 0 \quad \text{or} \quad x - 3 = 0$$
$$x = -6 \qquad\qquad x = 3$$

2.
$$x(6x + 19) = -15$$
$$6x^2 + 19x + 15 = 0$$
$$(2x + 3)(3x + 5) = 0$$
$$2x + 3 = 0 \quad \text{or} \quad 3x + 5 = 0$$
$$x = -\tfrac{3}{2} \qquad\qquad x = -\tfrac{5}{3}$$

3. $\left(\tfrac{1}{2} \cdot 24\right)^2 = 12^2 = 144$

4. $\left(\tfrac{1}{2} \cdot (-50)\right)^2 = (-25)^2 = 625$

5.
$$x^2 + 4x + 1 = 0$$
$$x^2 + 4x = -1$$
$$x^2 + 4x + 4 = -1 + 4$$
$$(x + 2)^2 = 3$$
$$x + 2 = \pm\sqrt{3}$$
$$x = -2 \pm \sqrt{3}$$

6.
$$x^2 - 5x - 3 = 0$$
$$x^2 - 5x = 3$$
$$x^2 - 5x + \frac{25}{4} = 3 + \frac{25}{4}$$
$$\left(x - \frac{5}{2}\right)^2 = \frac{37}{4}$$
$$x - \frac{5}{2} = \pm\sqrt{\frac{37}{4}}$$
$$x = \frac{5}{2} \pm \frac{\sqrt{37}}{2}$$

7.
$$2x^2 + 5x + 1 = 0$$
$$a = 2, b = 5, c = 1$$
$$x = \frac{-b \pm \sqrt{b^2 - 4ac}}{2a}$$
$$= \frac{-5 \pm \sqrt{5^2 - 4(2)(1)}}{2(2)}$$
$$= \frac{-5 \pm \sqrt{25 - 8}}{4}$$
$$= \frac{-5 \pm \sqrt{17}}{4} = -\frac{5}{4} \pm \frac{\sqrt{17}}{4}$$

8.
$$x^2 - x + 3 = 0$$
$$a = 1, b = -1, c = 3$$
$$x = \frac{-b \pm \sqrt{b^2 - 4ac}}{2a}$$
$$= \frac{-(-1) \pm \sqrt{(-1)^2 - 4(1)(3)}}{2(1)}$$
$$= \frac{1 \pm \sqrt{1 - 12}}{2}$$
$$= \frac{1 \pm \sqrt{-11}}{2} = \frac{1}{2} \pm \frac{\sqrt{11}}{2}i$$

9.
$$3x^2 + 5x + 17 = 0$$
$$a = 3, b = 5, c = 17$$
$$b^2 - 4ac = 5^2 - 4(3)(17)$$
$$= 25 - 208 = -183$$
nonreal solutions

10.
$$4x^2 - 2kx + k - 1 = 0$$
$$a = 4, b = -2k, c = k - 1$$
Set the discriminant equal to 0:
$$b^2 - 4ac = 0$$
$$(-2k)^2 - 4(4)(k - 1) = 0$$
$$4k^2 - 16k + 16 = 0$$
$$4(k - 2)(k - 2) = 0$$
$$k - 2 = 0 \quad \textbf{or} \quad k - 2 = 0$$
$$k = 2 \qquad\qquad k = 2$$

11. Let $x =$ the length of the shorter leg.
Then $x + 14 =$ the other length.
$$x^2 + (x + 14)^2 = 26^2$$
$$x^2 + x^2 + 28x + 196 = 676$$
$$2x^2 + 28x - 480 = 0$$
$$2(x + 24)(x - 10) = 0$$
$$x = -24 \quad \text{or} \quad x = 10$$
The shorter leg is 10 inches long.

12.
$$2y - 3y^{1/2} + 1 = 0$$
$$\left(2y^{1/2} - 1\right)\left(y^{1/2} - 1\right) = 0$$
$$2y^{1/2} - 1 = 0 \quad \textbf{or} \quad y^{1/2} - 1 = 0$$
$$2y^{1/2} = 1 \qquad\qquad y^{1/2} = 1$$
$$\left(y^{1/2}\right)^2 = \left(\tfrac{1}{2}\right)^2 \qquad \left(y^{1/2}\right)^2 = 1^2$$
$$y = \tfrac{1}{4} \qquad\qquad y = 1$$
Solution $\qquad\qquad$ Solution

13. $y = \dfrac{1}{2}x^2 - 4 = \dfrac{1}{2}(x - 0)^2 - 4$

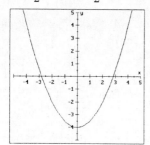

vertex: $(0, -4)$

14. $y \leq -x^2 + 3$

15. $x^2 - 2x - 8 > 0$

$(x + 2)(x - 4) > 0$

$x - 4 \quad --------0\,++++$
$x + 2 \quad ---\;\;0++++++\,++++$

$\overleftarrow{\hspace{1.5cm}} \underset{-2}{)} \rule{1cm}{0.4pt} \underset{4}{(} \overrightarrow{\hspace{1.5cm}}$

solution set: $(-\infty, -2) \cup (4, \infty)$

16. $\dfrac{x - 2}{x + 3} \leq 0$

$x - 2 \quad -------0 \;\; ++++$
$x + 3 \quad ---\,0++++++ \;\; ++++$

$\overleftarrow{\hspace{1cm}} \underset{-3}{(} \rule{1cm}{0.4pt} \underset{2}{]} \overrightarrow{\hspace{1cm}}$

solution set: $(-3, 2]$

17. $g + f = g(x) + f(x) = x - 1 + 4x$
$\qquad\qquad\qquad\qquad = 5x - 1$

18. $f - g = f(x) - g(x) = 4x - (x - 1)$
$\qquad\qquad\qquad\qquad = 3x + 1$

19. $g \cdot f = g(x)f(x) = (x - 1)4x = 4x^2 - 4x$

20. $g/f = \dfrac{g(x)}{f(x)} = \dfrac{x - 1}{4x}$

21. $(g \circ f)(1) = g(f(1)) = g(4(1))$
$\qquad\qquad\qquad\quad = g(4) = 4 - 1 = 3$

22. $(f \circ g)(0) = f(g(0)) = f(0 - 1)$
$\qquad\qquad\qquad\quad = f(-1) = 4(-1) = -4$

23. $(f \circ g)(-1) = f(g(-1)) = f(-1 - 1) = f(-2) = 4(-2) = -8$

24. $(g \circ f)(-2) = g(f(-2)) = g(4(-2)) = g(-8) = -8 - 1 = -9$

25. $(f \circ g)(x) = f(g(x)) = f(x - 1) = 4(x - 1)$

26. $(g \circ f)(x) = g(f(x)) = g(4x) = 4x - 1$

27. $3x + 2y = 12$
$3y + 2x = 12$
$\qquad 3y = -2x + 12$
$\qquad\; y = \dfrac{-2x + 12}{3}$

28. $y = 3x^2 + 4$
$x = 3y^2 + 4$
$x - 4 = 3y^2$
$\dfrac{x - 4}{3} = y^2$
$-\sqrt{\dfrac{x - 4}{3}} = y$

Cumulative Review Exercises (page 710)

1. $y = f(x) = 2x^2 - 3$
domain $= (-\infty, \infty)$
range $= [-3, \infty)$

2. $y = f(x) = -|x - 4|$
domain $= (-\infty, \infty)$
range $= (-\infty, 0]$

3. $y - y_1 = m(x - x_1)$
$y + 4 = 3(x + 2)$
$y = 3x + 2$

4. $2x + 3y = 6$
$3y = -2x + 6$
$y = -\dfrac{2}{3}x + 2$
$y - y_1 = m(x - x_1)$
$y + 2 = -\dfrac{2}{3}(x - 0)$
$y = -\dfrac{2}{3}x - 2$

5. $(2a^2 + 4a - 7) - 2(3a^2 - 4a) = 2a^2 + 4a - 7 - 6a^2 + 8a = -4a^2 + 12a - 7$

6. $(3x + 2)(2x - 3) = 6x^2 - 9x + 4x - 6 = 6x^2 - 5x - 6$

7. $x^4 - 16y^4 = (x^2 + 4y^2)(x^2 - 4y^2) = (x^2 + 4y^2)(x + 2y)(x - 2y)$

8. $15x^2 - 2x - 8 = (5x - 4)(3x + 2)$

9. $x^2 - 5x - 6 = 0$
$(x - 6)(x + 1) = 0$
$x - 6 = 0 \quad \textbf{or} \quad x + 1 = 0$
$x = 6 \qquad\qquad x = -1$

10. $6a^3 - 2a = a^2$
$6a^3 - a^2 - 2a = 0$
$a(6a^2 - a - 2) = 0$
$a(3a - 2)(2a + 1) = 0$
$a = 0 \quad \textbf{or} \quad 3a - 2 = 0 \quad \textbf{or} \quad 2a + 1 = 0$
$a = \frac{2}{3} \qquad\qquad a = -\frac{1}{2}$

11. $\sqrt{25x^4} = 5x^2$

12. $\sqrt{48t^3} = \sqrt{16t^2}\sqrt{3t} = 4t\sqrt{3t}$

13. $\sqrt[3]{-27x^3} = -3x$

14. $\sqrt[3]{\dfrac{128x^4}{2x}} = \sqrt[3]{64x^3} = 4x$

15. $8^{-1/3} = \dfrac{1}{8^{1/3}} = \dfrac{1}{2}$

16. $64^{2/3} = \left(64^{1/3}\right)^2 = 4^2 = 16$

17. $\dfrac{y^{2/3}y^{5/3}}{y^{1/3}} = \dfrac{y^{7/3}}{y^{1/3}} = y^{6/3} = y^2$

18. $\dfrac{x^{5/3}x^{1/2}}{x^{3/4}} = \dfrac{x^{13/6}}{x^{3/4}} = x^{17/12}$

19. $f(x) = \sqrt{x-2}$; Shift $y = \sqrt{x}$ right 2.

$D = [2, \infty), R = [0, \infty)$

20. $f(x) = -\sqrt{x+2}$; Reflect $y = \sqrt{x}$ about the x-axis and shift left 2.

$D = [-2, \infty), R = (-\infty, 0]$

21. $\left(x^{2/3} - x^{1/3}\right)\left(x^{2/3} + x^{1/3}\right) = x^{4/3} + x^{3/3} - x^{3/3} - x^{2/3} = x^{4/3} - x^{2/3}$

22. $\left(x^{-1/2} + x^{1/2}\right)^2 = \left(x^{-1/2} + x^{1/2}\right)\left(x^{-1/2} + x^{1/2}\right) = x^{-2/2} + x^0 + x^0 + x^{2/2} = x + 2 + \frac{1}{x}$

23. $\sqrt{50} - \sqrt{8} + \sqrt{32} = \sqrt{25}\sqrt{2} - \sqrt{4}\sqrt{2} + \sqrt{16}\sqrt{2} = 5\sqrt{2} - 2\sqrt{2} + 4\sqrt{2} = 7\sqrt{2}$

24. $-3\sqrt[4]{32} - 2\sqrt[4]{162} + 5\sqrt[4]{48} = -3\sqrt[4]{16}\sqrt[4]{2} - 2\sqrt[4]{81}\sqrt[4]{2} + 5\sqrt[4]{16}\sqrt[4]{3}$
$$= -3(2)\sqrt[4]{2} - 2(3)\sqrt[4]{2} + 5(2)\sqrt[4]{3}$$
$$= -6\sqrt[4]{2} - 6\sqrt[4]{2} + 10\sqrt[4]{3} = -12\sqrt[4]{2} + 10\sqrt[4]{3}$$

25. $3\sqrt{2}(2\sqrt{3} - 4\sqrt{12}) = 6\sqrt{6} - 12\sqrt{24} = 6\sqrt{6} - 12\sqrt{4}\sqrt{6} = 6\sqrt{6} - 24\sqrt{6} = -18\sqrt{6}$

26. $\dfrac{5}{\sqrt[3]{x}} = \dfrac{5}{\sqrt[3]{x}} \cdot \dfrac{\sqrt[3]{x^2}}{\sqrt[3]{x^2}} = \dfrac{5\sqrt[3]{x^2}}{\sqrt[3]{x^3}} = \dfrac{5\sqrt[3]{x^2}}{x}$

27. $\dfrac{\sqrt{x}+2}{\sqrt{x}-1} = \dfrac{\sqrt{x}+2}{\sqrt{x}-1} \cdot \dfrac{\sqrt{x}+1}{\sqrt{x}+1} = \dfrac{x+3\sqrt{x}+2}{x-1}$

28. $\sqrt[6]{x^3 y^3} = (x^3 y^3)^{1/6} = x^{3/6} y^{3/6} = x^{1/2} y^{1/2} = \sqrt{xy}$

29.
$$5\sqrt{x+2} = x+8$$
$$\left(5\sqrt{x+2}\right)^2 = (x+8)^2$$
$$25(x+2) = x^2 + 16x + 64$$
$$25x + 50 = x^2 + 16x + 64$$
$$0 = x^2 - 9x + 14$$
$$0 = (x-7)(x-2)$$
$$x = 7 \quad \text{or} \quad x = 2 \quad \text{(Both check.)}$$

30.
$$\sqrt{x} + \sqrt{x+2} = 2$$
$$\sqrt{x} = 2 - \sqrt{x+2}$$
$$\left(\sqrt{x}\right)^2 = \left(2 - \sqrt{x+2}\right)^2$$
$$x = 4 - 4\sqrt{x+2} + x + 2$$
$$4\sqrt{x+2} = 6$$
$$\left(4\sqrt{x+2}\right)^2 = 6^2$$
$$16(x+2) = 36$$
$$16x + 32 = 36$$
$$16x = 4$$
$$x = \frac{4}{16} = \frac{1}{4}$$

31. hypotenuse $= 3\sqrt{2}$ in.

32. hypotenuse $= 2 \cdot \dfrac{3}{\sqrt{3}} = \dfrac{6\sqrt{3}}{3} = 2\sqrt{3}$ in.

33. $d = \sqrt{(-2-4)^2 + (6-14)^2} = \sqrt{(-6)^2 + (-8)^2} = \sqrt{36+64} = \sqrt{100} = 10$

34. $\left(\frac{1}{2} \cdot 6\right)^2 = 3^2 = 9$

35.
$$2x^2 + x - 3 = 0$$
$$x^2 + \frac{1}{2}x - \frac{3}{2} = 0$$
$$x^2 + \frac{1}{2}x = \frac{3}{2}$$
$$x^2 + \frac{1}{2}x + \frac{1}{16} = \frac{3}{2} + \frac{1}{16}$$
$$\left(x + \frac{1}{4}\right)^2 = \frac{25}{16}$$
$$x + \frac{1}{4} = \pm\frac{5}{4}$$
$$x = -\frac{1}{4} \pm \frac{5}{4}$$
$$x = \frac{4}{4} = 1 \quad \text{or} \quad x = -\frac{6}{4} = -\frac{3}{2}$$

36.
$$3x^2 + 4x - 1 = 0$$
$$a = 3, b = 4, c = -1$$
$$x = \frac{-b \pm \sqrt{b^2 - 4ac}}{2a}$$
$$= \frac{-4 \pm \sqrt{4^2 - 4(3)(-1)}}{2(3)}$$
$$= \frac{-4 \pm \sqrt{16 + 12}}{6}$$
$$= \frac{-4 \pm \sqrt{28}}{6}$$
$$= \frac{-4 \pm 2\sqrt{7}}{6} = -\frac{2}{3} \pm \frac{\sqrt{7}}{3}$$

37. $y = \dfrac{1}{2}x^2 + 5 = \dfrac{1}{2}(x - 0)^2 + 5$

vertex: $(0, 5)$

38. $y \le -x^2 + 3$

vertex: $(0, 3)$

39. $(3 + 5i) + (4 - 3i) = 3 + 5i + 4 - 3i = 7 + 2i$

40. $(7 - 4i) - (12 + 3i) = 7 - 4i - 12 - 3i = -5 - 7i$

41. $(2 - 3i)(2 + 3i) = 4 + 6i - 6i - 9i^2 = 4 + 9 = 13 = 13 + 0i$

42. $(3 + i)(3 - 3i) = 9 - 9i + 3i - 3i^2 = 9 - 6i + 3 = 12 - 6i$

43. $(3 - 2i) - (4 + i)^2 = 3 - 2i - (16 + 8i + i^2) = 3 - 2i - (15 + 8i) = 3 - 2i - 15 - 8i$
$$= -12 - 10i$$

44. $\dfrac{5}{3 - i} = \dfrac{5}{3 - i} \cdot \dfrac{3 + i}{3 + i} = \dfrac{5(3 + i)}{9 - i^2} = \dfrac{5(3 + i)}{10} = \dfrac{3 + i}{2} = \dfrac{3}{2} + \dfrac{1}{2}i$

45. $|3 + 2i| = \sqrt{3^2 + 2^2} = \sqrt{9 + 4} = \sqrt{13}$

46. $|5 - 6i| = \sqrt{5^2 + (-6)^2} = \sqrt{25 + 36} = \sqrt{61}$

47.
$$2x^2 + 4x = k$$
$$2x^2 + 4x - k = 0$$
$$a = 2, b = 4, c = -k$$
Set the discriminant equal to 0:
$$b^2 - 4ac = 0$$
$$4^2 - 4(2)(-k) = 0$$
$$16 + 8k = 0$$
$$8k = -16$$
$$k = -2$$

48.
$$a - 7a^{1/2} + 12 = 0$$
$$\left(a^{1/2} - 3\right)\left(a^{1/2} - 4\right) = 0$$
$$a^{1/2} - 3 = 0 \quad \text{or} \quad a^{1/2} - 4 = 0$$
$$a^{1/2} = 3 \qquad\qquad a^{1/2} = 4$$
$$\left(a^{1/2}\right)^2 = 3^2 \qquad \left(a^{1/2}\right)^2 = 4^2$$
$$a = 9 \qquad\qquad a = 16$$
$$\text{Solution} \qquad\qquad \text{Solution}$$

49. $x^2 - x - 6 > 0$

$(x+2)(x-3) > 0$

$x - 3 \quad \text{-------} 0 +++$

$x + 2 \quad \text{---} 0++++++ +++$

$\longleftarrow \quad) \quad\quad (\longrightarrow$

$\quad\quad -2 \quad\quad 3$

solution set: $(-\infty, -2) \cup (3, \infty)$

50. $x^2 - x - 6 \leq 0$

$(x+2)(x-3) \leq 0$

$x - 3 \quad \text{-------} 0 \quad +++$

$x + 2 \quad \text{---} 0++++++ \quad +++$

$\longleftarrow \quad [\quad\quad] \longrightarrow$

$\quad\quad -2 \quad\quad 3$

solution set: $[-2, 3]$

51. $f(-1) = 3(-1)^2 + 2 = 3(1) + 2 = 3 + 2 = 5$

52. $(g \circ f)(2) = g(f(2)) = g(3(2)^2 + 2) = g(14) = 2(14) - 1 = 27$

53. $(f \circ g)(x) = f(g(x)) = f(2x - 1) = 3(2x - 1)^2 + 2 = 3(4x^2 - 4x + 1) + 2 = 12x^2 - 12x + 5$

54. $(g \circ f)(x) = g(f(x)) = g(3x^2 + 2) = 2(3x^2 + 2) - 1 = 6x^2 + 3$

55.
$y = 3x + 2$

$x = 3y + 2$

$x - 2 = 3y$

$\dfrac{x - 2}{3} = y$, or $y = f^{-1}(x) = \dfrac{x - 2}{3}$

56.
$y = x^3 + 4$

$x = y^3 + 4$

$x - 4 = y^3$

$\sqrt[3]{x - 4} = y$, or $y = f^{-1}(x) = \sqrt[3]{x - 4}$

Exercise 11.1 (page 722)

1. $3x + 2x - 20 = 180$

$5x = 200$

$x = 40$

3. $m(\angle 2) = 3x = 3(40) = 120°$

5. exponential

7. $(0, \infty)$

9. increasing

11. $P\left(1 + \frac{r}{k}\right)^{kt}$

Problems 13-15 are to be solved using a calculator. The keystrokes needed to solve each problem using a TI-83 graphing calculator appear in each solution. There may be other solutions. Keystrokes for other calculators may be slightly different.

13. $2^{\sqrt{2}} \Rightarrow \boxed{2} \boxed{\wedge} \boxed{\sqrt{}} \boxed{2} \boxed{\text{ENTER}}$

$\{2.6651\}$

15. $5^{\sqrt{5}} \Rightarrow \boxed{5} \boxed{\wedge} \boxed{\sqrt{}} \boxed{5} \boxed{\text{ENTER}}$

$\{36.5548\}$

17. $\left(2^{\sqrt{3}}\right)^{\sqrt{3}} = 2^{(\sqrt{3})(\sqrt{3})} = 2^3 = 8$

19. $7^{\sqrt{3}} 7^{\sqrt{12}} = 7^{\sqrt{3} + \sqrt{12}} = 7^{\sqrt{3} + 2\sqrt{3}} = 7^{3\sqrt{3}}$

21. $y = f(x) = 3^x$
through $(0, 1)$ and $(1, 3)$

23. $y = f(x) = \left(\frac{1}{3}\right)^x$
through $(0, 1)$ and $\left(1, \frac{1}{3}\right)$

25. $f(x) = 3^x - 2$
Shift $y = 3^x$ down 2.

27. $f(x) = 3^{x-1}$
Shift $y = 3^x$ right 1.

29. $y = b^x$
$\frac{1}{2} = b^1$
$\frac{1}{2} = b$

31. $y = b^x$
$2 = b^0$
no such value of b

33. $y = b^x$
$2 = b^1$
$2 = b$

35. $y = b^x$
$9 = b^2$
$3 = b$

37. $y = f(x) = \frac{1}{2}\left(3^{x/2}\right)$

increasing

39. $y = f(x) = 2\left(3^{-x/2}\right)$

decreasing

41. $A = P\left(1 + \dfrac{r}{k}\right)^{kt}$

$$= 10{,}000\left(1 + \dfrac{0.08}{4}\right)^{4(10)}$$

$$= 10{,}000(1.02)^{40}$$

$$\approx \$22{,}080.40$$

43. $A = P\left(1 + \dfrac{r}{k}\right)^{kt}$

$$= 1000\left(1 + \dfrac{0.05}{4}\right)^{4(5)}$$

$$= 1000(1.0125)^{20}$$

$$\approx \$1282.040$$

$A = P\left(1 + \dfrac{r}{k}\right)^{kt}$

$$= 1000\left(1 + \dfrac{0.055}{4}\right)^{4(5)}$$

$$= 1000(1.01375)^{20}$$

$$\approx \$1314.07$$

$\text{difference} = \$1314.07 - \$1282.04$

$$= \$32.03$$

45. $A = P\left(1 + \dfrac{r}{k}\right)^{kt}$

$$= 1\left(1 + \dfrac{0.05}{1}\right)^{1(300)}$$

$$= 1(1.05)^{300}$$

$$\approx \$2{,}273{,}996.13$$

47. $A = A_0\left(\dfrac{2}{3}\right)^{t}$

$$A = A_0\left(\dfrac{2}{3}\right)^{5}$$

$$A = \dfrac{32}{243}A_0$$

49. $C = (3 \times 10^{-4})(0.7)^t = (3 \times 10^{-4})(0.7)^5 \approx 5.0421 \times 10^{-5}$ coulombs

51. $A = P\left(1 + \dfrac{r}{k}\right)^{kt}$

$$= 4700\left(1 + \dfrac{-0.25}{1}\right)^{1(5)}$$

$$= 4700(0.75)^5$$

$$\approx \$1115.33$$

53. **Answers may vary.**

55. If the base were 0, then the function would not be defined for $x = 0 \Rightarrow y = 0^0$.

Exercise 11.2 (page 729)

1. $\sqrt{240x^5} = \sqrt{16x^4}\sqrt{15x} = 4x^2\sqrt{15x}$

3. $4\sqrt{48y^3} - 3y\sqrt{12y} = 4\sqrt{16y^2}\sqrt{3y} - 3y\sqrt{4}\sqrt{3y} = 4(4y)\sqrt{3y} - 3y(2)\sqrt{3y}$
$$= 16y\sqrt{3y} - 6y\sqrt{3y} = 10y\sqrt{3y}$$

5. 2.72

7. increasing

9. $A = Pe^{rt}$

11. $y = f(x) = e^x + 1$
Shift $y = e^x$ up 1.

13. $y = f(x) = e^{(x+3)}$
Shift $y = e^x$ left 3.

15. $y = f(x) = -e^x$; Reflect $y = e^x$ about the x-axis.

17. $y = f(x) = 2e^x$; Stretch $y = e^x$ vertically by a factor of 2.

19. The graph should be increasing. The graph could not look like this.

21. The graph should go through the point $(0, 1)$. The graph could not look like this.

23. $A = Pe^{rt}$
$= 5000e^{0.06(12)}$
$= 5000e^{0.72}$
$\approx \$10,272.17$

25. $A = Pe^{rt}$
$12000 = Pe^{0.07(9)}$
$12000 = Pe^{0.63}$
$12000 \approx P(1.8776106)$
$P \approx \dfrac{12000}{1.8776106}$
$P \approx \$6,391.10$

27. $A = Pe^{rt}$
$= 5000e^{0.085(5)}$
$= 5000e^{0.425}$
$\approx \$7,647.95$ (continuous)

$A = P\left(1 + \dfrac{r}{k}\right)^{kt}$
$= 5000\left(1 + \dfrac{0.085}{1}\right)^{1(5)}$
$= 5000(1.085)^5$
$\approx \$7,518.28$ (annual)

29. $A = Pe^{rt}$
$= 6e^{0.019(30)}$
$= 6e^{0.57}$
≈ 10.6 billion people

31. $A = Pe^{rt}$
$= 6e^{0.019(50)}$
$= 6e^{0.95}$
$\approx 6(2.6)$
It will increase by a
factor of about 2.6.

33. $A = 8000e^{-0.008t}$
$= 8000e^{-0.008(20)}$
$= 8000e^{-0.16}$
≈ 6817

35. $P = 0.3(1 - e^{-0.05t})$
$= 0.3(1 - e^{-0.05(15)})$
$= 0.3(1 - e^{-0.75})$
$\approx 0.3(1 - 0.47237)$
$\approx 0.3(0.52763)$
≈ 0.16

37. $x = 0.08(1 - e^{-0.1t})$
$= 0.08(1 - e^{-0.1(0)})$
$= 0.08(1 - e^{0})$
$= 0.08(1 - 1)$
$= 0.08(0) = 0$

39. $v = 50(1 - e^{-0.2t})$
$= 50(1 - e^{-0.2(20)})$
$= 50(1 - e^{-4})$
$\approx 50(1 - 0.01832)$
$\approx 50(0.98168) \approx 49$ mps

41. $A = Pe^{rt}$
$= 4570e^{-0.06(6.5)}$
$\approx \$3094.15$

43. $y = 1000e^{0.02x}$
$y = 31x + 2000$

about 72 years

45. **Answers may vary.**

47. $e \approx 2.7182$;
$1 + 1 + \dfrac{1}{2} + \dfrac{1}{2 \cdot 3} + \dfrac{1}{2 \cdot 3 \cdot 4} + \dfrac{1}{2 \cdot 3 \cdot 4 \cdot 5}$
≈ 2.7167

49. $e^{t+5} = ke^{t}$
$e^{t} \cdot e^{5} = ke^{t}$
$e^{5}e^{t} = ke^{t}$
$k = e^{5}$

Exercise 11.3 (page 739)

1. $\sqrt[3]{6x + 4} = 4$
$6x + 4 = 4^{3}$
$6x + 4 = 64$
$6x = 60$
$x = 10$

3. $\sqrt{a + 1} - 1 = 3a$
$\sqrt{a + 1} = 3a + 1$
$a + 1 = (3a + 1)^{2}$
$a + 1 = 9a^{2} + 6a + 1$
$0 = 9a^{2} + 5a$
$0 = a(9a + 5)$
$a = 0$ or $a = -\tfrac{5}{9}$ $\left(-\tfrac{5}{9} \text{ does not check.}\right)$

5. $x = b^y$

7. range

9. inverse

11. exponent

13. $(b, 1); (1, 0)$

15. $20 \log \dfrac{E_O}{E_I}$

17. $\log_3 27 = 3 \Rightarrow 3^3 = 27$

19. $\log_{1/2} \dfrac{1}{4} = 2 \Rightarrow \left(\dfrac{1}{2}\right)^2 = \dfrac{1}{4}$

21. $\log_4 \dfrac{1}{64} = -3 \Rightarrow 4^{-3} = \dfrac{1}{64}$

23. $\log_{1/2} \dfrac{1}{8} = 3 \Rightarrow \left(\dfrac{1}{2}\right)^3 = \dfrac{1}{8}$

25. $6^2 = 36 \Rightarrow \log_6 36 = 2$

27. $5^{-2} = \dfrac{1}{25} \Rightarrow \log_5 \dfrac{1}{25} = -2$

29. $\left(\dfrac{1}{2}\right)^{-5} = 32 \Rightarrow \log_{1/2} 32 = -5$

31. $x^y = z \Rightarrow \log_x z = y$

33. $\log_2 16 = x \Rightarrow 2^x = 16 \Rightarrow x = 4$

35. $\log_4 16 = x \Rightarrow 4^x = 16 \Rightarrow x = 2$

37. $\log_{1/2} \dfrac{1}{8} = x \Rightarrow \left(\dfrac{1}{2}\right)^x = \dfrac{1}{8} \Rightarrow x = 3$

39. $\log_9 3 = x \Rightarrow 9^x = 3 \Rightarrow x = \dfrac{1}{2}$

41. $\log_{1/2} 8 = x \Rightarrow \left(\dfrac{1}{2}\right)^x = 8 \Rightarrow x = -3$

43. $\log_7 x = 2 \Rightarrow 7^2 = x \Rightarrow x = 49$

45. $\log_6 x = 1 \Rightarrow 6^1 = x \Rightarrow x = 6$

47. $\log_{25} x = \dfrac{1}{2} \Rightarrow 25^{1/2} = x \Rightarrow x = 5$

49. $\log_5 x = -2 \Rightarrow 5^{-2} = x \Rightarrow x = \dfrac{1}{25}$

51. $\log_{36} x = -\dfrac{1}{2} \Rightarrow 36^{-1/2} = x \Rightarrow x = \dfrac{1}{6}$

53. $\log_{100} \dfrac{1}{1000} = x \Rightarrow 100^x = \dfrac{1}{1000} \Rightarrow$
$x = -\dfrac{3}{2}$

55. $\log_{27} 9 = x \Rightarrow 27^x = 9 \Rightarrow x = \dfrac{2}{3}$

57. $\log_x 5^3 = 3 \Rightarrow x^3 = 5^3 \Rightarrow x = 5$

59. $\log_x \dfrac{9}{4} = 2 \Rightarrow x^2 = \dfrac{9}{4} \Rightarrow x = \dfrac{3}{2}$

61. $\log_x \dfrac{1}{64} = -3 \Rightarrow x^{-3} = \dfrac{1}{64} \Rightarrow x = 4$

63. $\log_{2\sqrt{2}} x = 2 \Rightarrow (2\sqrt{2})^2 = x \Rightarrow x = 8$

65. $2^{\log_2 4} = x \Rightarrow x = 4$

67. $x^{\log_4 6} = 6 \Rightarrow x = 4$

69. $\log 10^3 = x \Rightarrow 10^x = 10^3 \Rightarrow x = 3$

71. $10^{\log x} = 100 \Rightarrow \log x = 2 \Rightarrow x = 100$

73. $\log 8.25 \approx 0.9165$

75. $\log 0.00867 \approx -2.0620$

77. $\log y = 1.4023 \Rightarrow y = 25.25$

79. $\log y = 4.24 \Rightarrow y = 17,378.01$

81. $\log y = -3.71 \Rightarrow y = 0.00$

83. $\log y = \log 8 \Rightarrow \log y = 0.9030 \Rightarrow y = 8$

85. $y = f(x) = \log_3 x$
through $(1, 0)$ and $(3, 1)$

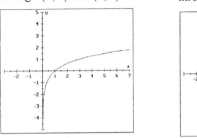

increasing

87. $y = f(x) = \log_{1/2} x$
through $(1, 0)$ and $\left(\frac{1}{2}, 1\right)$

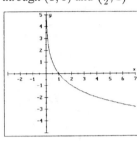

decreasing

89. $y = f(x) = 3 + \log_3 x$
Shift $y = \log_3 x$ up 3.

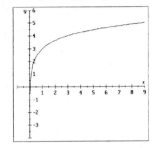

91. $y = f(x) = \log_{1/2}(x - 2)$
Shift $y = \log_{1/2} x$ right 2.

93. $y = f(x) = 2^x$
$y = g(x) = \log_2 x$

95. $y = f(x) = \left(\frac{1}{4}\right)^x$
$y = g(x) = \log_{1/4} x$

97. $\log_b 2 = 0 \Rightarrow b^0 = 2$; No such value exists.

99. $\log_b 9 = 2 \Rightarrow b^2 = 9 \Rightarrow b = 3$

101. dB gain $= 20 \log \dfrac{E_O}{E_I} = 20 \log \dfrac{20}{0.71} = 20 \log 28.169 \approx 29.0$ dB

103. dB gain $= 20 \log \dfrac{E_O}{E_I} = 20 \log \dfrac{30}{0.1} = 20 \log 300 \approx 49.5$ dB

105. $R = \log \dfrac{A}{P} = \log \dfrac{5000}{0.2} = \log 25,000$
≈ 4.4

107. $R = \log \dfrac{A}{P} = \log \dfrac{2500}{0.25} = \log 10000$
$= 4$

109. $n = \dfrac{\log V - \log C}{\log\left(1 - \frac{2}{N}\right)} = \dfrac{\log 2000 - \log 17000}{\log\left(1 - \frac{2}{5}\right)} \approx \dfrac{-0.929419}{-0.221849} \approx 4.2$ years old

111. $n = \dfrac{\log\left[\frac{Ar}{P}+1\right]}{\log(1+r)} = \dfrac{\log\left[\frac{20{,}000(0.12)}{1000}+1\right]}{\log(1+0.12)} = \dfrac{\log 3.4}{\log 1.12} \approx 10.8$ years

113. Answers may vary. **115.** Answers may vary. **117.** Answers may vary.

Exercise 11.4 (page 746)

1. $y = 5x + 8 \Rightarrow m = 5$
Use the parallel slope:
$y - y_1 = m(x - x_1)$
$y - 0 = 5(x - 0)$
$y = 5x$

3. $y = \dfrac{2}{3}x - 12 \Rightarrow m = \dfrac{2}{3}$
Use the perpendicular slope:
$y - y_1 = m(x - x_1)$
$y - 2 = -\dfrac{3}{2}(x - 3)$
$y - 2 = -\dfrac{3}{2}x + \dfrac{9}{2}$
$y = -\dfrac{3}{2}x + \dfrac{13}{2}$

5. $x = 5$

7. $\dfrac{2x+3}{4x^2-9} = \dfrac{2x+3}{(2x+3)(2x-3)} = \dfrac{1}{2x-3}$

9. $\dfrac{x^2+3x+2}{3x+12} \cdot \dfrac{x+4}{x^2-4} = \dfrac{(x+2)(x+1)}{3(x+4)} \cdot \dfrac{x+4}{(x+2)(x-2)} = \dfrac{x+1}{3(x-2)}$

11. $\log_e x$ **13.** $(-\infty, \infty)$ **15.** 10 **17.** $\dfrac{\ln 2}{r} \cdot 1$

19. $\ln 25.25 \approx 3.2288$ **21.** $\ln 9.89 \approx 2.2915$

23. $\log(\ln 2) \approx \log(0.6931) \approx -0.1592$ **25.** $\ln(\log 0.5) = \ln(-0.3010) \Rightarrow$ impossible

27. $\ln y = 2.3015 \Rightarrow y = 9.9892$ **29.** $\ln y = 3.17 \Rightarrow y = 23.8075$

31. $\ln y = -4.72 \Rightarrow y = 0.0089$

33. $\log y = \ln 6 \Rightarrow \log y \approx 1.7918 \Rightarrow y \approx 61.9098$
(The answer will vary if rounding is used on the calculator.)

35. The graph must be increasing. The graph could not look like this.

37. The graph must go through $(1, 0)$. The graph could not look like this.

39. $y = -\ln x$

41. $y = \ln(-x)$

43. $t = \dfrac{\ln 2}{r} = \dfrac{\ln 2}{0.12} \approx 5.8$ years

45. $t = \dfrac{\ln 3}{r} = \dfrac{\ln 3}{0.12} \approx 9.2$ years

47. $t = -\dfrac{1}{0.9} \ln \dfrac{50 - T_r}{200 - T_r} = -\dfrac{1}{0.9} \ln \dfrac{50 - 38}{200 - 38} = -\dfrac{1}{0.9} \ln \dfrac{12}{162} \approx -\dfrac{1}{0.9}(-2.6027) \approx 2.9$ hours

49. **Answers may vary.**

51. $P = P_0 e^{rt} = P_0 e^{r \frac{\ln 2}{r}} = P_0 e^{\ln 2} = 2P_0$

53. Let $t = \dfrac{\ln 4}{r} \Rightarrow P = P_0 e^{rt} = P_0 e^{r \frac{\ln 4}{r}} = P_0 e^{\ln 4} = 4P_0$

Exercise 11.5 (page 755)

1. $m = \dfrac{\Delta y}{\Delta x} = \dfrac{3 - (-4)}{-2 - 4} = \dfrac{7}{-6} = -\dfrac{7}{6}$

3. $x = \dfrac{-2+4}{2} = \dfrac{2}{2} = 1;\ y = \dfrac{3+(-4)}{2} = \dfrac{-1}{2} = -\dfrac{1}{2}$
midpoint: $\left(1, -\dfrac{1}{2}\right)$

5. 0

7. $M;\ N$

9. $x;\ y$

11. x

13. $\neq$

15. 0

17. 7

19. 10

21. 1

23. 0

25. 7

27. 10

29. 1

Problems 31-35 are to be solved using a calculator. The keystrokes needed to solve each problem using a TI-83 graphing calculator appear in each solution. There may be other solutions. Keystrokes for other calculators may be slightly different.

31. | log | 2 | . | 5 | × | 3 | . | 7 | ENTER | {0.96614}
| log | 2 | . | 5 |) | + | log | 3 | . | 7 |) | ENTER | {0.96614}

33. | ln | 2 | . | 2 | 5 | ^ | 4 | ENTER | {3.24372}
| 4 | ln | 2 | . | 2 | 5 | ENTER | {3.24372}

35. [log] [√] [2] [4] [.] [3] [ENTER] {0.69280}
[.] [5] [log] [2] [4] [.] [3] [ENTER] {0.69280}

37. $\log_b xyz = \log_b x + \log_b y + \log_b z$

39. $\log_b \dfrac{2x}{y} = \log_b 2x - \log_b y$
$$= \log_b 2 + \log_b x - \log_b y$$

41. $\log_b x^3 y^2 = \log_b x^3 + \log_b y^2 = 3\log_b x + 2\log_b y$

43. $\log_b (xy)^{1/2} = \dfrac{1}{2}\log_b xy = \dfrac{1}{2}(\log_b x + \log_b y) = \dfrac{1}{2}\log_b x + \dfrac{1}{2}\log_b y$

45. $\log_b x\sqrt{z} = \log_b xz^{1/2} = \log_b x + \log_b z^{1/2} = \log_b x + \dfrac{1}{2}\log_b z$

47. $\log_b \dfrac{\sqrt[3]{x}}{\sqrt[4]{yz}} = \log_b \dfrac{x^{1/3}}{(yz)^{1/4}} = \log_b x^{1/3} - \log_b (yz)^{1/4} = \frac{1}{3}\log_b x - \frac{1}{4}\log_b yz$
$$= \tfrac{1}{3}\log_b x - \tfrac{1}{4}(\log_b y + \log_b z)$$
$$= \tfrac{1}{3}\log_b x - \tfrac{1}{4}\log_b y - \tfrac{1}{4}\log_b z$$

49. $\log_b (x+1) - \log_b x = \log_b \dfrac{x+1}{x}$

51. $2\log_b x + \dfrac{1}{2}\log_b y = \log_b x^2 + \log_b y^{1/2}$
$$= \log_b x^2 y^{1/2}$$

53. $-3\log_b x - 2\log_b y + \dfrac{1}{2}\log_b z = \log_b x^{-3} + \log_b y^{-2} + \log_b z^{1/2} = \log_b x^{-3}y^{-2}z^{1/2} = \log_b \dfrac{z^{1/2}}{x^3 y^2}$

55. $\log_b \left(\dfrac{x}{z} + x\right) - \log_b \left(\dfrac{y}{z} + y\right) = \log_b \dfrac{\frac{x}{z} + x}{\frac{y}{z} + y} = \log_b \dfrac{x + xz}{y + yz} = \log_b \dfrac{x(1+z)}{y(1+z)} = \log_b \dfrac{x}{y}$

57. $\log_b 0 = 1 \Rightarrow b^1 = 0 \Rightarrow b = 0$
$\Rightarrow$ FALSE $(b \neq 0)$

59. $\log_b xy = (\log_b x)(\log_b y) \Rightarrow$ FALSE
$(\log_b xy = \log_b x + \log_b y)$

61. $\log_7 7^7 = 7 \Rightarrow 7^7 = 7^7 \Rightarrow$ TRUE

63. $\dfrac{\log_b A}{\log_b B} = \log_b A - \log_b B$
$$\Rightarrow \text{FALSE} \left(\log_b \dfrac{A}{B} = \log_b A - \log_b B\right)$$

65. $3\log_b \sqrt[3]{a} = 3\log_b a^{1/3} = \dfrac{1}{3}\cdot 3\log_b a = \log_b a \Rightarrow$ TRUE

67. $\log_b \dfrac{1}{a} = \log_b 1 - \log_b a = 0 - \log_b a = -\log_b a \Rightarrow$ TRUE

69. $\log 28 = \log 4\cdot 7 = \log 4 + \log 7 = 0.6021 + 0.8451 = 1.4472$

71. $\log 2.25 = \log \dfrac{9}{4} = \log 9 - \log 4 = 0.9542 - 0.6021 = 0.3521$

73. $\log \dfrac{63}{4} = \log \dfrac{7 \cdot 9}{4} = \log 7 + \log 9 - \log 4 = 0.8451 + 0.9542 - 0.6021 = 1.1972$

75. $\log 252 = \log 4 \cdot 7 \cdot 9 = \log 4 + \log 7 + \log 9 = 0.6021 + 0.8451 + 0.9542 = 2.4014$

77. $\log 112 = \log 4^2 \cdot 7 = \log 4^2 + \log 7 = 2 \log 4 + \log 7 = 2(0.6021) + 0.8451 = 2.0493$

79. $\log \dfrac{144}{49} = \log \dfrac{16 \cdot 9}{7^2} = \log \dfrac{4^2 \cdot 9}{7^2} = \log 4^2 + \log 9 - \log 7^2 = 2 \log 4 + \log 9 - 2 \log 7$

$$= 2(0.6021) + 0.9542 - 2(0.8451)$$
$$= 0.4682$$

81. $\log_3 7 = \dfrac{\log 7}{\log 3} \approx 1.7712$

83. $\log_{1/3} 3 = \dfrac{\log 3}{\log \frac{1}{3}} \approx -1.0000$

85. $\log_3 8 = \dfrac{\log 8}{\log 3} \approx 1.8928$

87. $\log_{\sqrt 2} \sqrt 5 = \dfrac{\log \sqrt 5}{\log \sqrt 2} \approx 2.3219$

89. $\text{pH} = -\log [\text{H}^+] = -\log (1.7 \times 10^{-5}) \approx 4.77$

91. low pH:

  $\text{pH} = -\log [\text{H}^+]$
  $6.8 = -\log [\text{H}^+]$
  $-6.8 = \log [\text{H}^+]$
  $[\text{H}^+] = 1.5849 \times 10^{-7}$

  high pH:

  $\text{pH} = -\log [\text{H}^+]$
  $7.6 = -\log [\text{H}^+]$
  $-7.6 = \log [\text{H}^+]$
  $[\text{H}^+] = 2.5119 \times 10^{-8}$

93. $k \ln 2I = k(\ln 2 + \ln I)$
$\qquad\qquad = k \ln 2 + k \ln I$
$\qquad\qquad = k \ln 2 + L$
The loudness increases by $k \ln 2$.

95. $L = 3k \ln I = k \cdot 3 \ln I$
$\qquad\qquad\quad = k \ln I^3$
The intensity must be cubed.

97. Answers may vary.

99. $\ln(e^x) = \log_e(e^x) = x$

101. Let $\log_{b^2} x = y$. Then
$$\left(b^2\right)^y = x$$
$$b^{2y} = x$$
$$\left(b^{2y}\right)^{1/2} = x^{1/2}$$
$$b^y = x^{1/2}$$
$$\log_b x^{1/2} = y$$
$$\frac{1}{2} \log_b x = y$$

Exercise 11.6 (page 764)

1. $5x^2 - 25x = 0$
$5x(x - 5) = 0$
$5x = 0$ **or** $x - 5 = 0$
$x = 0$ $\qquad x = 5$

3. $3p^2 + 10p = 8$
$3p^2 + 10p - 8 = 0$
$(3p - 2)(p + 4) = 0$
$3p - 2 = 0$ **or** $p + 4 = 0$
$p = \frac{2}{3}$ $\qquad p = -4$

5. exponential

7. $A_0 2^{-t/h}$

9. $4^x = 5$
$\log 4^x = \log 5$
$x \log 4 = \log 5$
$x = \dfrac{\log 5}{\log 4}$
$x \approx 1.1610$

11. $e^t = 50$
$\ln e^t = \ln 50$
$t \ln e = \ln 50$
$t = \ln 50$
$t \approx 3.9120$

13. $5 = 2.1(1.04)^t$
$\dfrac{5}{2.1} = (1.04)^t$
$\log \dfrac{5}{2.1} = \log (1.04)^t$
$\log \dfrac{5}{2.1} = t \log 1.04$
$\dfrac{\log \frac{5}{2.1}}{\log 1.04} = t$
$22.1184 \approx t$

15. $13^{x-1} = 2$
$\log 13^{x-1} = \log 2$
$(x - 1) \log 13 = \log 2$
$x - 1 = \dfrac{\log 2}{\log 13}$
$x = \dfrac{\log 2}{\log 13} + 1$
$x \approx 1.2702$

17. $2^{x+1} = 3^x$
$\log 2^{x+1} = \log 3^x$
$(x + 1) \log 2 = x \log 3$
$x \log 2 + \log 2 = x \log 3$
$\log 2 = x \log 3 - x \log 2$
$\log 2 = x(\log 3 - \log 2)$
$\dfrac{\log 2}{\log 3 - \log 2} = x$
$1.7095 \approx x$

19. $2^x = 3^x$
$\log 2^x = \log 3^x$
$x \log 2 = x \log 3$
$0 = x \log 3 - x \log 2$
$0 = x(\log 3 - \log 2)$
$\dfrac{0}{\log 3 - \log 2} = x$
$0 = x$

21. $7^{x^2} = 10$
$\log 7^{x^2} = \log 10$
$x^2 \log 7 = \log 10$
$x^2 = \dfrac{\log 10}{\log 7}$
$x^2 \approx 1.1833$
$x \approx \pm 1.0878$

23.
$$8^{x^2} = 9^x$$
$$\log 8^{x^2} = \log 9^x$$
$$x^2 \log 8 = x \log 9$$
$$x^2 \log 8 - x \log 9 = 0$$
$$x(x \log 8 - \log 9) = 0$$
$$x = 0 \quad \text{or} \quad x \log 8 - \log 9 = 0$$
$$x \log 8 = \log 9$$
$$x = \frac{\log 9}{\log 8}$$
$$x \approx 1.0566$$

25.
$$2^{x^2 - 2x} = 8$$
$$2^{x^2 - 2x} = 2^3$$
$$x^2 - 2x = 3$$
$$x^2 - 2x - 3 = 0$$
$$(x + 1)(x - 3) = 0$$
$$x + 1 = 0 \quad \text{or} \quad x - 3 = 0$$
$$x = -1 \qquad\qquad x = 3$$

27.
$$3^{x^2 + 4x} = \frac{1}{81}$$
$$3^{x^2 + 4x} = 3^{-4}$$
$$x^2 + 4x = -4$$
$$x^2 + 4x + 4 = 0$$
$$(x + 2)(x + 2) = 0$$
$$x + 2 = 0 \quad \text{or} \quad x + 2 = 0$$
$$x = -2 \qquad\qquad x = -2$$

29.
$$4^{x+2} - 4^x = 15$$
$$4^x 4^2 - 4^x = 15$$
$$16 \cdot 4^x - 4^x = 15$$
$$15 \cdot 4^x = 15$$
$$4^x = 1$$
$$x = 0$$

31.
$$2(3^x) = 6^{2x}$$
$$\log 2(3^x) = \log 6^{2x}$$
$$\log 2 + \log 3^x = 2x \log 6$$
$$\log 2 + x \log 3 = 2x \log 6$$
$$\log 2 = 2x \log 6 - x \log 3$$
$$\log 2 = x(2 \log 6 - \log 3)$$
$$\frac{\log 2}{2 \log 6 - \log 3} = x$$
$$0.2789 \approx x$$

33. $2^{x+1} = 7 \Rightarrow$ Graph $y = 2^{x+1}$ and $y = 7$.

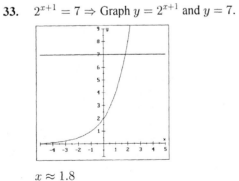

$$x \approx 1.8$$

35. $2^{x^2 - 2x} - 8 = 0 \Rightarrow$ Graph $y = 2^{x^2 - 2x} - 8$ and find any x-intercept(s).

$$x = 3 \text{ or } x = -1$$

37. $\log 2x = \log 4$

$\qquad 2x = 4$

$\qquad x = 2$

39. $\log (3x + 1) = \log (x + 7)$

$\qquad 3x + 1 = x + 7$

$\qquad 2x = 6$

$\qquad x = 3$

41. $\log (3 - 2x) - \log (x + 24) = 0$

$\qquad \log (3 - 2x) = \log (x + 24)$

$\qquad 3 - 2x = x + 24$

$\qquad -21 = 3x$

$\qquad -7 = x$

43. $\log \dfrac{4x + 1}{2x + 9} = 0$

$\qquad 10^0 = \dfrac{4x + 1}{2x + 9}$

$\qquad 1 = \dfrac{4x + 1}{2x + 9}$

$\qquad 2x + 9 = 4x + 1$

$\qquad 8 = 2x$

$\qquad 4 = x$

45. $\log x^2 = 2$

$\qquad 10^2 = x^2$

$\qquad 100 = x^2$

$\qquad \pm 10 = x$

47. $\log x + \log (x - 48) = 2$

$\qquad \log x(x - 48) = 2$

$\qquad 10^2 = x(x - 48)$

$\qquad 0 = x^2 - 48x - 100$

$\qquad 0 = (x - 50)(x + 2)$

$x - 50 = 0 \quad$ **or** $\quad x + 2 = 0$

$\qquad x = 50 \qquad\qquad x = -2$

$\qquad\qquad\qquad\qquad$ Does not check.

49. $\log x + \log (x - 15) = 2$

$\qquad \log x(x - 15) = 2$

$\qquad 10^2 = x(x - 15)$

$\qquad 0 = x^2 - 15x - 100$

$\qquad 0 = (x - 20)(x + 5)$

$x - 20 = 0 \quad$ **or** $\quad x + 5 = 0$

$\qquad x = 20 \qquad\qquad x = -5$

$\qquad\qquad\qquad\qquad$ Does not check.

51. $\qquad \log (x + 90) = 3 - \log x$

$\log x + \log (x + 90) = 3$

$\qquad \log x(x + 90) = 3$

$\qquad 10^3 = x(x + 90)$

$\qquad 0 = x^2 + 90x - 1000$

$\qquad 0 = (x - 10)(x + 100)$

$x - 10 = 0 \quad$ **or** $\quad x + 100 = 0$

$\qquad x = 10 \qquad\qquad x = -100$

$\qquad\qquad\qquad\qquad$ Does not check.

53. $\log (x - 6) - \log (x - 2) = \log \dfrac{5}{x}$

$$\log \dfrac{x - 6}{x - 2} = \log \dfrac{5}{x}$$

$$\dfrac{x - 6}{x - 2} = \dfrac{5}{x}$$

$$x(x - 6) = 5(x - 2)$$

$$x^2 - 6x = 5x - 10$$

$$x^2 - 11x + 10 = 0$$

$$(x - 10)(x - 1) = 0$$

$$x - 10 = 0 \quad \textbf{or} \quad x - 1 = 0$$

$$x = 10 \qquad\qquad x = 1$$

Does not check.

55. $\log x^2 = (\log x)^2$

$$2 \log x = (\log x)^2$$

$$0 = (\log x)^2 - 2 \log x$$

$$0 = \log x \,(\log x - 2)$$

$$\log x = 0 \quad \textbf{or} \quad \log x - 2 = 0$$

$$x = 1 \qquad\qquad \log x = 2$$

$$x = 100$$

57. $\dfrac{\log (3x - 4)}{\log x} = 2$

$$\log (3x - 4) = 2 \log x$$

$$\log (3x - 4) = \log x^2$$

$$3x - 4 = x^2$$

$$0 = x^2 - 3x + 4$$

$$b^2 - 4ac = (-3)^2 - 4(1)(4) = -7 \Rightarrow$$

solutions are nonreal $\Rightarrow$ no solution

59. $\dfrac{\log (5x + 6)}{2} = \log x$

$$\log (5x + 6) = 2 \log x$$

$$\log (5x + 6) = \log x^2$$

$$5x + 6 = x^2$$

$$0 = x^2 - 5x - 6$$

$$0 = (x - 6)(x + 1)$$

$$x - 6 = 0 \quad \textbf{or} \quad x + 1 = 0$$

$$x = 6 \qquad\qquad x = -1$$

Does not check.

61. $\log_3 x = \log_3 \left(\dfrac{1}{x} \right) + 4$

$$\log_3 x = \log_3 \left(\dfrac{1}{x} \right) + \log_3 81$$

$$\log_3 x = \log_3 \left(\dfrac{81}{x} \right)$$

$$x = \dfrac{81}{x}$$

$$x^2 = 81$$

$$x = 9 \ (-9 \text{ does not check.})$$

63. $2 \log_2 x = 3 + \log_2 (x - 2)$

$$\log_2 x^2 - \log_2 (x - 2) = 3$$

$$\log_2 \dfrac{x^2}{x - 2} = 3$$

$$\dfrac{x^2}{x - 2} = 8$$

$$x^2 = 8(x - 2)$$

$$x^2 = 8x - 16$$

$$x^2 - 8x + 16 = 0$$

$$(x - 4)(x - 4) = 0$$

$$x - 4 = 0 \quad \textbf{or} \quad x - 4 = 0$$

$$x = 4 \qquad\qquad x = 4$$

65. $\log (7y + 1) = 2 \log (y + 3) - \log 2$

$\log (7y + 1) = \log (y + 3)^2 - \log 2$

$\log (7y + 1) = \log \dfrac{y^2 + 6y + 9}{2}$

$7y + 1 = \dfrac{y^2 + 6y + 9}{2}$

$2(7y + 1) = y^2 + 6y + 9$

$14y + 2 = y^2 + 6y + 9$

$0 = y^2 - 8y + 7$

$0 = (y - 7)(y - 1)$

$y - 7 = 0$ **or** $y - 1 = 0$

$ y = 7 y = 1$

67. $\log x + \log (x - 15) = 2$

Graph $y = \log x + \log (x - 15)$ and $y = 2$.

$x = 20$

69. $\ln (2x + 5) - \ln 3 = \ln (x - 1) \Rightarrow$ Graph
$y = \ln (2x + 5) - \ln 3$ and $y = \ln (x - 1)$.

$x = 8$

71.
$$A = A_0 2^{-t/h}$$
$$0.75 A_0 = A_0 2^{-t/12.4}$$
$$0.75 = 2^{-t/12.4}$$
$$\log 0.75 = \log 2^{-t/12.4}$$
$$\log 0.75 = -\dfrac{t}{12.4} \log 2$$
$$\dfrac{\log 0.75}{-\log 2} = \dfrac{t}{12.4}$$
$$-12.4 \dfrac{\log 0.75}{\log 2} = t$$
$$5.1 \text{ years} \approx t$$

73.
$$A = A_0 2^{-t/h}$$
$$0.20 A_0 = A_0 2^{-t/18.4}$$
$$0.20 = 2^{-t/18.4}$$
$$\log 0.20 = \log 2^{-t/18.4}$$
$$\log 0.20 = -\dfrac{t}{18.4} \log 2$$
$$\dfrac{\log 0.20}{-\log 2} = \dfrac{t}{18.4}$$
$$-18.4 \dfrac{\log 0.20}{\log 2} = t$$
$$42.7 \text{ days} \approx t$$

75.
$$A = A_0 2^{-t/h}$$
$$0.60 A_0 = A_0 2^{-t/5700}$$
$$0.60 = 2^{-t/5700}$$
$$\log 0.60 = \log 2^{-t/5700}$$
$$\log 0.60 = -\dfrac{t}{5700} \log 2$$
$$\dfrac{\log 0.60}{-\log 2} = \dfrac{t}{5700}$$
$$-5700 \dfrac{\log 0.60}{\log 2} = t$$
$$4200 \text{ years} \approx t$$

77.
$$A = P\left(1 + \frac{r}{k}\right)^{kt}$$
$$800 = 500\left(1 + \frac{0.085}{2}\right)^{2t}$$
$$1.6 = (1.0425)^{2t}$$
$$\log 1.6 = \log (1.0425)^{2t}$$
$$\log 1.6 = 2t \log (1.0425)$$
$$\frac{\log 1.6}{2 \log 1.0425} = t$$
$$5.6 \text{ years} \approx t$$

79.
$$A = P\left(1 + \frac{r}{k}\right)^{kt}$$
$$2100 = 1300\left(1 + \frac{0.09}{4}\right)^{4t}$$
$$\frac{2100}{1300} = (1.0225)^{4t}$$
$$\log \frac{2100}{1300} = \log (1.0225)^{4t}$$
$$\log \frac{21}{13} = 4t \log (1.0225)$$
$$\frac{\log \frac{21}{13}}{4 \log 1.0225} = t$$
$$5.4 \text{ years} \approx t$$

81. doubling time $= t = \dfrac{\ln 2}{r} = \dfrac{100 \ln 2}{100r} \approx \dfrac{70}{100r} = \dfrac{70}{r}$, written as a %

83.
$$P = P_0 e^{kt}$$
$$2P_0 = P_0 e^{5k}$$
$$2 = e^{5k}$$
$$\ln 2 = \ln e^{5k}$$
$$\ln 2 = 5k$$
$$\frac{\ln 2}{5} = k$$

$$P = P_0 e^{kt}$$
$$1,000,000 = 30,000 e^{\frac{\ln 2}{5}t}$$
$$33.333 \approx e^{\frac{\ln 2}{5}t}$$
$$\ln 33.333 \approx \ln e^{\frac{\ln 2}{5}t}$$
$$\ln 33.333 \approx \frac{\ln 2}{5}t$$
$$\frac{5 \ln 33.333}{\ln 2} \approx t$$
$$25.3 \text{ years} \approx t$$

85
$$P = P_0 e^{kt}$$
$$2P_0 = P_0 e^{24k}$$
$$2 = e^{24k}$$
$$\ln 2 = \ln e^{24k}$$
$$\ln 2 = 24k$$
$$\frac{\ln 2}{24} = k$$

$$P = P_0 e^{kt}$$
$$P = P_0 e^{\frac{\ln 2}{24}(36)}$$
$$P = P_0 e^{\frac{3\ln 2}{2}}$$
$$P = P_0(2.828)$$
It will be about 2.828 times larger.

87. $n = \dfrac{1}{\log 2}\left(\log \dfrac{B}{b}\right) = \dfrac{1}{\log 2}\log \dfrac{5 \times 10^6}{500} = \dfrac{1}{\log 2}\log 10{,}000 \approx 13.3 \text{ generations}$

89. Answers may vary.

91. Since the logarithm of a negative number is not defined (as a real number), the values $x - 3$ and $x^2 + 2$ must be nonnegative. Since $x^2 + 2$ is always greater than 0, the only restriction is that $x - 3 > 0$, or $x > 3$. Thus, x cannot be a solution if $x \le 3$.

Chapter 11 Summary (page 768)

1. $5^{\sqrt{2}} \cdot 5^{\sqrt{2}} = 5^{\sqrt{2}+\sqrt{2}} = 5^{2\sqrt{2}}$

2. $\left(2^{\sqrt{5}}\right)^{\sqrt{2}} = 2^{\left(\sqrt{5}\right)\left(\sqrt{2}\right)} = 2^{\sqrt{10}}$

3. $y = 3^x$; through $(0, 1)$ and $(1, 3)$

4. $y = \left(\frac{1}{3}\right)^x$; through $(0, 1)$ and $\left(1, \frac{1}{3}\right)$

5. The graph will go through $(0, 1)$ and $(1, 6)$, so $x = 1$ and $y = 6$.

6. domain $= (-\infty, \infty)$; range $= (0, \infty)$

7. $y = f(x) = \left(\frac{1}{2}\right)^x - 2$

Shift $y = \left(\frac{1}{2}\right)^x$ down 2.

8. $y = f(x) = \left(\frac{1}{2}\right)^{x+2}$

Shift $y = \left(\frac{1}{2}\right)^x$ left 2.

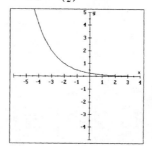

9. $A = P\left(1 + \dfrac{r}{k}\right)^{kt} = 10500\left(1 + \dfrac{0.09}{4}\right)^{4\cdot60} = 10500(1.0225)^{240} \approx \$2,189,703.45$

10. $A = Pe^{rt} = 10500e^{0.09(60)} = 10500e^{5.4} \approx \$2,324,767.37$

11. $y = f(x) = e^x + 1$; Shift $y = e^x$ up 1.

12. $y = f(x) = e^{x-3}$; Shift $y = e^x$ right 3.

13. $P = P_0 e^{kt} = 275,000,000 e^{0.015(50)}$
$\approx 582,000,000$

14. domain $= (0, \infty)$; range $= (-\infty, \infty)$

15. $\log_3 9 = 2$

16. $\log_9 \dfrac{1}{3} = -\dfrac{1}{2}$

17. $\log_\pi 1 = 0$

18. $\log_5 0.04 = \log_5 \dfrac{1}{25} = -2$

19. $\log_a \sqrt{a} = \log_a a^{1/2} = \dfrac{1}{2}$

20. $\log_a \sqrt[3]{a} = \log_a a^{1/3} = \dfrac{1}{3}$

21. $\log_2 x = 5 \Rightarrow 2^5 = x \Rightarrow x = 32$

22. $\log_{\sqrt{3}} x = 4 \Rightarrow \left(\sqrt{3}\right)^4 = x \Rightarrow x = 9$

23. $\log_{\sqrt{3}} x = 6 \Rightarrow \left(\sqrt{3}\right)^6 = x \Rightarrow x = 27$

24. $\log_{0.1} 10 = x \Rightarrow (0.1)^x = 10$
$\left(\frac{1}{10}\right)^x = 10 \Rightarrow x = -1$

25. $\log_x 2 = -\dfrac{1}{3} \Rightarrow x^{-1/3} = 2$
$\left(x^{-1/3}\right)^{-3} = 2^{-3} \Rightarrow x = \dfrac{1}{8}$

26. $\log_x 32 = 5 \Rightarrow x^5 = 32 \Rightarrow x = 2$

27. $\log_{0.25} x = -1 \Rightarrow (0.25)^{-1} = x$
$\left(\dfrac{1}{4}\right)^{-1} = x \Rightarrow x = 4$

28. $\log_{0.125} x = -\dfrac{1}{3} \Rightarrow (0.125)^{-1/3} = x$
$\left(\dfrac{1}{8}\right)^{-1/3} = x \Rightarrow x = 2$

29. $\log_{\sqrt{2}} 32 = x \Rightarrow \left(\sqrt{2}\right)^x = 32$
$\left(2^{1/2}\right)^x = 2^5 \Rightarrow \dfrac{1}{2}x = 5 \Rightarrow x = 10$

30. $\log_{\sqrt{5}} x = -4 \Rightarrow \left(\sqrt{5}\right)^{-4} = x$
$\left(5^{1/2}\right)^{-4} = x \Rightarrow 5^{-2} = x \Rightarrow x = \dfrac{1}{25}$

31. $\log_{\sqrt{3}} 9\sqrt{3} = x \Rightarrow \left(\sqrt{3}\right)^x = 9\sqrt{3}$
$\left(3^{1/2}\right)^x = 3^{5/2} \Rightarrow x = 5$

32. $\log_{\sqrt{5}} 5\sqrt{5} = x \Rightarrow \left(\sqrt{5}\right)^x = 5\sqrt{5}$
$\left(5^{1/2}\right)^x = 5^{3/2} \Rightarrow x = 3$

33. $y = f(x) = \log(x - 2)$
Shift $y = \log x$ right 2.

34. $y = f(x) = 3 + \log x$
Shift $y = \log x$ up 3.

35. $y = 4^x$
$y = \log_4 x$

36. $y = \left(\frac{1}{3}\right)^x$
$y = \log_{1/3} x$

37. dB gain $= 20 \log \frac{E_O}{E_I} = 20 \log \frac{18}{0.04} = 20 \log 450 \approx 53$ dB

38. $R = \log \frac{A}{P} = \log \frac{7500}{0.3} = \log 25{,}000 \approx 4.4$

39. $\ln 452 \approx 6.1137$

40. $\ln (\log 7.85) \approx \ln 0.8949 \approx -0.1111$

41. $\ln x = 2.336 \Rightarrow x = 10.3398$

42. $\ln x = \log 8.8 \Rightarrow x = 2.5715$

43. $y = f(x) = 1 + \ln x$
Shift $y = \ln x$ up 1.

44. $y = f(x) = \ln (x + 1)$
Shift $y = \ln x$ left 1.

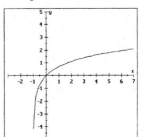

45. $t = \dfrac{\ln 2}{r} = \dfrac{\ln 2}{0.03} \approx 23$ years

46. $\log_7 1 = 0$

47. $\log_7 7 = 1$

48. $\log_7 7^3 = 3$

49. $7^{\log_7 4} = 4$

50. $\ln e^4 = 4$

51. $\ln 1 = 0$

52. $10^{\log_{10} 7} = 7$

53. $e^{\ln 3} = 3$

54. $\log_b b^4 = 4$

55. $\ln e^9 = 9$

56. $\log_b \dfrac{x^2 y^3}{z^4} = \log_b x^2 + \log_b y^3 - \log_b z^4 = 2 \log_b x + 3 \log_b y - 4 \log_b z$

57. $\log_b \sqrt{\dfrac{x}{yz^2}} = \log_b \left(\dfrac{x}{yz^2}\right)^{1/2} = \dfrac{1}{2}\log_b \dfrac{x}{yz^2} = \dfrac{1}{2}(\log_b x - \log_b y - \log_b z^2)$

$$= \dfrac{1}{2}(\log_b x - \log_b y - 2\log_b z)$$

58. $3\log_b x - 5\log_b y + 7\log_b z = \log_b x^3 - \log_b y^5 + \log_b z^7 = \log_b \dfrac{x^3 z^7}{y^5}$

59. $\dfrac{1}{2}\log_b x + 3\log_b y - 7\log_b z = \log_b x^{1/2} + \log_b y^3 - \log_b z^7 = \log_b \dfrac{y^3 \sqrt{x}}{z^7}$

60. $\log abc = \log a + \log b + \log c = 0.6 + 0.36 + 2.4 = 3.36$

61. $\log a^2 b = \log a^2 + \log b = 2\log a + \log b = 2(0.6) + 0.36 = 1.56$

62. $\log \dfrac{ac}{b} = \log a + \log c - \log b = 0.6 + 2.4 - 0.36 = 2.64$

63. $\log \dfrac{a^2}{c^3 b^2} = \log a^2 - \log c^3 - \log b^2 = 2\log a - 3\log c - 2\log b$

$$= 2(0.6) - 3(2.4) - 2(0.36) = -6.72$$

64. $\log_5 17 = \dfrac{\log 17}{\log 5} \approx 1.7604$

65. $\mathrm{pH} = -\log [\mathrm{H}^+]$
$3.1 = -\log [\mathrm{H}^+]$
$-3.1 = \log [\mathrm{H}^+]$
$[\mathrm{H}^+] = 7.94 \times 10^{-4}$ gram-ions per liter

66. $k\ln\left(\tfrac{1}{2}I\right) = k\left(\ln \tfrac{1}{2} + \ln I\right)$
$= k\ln \tfrac{1}{2} + k\ln I$
$= k\ln 2^{-1} + L$
$= -k\ln 2 + L$

The loudness decreases by $k\ln 2$.

67. $3^x = 7$
$\log 3^x = \log 7$
$x\log 3 = \log 7$
$x = \dfrac{\log 7}{\log 3} \approx 1.7712$

68. $5^{x+2} = 625$
$5^{x+2} = 5^4$
$x + 2 = 4$
$x = 2$

69.
$$25 = 5.5(1.05)^t$$
$$\frac{25}{5.5} = (1.05)^t$$
$$\log \frac{25}{5.5} = \log (1.05)^t$$
$$\log \frac{25}{5.5} = t \log 1.05$$
$$\frac{\log \frac{25}{5.5}}{\log 1.05} = t$$
$$31.0335 \approx t$$

70.
$$4^{2t-1} = 64$$
$$4^{2t-1} = 4^3$$
$$2t - 1 = 3$$
$$2t = 4$$
$$t = 2$$

71.
$$2^x = 3^{x-1}$$
$$\log 2^x = \log 3^{x-1}$$
$$x \log 2 = (x - 1) \log 3$$
$$x \log 2 = x \log 3 - \log 3$$
$$\log 3 = x \log 3 - x \log 2$$
$$\log 3 = x(\log 3 - \log 2)$$
$$\frac{\log 3}{\log 3 - \log 2} = x$$
$$2.7095 \approx x$$

72.
$$2^{x^2+4x} = \frac{1}{8}$$
$$2^{x^2+4x} = 2^{-3}$$
$$x^2 + 4x = -3$$
$$x^2 + 4x + 3 = 0$$
$$(x + 3)(x + 1) = 0$$
$$x + 3 = 0 \quad \textbf{or} \quad x + 1 = 0$$
$$x = -3 \qquad\qquad x = -1$$

73.
$$\log x + \log (29 - x) = 2$$
$$\log x(29 - x) = 2$$
$$10^2 = x(29 - x)$$
$$100 = 29x - x^2$$
$$x^2 - 29x + 100 = 0$$
$$(x - 25)(x - 4) = 0$$
$$x - 25 = 0 \quad \textbf{or} \quad x - 4 = 0$$
$$x = 25 \qquad\qquad x = 4$$

74.
$$\log_2 x + \log_2 (x - 2) = 3$$
$$\log_2 x(x - 2) = 3$$
$$x(x - 2) = 2^3$$
$$x^2 - 2x = 8$$
$$x^2 - 2x - 8 = 0$$
$$(x - 4)(x + 2) = 0$$
$$x - 4 = 0 \quad \textbf{or} \quad x + 2 = 0$$
$$x = 4 \qquad\qquad x = -2$$
Does not check.

75.
$$\log_2 (x + 2) + \log_2 (x - 1) = 2$$
$$\log_2 (x + 2)(x - 1) = 2$$
$$(x + 2)(x - 1) = 2^2$$
$$x^2 + x - 2 = 4$$
$$x^2 + x - 6 = 0$$
$$(x - 2)(x + 3) = 0$$
$$x - 2 = 0 \quad \textbf{or} \quad x + 3 = 0$$
$$x = 2 \qquad\qquad x = -3$$
Does not check.

76.
$$\frac{\log (7x - 12)}{\log x} = 2$$
$$\log (7x - 12) = 2 \log x$$
$$\log (7x - 12) = \log x^2$$
$$7x - 12 = x^2$$
$$0 = x^2 - 7x + 12$$
$$0 = (x - 4)(x - 3)$$
$$x - 4 = 0 \quad \textbf{or} \quad x - 3 = 0$$
$$x = 4 \qquad\qquad x = 3$$

77. $\log x + \log (x - 5) = \log 6$

$\log x(x - 5) = \log 6$

$x(x - 5) = 6$

$x^2 - 5x - 6 = 0$

$(x - 6)(x + 1) = 0$

$x - 6 = 0 \quad \textbf{or} \quad x + 1 = 0$

$x = 6 \qquad\qquad x = -1$

Does not check.

78. $\log 3 - \log (x - 1) = -1$

$\log \dfrac{3}{x - 1} = -1$

$\dfrac{3}{x - 1} = 10^{-1}$

$\dfrac{3}{x - 1} = \dfrac{1}{10}$

$30 = x - 1$

$31 = x$

79. $e^{x \ln 2} = 9$

$e^{\ln 2^x} = 9$

$2^x = 9$

$\ln 2^x = \ln 9$

$x \ln 2 = \ln 9$

$x = \dfrac{\ln 9}{\ln 2} \approx 3.1699$

80. $\ln x = \ln (x - 1)$

$x = x - 1$

$0 = -1$

There is no solution.

81. $\ln x = \ln (x - 1) + 1$

$\ln x - \ln (x - 1) = 1$

$\ln \dfrac{x}{x - 1} = 1$

$\dfrac{x}{x - 1} = e^1$

$x = e(x - 1)$

$x = ex - e$

$e = ex - x$

$e = x(e - 1)$

$\dfrac{e}{e - 1} = x$

$1.5820 \approx x$

82. $\ln x = \log_{10} x$

$\ln x = \dfrac{\ln x}{\ln 10}$

$\ln x \ln 10 = \ln x$

$\ln x \ln 10 - \ln x = 0$

$\ln x (\ln 10 - 1) = 0$

$\ln x = 0$

$x = 1$

83. $A = A_0 2^{-t/h}$

$\dfrac{2}{3} A_0 = A_0 2^{-t/5700}$

$\dfrac{2}{3} = 2^{-t/5700}$

$\log \dfrac{2}{3} = \log 2^{-t/5700}$

$\log \dfrac{2}{3} = -\dfrac{t}{5700} \log 2$

$\dfrac{\log \frac{2}{3}}{-\log 2} = \dfrac{t}{5700}$

$-5700 \dfrac{\log \frac{2}{3}}{\log 2} = t$

$3300 \text{ years} \approx t$

Chapter 11 Test (page 773)

1. $f(x) = 2^x + 1$; Shift $y = 2^x$ up 1.

2. $f(x) = 2^{-x}$; Reflect $y = 2^x$ about y-axis.

3. $A = A_0(2)^{-t} = 3(2)^{-6} = \dfrac{3}{2^6} = \dfrac{3}{64}$ gram

4. $A = A_0\left(1 + \dfrac{r}{k}\right)^{kt} = 1000\left(1 + \dfrac{0.06}{2}\right)^{2(1)} = 1000(1.03)^2 \approx \$1060.90$

5. $f(x) = e^x$

6. $\begin{aligned} A = A_0 e^{rt} &= 2000 e^{(0.08)10} \\ &= 2000 e^{0.8} \\ &\approx \$4451.08 \end{aligned}$

7. $\log_4 16 = x \Rightarrow 4^x = 16 \Rightarrow x = 2$

8. $\log_x 81 = 4 \Rightarrow x^4 = 81 \Rightarrow x = 3$

9. $\log_3 x = -3 \Rightarrow 3^{-3} = x \Rightarrow x = \dfrac{1}{27}$

10. $\log_x 100 = 2 \Rightarrow x^2 = 100 \Rightarrow x = 10$

11. $\log_{3/2} \dfrac{9}{4} = x \Rightarrow \left(\dfrac{3}{2}\right)^x = \dfrac{9}{4} \Rightarrow x = 2$

12. $\log_{2/3} x = -3 \Rightarrow \left(\dfrac{2}{3}\right)^{-3} = x \Rightarrow x = \dfrac{27}{8}$

13. $f(x) = -\log_3 x$

14. $f(x) = \ln x$

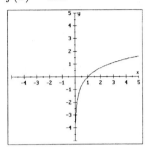

15. $\log a^2bc^3 = \log a^2 + \log b + \log c^3 = 2\log a + \log b + 3\log c$

16. $\ln\sqrt{\dfrac{a}{b^2c}} = \ln\left(\dfrac{a}{b^2c}\right)^{1/2} = \dfrac{1}{2}\ln\dfrac{a}{b^2c} = \dfrac{1}{2}(\ln a - \ln b^2 - \ln c) = \dfrac{1}{2}\ln a - \ln b - \dfrac{1}{2}\ln c$

17. $\dfrac{1}{2}\log(a+2) + \log b - 3\log c = \log(a+2)^{1/2} + \log b - \log c^3 = \log\dfrac{b\sqrt{a+2}}{c^3}$

18. $\dfrac{1}{3}(\log a - 2\log b) - \log c = \dfrac{1}{3}(\log a - \log b^2) - \log c = \dfrac{1}{3}\log\dfrac{a}{b^2} - \log c = \log\sqrt[3]{\dfrac{a}{b^2}} - \log c$

$$= \log\dfrac{\sqrt[3]{a}}{c\sqrt[3]{b^2}}$$

19. $\log 24 = \log 8 \cdot 3 = \log 2^3 \cdot 3 = \log 2^3 + \log 3 = 3\log 2 + \log 3 = 3(0.3010) + 0.4771 = 1.3801$

20. $\log\dfrac{8}{3} = \log\dfrac{2^3}{3} = \log 2^3 - \log 3 = 3\log 2 - \log 3 = 3(0.3010) - 0.4771 = 0.4259$

21. $\log_7 3 = \dfrac{\log 3}{\log 7}$ or $\dfrac{\ln 3}{\ln 7}$

22. $\log_\pi e = \dfrac{\log e}{\log \pi}$ or $\dfrac{\ln e}{\ln \pi}$

23. $\log_a ab = \log_a a + \log_a b = 1 + \log_a b \Rightarrow$ TRUE

24. $\dfrac{\log a}{\log b} = \log a - \log b \Rightarrow$ FALSE $\left(\log\dfrac{a}{b} = \log a - \log b\right)$

25. $\log a^{-3} = -3\log a \neq \dfrac{1}{3\log a} \Rightarrow$ FALSE

26. $\ln(-x) = -\ln x \Rightarrow$ FALSE (This implies one of the logarithms is negative, which is impossible.)

27. $\text{pH} = -\log[\text{H}^+] = -\log(3.7 \times 10^{-7}) \approx 6.4$

28. $\text{dB gain} = 20\log\dfrac{E_O}{E_I} = 20\log\dfrac{60}{0.3} = 20\log 200 \approx 46$

29.
$$5^x = 3$$
$$\log 5^x = \log 3$$
$$x \log 5 = \log 3$$
$$x = \frac{\log 3}{\log 5}$$

30.
$$3^{x-1} = 100^x$$
$$\log 3^{x-1} = \log 100^x$$
$$(x-1)\log 3 = x \log 100$$
$$x \log 3 - \log 3 = 2x$$
$$x \log 3 - 2x = \log 3$$
$$x(\log 3 - 2) = \log 3$$
$$x = \frac{\log 3}{\log 3 - 2}$$

31.
$$\log(5x + 2) = \log(2x + 5)$$
$$5x + 2 = 2x + 5$$
$$3x = 3$$
$$x = 1$$

32.
$$\log x + \log(x - 9) = 1$$
$$\log x(x - 9) = 1$$
$$x(x - 9) = 10$$
$$x^2 - 9x - 10 = 0$$
$$(x - 10)(x + 1) = 0$$
$$x - 10 = 0 \quad \textbf{or} \quad x + 1 = 0$$
$$x = 10 \qquad\qquad x = -1$$
$$\text{Does not check.}$$

Exercise 12.1 (page 785)

1.
$$|3x - 4| = 11$$
$$3x - 4 = 11 \quad \textbf{or} \quad 3x - 4 = -11$$
$$3x = 15 \qquad\qquad 3x = -7$$
$$x = 5 \qquad\qquad x = -\tfrac{7}{3}$$

3.
$$|3x + 4| = |5x - 2|$$
$$3x + 4 = 5x - 2 \quad \textbf{or} \quad 3x + 4 = -(5x - 2)$$
$$-2x = -6 \qquad\qquad 3x + 4 = -5x + 2$$
$$x = 3 \qquad\qquad 8x = -2$$
$$\qquad\qquad\qquad x = -\tfrac{1}{4}$$

5. circle; plane

7. $r^2 < 0$

9. parabola; $(3, 2)$; right

11. $x^2 + y^2 = 9$
C $(0,0)$; $r = \sqrt{9} = 3$

13. $(x - 2)^2 + y^2 = 9$
C $(2,0)$; $r = \sqrt{9} = 3$

15. $(x - 2)^2 + (y - 4)^2 = 4$
C $(2,4)$; $r = \sqrt{4} = 2$

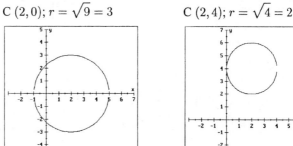

17. $(x+3)^2 + (y-1)^2 = 16$

C $(-3, 1)$; $r = \sqrt{16} = 4$

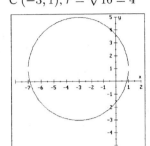

19. $x^2 + (y+3)^2 = 1$

C $(0, -3)$; $r = \sqrt{1} = 1$

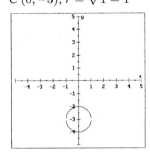

21. $3x^2 + 3y^2 = 16$

$$3y^2 = 16 - 3x^2$$

$$y^2 = \frac{16 - 3x^2}{3}$$

$$y = \pm\sqrt{\frac{16 - 3x^2}{3}}$$

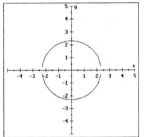

23. $(x+1)^2 + y^2 = 16$

$$y^2 = 16 - (x+1)^2$$

$$y = \pm\sqrt{16 - (x+1)^2}$$

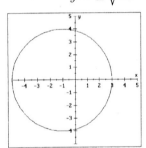

25. $(x-h)^2 + (y-k)^2 = r^2$

$$(x-0)^2 + (y-0)^2 = 1^2$$

$$x^2 + y^2 = 1$$

27. $(x-h)^2 + (y-k)^2 = r^2$

$$(x-6)^2 + (y-8)^2 = 5^2$$

$$(x-6)^2 + (y-8)^2 = 25$$

29. $(x-h)^2 + (y-k)^2 = r^2$

$$(x-(-2))^2 + (y-6)^2 = 12^2$$

$$(x+2)^2 + (y-6)^2 = 144$$

31. $(x-h)^2 + (y-k)^2 = r^2$

$$(x-0)^2 + (y-0)^2 = \left(\sqrt{2}\right)^2$$

$$x^2 + y^2 = 2$$

33. $x^2 + y^2 + 2x - 8 = 0$
$$x^2 + 2x + y^2 = 8$$
$$x^2 + 2x + 1 + y^2 = 8 + 1$$
$$(x+1)^2 + y^2 = 9$$
$$C(-1, 0);\ r = \sqrt{9} = 3$$

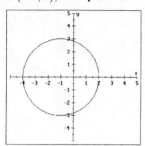

35. $9x^2 + 9y^2 - 12y = 5$
$$x^2 + y^2 - \frac{4}{3}y = \frac{5}{9}$$
$$x^2 + y^2 - \frac{4}{3}y + \frac{4}{9} = \frac{5}{9} + \frac{4}{9}$$
$$x^2 + \left(y - \frac{2}{3}\right)^2 = 1$$
$$C\left(0, \frac{2}{3}\right);\ r = \sqrt{1} = 1$$

37. $x^2 + y^2 - 2x + 4y = -1$
$$x^2 - 2x + y^2 + 4y = -1$$
$$x^2 - 2x + 1 + y^2 + 4y + 4 = -1 + 1 + 4$$
$$(x-1)^2 + (y+2)^2 = 4$$
$$C(1, -2);\ r = \sqrt{4} = 2$$

39. $x^2 + y^2 + 6x - 4y = -12$
$$x^2 + 6x + y^2 - 4y = -12$$
$$x^2 + 6x + 9 + y^2 - 4y + 4 = -12 + 9 + 4$$
$$(x+3)^2 + (y-2)^2 = 1$$
$$C(-3, 2);\ r = \sqrt{1} = 1$$

41. $x = y^2$

$x = (y - 0)^2 + 0$

V $(0, 0)$; opens R

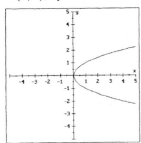

43. $x = -\dfrac{1}{4}y^2$

$x = -\dfrac{1}{4}(y - 0)^2 + 0$

V $(0, 0)$; opens L

45. $y = x^2 + 4x + 5$

$y = x^2 + 4x + 4 + 5 - 4$

$y = (x + 2)^2 + 1$

V $(-2, 1)$; opens U

47. $y = -x^2 - x + 1$

$y = -(x^2 + x) + 1$

$y = -\left(x^2 + x + \dfrac{1}{4}\right) + 1 + \dfrac{1}{4}$

$y = -\left(x + \dfrac{1}{2}\right)^2 + \dfrac{5}{4}$

V $\left(-\dfrac{1}{2}, \dfrac{5}{4}\right)$; opens D

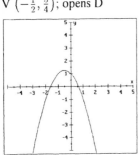

49. $y^2 + 4x - 6y = -1$

$4x = -y^2 + 6y - 1$

$x = -\dfrac{1}{4}y^2 + \dfrac{3}{2}y - \dfrac{1}{4}$

$x = -\dfrac{1}{4}(y^2 - 6y) - \dfrac{1}{4}$

$x = -\dfrac{1}{4}(y^2 - 6y + 9) - \dfrac{1}{4} + \dfrac{9}{4}$

$x = -\dfrac{1}{4}(y - 3)^2 + 2$

V $(2, 3)$; opens L

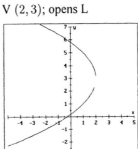

51. $y = 2(x-1)^2 + 3$
V $(1,3)$; opens U

53. $x = 2y^2$

$y^2 = \dfrac{x}{2}$

$y = \pm\sqrt{\dfrac{x}{2}}$

55. $x^2 - 2x + y = 6$

$y = 6 - x^2 + 2x$

57. The radius of the larger gear is $\sqrt{16} = 4$. Centers: 7 units apart $\Rightarrow$ smaller gear $r = 3$.
$(x-h)^2 + (y-k)^2 = r^2$
$(x-7)^2 + (y-0)^2 = 3^2 \Rightarrow (x-7)^2 + y^2 = 9$

59.
$$x^2 + y^2 - 8x - 20y + 16 = 0$$
$$x^2 - 8x + y^2 - 20y = -16$$
$$x^2 - 8x + 16 + y^2 - 20y + 100 = -16 + 16 + 100$$
$$(x-4)^2 + (y-10)^2 = 100$$
center: $(4, 10)$; radius $= 10$
$$x^2 + y^2 + 2x + 4y - 11 = 0$$
$$x^2 + 2x + 1 + y^2 + 4y + 4 = 11 + 1 + 4$$
$$(x+1)^2 + (y+2)^2 = 16$$
center: $(-1, -2)$; radius $= 4$

Since the ranges overlap (see graph), they can not be licensed for the same frequency.

61. Set $y = 0$:
$y = 30x - x^2$
$0 = 30x - x^2$
$0 = x(30 - x)$
$x = 0$ or $x = 30$
It lands 30 feet away.

63. Find the vertex:
$2y^2 - 9x = 18$
$-9x = -2y^2 + 18$
$x = \dfrac{2}{9}y^2 - 2$
$x = \dfrac{2}{9}(y-0)^2 - 2$
vertex: $(-2, 0) \Rightarrow$ distance $= 2$ AU

65. **Answers may vary.**

67. **Answers may vary.**

Exercise 12.2 (page 796)

1. $3x^{-2}y^2(4x^2 + 3y^{-2}) = 12x^0y^2 + 9x^{-2}y^0 = 12y^2 + \dfrac{9}{x^2}$

3. $\dfrac{x^{-2} + y^{-2}}{x^{-2} - y^{-2}} = \dfrac{x^{-2} + y^{-2}}{x^{-2} - y^{-2}} \cdot \dfrac{x^2y^2}{x^2y^2} = \dfrac{y^2 + x^2}{y^2 - x^2}$

5. ellipse; sum **7.** center **9.** $(0,0)$

11. $\dfrac{x^2}{4} + \dfrac{y^2}{9} = 1$

C $(0,0)$; move 2 horiz. and 3 vert.

13. $x^2 + 9y^2 = 9$

$\dfrac{x^2}{9} + \dfrac{9y^2}{9} = \dfrac{9}{9}$

$\dfrac{x^2}{9} + \dfrac{y^2}{1} = 1$

C $(0,0)$; move 3 horiz. and 1 vert.

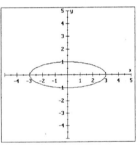

15. $16x^2 + 4y^2 = 64$

$\dfrac{16x^2}{64} + \dfrac{4y^2}{64} = \dfrac{64}{64}$

$\dfrac{x^2}{4} + \dfrac{y^2}{16} = 1$

C $(0,0)$; move 2 horiz. and 4 vert.

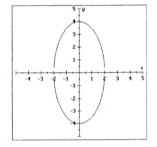

17. $\dfrac{(x-2)^2}{9} + \dfrac{(y-1)^2}{4} = 1$

C $(2,1)$; move 3 horiz. and 2 vert.

19. $(x+1)^2 + 4(y+2)^2 = 4$

$\dfrac{(x+1)^2}{4} + \dfrac{4(y+2)^2}{4} = \dfrac{4}{4}$

$\dfrac{(x+1)^2}{4} + \dfrac{(y+2)^2}{1} = 1$

C $(-1,-2)$; move 2 horiz. and 1 vert.

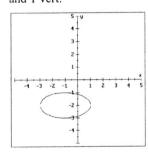

21. $\dfrac{x^2}{9} + \dfrac{y^2}{4} = 1$

$\dfrac{y^2}{4} = 1 - \dfrac{x^2}{9}$

$y^2 = 4\left(1 - \dfrac{x^2}{9}\right)$

$y = \pm\sqrt{4\left(1 - \dfrac{x^2}{9}\right)}$

23. $\dfrac{x^2}{4} + \dfrac{(y-1)^2}{9} = 1$

$\dfrac{(y-1)^2}{9} = 1 - \dfrac{x^2}{4}$

$(y-1)^2 = 9\left(1 - \dfrac{x^2}{4}\right)$

$y - 1 = \pm\sqrt{9\left(1 - \dfrac{x^2}{4}\right)}$

$y = 1 \pm \sqrt{9\left(1 - \dfrac{x^2}{4}\right)}$

25.
$$x^2 + 4y^2 - 4x + 8y + 4 = 0$$
$$x^2 - 4x + 4\left(y^2 + 2y\right) = -4$$
$$x^2 - 4x + 4 + 4\left(y^2 + 2y + 1\right) = -4 + 4 + 4$$
$$(x-2)^2 + 4(y+1)^2 = 4$$
$$\dfrac{(x-2)^2}{4} + \dfrac{(y+1)^2}{1} = 1$$
C $(2, -1)$; move 2 horiz. and 1 vert.

27.
$$9x^2 + 4y^2 - 18x + 16y = 11$$
$$9\left(x^2 - 2x\right) + 4\left(y^2 + 4y\right) = 11$$
$$9\left(x^2 - 2x + 1\right) + 4\left(y^2 + 4y + 4\right) = 11 + 9 + 16$$
$$9(x-1)^2 + 4(y+2)^2 = 36$$
$$\dfrac{(x-1)^2}{4} + \dfrac{(y+2)^2}{9} = 1$$
C $(1, -2)$; move 2 horiz. and 3 vert.

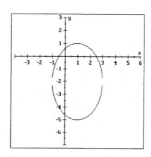

29. Note: $a = 40/2 = 20$, $b = 10$

$$\frac{x^2}{a^2} + \frac{y^2}{b^2} = 1$$

$$\frac{x^2}{400} + \frac{y^2}{100} = 1$$

$$x^2 + 4y^2 = 400$$

$$4y^2 = 400 - x^2$$

$$y^2 = \tfrac{1}{4}(400 - x^2)$$

$$y = \tfrac{1}{2}\sqrt{400 - x^2}$$

31. $9x^2 + 16y^2 = 144$

$$\frac{9x^2}{144} + \frac{16y^2}{144} = \frac{144}{144}$$

$$\frac{x^2}{16} + \frac{y^2}{9} = 1$$

area $= \pi a b$

$= \pi(4)(3) = 12\pi$ square units

33. **Answers may vary.**

35. It is a circle.

Exercise 12.3 (page 805)

1. $-6x^4 + 9x^3 - 6x^2 = -3x^2(2x^2 - 3x + 2)$ **3.** $15a^2 - 4ab - 4b^2 = (3a - 2b)(5a + 2b)$

5. hyperbola; difference   **7.** center   **9.** $(0, 0)$

11. $\dfrac{x^2}{9} - \dfrac{y^2}{4} = 1$

C $(0,0)$; open horiz.;
move 3 horiz. and 2 vert.

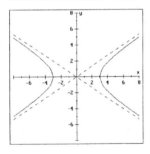

13. $\dfrac{y^2}{4} - \dfrac{x^2}{9} = 1$

C $(0,0)$; open vert.;
move 3 horiz. and 2 vert.

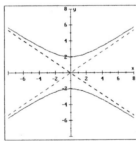

15. $25x^2 - y^2 = 25$

$$\frac{25x^2}{25} - \frac{y^2}{25} = 1$$

$$\frac{x^2}{1} - \frac{y^2}{25} = 1$$

C $(0,0)$; open horiz.;
move 1 horiz. and 5 vert.

17. $\dfrac{(x-2)^2}{9} - \dfrac{y^2}{16} = 1$

C $(2,0)$; open horiz.;
move 3 horiz. and 4 vert.

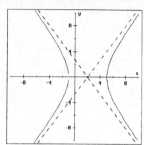

19. $4(x+3)^2 - (y-1)^2 = 4$

$$\dfrac{4(x+3)^2}{4} - \dfrac{(y-1)^2}{4} = \dfrac{4}{4}$$

$$\dfrac{(x+3)^2}{1} - \dfrac{(y-1)^2}{4} = 1$$

C $(-3,1)$; open horiz.;
move 1 horiz. and 2 vert.

21. $xy = 8$

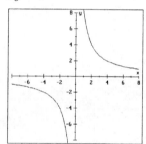

23. $\dfrac{x^2}{9} - \dfrac{y^2}{4} = 1$

$$\dfrac{y^2}{4} = \dfrac{x^2}{9} - 1$$

$$y^2 = 4\left(\dfrac{x^2}{9} - 1\right)$$

$$y = \pm\sqrt{4\left(\dfrac{x^2}{9} - 1\right)}$$

363

SECTION 12.3

25.
$$\frac{x^2}{4} - \frac{(y-1)^2}{9} = 1$$
$$\frac{(y-1)^2}{9} = \frac{x^2}{4} - 1$$
$$(y-1)^2 = 9\left(\frac{x^2}{4} - 1\right)$$
$$y - 1 = \pm\sqrt{9\left(\frac{x^2}{4} - 1\right)}$$
$$y = 1 \pm \sqrt{9\left(\frac{x^2}{4} - 1\right)}$$

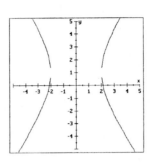

27.
$$4x^2 - y^2 + 8x - 4y = 4$$
$$4x^2 + 8x - y^2 - 4y = 4$$
$$4(x^2 + 2x) - (y^2 + 4y) = 4$$
$$4(x^2 + 2x + 1) - (y^2 + 4y + 4) = 4 + 4 - 4$$
$$4(x+1)^2 - (y+2)^2 = 4$$
$$\frac{(x+1)^2}{1} - \frac{(y+2)^2}{4} = 1$$
C $(-1, -2)$; opens horiz.; move 1 horiz. and 2 vert.

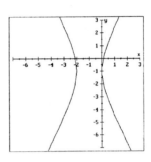

29.
$$4y^2 - x^2 + 8y + 4x = 4$$
$$4y^2 + 8y - x^2 + 4x = 4$$
$$4(y^2 + 2y) - (x^2 - 4x) = 4$$
$$4(y^2 + 2y + 1) - (x^2 - 4x + 4) = 4 + 4 - 4$$
$$4(y+1)^2 - (x-2)^2 = 4$$
$$\frac{(y+1)^2}{1} - \frac{(x-2)^2}{4} = 1$$
C $(2, -1)$; opens vert.; move 2 horiz. and 1 vert.

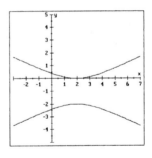

31.
$$9y^2 - x^2 = 81$$
$$\frac{9y^2}{81} - \frac{x^2}{81} = \frac{81}{81}$$
$$\frac{y^2}{9} - \frac{x^2}{81} = 1$$
distance $= \sqrt{9} = 3$ units

33.
$$y^2 - x^2 = 25$$
$$\frac{y^2}{25} - \frac{x^2}{25} = 1$$
vertex: $(0, 5)$
Let $y = 10$:
$$10^2 - x^2 = 25$$
$$-x^2 = -75$$
$$x = \sqrt{75} = 5\sqrt{3}$$
width $= 2(5\sqrt{3}) = 10\sqrt{3}$ miles

35. **Answers may vary.**

37. If $a = b$, the rectangle is a square.

Exercise 12.4 (page 811)

1. $(6x - 10) + (3x + 10) = 180$
$$9x = 180$$
$$x = 20$$

3. domains

5. constant; $f(x)$

7. step

9. increasing on $(-\infty, 0)$, decreasing on $(0, \infty)$

11. decreasing on $(-\infty, 0)$, constant on $(0, 2)$, increasing on $(2, \infty)$

13. $f(x) = \begin{cases} -1 & \text{if } x \le 0 \\ x & \text{if } x > 0 \end{cases}$

15. $f(x) = \begin{cases} -x & \text{if } x \le 0 \\ x & \text{if } 0 < x < 2 \\ -x & \text{if } x \ge 2 \end{cases}$

17. $f(x) = -[[x]]$

constant on $(-\infty, 0)$
increasing on $(0, \infty)$

decreasing on $(-\infty, 0)$
increasing on $(0, 2)$
decreasing on $(2, \infty)$

19. $f(x) = 2[[x]]$

21. $f(x) = \begin{cases} -1 & \text{if } x < 0 \\ 0 & \text{if } x = 0 \\ 1 & \text{if } x > 0 \end{cases}$

23. Find y when $x = 2.5$.
cost = \$30

25. After 2 hours, B is cheaper.

27. Answers may vary.

29. $f(x) = \begin{cases} x & \text{if } x < -2 \\ -x & \text{if } x > -2 \end{cases}$

Chapter 12 Summary (page 814)

1. $(x-1)^2 + (y+2)^2 = 9$

$C(1, -2); r = \sqrt{9} = 3$

2. $x^2 + y^2 = 16$

$C(0,0); r = \sqrt{16} = 4$

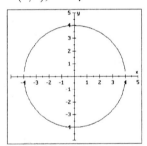

3.
$$x^2 + y^2 + 4x - 2y = 4$$
$$x^2 + 4x + y^2 - 2y = 4$$
$$x^2 + 4x + 4 + y^2 - 2y + 1 = 4 + 4 + 1$$
$$(x+2)^2 + (y-1)^2 = 9$$
$$C(-2, 1); r = \sqrt{9} = 3$$

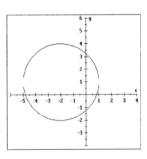

4. $x = -3(y-2)^2 + 5$
V $(5, 2)$; opens L

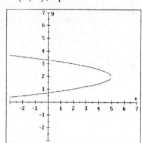

5. $x = 2(y+1)^2 - 2$
V $(-2, -1)$; opens R

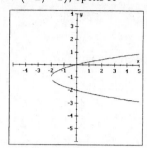

6. $9x^2 + 16y^2 = 144$

$$\dfrac{9x^2}{144} + \dfrac{16y^2}{144} = \dfrac{144}{144}$$

$$\dfrac{x^2}{16} + \dfrac{y^2}{9} = 1$$

C $(0, 0)$; move 4 horiz. and 3 vert.

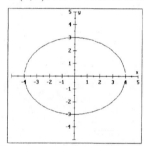

7. $\dfrac{(x-2)^2}{4} + \dfrac{(y-1)^2}{9} = 1$

C $(2, 1)$; move 2 horiz. and 3 vert.

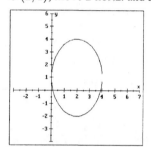

8.

$$4x^2 + 9y^2 + 8x - 18y = 23$$
$$4x^2 + 8x + 9y^2 - 18y = 23$$
$$4(x^2 + 2x) + 9(y^2 - 2y) = 23$$
$$4(x^2 + 2x + 1) + 9(y^2 - 2y + 1) = 23 + 4 + 9$$
$$4(x+1)^2 + 9(y-1)^2 = 36$$
$$\dfrac{(x+1)^2}{9} + \dfrac{(y-1)^2}{4} = 1$$

C $(-1, 1)$; move 3 horiz. and 2 vert.

9.
$$9x^2 - y^2 = -9$$
$$\frac{9x^2}{-9} - \frac{y^2}{-9} = \frac{-9}{-9}$$
$$\frac{y^2}{9} - \frac{x^2}{1} = 1$$
C $(0,0)$; opens vert.; move 1 horiz. and 3 vert.

10. $xy = 9$

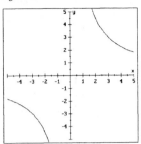

11.
$$4x^2 - 2y^2 + 8x - 8y = 8$$
$$4x^2 + 8x - 2y^2 - 8y = 8$$
$$4(x^2 + 2x) - 2(y^2 + 4y) = 8$$
$$4(x^2 + 2x + 1) - 2(y^2 + 4y + 4) = 8$$
$$4(x+1)^2 - 2(y+2)^2 = 8$$
$$\frac{(x+1)^2}{2} - \frac{(y+2)^2}{4} = 1 \Rightarrow \text{hyperbola}$$

12.
$$9x^2 - 4y^2 - 18x - 8y = 31$$
$$9x^2 - 18x - 4y^2 - 8y = 31$$
$$9(x^2 - 2x) - 4(y^2 + 2y) = 31$$
$$9(x^2 - 2x + 1) - 4(y^2 + 2y + 1) = 31 + 9 - 4$$
$$9(x-1)^2 - 4(y+1)^2 = 36$$
$$\frac{(x-1)^2}{4} - \frac{(y+1)^2}{9} = 1$$
C $(1,-1)$; opens horiz.; move 2 horiz. and 3 vert.

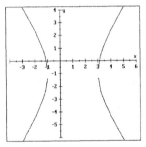

13. increasing on $(-\infty, -2)$; constant on $(-2, 1)$; decreasing on $(1, \infty)$

14. $f(x) = \begin{cases} x & \text{if } x \leq 1 \\ -x^2 & \text{if } x > 1 \end{cases}$

15. $f(x) = 3[[x]]$

Chapter 12 Test (page 817)

1. $(x-2)^2 + (y+3)^2 = 4$
Center: $(2, -3)$; radius $= 2$

2.
$$x^2 + y^2 + 4x - 6y = 3$$
$$x^2 + 4x + y^2 - 6y = 3$$
$$x^2 + 4x + 4 + y^2 - 6y + 9 = 3 + 4 + 9$$
$$(x+2)^2 + (y-3)^2 = 16$$
Center: $(-2, 3)$; radius $= 4$

3. $(x+1)^2 + (y-2)^2 = 9$
$C(-1, 2); r = \sqrt{9} = 3$

4. $x = (y-2)^2 - 1$
$V(-1, 2)$; opens R

5.
$$9x^2 + 4y^2 = 36$$
$$\frac{9x^2}{36} + \frac{4y^2}{36} = \frac{36}{36}$$
$$\frac{x^2}{4} + \frac{y^2}{9} = 1$$
C $(0,0)$; move 2 horiz. and 3 vert.

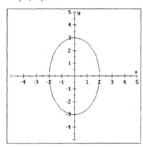

6.
$$\frac{(x-2)^2}{9} - y^2 = 1$$
C $(2,0)$; opens horiz; move 3 horiz and 1 **vert**

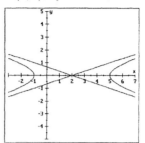

7.
$$4x^2 + y^2 - 24x + 2y = -33$$
$$4(x^2 - 6x) + (y^2 + 2y) = -33$$
$$4(x^2 - 6x + 9) + (y^2 + 2y + 1) = -33 + 36 + 1$$
$$4(x-3)^2 + (y+1)^2 = 4$$
$$\frac{(x-3)^2}{1} + \frac{(y+1)^2}{4} = 1$$
C $(3,-1)$; move 1 horiz. and 2 vert.

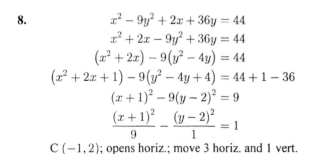

8.
$$x^2 - 9y^2 + 2x + 36y = 44$$
$$x^2 + 2x - 9y^2 + 36y = 44$$
$$(x^2 + 2x) - 9(y^2 - 4y) = 44$$
$$(x^2 + 2x + 1) - 9(y^2 - 4y + 4) = 44 + 1 - 36$$
$$(x+1)^2 - 9(y-2)^2 = 9$$
$$\frac{(x+1)^2}{9} - \frac{(y-2)^2}{1} = 1$$
C $(-1,2)$; opens horiz.; move 3 horiz. and 1 vert.

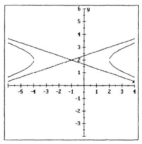

9. increasing: $(-3, 0)$; decreasing: $(0, 3)$

10. $f(x) = \begin{cases} -x^2 & \text{when } x < 0 \\ -x & \text{when } x \geq 0 \end{cases}$

Cumulative Review Exercises (page 818)

1. $(4x - 3y)(3x + y) = 12x^2 + 4xy - 9xy - 3y^2 = 12x^2 - 5xy - 3y^2$

2. $(a^n + 1)(a^n - 3) = a^n a^n - 3a^n + a^n - 3 = a^{2n} - 2a^n - 3$

3. $\dfrac{5a - 10}{a^2 - 4a + 4} = \dfrac{5(a - 2)}{(a - 2)(a - 2)} = \dfrac{5}{a - 2}$

4. $\dfrac{a^4 - 5a^2 + 4}{a^2 + 3a + 2} = \dfrac{(a^2 - 4)(a^2 - 1)}{(a + 2)(a + 1)} = \dfrac{(a + 2)(a - 2)(a + 1)(a - 1)}{(a + 2)(a + 1)} = (a - 2)(a - 1)$

$$= a^2 - 3a + 2$$

5. $\dfrac{a^2 - a - 6}{a^2 - 4} \div \dfrac{a^2 - 9}{a^2 + a - 6} = \dfrac{a^2 - a - 6}{a^2 - 4} \cdot \dfrac{a^2 + a - 6}{a^2 - 9} = \dfrac{(a - 3)(a + 2)}{(a + 2)(a - 2)} \cdot \dfrac{(a + 3)(a - 2)}{(a + 3)(a - 3)} = 1$

6. $\dfrac{2}{a - 2} + \dfrac{3}{a + 2} - \dfrac{a - 1}{a^2 - 4} = \dfrac{2}{a - 2} + \dfrac{3}{a + 2} - \dfrac{a - 1}{(a + 2)(a - 2)}$

$$= \dfrac{2(a + 2)}{(a - 2)(a + 2)} + \dfrac{3(a - 2)}{(a + 2)(a - 2)} - \dfrac{a - 1}{(a + 2)(a - 2)}$$

$$= \dfrac{2(a + 2) + 3(a - 2) - (a - 1)}{(a + 2)(a - 2)}$$

$$= \dfrac{2a + 4 + 3a - 6 - a + 1}{(a + 2)(a - 2)} = \dfrac{4a - 1}{(a + 2)(a - 2)}$$

7. $3x - 4y = 12$ $y = \dfrac{3}{4}x - 5$

$-4y = -3x + 12$

$y = \dfrac{3}{4}x - 3$ $m = \dfrac{3}{4}$

$m = \dfrac{3}{4}$

Parallel

8. $y = 3x + 4$ $x = -3y + 4$

$m = 3$ $3y = -x + 4$

$y = -\dfrac{1}{3}x + \dfrac{4}{3}$

$m = -\dfrac{1}{3}$

Perpendicular

9.
$$y - y_1 = m(x - x_1)$$
$$y - 5 = -2(x - 0)$$
$$y - 5 = -2x$$
$$y = -2x + 5$$

10.
$$m = \frac{y_2 - y_1}{x_2 - x_1} = \frac{4 - (-5)}{-5 - 8} = \frac{9}{-13} = -\frac{9}{13}$$
$$y - y_1 = m(x - x_1)$$
$$y - (-5) = -\frac{9}{13}(x - 8)$$
$$y + 5 = -\frac{9}{13}x + \frac{72}{13}$$
$$y = -\frac{9}{13}x + \frac{7}{13}$$

11. $2x - 3y < 6$

12. $y \geq x^2 - 4$

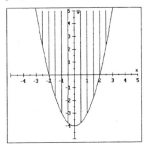

13. $\sqrt{98} + \sqrt{8} - \sqrt{32} = \sqrt{49}\sqrt{2} + \sqrt{4}\sqrt{2} - \sqrt{16}\sqrt{2} = 7\sqrt{2} + 2\sqrt{2} - 4\sqrt{2} = 5\sqrt{2}$

14. $12\sqrt[3]{648x^4} + 3\sqrt[3]{81x^4} = 12\sqrt[3]{216x^3}\sqrt[3]{3x} + 3\sqrt[3]{27x^3}\sqrt[3]{3x} = 12(6x)\sqrt[3]{3x} + 3(3x)\sqrt[3]{3x}$
$$= 72x\sqrt[3]{3x} + 9x\sqrt[3]{3x} = 81x\sqrt[3]{3x}$$

15.
$$\sqrt{3a + 1} = a - 1$$
$$\left(\sqrt{3a + 1}\right)^2 = (a - 1)^2$$
$$3a + 1 = a^2 - 2a + 1$$
$$0 = a^2 - 5a$$
$$0 = a(a - 5)$$
$$a = 0 \quad \textbf{or} \quad a - 5 = 0$$
$$\text{doesn't check} \qquad a = 5$$

16.
$$\sqrt{x + 3} - \sqrt{3} = \sqrt{x}$$
$$\sqrt{x + 3} = \sqrt{x} + \sqrt{3}$$
$$\left(\sqrt{x + 3}\right)^2 = \left(\sqrt{x} + \sqrt{3}\right)^2$$
$$x + 3 = x + 2\sqrt{3x} + 3$$
$$0 = 2\sqrt{3x}$$
$$0^2 = \left(2\sqrt{3x}\right)^2$$
$$0 = 4(3x)$$
$$0 = 12x$$
$$0 = x$$

17.
$$6a^2 + 5a - 6 = 0$$
$$(2a + 3)(3a - 2) = 0$$
$$2a + 3 = 0 \quad \textbf{or} \quad 3a - 2 = 0$$
$$a = -\frac{3}{2} \qquad\qquad a = \frac{2}{3}$$

18.
$$3x^2 + 8x - 1 = 0$$
$$a = 3, b = 8, c = -1$$
$$x = \frac{-b \pm \sqrt{b^2 - 4ac}}{2a}$$
$$= \frac{-8 \pm \sqrt{8^2 - 4(3)(-1)}}{2(3)}$$
$$= \frac{-8 \pm \sqrt{64 + 12}}{6}$$
$$= \frac{-8 \pm \sqrt{76}}{6}$$
$$= -\frac{8}{6} \pm \frac{2\sqrt{19}}{6} = -\frac{4}{3} \pm \frac{\sqrt{19}}{3}$$

19. $(f \circ g)(x) = f(g(x)) = f(2x + 1) = (2x + 1)^2 - 2 = 4x^2 + 4x + 1 - 2 = 4x^2 + 4x - 1$

20.
$$y = 2x^3 - 1$$
$$x = 2y^3 - 1$$
$$x + 1 = 2y^3$$
$$\frac{x + 1}{2} = y^3$$
$$\sqrt[3]{\frac{x + 1}{2}} = y$$
$$y = f^{-1}(x) = \sqrt[3]{\frac{x + 1}{2}}$$

21. $y = \left(\dfrac{1}{2}\right)^x$

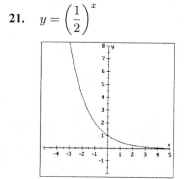

22. $y = \log_2 x \Rightarrow 2^y = x$

23.
$$2^{x+2} = 3^x$$
$$\log 2^{x+2} = \log 3^x$$
$$(x + 2)\log 2 = x \log 3$$
$$x \log 2 + 2 \log 2 = x \log 3$$
$$2 \log 2 = x \log 3 - x \log 2$$
$$2 \log 2 = x(\log 3 - \log 2)$$
$$\frac{2 \log 2}{\log 3 - \log 2} = x$$

24.
$$2 \log 5 + \log x - \log 4 = 2$$
$$\log 5^2 + \log x - \log 4 = 2$$
$$\log \frac{25x}{4} = 2$$
$$10^2 = \frac{25x}{4}$$
$$400 = 25x$$
$$16 = x$$

25. $x^2 + (y+1)^2 = 9$

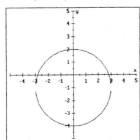

26. $x^2 - 9(y+1)^2 = 9$

$$\frac{x^2}{9} - \frac{(y+1)^2}{1} = 1$$

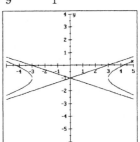

Exercise 13.1 (page 831)

1. $\left(a^2 a^3\right)^2 \left(a^4 a^2\right)^2 = \left(a^5\right)^2 \left(a^6\right)^2$
$$= a^{10} a^{12} = a^{22}$$

3. $\left(\dfrac{-3x^3 y^4}{x^{-5} y^3}\right)^{-4} = \left(\dfrac{x^{-5} y^3}{-3x^3 y^4}\right)^4$
$$= \left(\frac{1}{-3x^8 y}\right)^4 = \frac{1}{81 x^{32} y^4}$$

5. $\quad A = p + prt$
$A - p = prt$
$\dfrac{A-p}{pt} = r$, or $r = \dfrac{A-p}{pt}$

7. $\quad \dfrac{1}{r} = \dfrac{1}{r_1} + \dfrac{1}{r_2}$
$\dfrac{1}{r} \cdot rr_1 r_2 = \left(\dfrac{1}{r_1} + \dfrac{1}{r_2}\right) rr_1 r_2$
$r_1 r_2 = rr_2 + rr_1$
$r_1 r_2 = r(r_2 + r_1)$
$\dfrac{r_1 r_2}{r_2 + r_1} = r$, or $r = \dfrac{r_1 r_2}{r_2 + r_1}$

9. consistent

11. independent

13. $\begin{cases} x - y = 4 \\ 2x + y = 5 \end{cases}$

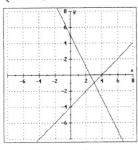

$(3, -1)$ is the solution.

15. $\begin{cases} x = 13 - 4y \\ 3x = 4 + 2y \end{cases}$

$\left(3, \frac{5}{2}\right)$ is the solution.

17. $\begin{cases} 2x + 3y = 0 \\ 2x + y = 4 \end{cases}$

$(3, -2)$ is the solution.

19. $\begin{cases} x = 3 - 2y \\ 2x + 4y = 6 \end{cases}$

dependent equations

21. $\begin{cases} y = 3 \\ x = 2 \end{cases}$

$(2, 3)$ is the solution.

23. $\begin{cases} x = \dfrac{11 - 2y}{3} \\ y = \dfrac{11 - 6x}{4} \end{cases}$

inconsistent system

25. $\begin{cases} \dfrac{5}{2}x + y = \dfrac{1}{2} \\ 2x - \dfrac{3}{2}y = 5 \end{cases}$

$(1, -2)$ is the solution.

27. $\begin{cases} y = 3.2x - 1.5 \\ y = -2.7x - 3.7 \end{cases}$

solution: $(-0.37, -2.69)$

29. $\begin{cases} 1.7x + 2.3y = 3.2 \\ y = 0.25x + 8.95 \end{cases}$

solution: $(-7.64, 7.04)$

31. $\begin{cases} (1) \quad y = x \\ (2) \quad x + y = 4 \end{cases}$

Substitute $y = x$ from (1) into (2):

$x + \boldsymbol{y} = 4$

$x + \boldsymbol{x} = 4$

$\quad 2x = 4$

$\quad\ \ x = 2$

Substitute this and solve for y:

$\quad y = x = 2$

Solution: $(2, 2)$

33. $\begin{cases} (1) \quad x - y = 2 \\ (2) \quad 2x + y = 13 \end{cases}$

Substitute $x = y + 2$ from (1) into (2):

$\quad 2x + y = 13$

$2(\boldsymbol{y + 2}) + y = 13$

$\quad 2y + 4 + y = 13$

$\quad\quad\quad\ 3y = 9$

$\quad\quad\quad\ \ y = 3$

Substitute this and solve for x:

$\quad x = y + 2 = 3 + 2 = 5$

Solution: $(5, 3)$

35. $\begin{cases} (1) & x + 2y = 6 \\ (2) & 3x - y = -10 \end{cases}$

Substitute $x = -2y + 6$ from (1) into (2):
$$3x - y = -10$$
$$3(-2y + 6) - y = -10$$
$$-6y + 18 - y = -10$$
$$-7y = -28$$
$$y = 4$$

Substitute this and solve for x:
$$x = -2y + 6$$
$$x = -2(4) + 6$$
$$x = -8 + 6 = -2$$

Solution: $(-2, 4)$

37. $\begin{cases} (1) & 3x = 2y - 4 \\ (2) & 6x - 4y = -4 \end{cases}$

Substitute $x = \dfrac{2y - 4}{3}$ from (1) into (2):
$$6x - 4y = -4$$
$$6\left(\frac{2y - 4}{3}\right) - 4y = -4$$
$$2(2y - 4) - 4y = -4$$
$$4y - 8 - 4y = -4$$
$$-8 = -4$$

Impossible $\Rightarrow$ no solution

39. $\begin{cases} (1) & 3x - 4y = 9 \\ (2) & x + 2y = 8 \end{cases}$

Substitute $x = -2y + 8$ from (2) into (1):
$$3x - 4y = 9$$
$$3(-2y + 8) - 4y = 9$$
$$-6y + 24 - 4y = 9$$
$$-10y = -15$$
$$y = \tfrac{3}{2}$$

Substitute this and solve for x:
$$x = -2y + 8$$
$$x = -2\left(\frac{3}{2}\right) + 8$$
$$x = -3 + 8 = 5$$

Solution: $\left(5, \tfrac{3}{2}\right)$

41. $\begin{cases} (1) & 2x + 2y = -1 \\ (2) & 3x + 4y = 0 \end{cases}$

Substitute $y = \dfrac{-2x - 1}{2}$ from (1) into (2):
$$3x + 4y = 0$$
$$3x + 4\left(\frac{-2x - 1}{2}\right) = 0$$
$$3x + 2(-2x - 1) = 0$$
$$3x - 4x - 2 = 0$$
$$-x = 2$$
$$x = -2$$

Substitute this and solve for y:
$$y = \frac{-2x - 1}{2}$$
$$y = \frac{-2(-2) - 1}{2}$$
$$y = \frac{4 - 1}{2} = \frac{3}{2}$$

Solution: $\left(-2, \dfrac{3}{2}\right)$

43.

$$x - y = \ 3$$
$$\underline{x + y = \ 7}$$
$$2x \quad\ = 10$$
$$x \quad\ = \ 5$$

Substitute and solve for y:

$$x + y = 7$$
$$5 + y = 7$$
$$y = 2$$

The solution is $(5, 2)$.

45.

$$2x + y = -10$$
$$\underline{2x - y = \ -6}$$
$$4x \quad\ = -16$$
$$x \quad\ = \ -4$$

Substitute and solve for y:

$$2x + y = -10$$
$$2(-4) + y = -10$$
$$-8 + y = -10$$
$$y = -2$$

The solution is $(-4, -2)$.

47.

$$\begin{aligned}2x + 3y = \ \ 8 &\Rightarrow \times (2)\\ \underline{3x - 2y = -1} &\Rightarrow \times (3)\end{aligned} \quad \begin{aligned}4x + 6y &= \ 16\\ \underline{9x - 6y} &= -3\\ 13x \quad\ &= \ 13\\ x \quad\ &= \ \ 1\end{aligned} \quad \begin{aligned}2x + 3y &= 8\\ 2(1) + 3y &= 8\\ 3y &= 6\\ y &= 2\end{aligned}$$

Solution: $\boxed{(1, 2)}$

49.

$$\begin{aligned}4x + 9y = \ \ 8\\ \underline{2x - 6y = -3} \Rightarrow \times (-2)\end{aligned} \quad \begin{aligned}4x + \ 9y &= \ 8\\ \underline{-4x + 12y} &= \ 6\\ 21y &= 14\\ y &= \tfrac{14}{21} = \tfrac{2}{3}\end{aligned} \quad \begin{aligned}2x - 6y &= -3\\ 2x - 6\left(\tfrac{2}{3}\right) &= -3\\ 2x - 4 &= -3\\ 2x &= 1\\ x &= \tfrac{1}{2}\end{aligned}$$

Solution: $\boxed{\left(\tfrac{1}{2}, \tfrac{2}{3}\right)}$

51.

$$\begin{aligned}8x - 4y = 16 &\Rightarrow 8x - 4y = 16 \Rightarrow\\ \underline{2x - \ 4 = \ y} &\Rightarrow \underline{2x - \ y = \ \ 4} \Rightarrow \times (-4)\end{aligned} \quad \begin{aligned}8x - 4y &= \ \ 16\\ \underline{-8x + 4y} &= -16\\ 0 &= \ \ \ \ 0 \Rightarrow \boxed{\text{Dependent equations}}\end{aligned}$$

53.

$$x = \frac{3}{2}y + 5 \Rightarrow \times (2) \quad 2x = 3y + 10 \Rightarrow 2x - 3y = 10 \Rightarrow \qquad 2x - 3y = \ \ 10$$
$$\underline{2x - \ 3y = 8} \Rightarrow \qquad \underline{2x - 3y = \ \ 8} \Rightarrow 2x - 3y = \ \ 8 \Rightarrow \times (-1) \quad \underline{-2x + 3y = -8}$$
$$0 \neq \ \ 2 \Rightarrow \boxed{\begin{array}{c}\text{No}\\ \text{solution}\end{array}}$$

55.

$$\begin{aligned}\tfrac{x}{2} + \tfrac{y}{2} = \ \ 6 &\Rightarrow \times 2\\ \tfrac{x}{2} - \tfrac{y}{2} = -2 &\Rightarrow \times 2\end{aligned} \quad \begin{aligned}x + y &= \ 12\\ \underline{x - y} &= -4\\ 2x \quad\ &= \ \ 8\\ x \quad\ &= \ \ 4\end{aligned} \quad \begin{aligned}x + y &= 12\\ 4 + y &= 12\\ y &= 8\end{aligned}$$

Solution: $\boxed{(4, 8)}$

57.

$$\begin{aligned}\tfrac{3}{4}x + \tfrac{2}{3}y = \ \ 7 &\Rightarrow \times 12 \quad 9x + 8y = \ \ 84 \Rightarrow \times 2\\ \underline{\tfrac{3}{5}x - \tfrac{1}{2}y = 18} &\Rightarrow \times 10 \quad \underline{6x - 5y = 180} \Rightarrow \times (-3)\end{aligned} \quad \begin{aligned}18x + 16y &= \ \ 168\\ \underline{-18x + 15y} &= -540\\ 31y &= -372\\ y &= \ -12\end{aligned} \quad \begin{aligned}6x - 5y &= 180\\ 6x - 5(-12) &= 180\\ 6x &= 120\\ x &= 20\end{aligned}$$

Solution: $\boxed{(20, -12)}$

59. $\begin{cases} y < 3x + 2 \\ y < -2x + 3 \end{cases}$

61. $\begin{cases} 3x + 2y > 6 \\ x + 3y \le 2 \end{cases}$

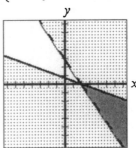

63. **a.** The point $(15, 2.0)$ is on the graph of the cost function, so it costs $2 million to manufacture 15,000 cameras.

b. The point $(20, 3.0)$ is on the graph of the revenue function, so there is a revenue of $3 million when 20,000 cameras are sold.

c. The graphs of the cost function and the revenue function meet at the point $(10, 1.5)$, so the revenue and cost functions are equal for 10,000 cameras.

65. Let $x =$ the cost of the pair of shoes and $y =$ the cost of the sweater.

(1) $\quad x + y = 98 \qquad\qquad x + y = 98 \quad y = x + 16 \qquad$ The sweater cost $57.
(2) $\qquad y = x + 16 \qquad x + x + 16 = 98 \quad y = 41 + 16$
$\qquad\qquad\qquad\qquad\qquad\qquad 2x = 82 \qquad y = 57$
$\qquad\qquad\qquad\qquad\qquad\qquad\quad x = 41$

67. (1) $\quad R_1 + R_2 = 1375 \qquad\qquad R_1 + R_2 = 1375 \quad R_1 = R_2 + 125 \qquad$ The resistances
(2) $\qquad R_1 = R_2 + 125 \quad R_2 + 125 + R_2 = 1375 \quad R_1 = 625 + 125 \qquad$ are $R_1 = 750$
$\qquad\qquad\qquad\qquad\qquad\qquad\qquad 2R_2 = 1250 \qquad R_1 = 750 \qquad$ ohms and
$\qquad\qquad\qquad\qquad\qquad\qquad\qquad\quad R_2 = 625 \qquad\qquad\qquad\qquad\qquad R_2 = 625$ ohms.

69. Let $l =$ the length of the field and $w =$ the width of the field.

$2w + 2l = 72 \Rightarrow \times(-1) \quad -2w - 2l = -72 \qquad 2w + 2l = 72$
$\underline{3w + 2l = 88} \Rightarrow \qquad\qquad \underline{3w + 2l = 88} \qquad 2(16) + 2l = 72$
$\qquad\qquad\qquad\qquad\qquad\quad w \quad = \quad 16 \qquad\quad 32 + 2l = 72$
$\qquad\qquad\qquad\qquad\qquad\qquad\qquad\qquad\qquad\qquad 2l = 40$
$\qquad\qquad\qquad\qquad\qquad\qquad\qquad\qquad\qquad\quad l = 20$

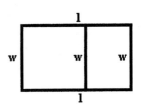

The dimensions of the field are 20 meters by 16 meters.

71. **Answers may vary.**   **73.** **Answers may vary.**   **75.** **Answers may vary.**

77. No.

Exercise 13.2 (page 841)

1. $m = \dfrac{\Delta y}{\Delta x} = \dfrac{-4-5}{-2-3} = \dfrac{-9}{-5} = \dfrac{9}{5}$

3. $f(0) = 2(0)^2 + 1 = 2(0) + 1 = 0 + 1 = 1$

5. $f(s) = 2s^2 + 1$

7. plane

9. infinitely

11.

$x - y + z = 2$	$2x + y - z = 4$	$2x - 3y + z = 2$	$(2, 1, 1)$ is a solution to
$2 - 1 + 1 \overset{?}{=} 2$	$2(2) + 1 - 1 \overset{?}{=} 4$	$2(2) - 3(1) + 1 \overset{?}{=} 2$	the system.
$2 = 2$	$4 + 1 - 1 \overset{?}{=} 4$	$4 - 3 + 1 \overset{?}{=} 2$	
	$4 = 4$	$2 = 2$	

13.

(1) $x + y + z = 4$ (1) $x + y + z = 4$ (2) $2x + y - z = 1$

(2) $2x + y - z = 1$ (2) $2x + y - z = 1$ (3) $2x - 3y + z = 1$

(3) $2x - 3y + z = 1$ (4) $\overline{3x + 2y \quad\;\; = 5}$ (5) $\overline{4x - 2y \quad\;\; = 2}$

(4) $3x + 2y = 5$ $3x + 2y = 5$ $x + y + z = 4$

(5) $\dfrac{4x - 2y = 2}{7x \quad\;\; = 7}$ $3(1) + 2y = 5$ $1 + 1 + z = 4$

$\quad x \quad\;\; = 1$ $3 + 2y = 5$ $2 + z = 4$

 $2y = 2$ $z = 2$ The solution is $(1, 1, 2)$.

 $y = 1$

15.

(1) $2x + 2y + 3z = 10$ (1) $2x + 2y + 3z = 10$ (3) $x + y + 2z = 6$

(2) $3x + y - z = 0$ $3 \cdot$ (2) $9x + 3y - 3z = 0$ $2 \cdot$ (2) $6x + 2y - 2z = 0$

(3) $x + y + 2z = 6$ (4) $\overline{11x + 5y \quad\;\; = 10}$ (5) $\overline{7x + 3y \quad\;\; = 6}$

$11x + 5y = 10 \Rightarrow \times 3$ $33x + 15y = 30$ $11x + 5y = 10$ $x + y + 2z = 6$

$\dfrac{7x + 3y = 6 \Rightarrow \times(-5)}{}$ $\dfrac{-35x - 15y = -30}{-2x \quad\quad\;\; = 0}$ $11(0) + 5y = 10$ $0 + 2 + 2z = 6$

 $x \quad\quad\;\; = 0$ $0 + 5y = 10$ $2 + 2z = 6$

 $5y = 10$ $2z = 4$

 $y = 2$ $z = 2$

Solution: $(0, 2, 2)$

17.

(1) $a + b + 2c = 7$ (1) $a + b + 2c = 7$ (1) $a + b + 2c = 7$

(2) $a + 2b + c = 8$ $-2 \cdot$ (2) $-2a - 4b - 2c = -16$ $-2 \cdot$ (3) $-4a - 2b - 2c = -18$

(3) $2a + b + c = 9$ (4) $\overline{-a - 3b \quad\;\; = -9}$ (5) $\overline{-3a - b \quad\;\; = -11}$

$-a - 3b = -9 \Rightarrow \times(-3)$ $3a + 9b = 27$ $-a - 3b = -9$ $2a + b + c = 9$

$\dfrac{-3a - b = -11 \Rightarrow}{}$ $\dfrac{-3a - b = -11}{8b = 16}$ $-a - 3(2) = -9$ $2(3) + 2 + c = 9$

 $b = 2$ $-a - 6 = -9$ $6 + 2 + c = 9$

 $-a = -3$ $8 + c = 9$

 $a = 3$ $c = 1$

Solution: $(3, 2, 1)$

19. (1) $2x + y - z = 1$ (2) $x + 2y + 2z = 2$ (3) $4x + 5y + 3z = 3$
(2) $x + 2y + 2z = 2$ $2 \cdot (1)$ $4x + 2y - 2z = 2$ $3 \cdot (1)$ $6x + 3y - 3z = 3$
(3) $4x + 5y + 3z = 3$ (4) $\overline{5x + 4y \qquad = 4}$ (5) $\overline{10x + 8y \qquad = 6}$

$5x + 4y = 4 \Rightarrow \times(-2)$ $-10x - 8y = -8$ Since this equation is always false, there is no
$\underline{10x + 8y = 6 \Rightarrow}$ $\underline{10x + 8y = \quad 6}$ solution. The system is inconsistent.
$\qquad\qquad\qquad\qquad\qquad 0 = -2$

21. (1) $4x + 3z = 4$ (2) $2y - 6z = -1$ (2) $2y - 6z = -1$
(2) $2y - 6z = -1$ $2 \cdot (1)$ $8x \qquad + 6z = 8$ $2 \cdot (3)$ $16x + 8y + 6z = 18$
(3) $8x + 4y + 3z = 9$ (4) $\overline{8x + 2y \qquad = 7}$ (5) $\overline{16x + 10y \qquad = 17}$

$8x + 2y = 7 \Rightarrow \times(-2)$ $-16x - 4y = -14$ $8x + 2y = 7$ $4x + 3z = 4$
$\underline{16x + 10y = 17 \Rightarrow}$ $\underline{16x + 10y = \quad 17}$ $8x + 2\left(\frac{1}{2}\right) = 7$ $4\left(\frac{3}{4}\right) + 3z = 4$
$\qquad\qquad\qquad\qquad\qquad 6y = \quad 3$ $8x + 1 = 7$ $3 + 3z = 4$
$\qquad\qquad\qquad\qquad\qquad y = \quad \frac{1}{2}$ $8x = 6$ $3z = 1$
$\qquad\qquad\qquad\qquad\qquad\qquad\qquad\qquad x = \frac{3}{4}$ $z = \frac{1}{3}$

Solution: $\boxed{\left(\frac{3}{4}, \frac{1}{2}, \frac{1}{3}\right)}$

23. (1) $2x + 3y + 4z = 6$ (3) $4x + 6y + 8z = 12$
(2) $2x - 3y - 4z = -4$ $-2 \cdot (1)$ $-4x - 6y - 8z = -12$
(3) $4x + 6y + 8z = 12$ (4) $\overline{\qquad\qquad\quad 0 = \quad 0}$

Since equation (4) is always true, the equations are dependent.

25. $x + \frac{1}{3}y + z = 13$ $\Rightarrow \times 3$ (1) $3x + y + 3z = 39$
$\frac{1}{2}x - y + \frac{1}{3}z = -2$ $\Rightarrow \times 6$ (2) $3x - 6y + 2z = -12$
$x + \frac{1}{2}y - \frac{1}{3}z = 2$ $\Rightarrow \times 6$ (3) $6x + 3y - 2z = 12$

(2) $3x - 6y + 2z = -12$ (2) $3x - 6y + 2z = -12$
$6 \cdot (1)$ $18x + 6y + 18z = 234$ $2 \cdot (3)$ $12x + 6y - 4z = 24$
(4) $\overline{21x \qquad + 20z = 222}$ (5) $\overline{15x \qquad - 2z = 12}$

$21x + 20z = 222 \Rightarrow$ $21x + 20z = 222$ $15x - 2z = 12$ $3x + y + 3z = 39$
$\underline{15x - 2z = 12 \Rightarrow \times 10}$ $\underline{150x - 20z = 120}$ $15(2) - 2z = 12$ $3(2) + y + 3(9) = 39$
$\qquad\qquad\qquad\qquad\qquad 171x \qquad = 342$ $30 - 2z = 12$ $6 + y + 27 = 39$
$\qquad\qquad\qquad\qquad\qquad x \qquad = 2$ $-2z = -18$ $y = 6$
$\qquad\qquad\qquad\qquad\qquad\qquad\qquad\qquad\qquad z = 9$

Solution: $\boxed{(2, 6, 9)}$

27. Let $x =$ the first integer, $y =$ the second integer and $z =$ the third integer.

$$
\begin{aligned}
x + y + z &= 18 \\
z &= 4y \\
y &= x + 6
\end{aligned}
\quad
\begin{aligned}
&\Rightarrow \\
&\Rightarrow \\
&\Rightarrow
\end{aligned}
\quad
\begin{aligned}
(1)\ & x + y + z = 18 \\
(2)\ & -4y + z = 0 \\
(3)\ & -x + y = 6
\end{aligned}
\qquad
\begin{aligned}
(1)\ & \\
(3)\ & \\
(4)\ &
\end{aligned}
\quad
\begin{aligned}
x + y + z &= 18 \\
-x + y &= 6 \\
\hline
2y + z &= 24
\end{aligned}
$$

$$
\begin{aligned}
2y + z = 24 &\Rightarrow \times 2 \\
-4y + z = 0 &\Rightarrow
\end{aligned}
\quad
\begin{aligned}
4y + 2z &= 48 \\
-4y + z &= 0 \\
\hline
3z &= 48 \\
z &= 16
\end{aligned}
\qquad
\begin{aligned}
2y + z &= 24 \\
2y + 16 &= 24 \\
2y &= 8 \\
y &= 4
\end{aligned}
\qquad
\begin{aligned}
x + y + z &= 18 \\
x + 4 + 16 &= 18 \\
x &= -2
\end{aligned}
$$

The integers are $-2, 4$ and 16.

29. Let A, B and C represent the measures of the three angles.

$$
\begin{aligned}
A + B + C &= 180 \\
A &= B + C - 100 \\
C &= 2B - 40
\end{aligned}
\quad
\begin{aligned}
&\Rightarrow \\
&\Rightarrow \\
&\Rightarrow
\end{aligned}
\quad
\begin{aligned}
(1)\ & A + B + C = 180 \\
(2)\ & A - B - C = -100 \\
(3)\ & -2B + C = -40
\end{aligned}
\qquad
\begin{aligned}
(1)\ & \\
(2)\ & \\
(4)\ &
\end{aligned}
\quad
\begin{aligned}
A + B + C &= 180 \\
A - B - C &= -100 \\
\hline
2A &= 80 \\
A &= 40
\end{aligned}
$$

$$
\begin{aligned}
A + B + C &= 180 \\
40 + B + C &= 180 \\
(5)\quad B + C &= 140
\end{aligned}
\qquad
\begin{aligned}
(-1) \cdot 3 \\
(5)
\end{aligned}
\quad
\begin{aligned}
2B - C &= 40 \\
B + C &= 140 \\
\hline
3B &= 180 \\
B &= 60
\end{aligned}
\qquad
\begin{aligned}
B + C &= 140 \\
60 + C &= 140 \\
C &= 80
\end{aligned}
$$

The angles have measures of $40°, 60°$ and $80°$.

31. Let $A =$ the units of food A, $B =$ the units of food B and $C =$ the units of food C.

$$
\begin{aligned}
(1)\ & A + 2B + 2C = 11 \quad \text{(fat)} \\
(2)\ & A + B + C = 6 \quad \text{(carbohydrate)} \\
(3)\ & 2A + B + 2C = 10 \quad \text{(protein)}
\end{aligned}
\qquad
\begin{aligned}
(1)\ & \\
-2 \cdot (2)\ & \\
(4)\ &
\end{aligned}
\quad
\begin{aligned}
A + 2B + 2C &= 11 \\
-2A - 2B - 2C &= -12 \\
\hline
-A &= -1 \\
A &= 1
\end{aligned}
$$

$$
\begin{aligned}
(1) \\
-2 \cdot (3) \\
(5)
\end{aligned}
\quad
\begin{aligned}
A + 2B + 2C &= 11 \\
-4A - 2B - 4C &= -20 \\
\hline
-3A - 2C &= -9
\end{aligned}
\qquad
\begin{aligned}
-3A - 2C &= -9 \\
-3(1) - 2C &= -9 \\
-3 - 2C &= -9 \\
-2C &= -6 \\
C &= 3
\end{aligned}
\qquad
\begin{aligned}
A + B + C &= 6 \\
1 + B + 3 &= 6 \\
B + 4 &= 6 \\
B &= 2
\end{aligned}
$$

1 unit of food A, 2 units of food B and 3 units of food C should be used.

33. Let $x =$ the number of \$5 statues, $y =$ the number of \$4 statues and $z =$ the number of \$3 statues.

$$
\begin{aligned}
(1)\ & x + y + z = 180 \quad \text{(total number made)} \\
(2)\ & 5x + 4y + 3z = 650 \quad \text{(total cost)} \\
(3)\ & 20x + 12y + 9z = 2100 \quad \text{(total revenue)}
\end{aligned}
\qquad
\begin{aligned}
-3 \cdot (1)\ & \\
(2)\ & \\
(4)\ &
\end{aligned}
\quad
\begin{aligned}
-3x - 3y - 3z &= -540 \\
5x + 4y + 3z &= 650 \\
\hline
2x + y &= 110
\end{aligned}
$$

$$
\begin{aligned}
-9 \cdot (1) \\
(3) \\
(5)
\end{aligned}
\quad
\begin{aligned}
-9x - 9y - 9z &= -1620 \\
20x + 12y + 9z &= 2100 \\
\hline
11x + 3y &= 480
\end{aligned}
\qquad
\begin{aligned}
2x + y = 110 &\Rightarrow \times(-3) \\
11x + 3y = 480 &\Rightarrow
\end{aligned}
\quad
\begin{aligned}
-6x - 3y &= -330 \\
11x + 3y &= 480 \\
\hline
5x &= 150 \\
x &= 30
\end{aligned}
$$

$$
\begin{aligned}
2x + y &= 110 \\
2(30) + y &= 110 \\
60 + y &= 110 \\
y &= 50
\end{aligned}
\qquad
\begin{aligned}
x + y + z &= 180 \\
30 + 50 + z &= 180 \\
z &= 100
\end{aligned}
$$

30 of the \$5, 50 of the \$4 and 100 of the \$3 statues should be made.

35. Let $x =$ the number of \$5 tickets, $y =$ the number of \$3 tickets and $z =$ the number of \$2 tickets.

$$
\begin{array}{lll}
(1) & x + y + z = 750 & \text{(total sold)} \\
(2) & x = 2z \Rightarrow x - 2z = 0 & \text{(twice as many)} \\
(3) & 5x + 3y + 2z = 2625 & \text{(total revenue)}
\end{array}
$$

$$
\begin{array}{lll}
2 \cdot (1) & 2x + 2y + 2z = 1500 \\
(2) & x \quad\;\; - 2z = \quad 0 \\
(4) & 3x + 2y \quad\;\; = 1500
\end{array}
$$

$$
\begin{array}{ll}
(2) & x \quad - 2z = \quad 0 \\
(3) & 5x + 3y + 2z = 2625 \\
(5) & 6x + 3y \quad\;\; = 2625
\end{array}
$$

$$
\begin{array}{l}
3x + 2y = 1500 \Rightarrow \times(-2) \quad -6x - 4y = -3000 \\
6x + 3y = 2625 \Rightarrow \qquad\quad\;\; 6x + 3y = \quad 2625 \\
\hline
\qquad\qquad\qquad\qquad\qquad\qquad -y = \;\; -375 \\
\qquad\qquad\qquad\qquad\qquad\qquad\;\; y = \qquad 375
\end{array}
$$

$$
\begin{array}{ll}
3x + 2y = 1500 & x + y + z = 750 \\
3x + 2(375) = 1500 & 250 + 375 + z = 750 \\
3x + 750 = 1500 & 625 + z = 750 \\
3x = 750 & z = 125 \\
x = 250 &
\end{array}
$$

250 of the \$5, 375 of the \$3 and 125 of the \$2 tickets were sold.

37. Let $x =$ the number of totem poles, $y =$ the number of bears and $z =$ the number of deer.

$$
\begin{array}{lll}
(1) & 2x + 2y + z = 14 & \text{(carving)} \\
(2) & x + 2y + 2z = 15 & \text{(sanding)} \\
(3) & 3x + 2y + 2z = 21 & \text{(painting)}
\end{array}
$$

$$
\begin{array}{ll}
-2 \cdot (1) & -4x - 4y - 2z = -28 \\
(2) & x + 2y + 2z = \quad 15 \\
(4) & -3x - 2y \qquad = -13
\end{array}
$$

$$
\begin{array}{ll}
-2 \cdot (1) & -4x - 4y - 2z = -28 \\
(3) & 3x + 2y + 2z = \quad 21 \\
\hline
(5) & -x - 2y \qquad\; = \quad -7
\end{array}
$$

$$
\begin{array}{l}
-3x - 2y = -13 \Rightarrow \times(-1) \quad 3x + 2y = \quad 13 \\
-x - 2y = \;\; -7 \Rightarrow \qquad\qquad\;\; -x - 2y = -7 \\
\hline
\qquad\qquad\qquad\qquad\qquad\qquad\quad 2x \qquad = \quad 6 \\
\qquad\qquad\qquad\qquad\qquad\qquad\qquad x \qquad = \quad 3
\end{array}
$$

$$
\begin{array}{ll}
3x + 2y = 13 & 2x + 2y + z = 14 \\
3(3) + 2y = 13 & 2(3) + 2(2) + z = 14 \\
9 + 2y = 13 & 6 + 4 + z = 14 \\
2y = 4 & z = 4 \\
y = 2 &
\end{array}
$$

3 totem poles, 2 bears and 4 deer should be made.

39. Substitute the coordinates of each point for x and y in the equation $y = ax^2 + bx + c$.

$$
\begin{array}{lll}
y = ax^2 + bx + c & y = ax^2 + bx + c & y = ax^2 + bx + c \\
0 = a(0)^2 + b(0) + c & -4 = a(2)^2 + b(2) + c & 0 = a(4)^2 + b(4) + c \\
0 = c & -4 = 4a + 2b + c & 0 = 16a + 4b + c
\end{array}
$$

Solve the system of equations formed from the three equations:

$$
\begin{array}{ll}
(1) & c = 0 \\
(2) & 4a + 2b + c = -4 \\
(3) & 16a + 4b + c = 0
\end{array}
$$

$$
\begin{array}{l}
4a + 2b = -4 \Rightarrow \times(-2) \quad -8a - 4b = 8 \\
16a + 4b = \quad 0 \Rightarrow \qquad\qquad\;\; 16a + 4b = 0 \\
\hline
\qquad\qquad\qquad\qquad\qquad\qquad\quad 8a \qquad = 8 \\
\qquad\qquad\qquad\qquad\qquad\qquad\quad\;\; a \qquad = 1
\end{array}
$$

$$
\begin{array}{l}
16a + 4b = 0 \\
16(1) + 4b = 0 \\
16 + 4b = 0 \\
4b = -16 \\
b = -4
\end{array}
$$

The equation is $y = x^2 - 4x$.

41. Substitute the coordinates of each point for x and y in the equation $x^2 + y^2 + cx + dy + e = 0$.

$$x^2 + y^2 + cx + dy + e = 0 \qquad\qquad x^2 + y^2 + cx + dy + e = 0$$
$$(1)^2 + (3)^2 + c(1) + d(3) + e = 0 \qquad (3)^2 + (1)^2 + c(3) + d(1) + e = 0$$
$$1 + 9 + c + 3d + e = 0 \qquad\qquad 9 + 1 + 3c + d + e = 0$$
$$c + 3d + e = -10 \qquad\qquad 3c + d + e = -10$$

$$x^2 + y^2 + cx + dy + e = 0 \qquad (1)\quad c + 3d + e = -10$$
$$(1)^2 + (-1)^2 + c(1) + d(-1) + e = 0 \qquad (2)\quad 3c + d + e = -10$$
$$1 + 1 + c - d + e = 0 \qquad\qquad (3)\quad c - d + e = -2$$
$$c - d + e = -2$$

$$
\begin{array}{ll}
(1) \quad c + 3d + e = -10 & (1) \quad c + 3d + e = -10 \qquad -2c + 2d = 0 \\
-1 \cdot (2) \quad \underline{-3c - d - e = 10} & -1 \cdot (3) \quad \underline{-c + d - e = 2} \qquad -2c + 2(-2) = 0 \\
(4) \quad \overline{-2c + 2d = 0} & (5) \quad \overline{4d = -8} \qquad\quad -2c - 4 = 0 \\
& d = -2 \qquad\qquad -2c = 4 \\
& \phantom{(5) \quad d = -2 \qquad} c = -2
\end{array}
$$

$$c + 3d + e = -10 \quad \text{The equation is } x^2 + y^2 - 2x - 2y - 2 = 0.$$
$$-2 + 3(-2) + e = -10$$
$$-2 - 6 + e = -10$$
$$-8 + e = -10$$
$$e = -2$$

43. Answers may vary.

45.

$$
\begin{array}{ll}
(1) \quad x + y + z + w = 3 & (1) \quad x + y + z + w = 3 \qquad (1) \quad x + y + z + w = 3 \\
(2) \quad x - y - z - w = -1 & (2) \quad \underline{x - y - z - w = -1} \qquad (3) \quad \underline{x + y - z - w = 1} \\
(3) \quad x + y - z - w = 1 & \quad 2x = 2 \qquad \quad 2x + 2y = 4 \\
(4) \quad x + y - z + w = 3 & \quad x = 1
\end{array}
$$

$$
\begin{array}{llll}
(1) \quad x + y + z + w = 3 & 2x + 2y = 4 & 2x + 2y + 2w = 6 & x + y + z + w = 3 \\
(4) \quad \underline{x + y - z + w = 3} & 2(1) + 2y = 4 & 2(1) + 2(1) + 2w = 6 & 1 + 1 + z + 1 = 3 \\
 \quad 2x + 2y + 2w = 6 & 2y = 2 & 4 + 2w = 6 & 3 + z = 3 \\
& y = 1 & 2w = 2 & z = 0 \\
& & w = 1 &
\end{array}
$$

The solution is $(1, 1, 0, 1)$.

Exercise 13.3 (page 851)

Note: The notation $3R_1 + R_3 \Rightarrow R_2$ means to multiply Row #1 of the previous matrix *by* 3, add that result to Row #3 of the previous matrix, and write the final result in Row #2 of the current matrix.

1. $93{,}000{,}000 = 9.3 \times 10^7$ **3.** $63 \times 10^3 = 6.3 \times 10^1 \times 10^3 = 6.3 \times 10^4$

5. matrix **7.** 3; columns **9.** augmented **11.** type 1

13. nonzero

15.
$$R_2 + (-R_1) \Rightarrow R_2$$
$$\begin{bmatrix} 2 & 1 & 1 \\ 5 & 4 & 1 \end{bmatrix} \Rightarrow \begin{bmatrix} 2 & 1 & 1 \\ 3 & 3 & \boxed{0} \end{bmatrix}$$

17.
$$2R_2 \Rightarrow R_2$$
$$\begin{bmatrix} 3 & -2 & 1 \\ -1 & 2 & 4 \end{bmatrix} \Rightarrow \begin{bmatrix} 3 & -2 & 1 \\ -2 & 4 & \boxed{8} \end{bmatrix}$$

19.
$$R_1 + (-R_2) \Rightarrow R_2 \qquad \tfrac{1}{2}R_2 \Rightarrow R_2$$
$$\begin{bmatrix} 1 & 1 & | & 2 \\ 1 & -1 & | & 0 \end{bmatrix} \Rightarrow \begin{bmatrix} 1 & 1 & | & 2 \\ 0 & 2 & | & 2 \end{bmatrix} \Rightarrow \begin{bmatrix} 1 & 1 & | & 2 \\ 0 & 1 & | & 1 \end{bmatrix}$$
From R_2, $y = 1$. From R_1:   The solution is $(1,1)$.
$$x + y = 2$$
$$x + 1 = 2 \Rightarrow x = 1$$

21.
$$-2R_1 + R_2 \Rightarrow R_2 \qquad -\tfrac{1}{3}R_2 \Rightarrow R_2$$
$$\begin{bmatrix} 1 & 2 & | & -4 \\ 2 & 1 & | & 1 \end{bmatrix} \Rightarrow \begin{bmatrix} 1 & 2 & | & -4 \\ 0 & -3 & | & 9 \end{bmatrix} \Rightarrow \begin{bmatrix} 1 & 2 & | & -4 \\ 0 & 1 & | & -3 \end{bmatrix}$$
From R_2, $y = -3$. From R_1:   The solution is $(2,-3)$.
$$x + 2y = -4$$
$$x + 2(-3) = -4$$
$$x - 6 = -4 \Rightarrow x = 2$$

23.
$$-3R_1 + R_2 \Rightarrow R_2 \qquad -\tfrac{1}{14}R_2 \Rightarrow R_2$$
$$\begin{bmatrix} 3 & 4 & | & -12 \\ 9 & -2 & | & 6 \end{bmatrix} \Rightarrow \begin{bmatrix} 3 & 4 & | & -12 \\ 0 & -14 & | & 42 \end{bmatrix} \Rightarrow \begin{bmatrix} 3 & 4 & | & -12 \\ 0 & 1 & | & -3 \end{bmatrix}$$
From R_2, $y = -3$. From R_1:   The solution is $(0,-3)$.
$$3x + 4y = -12$$
$$3x + 4(-3) = -12$$
$$3x - 12 = -12$$
$$3x = 0 \Rightarrow x = 0$$

25.
$$\begin{cases} 5a = 24 + 2b \\ 5b = 3a + 16 \end{cases} \Rightarrow \begin{cases} 5a - 2b = 24 \\ -3a + 5b = 16 \end{cases}$$
$$2R_2 + R_1 \Rightarrow R_1 \qquad -R_1 \Rightarrow R_1 \qquad 3R_1 + R_2 \Rightarrow R_2$$
$$\begin{bmatrix} 5 & -2 & | & 24 \\ -3 & 5 & | & 16 \end{bmatrix} \Rightarrow \begin{bmatrix} -1 & 8 & | & 56 \\ -3 & 5 & | & 16 \end{bmatrix} \Rightarrow \begin{bmatrix} 1 & -8 & | & -56 \\ -3 & 5 & | & 16 \end{bmatrix} \Rightarrow \begin{bmatrix} 1 & -8 & | & -56 \\ 0 & -19 & | & -152 \end{bmatrix}$$
$$-\tfrac{1}{19}R_2 \Rightarrow R_2 \qquad \text{From } R_2, b = 8. \quad \text{From } R_1 : \qquad \text{The solution is } (8,8).$$
$$\Rightarrow \begin{bmatrix} 1 & -8 & | & -56 \\ 0 & 1 & | & 8 \end{bmatrix}$$
$$a - 8b = -56$$
$$a - 8(8) = -56$$
$$a - 64 = -56 \Rightarrow a = 8$$

27.
$$-R_1 + R_2 \Rightarrow R_2$$
$$-R_1 + R_3 \Rightarrow R_3$$
$$\begin{bmatrix} 1 & 1 & 1 & | & 6 \\ 1 & 2 & 1 & | & 8 \\ 1 & 1 & 2 & | & 9 \end{bmatrix} \Rightarrow \begin{bmatrix} 1 & 1 & 1 & | & 6 \\ 0 & 1 & 0 & | & 2 \\ 0 & 0 & 1 & | & 3 \end{bmatrix}$$

From R_3, $z = 3$.　From R_2, $y = 2$.　From R_1:　　The solution is $(1, 2, 3)$.
$$x + y + z = 6$$
$$x + 2 + 3 = 6$$
$$x = 1$$

29.
$$R_1 + R_2 \Rightarrow R_2 \qquad\quad -\tfrac{1}{4}R_3 \Rightarrow R_2$$
$$-2R_1 + R_3 \Rightarrow R_3 \qquad \tfrac{1}{4}R_2 \Rightarrow R_3$$
$$\begin{bmatrix} 2 & 1 & 3 & | & 3 \\ -2 & -1 & 1 & | & 5 \\ 4 & -2 & 2 & | & 2 \end{bmatrix} \Rightarrow \begin{bmatrix} 2 & 1 & 3 & | & 3 \\ 0 & 0 & 4 & | & 8 \\ 0 & -4 & -4 & | & -4 \end{bmatrix} \Rightarrow \begin{bmatrix} 2 & 1 & 3 & | & 3 \\ 0 & 1 & 1 & | & 1 \\ 0 & 0 & 1 & | & 2 \end{bmatrix}$$

From R_3, $z = 2$.　From R_2:　　From R_1:　　　　The solution is $(-1, -1, 2)$.
$$y + z = 1 \qquad\qquad 2x + y + 3z = 3$$
$$y + 2 = 1 \qquad\qquad 2x + (-1) + 3(2) = 3$$
$$y = -1 \qquad\qquad 2x - 1 + 6 = 3$$
$$2x + 5 = 3$$
$$2x = -2$$
$$x = -1$$

31.
$$-3R_2 + R_1 \Rightarrow R_2$$
$$-2R_1 + R_3 \Rightarrow R_3 \qquad\quad \tfrac{2}{5}R_2 + R_3 \Rightarrow R_3$$
$$\begin{bmatrix} 3 & -2 & 4 & | & 4 \\ 1 & 1 & 1 & | & 3 \\ 6 & -2 & -3 & | & 10 \end{bmatrix} \Rightarrow \begin{bmatrix} 3 & -2 & 4 & | & 4 \\ 0 & -5 & 1 & | & -5 \\ 0 & 2 & -11 & | & 2 \end{bmatrix} \Rightarrow \begin{bmatrix} 3 & -2 & 4 & | & 4 \\ 0 & -5 & 1 & | & -5 \\ 0 & 0 & -\frac{53}{5} & | & 0 \end{bmatrix}$$

From R_3, $z = 0$.　From R_2:　　　From R_1:　　　　The solution is $(2, 1, 0)$.
$$-5y + z = -5 \qquad\quad 3x - 2y + 4z = 4$$
$$-5y + 0 = -5 \qquad\quad 3x - 2(1) + 4(0) = 4$$
$$-5y = -5 \qquad\qquad 3x - 2 = 4$$
$$y = 1 \qquad\qquad\quad 3x = 6$$
$$x = 2$$

33.
$$-3R_1 + R_2 \Rightarrow R_2 \qquad -\tfrac{1}{4}R_2 \Rightarrow R_2$$
$$-2R_1 + R_3 \Rightarrow R_3 \quad -4R_3 + R_2 \Rightarrow R_3$$
$$\begin{bmatrix} 1 & 1 & | & 3 \\ 3 & -1 & | & 1 \\ 2 & 1 & | & 4 \end{bmatrix} \Rightarrow \begin{bmatrix} 1 & 1 & | & 3 \\ 0 & -4 & | & -8 \\ 0 & -1 & | & -2 \end{bmatrix} \Rightarrow \begin{bmatrix} 1 & 1 & | & 3 \\ 0 & 1 & | & 2 \\ 0 & 0 & | & 0 \end{bmatrix}$$

From R_2, $y = 2$.　From R_1:　The solution is $(1, 2)$.
$$x + y = 3$$
$$x + 2 = 3$$
$$x = 1$$

35.
$$-2R_2 + R_1 \Rightarrow R_2 \qquad -\tfrac{1}{7}R_2 \Rightarrow R_2$$
$$2R_3 + R_1 \Rightarrow R_3 \qquad -\tfrac{7}{9}R_2 + R_3 \Rightarrow R_3$$

$$\begin{bmatrix} 2 & -1 & 4 \\ 1 & 3 & 2 \\ -1 & -4 & -2 \end{bmatrix} \Rightarrow \begin{bmatrix} 2 & -1 & 4 \\ 0 & -7 & 0 \\ 0 & -9 & 0 \end{bmatrix} \Rightarrow \begin{bmatrix} 2 & -1 & 4 \\ 0 & 1 & 0 \\ 0 & 0 & 0 \end{bmatrix}$$

From R_2, $y = 0$.   From R_1:   The solution is $(2, 0)$.
$$2x - y = 4$$
$$2x - 0 = 4$$
$$2x = 4$$
$$x = 2$$

37.
$$-2R_2 + R_1 \Rightarrow R_2 \qquad \tfrac{1}{3}R_2 \Rightarrow R_2$$
$$2R_3 + R_1 \Rightarrow R_3 \qquad -\tfrac{7}{3}R_2 + R_3 \Rightarrow R_3$$

$$\begin{bmatrix} 2 & 1 & 7 \\ 1 & -1 & 2 \\ -1 & 3 & -2 \end{bmatrix} \Rightarrow \begin{bmatrix} 2 & 1 & 7 \\ 0 & 3 & 3 \\ 0 & 7 & 3 \end{bmatrix} \Rightarrow \begin{bmatrix} 2 & 1 & 7 \\ 0 & 1 & 1 \\ 0 & 0 & -4 \end{bmatrix}$$

R_3: $0x + 0y = -4$, or $0 = -4$, which is an impossible equation.   NO SOLUTION

39.
$$-R_2 + R_1 \Rightarrow R_2 \qquad \tfrac{1}{2}R_2 \Rightarrow R_2$$
$$-3R_1 + R_3 \Rightarrow R_3 \quad 4R_2 + R_3 \Rightarrow R_3$$

$$\begin{bmatrix} 1 & 3 & 7 \\ 1 & 1 & 3 \\ 3 & 1 & 5 \end{bmatrix} \Rightarrow \begin{bmatrix} 1 & 3 & 7 \\ 0 & 2 & 4 \\ 0 & -8 & -16 \end{bmatrix} \Rightarrow \begin{bmatrix} 1 & 3 & 7 \\ 0 & 1 & 2 \\ 0 & 0 & 0 \end{bmatrix}$$

From R_2, $y = 2$.   From R_1:   The solution is $(1, 2)$.
$$x + 3y = 7$$
$$x + 3(2) = 7$$
$$x + 6 = 7 \Rightarrow x = 1$$

41.
$$R_1 + R_2 \Rightarrow R_2$$
$$\begin{bmatrix} 1 & 2 & 3 & -2 \\ -1 & -1 & -2 & 4 \end{bmatrix} \Rightarrow \begin{bmatrix} 1 & 2 & 3 & -2 \\ 0 & 1 & 1 & 2 \end{bmatrix}$$

From R_2:   From R_1:   The solution is
$$y + z = 2 \qquad x + 2y + 3z = -2 \qquad (-6 - z, 2 - z, z).$$
$$y = 2 - z \qquad x + 2(2 - z) + 3z = -2$$
$$x + 4 - 2z + 3z = -2$$
$$x + z = -6$$
$$x = -6 - z$$

43.

$$\begin{array}{c} \\ \begin{bmatrix} 1 & -1 & 0 & | & 1 \\ 0 & 1 & 1 & | & 1 \\ 1 & 0 & 1 & | & 2 \end{bmatrix} \end{array} \overset{-R_1 + R_3 \Rightarrow R_3}{\Rightarrow} \begin{bmatrix} 1 & -1 & 0 & | & 1 \\ 0 & 1 & 1 & | & 1 \\ 0 & 1 & 1 & | & 1 \end{bmatrix} \overset{-R_2 + R_3 \Rightarrow R_3}{\Rightarrow} \begin{bmatrix} 1 & -1 & 0 & | & 1 \\ 0 & 1 & 1 & | & 1 \\ 0 & 0 & 0 & | & 0 \end{bmatrix}$$

From R_2:

$$y + z = 1$$
$$y = 1 - z$$

From R_1:

$$x - y = 1$$
$$x - (1 - z) = 1$$
$$x - 1 + z = 1$$
$$x + z = 2$$
$$x = 2 - z$$

The solution is

$$(2 - z, 1 - z, z).$$

45. Let $x =$ the measure of the first angle and $y =$ the measure of the second angle. Form and solve this

system of equations: $\begin{cases} x + y = 90 \\ \quad\quad y = x + 46 \end{cases} \Rightarrow \begin{cases} x + y = 90 \\ -x + y = 46 \end{cases}$

$$\begin{bmatrix} 1 & 1 & | & 90 \\ -1 & 1 & | & 46 \end{bmatrix} \overset{R_1 + R_2 \Rightarrow R_2}{\Rightarrow} \begin{bmatrix} 1 & 1 & | & 90 \\ 0 & 2 & | & 136 \end{bmatrix} \overset{\frac{1}{2}R_2 \Rightarrow R_2}{\Rightarrow} \begin{bmatrix} 1 & 1 & | & 90 \\ 0 & 1 & | & 68 \end{bmatrix}$$

From R_2, $y = 68$.

From R_1:

$$x + y = 90$$
$$x + 68 = 90$$
$$x = 22$$

The angles have measures of $22°$ and $68°$.

47. Let A, B and C represent the measures of the three angles.

$$\begin{cases} A + B + C = 180 \\ \quad\quad B = A + 25 \\ \quad\quad C = 2A - 5 \end{cases} \Rightarrow \begin{cases} A + B + C = 180 \\ -A + B = 25 \\ -2A + C = -5 \end{cases}$$

$$\begin{bmatrix} 1 & 1 & 1 & | & 180 \\ -1 & 1 & 0 & | & 25 \\ -2 & 0 & 1 & | & -5 \end{bmatrix} \overset{R_1 + R_2 \Rightarrow R_2}{\underset{2R_1 + R_3 \Rightarrow R_3}{\Rightarrow}} \begin{bmatrix} 1 & 1 & 1 & | & 180 \\ 0 & 2 & 1 & | & 205 \\ 0 & 2 & 3 & | & 355 \end{bmatrix} \overset{-R_2 + R_3 \Rightarrow R_3}{\Rightarrow} \begin{bmatrix} 1 & 1 & 1 & | & 180 \\ 0 & 2 & 1 & | & 205 \\ 0 & 0 & 2 & | & 150 \end{bmatrix} \overset{\frac{1}{2}R_3 \Rightarrow R_3}{\Rightarrow} \begin{bmatrix} 1 & 1 & 1 & | & 180 \\ 0 & 2 & 1 & | & 205 \\ 0 & 0 & 1 & | & 75 \end{bmatrix}$$

From R_3, $C = 75$.

From R_2:

$$2B + C = 205$$
$$2B + 75 = 205$$
$$2B = 130$$
$$B = 65$$

From R_1:

$$A + B + C = 180$$
$$A + 65 + 75 = 180$$
$$A + 140 = 180$$
$$A = 40$$

The angles have measures of $40°$, $65°$ and $75°$.

49. Plug the coordinates of the points into the general equation to form and solve a system of equations.

$$y = ax^2 + bx + c \qquad\qquad y = ax^2 + bx + c \qquad\qquad y = ax^2 + bx + c$$
$$1 = a(0)^2 + b(0) + c \qquad 2 = a(1)^2 + b(1) + c \qquad 4 = a(-1)^2 + b(-1) + c$$
$$1 = c \qquad\qquad\qquad 2 = a + b + c \qquad\qquad 4 = a - b + c$$

$$\begin{cases} a + b + c = 2 \\ a - b + c = 4 \\ \phantom{a - b + {}} c = 1 \end{cases} \Rightarrow \begin{bmatrix} 1 & 1 & 1 & | & 2 \\ 1 & -1 & 1 & | & 4 \\ 0 & 0 & 1 & | & 1 \end{bmatrix} \Rightarrow \overset{-R_2 + R_1 \Rightarrow R_2}{\begin{bmatrix} 1 & 1 & 1 & | & 2 \\ 0 & 2 & 0 & | & -2 \\ 0 & 0 & 1 & | & 1 \end{bmatrix}} \Rightarrow \overset{\frac{1}{2}R_2 \Rightarrow R_2}{\begin{bmatrix} 1 & 1 & 1 & | & 2 \\ 0 & 1 & 0 & | & -1 \\ 0 & 0 & 1 & | & 1 \end{bmatrix}}$$

From R_3, $c = 1$. \qquad\qquad From R_2, $b = -1$. From R_1:

$$a + b + c = 2$$
$$a + (-1) + 1 = 2$$
$$a = 2$$

The equation is $y = 2x^2 - x + 1$.

51. Answers may vary.

53. The last equation represents the equation $0x + 0y + 0z = k$, or $0 = k$. If $k = 0$, then the system can be solved. However, if $k \neq 0$, the system will have no solution.

Exercise 13.4 (page 860)

1.
$$3(x + 2) - (2 - x) = x - 5$$
$$3x + 6 - 2 + x = x - 5$$
$$4x + 4 = x - 5$$
$$3x = -9$$
$$x = -3$$

3.
$$\frac{5}{3}(5x + 6) - 10 = 0$$
$$3 \cdot \frac{5}{3}(5x + 6) - 3 \cdot 10 = 3 \cdot 0$$
$$5(5x + 6) - 30 = 0$$
$$25x + 30 - 30 = 0$$
$$25x = 0$$
$$x = 0$$

5. number

7. $\begin{vmatrix} a_2 & c_2 \\ a_3 & c_3 \end{vmatrix}$

9. $\begin{vmatrix} 3 & 4 \\ 2 & -3 \end{vmatrix}$

11.
$$\begin{vmatrix} 2 & 3 \\ -2 & 1 \end{vmatrix} = 2(1) - 3(-2)$$
$$= 2 + 6 = 8$$

13.
$$\begin{vmatrix} -1 & 2 \\ 3 & -4 \end{vmatrix} = -1(-4) - 2(3)$$
$$= 4 - 6 = -2$$

15. $\begin{vmatrix} x & y \\ y & x \end{vmatrix} = x(x) - y(y) = x^2 - y^2$

17. $\begin{vmatrix} 1 & 0 & 1 \\ 0 & 1 & 0 \\ 1 & 1 & 1 \end{vmatrix} = 1\begin{vmatrix} 1 & 0 \\ 1 & 1 \end{vmatrix} - 0\begin{vmatrix} 0 & 0 \\ 1 & 1 \end{vmatrix} + 1\begin{vmatrix} 0 & 1 \\ 1 & 1 \end{vmatrix} = 1(1) - 0(0) + 1(-1) = 1 - 0 - 1 = 0$

19. $\begin{vmatrix} -1 & 2 & 1 \\ 2 & 1 & -3 \\ 1 & 1 & 1 \end{vmatrix} = -1\begin{vmatrix} 1 & -3 \\ 1 & 1 \end{vmatrix} - 2\begin{vmatrix} 2 & -3 \\ 1 & 1 \end{vmatrix} + 1\begin{vmatrix} 2 & 1 \\ 1 & 1 \end{vmatrix} = -1(4) - 2(5) + 1(1) = -4 - 10 + 1$

$$= -13$$

21. $\begin{vmatrix} 1 & -2 & 3 \\ -2 & 1 & 1 \\ -3 & -2 & 1 \end{vmatrix} = 1\begin{vmatrix} 1 & 1 \\ -2 & 1 \end{vmatrix} - (-2)\begin{vmatrix} -2 & 1 \\ -3 & 1 \end{vmatrix} + 3\begin{vmatrix} -2 & 1 \\ -3 & -2 \end{vmatrix} = 1(3) + 2(1) + 3(7) = 3 + 2 + 21$

$$= 26$$

23. $\begin{vmatrix} 1 & 2 & 3 \\ 4 & 5 & 6 \\ 7 & 8 & 9 \end{vmatrix} = 1\begin{vmatrix} 5 & 6 \\ 8 & 9 \end{vmatrix} - 2\begin{vmatrix} 4 & 6 \\ 7 & 9 \end{vmatrix} + 3\begin{vmatrix} 4 & 5 \\ 7 & 8 \end{vmatrix} = 1(-3) - 2(-6) + 3(-3) = -3 + 12 - 9 = 0$

25. $\begin{vmatrix} a & 2a & -a \\ 2 & -1 & 3 \\ 1 & 2 & -3 \end{vmatrix} = a\begin{vmatrix} -1 & 3 \\ 2 & -3 \end{vmatrix} - 2a\begin{vmatrix} 2 & 3 \\ 1 & -3 \end{vmatrix} + (-a)\begin{vmatrix} 2 & -1 \\ 1 & 2 \end{vmatrix} = a(-3) - 2a(-9) - a(5)$

$$= -3a + 18a - 5a = 10a$$

27. $\begin{vmatrix} 1 & a & b \\ 1 & 2a & 2b \\ 1 & 3a & 3b \end{vmatrix} = 1\begin{vmatrix} 2a & 2b \\ 3a & 3b \end{vmatrix} - a\begin{vmatrix} 1 & 2b \\ 1 & 3b \end{vmatrix} + b\begin{vmatrix} 1 & 2a \\ 1 & 3a \end{vmatrix} = 1(0) - a(b) + b(a) = 0 - ab + ab = 0$

29. $x = \dfrac{\begin{vmatrix} 6 & 1 \\ 2 & -1 \end{vmatrix}}{\begin{vmatrix} 1 & 1 \\ 1 & -1 \end{vmatrix}} = \dfrac{-6 - 2}{-1 - 1} = \dfrac{-8}{-2} = 4; \quad y = \dfrac{\begin{vmatrix} 1 & 6 \\ 1 & 2 \end{vmatrix}}{\begin{vmatrix} 1 & 1 \\ 1 & -1 \end{vmatrix}} = \dfrac{2 - 6}{-2} = \dfrac{-4}{-2} = 2;$ solution: $(4, 2)$

31. $x = \dfrac{\begin{vmatrix} 1 & 1 \\ -7 & -2 \end{vmatrix}}{\begin{vmatrix} 2 & 1 \\ 1 & -2 \end{vmatrix}} = \dfrac{-2 - (-7)}{-4 - 1} = \dfrac{5}{-5} = -1; \quad y = \dfrac{\begin{vmatrix} 2 & 1 \\ 1 & -7 \end{vmatrix}}{\begin{vmatrix} 2 & 1 \\ 1 & -2 \end{vmatrix}} = \dfrac{-14 - 1}{-5} = \dfrac{-15}{-5} = 3$

solution: $(-1, 3)$

33. $x = \dfrac{\begin{vmatrix} 0 & 3 \\ -4 & -6 \end{vmatrix}}{\begin{vmatrix} 2 & 3 \\ 4 & -6 \end{vmatrix}} = \dfrac{0 - (-12)}{-12 - 12} = \dfrac{12}{-24} = -\dfrac{1}{2}; \quad y = \dfrac{\begin{vmatrix} 2 & 0 \\ 4 & -4 \end{vmatrix}}{\begin{vmatrix} 2 & 3 \\ 4 & -6 \end{vmatrix}} = \dfrac{-8 - 0}{-24} = \dfrac{-8}{-24} = \dfrac{1}{3}$

solution: $\left(-\dfrac{1}{2}, \dfrac{1}{3}\right)$

35. $\begin{cases} y = \dfrac{-2x+1}{3} \\ 3x - 2y = 8 \end{cases} \Rightarrow \begin{cases} 3y = -2x+1 \\ 3x - 2y = 8 \end{cases} \Rightarrow \begin{cases} 2x + 3y = 1 \\ 3x - 2y = 8 \end{cases}$

$x = \dfrac{\begin{vmatrix} 1 & 3 \\ 8 & -2 \end{vmatrix}}{\begin{vmatrix} 2 & 3 \\ 3 & -2 \end{vmatrix}} = \dfrac{-2 - 24}{-4 - 9} = \dfrac{-26}{-13} = 2; \; y = \dfrac{\begin{vmatrix} 2 & 1 \\ 3 & 8 \end{vmatrix}}{\begin{vmatrix} 2 & 1 \\ 3 & -2 \end{vmatrix}} = \dfrac{16 - 3}{-13} = \dfrac{13}{-13} = -1;$ solution: $(2, -1)$

37. $\begin{cases} y = \dfrac{11 - 3x}{2} \\ x = \dfrac{11 - 4y}{6} \end{cases} \Rightarrow \begin{cases} 2y = 11 - 3x \\ 6x = 11 - 4y \end{cases} \Rightarrow \begin{cases} 3x + 2y = 11 \\ 6x + 4y = 11 \end{cases}$

$x = \dfrac{\begin{vmatrix} 11 & 2 \\ 11 & 4 \end{vmatrix}}{\begin{vmatrix} 3 & 2 \\ 6 & 4 \end{vmatrix}} = \dfrac{44 - 22}{12 - 12} = \dfrac{22}{0} \Rightarrow$ denominator $= 0$, numerator $\neq 0 \Rightarrow$ no solution

39. $\begin{cases} x = \dfrac{5y - 4}{2} \\ y = \dfrac{3x - 1}{5} \end{cases} \Rightarrow \begin{cases} 2x = 5y - 4 \\ 5y = 3x - 1 \end{cases} \Rightarrow \begin{cases} 2x - 5y = -4 \\ -3x + 5y = -1 \end{cases}$

$x = \dfrac{\begin{vmatrix} -4 & -5 \\ -1 & 5 \end{vmatrix}}{\begin{vmatrix} 2 & -5 \\ -3 & 5 \end{vmatrix}} = \dfrac{-20 - 5}{10 - 15} = \dfrac{-25}{-5} = 5; \; y = \dfrac{\begin{vmatrix} 2 & -4 \\ -3 & -1 \end{vmatrix}}{\begin{vmatrix} 2 & -5 \\ -3 & 5 \end{vmatrix}} = \dfrac{-2 - 12}{-5} = \dfrac{-14}{-5} = \dfrac{14}{5}$

solution: $\left(5, \frac{14}{5}\right)$

Note: In the following problems, D stands for the denominator determinant, while N_x, N_y and N_z stand for the numerator determinants for x, y and z, respectively.

41. $D = \begin{vmatrix} 1 & 1 & 1 \\ 1 & 1 & -1 \\ 1 & -1 & 1 \end{vmatrix} = 1\begin{vmatrix} 1 & -1 \\ -1 & 1 \end{vmatrix} - 1\begin{vmatrix} 1 & -1 \\ 1 & 1 \end{vmatrix} + 1\begin{vmatrix} 1 & 1 \\ 1 & -1 \end{vmatrix} = 1(0) - 1(2) + 1(-2) = -4$

$N_x = \begin{vmatrix} 4 & 1 & 1 \\ 0 & 1 & -1 \\ 2 & -1 & 1 \end{vmatrix} = 4\begin{vmatrix} 1 & -1 \\ -1 & 1 \end{vmatrix} - 1\begin{vmatrix} 0 & -1 \\ 2 & 1 \end{vmatrix} + 1\begin{vmatrix} 0 & 1 \\ 2 & -1 \end{vmatrix} = 4(0) - 1(2) + 1(-2) = -4$

$N_y = \begin{vmatrix} 1 & 4 & 1 \\ 1 & 0 & -1 \\ 1 & 2 & 1 \end{vmatrix} = 1\begin{vmatrix} 0 & -1 \\ 2 & 1 \end{vmatrix} - 4\begin{vmatrix} 1 & -1 \\ 1 & 1 \end{vmatrix} + 1\begin{vmatrix} 1 & 0 \\ 1 & 2 \end{vmatrix} = 1(2) - 4(2) + 1(2) = -4$

$N_z = \begin{vmatrix} 1 & 1 & 4 \\ 1 & 1 & 0 \\ 1 & -1 & 2 \end{vmatrix} = 1\begin{vmatrix} 1 & 0 \\ -1 & 2 \end{vmatrix} - 1\begin{vmatrix} 1 & 0 \\ 1 & 2 \end{vmatrix} + 4\begin{vmatrix} 1 & 1 \\ 1 & -1 \end{vmatrix} = 1(2) - 1(2) + 4(-2) = -8$

$x = \dfrac{N_x}{D} = \dfrac{-4}{-4} = 1; \; y = \dfrac{N_y}{D} = \dfrac{-4}{-4} = 1; \; z = \dfrac{N_z}{D} = \dfrac{-8}{-4} = 2 \Rightarrow$ solution: $(1, 1, 2)$

43. $D = \begin{vmatrix} 1 & 1 & 2 \\ 1 & 2 & 1 \\ 2 & 1 & 1 \end{vmatrix} = 1\begin{vmatrix} 2 & 1 \\ 1 & 1 \end{vmatrix} - 1\begin{vmatrix} 1 & 1 \\ 2 & 1 \end{vmatrix} + 2\begin{vmatrix} 1 & 2 \\ 2 & 1 \end{vmatrix} = 1(1) - 1(-1) + 2(-3) = -4$

$N_x = \begin{vmatrix} 7 & 1 & 2 \\ 8 & 2 & 1 \\ 9 & 1 & 1 \end{vmatrix} = 7\begin{vmatrix} 2 & 1 \\ 1 & 1 \end{vmatrix} - 1\begin{vmatrix} 8 & 1 \\ 9 & 1 \end{vmatrix} + 2\begin{vmatrix} 8 & 2 \\ 9 & 1 \end{vmatrix} = 7(1) - 1(-1) + 2(-10) = -12$

$N_y = \begin{vmatrix} 1 & 7 & 2 \\ 1 & 8 & 1 \\ 2 & 9 & 1 \end{vmatrix} = 1\begin{vmatrix} 8 & 1 \\ 9 & 1 \end{vmatrix} - 7\begin{vmatrix} 1 & 1 \\ 2 & 1 \end{vmatrix} + 2\begin{vmatrix} 1 & 8 \\ 2 & 9 \end{vmatrix} = 1(-1) - 7(-1) + 2(-7) = -8$

$N_z = \begin{vmatrix} 1 & 1 & 7 \\ 1 & 2 & 8 \\ 2 & 1 & 9 \end{vmatrix} = 1\begin{vmatrix} 2 & 8 \\ 1 & 9 \end{vmatrix} - 1\begin{vmatrix} 1 & 8 \\ 2 & 9 \end{vmatrix} + 7\begin{vmatrix} 1 & 2 \\ 2 & 1 \end{vmatrix} = 1(10) - 1(-7) + 7(-3) = -4$

$x = \dfrac{N_x}{D} = \dfrac{-12}{-4} = 3; \ y = \dfrac{N_y}{D} = \dfrac{-8}{-4} = 2; \ z = \dfrac{N_z}{D} = \dfrac{-4}{-4} = 1 \Rightarrow$ solution: $(3, 2, 1)$

45. $D = \begin{vmatrix} 2 & 1 & -1 \\ 1 & 2 & 2 \\ 4 & 5 & 3 \end{vmatrix} = 2\begin{vmatrix} 2 & 2 \\ 5 & 3 \end{vmatrix} - 1\begin{vmatrix} 1 & 2 \\ 4 & 3 \end{vmatrix} + (-1)\begin{vmatrix} 1 & 2 \\ 4 & 5 \end{vmatrix} = 2(-4) - 1(-5) - 1(-3) = 0$

$N_x = \begin{vmatrix} 1 & 1 & -1 \\ 2 & 2 & 2 \\ 3 & 5 & 3 \end{vmatrix} = 1\begin{vmatrix} 2 & 2 \\ 5 & 3 \end{vmatrix} - 1\begin{vmatrix} 2 & 2 \\ 3 & 3 \end{vmatrix} + (-1)\begin{vmatrix} 2 & 2 \\ 3 & 5 \end{vmatrix} = 1(-4) - 1(0) - 1(4) = -8$

$x = \dfrac{N_x}{D} = \dfrac{-8}{0} \Rightarrow$ denominator $= 0$, numerator $\neq 0 \Rightarrow$ no solution.

47. $D = \begin{vmatrix} 2 & 1 & 1 \\ 1 & -2 & 3 \\ 1 & 1 & -4 \end{vmatrix} = 2\begin{vmatrix} -2 & 3 \\ 1 & -4 \end{vmatrix} - 1\begin{vmatrix} 1 & 3 \\ 1 & -4 \end{vmatrix} + 1\begin{vmatrix} 1 & -2 \\ 1 & 1 \end{vmatrix} = 2(5) - 1(-7) + 1(3) = 20$

$N_x = \begin{vmatrix} 5 & 1 & 1 \\ 10 & -2 & 3 \\ -3 & 1 & -4 \end{vmatrix} = 5\begin{vmatrix} -2 & 3 \\ 1 & -4 \end{vmatrix} - 1\begin{vmatrix} 10 & 3 \\ -3 & -4 \end{vmatrix} + 1\begin{vmatrix} 10 & -2 \\ -3 & 1 \end{vmatrix}$

$\qquad = 5(5) - 1(-31) + 1(4) = 60$

$N_y = \begin{vmatrix} 2 & 5 & 1 \\ 1 & 10 & 3 \\ 1 & -3 & -4 \end{vmatrix} = 2\begin{vmatrix} 10 & 3 \\ -3 & -4 \end{vmatrix} - 5\begin{vmatrix} 1 & 3 \\ 1 & -4 \end{vmatrix} + 1\begin{vmatrix} 1 & 10 \\ 1 & -3 \end{vmatrix}$

$\qquad = 2(-31) - 5(-7) + 1(-13) = -40$

$N_z = \begin{vmatrix} 2 & 1 & 5 \\ 1 & -2 & 10 \\ 1 & 1 & -3 \end{vmatrix} = 2\begin{vmatrix} -2 & 10 \\ 1 & -3 \end{vmatrix} - 1\begin{vmatrix} 1 & 10 \\ 1 & -3 \end{vmatrix} + 5\begin{vmatrix} 1 & -2 \\ 1 & 1 \end{vmatrix} = 2(-4) - 1(-13) + 5(3) = 20$

$x = \dfrac{N_x}{D} = \dfrac{60}{20} = 3; \ y = \dfrac{N_y}{D} = \dfrac{-40}{20} = -2; \ z = \dfrac{N_z}{D} = \dfrac{20}{20} = 1 \Rightarrow$ solution: $(3, -2, 1)$

49. $D = \begin{vmatrix} 2 & 3 & 4 \\ 2 & -3 & -4 \\ 4 & 6 & 8 \end{vmatrix} = 2\begin{vmatrix} -3 & -4 \\ 6 & 8 \end{vmatrix} - 3\begin{vmatrix} 2 & -4 \\ 4 & 8 \end{vmatrix} + 4\begin{vmatrix} 2 & -3 \\ 4 & 6 \end{vmatrix} = 2(0) - 3(32) + 4(24) = 0$

$N_x = \begin{vmatrix} 6 & 3 & 4 \\ -4 & -3 & -4 \\ 12 & 6 & 8 \end{vmatrix} = 6\begin{vmatrix} -3 & -4 \\ 6 & 8 \end{vmatrix} - 3\begin{vmatrix} -4 & -4 \\ 12 & 8 \end{vmatrix} + 4\begin{vmatrix} -4 & -3 \\ 12 & 6 \end{vmatrix} = 6(0) - 3(16) + 4(12) = 0$

$x = \dfrac{N_x}{D} = \dfrac{0}{0} \Rightarrow$ denominator $= 0$, numerator $= 0 \Rightarrow$ dependent equations.

51. $\begin{cases} x + y = 1 \Rightarrow & x + y = 1 \\ \frac{1}{2}y + z = \frac{5}{2} \Rightarrow & y + 2z = 5 \\ x - z = -3 \Rightarrow & x - z = -3 \end{cases}$

$D = \begin{vmatrix} 1 & 1 & 0 \\ 0 & 1 & 2 \\ 1 & 0 & -1 \end{vmatrix} = 1\begin{vmatrix} 1 & 2 \\ 0 & -1 \end{vmatrix} - 1\begin{vmatrix} 0 & 2 \\ 1 & -1 \end{vmatrix} + 0\begin{vmatrix} 0 & 1 \\ 1 & 0 \end{vmatrix} = 1(-1) - 1(-2) + 0(-1) = 1$

$N_x = \begin{vmatrix} 1 & 1 & 0 \\ 5 & 1 & 2 \\ -3 & 0 & -1 \end{vmatrix} = 1\begin{vmatrix} 1 & 2 \\ 0 & -1 \end{vmatrix} - 1\begin{vmatrix} 5 & 2 \\ -3 & -1 \end{vmatrix} + 0\begin{vmatrix} 5 & 1 \\ -3 & 0 \end{vmatrix} = 1(-1) - 1(1) + 0(3) = -2$

$N_y = \begin{vmatrix} 1 & 1 & 0 \\ 0 & 5 & 2 \\ 1 & -3 & -1 \end{vmatrix} = 1\begin{vmatrix} 5 & 2 \\ -3 & -1 \end{vmatrix} - 1\begin{vmatrix} 0 & 2 \\ 1 & -1 \end{vmatrix} + 0\begin{vmatrix} 0 & 5 \\ 1 & -3 \end{vmatrix} = 1(1) - 1(-2) + 0(-5) = 3$

$N_z = \begin{vmatrix} 1 & 1 & 1 \\ 0 & 1 & 5 \\ 1 & 0 & -3 \end{vmatrix} = 1\begin{vmatrix} 1 & 5 \\ 0 & -3 \end{vmatrix} - 1\begin{vmatrix} 0 & 5 \\ 1 & -3 \end{vmatrix} + 1\begin{vmatrix} 0 & 1 \\ 1 & 0 \end{vmatrix} = 1(-3) - 1(-5) + 1(-1) = 1$

$x = \dfrac{N_x}{D} = \dfrac{-2}{1} = -2;\ y = \dfrac{N_y}{D} = \dfrac{3}{1} = 3;\ z = \dfrac{N_z}{D} = \dfrac{1}{1} = 1 \Rightarrow$ solution: $(-2, 3, 1)$

53. $\begin{cases} 2x - y + 4z + 2 = 0 \Rightarrow & 2x - y + 4z = -2 \\ 5x + 8y + 7z = -8 \Rightarrow & 5x + 8y + 7z = -8 \\ x + 3y + z + 3 = 0 \Rightarrow & x + 3y + z = -3 \end{cases}$

$D = \begin{vmatrix} 2 & -1 & 4 \\ 5 & 8 & 7 \\ 1 & 3 & 1 \end{vmatrix} = 2\begin{vmatrix} 8 & 7 \\ 3 & 1 \end{vmatrix} - (-1)\begin{vmatrix} 5 & 7 \\ 1 & 1 \end{vmatrix} + 4\begin{vmatrix} 5 & 8 \\ 1 & 3 \end{vmatrix} = 2(-13) + 1(-2) + 4(7) = 0$

$N_x = \begin{vmatrix} -2 & -1 & 4 \\ -8 & 8 & 7 \\ -3 & 3 & 1 \end{vmatrix} = -2\begin{vmatrix} 8 & 7 \\ 3 & 1 \end{vmatrix} - (-1)\begin{vmatrix} -8 & 7 \\ -3 & 1 \end{vmatrix} + 4\begin{vmatrix} -8 & 8 \\ -3 & 3 \end{vmatrix} = -2(-13) + 1(13) + 4(0)$

$= 39$

$x = \dfrac{N_x}{D} = \dfrac{39}{0} \Rightarrow$ denominator $= 0$, numerator $\neq 0 \Rightarrow$ no solution.

55. $\begin{vmatrix} x & 1 \\ 3 & 2 \end{vmatrix} = 1$

$2x - 3 = 1$

$2x = 4$

$x = 2$

57. $\begin{vmatrix} x & -2 \\ 3 & 1 \end{vmatrix} = \begin{vmatrix} 4 & 2 \\ x & 3 \end{vmatrix}$

$x + 6 = 12 - 2x$

$3x = 6$

$x = 2$

59. Let $x =$ the amount invested in HiTech, $y =$ the amount invested in SaveTel and $z =$ the amount invested in HiGas. Form and solve the following system of equations:

$$\begin{cases} x + y + z = 20000 \Rightarrow \\ 0.10x + 0.05y + 0.06z = 0.066(20000) \Rightarrow \\ y + z = 3x \Rightarrow \end{cases} \qquad \begin{aligned} x + y + z &= 20000 \\ 10x + 5y + 6x &= 132000 \\ -3x + y + z &= 0 \end{aligned}$$

$$D = \begin{vmatrix} 1 & 1 & 1 \\ 10 & 5 & 6 \\ -3 & 1 & 1 \end{vmatrix} = 1 \begin{vmatrix} 5 & 6 \\ 1 & 1 \end{vmatrix} - 1 \begin{vmatrix} 10 & 6 \\ -3 & 1 \end{vmatrix} + 1 \begin{vmatrix} 10 & 5 \\ -3 & 1 \end{vmatrix} = 1(-1) - 1(28) + 1(25) = -4$$

$$N_x = \begin{vmatrix} 20000 & 1 & 1 \\ 132000 & 5 & 6 \\ 0 & 1 & 1 \end{vmatrix} = 20000 \begin{vmatrix} 5 & 6 \\ 1 & 1 \end{vmatrix} - 1 \begin{vmatrix} 132000 & 6 \\ 0 & 1 \end{vmatrix} + 1 \begin{vmatrix} 132000 & 5 \\ 0 & 1 \end{vmatrix}$$

$$= 20000(-1) - 1(132000) + 1(132000) = -20000$$

$$N_y = \begin{vmatrix} 1 & 20000 & 1 \\ 10 & 132000 & 6 \\ -3 & 0 & 1 \end{vmatrix} = 1 \begin{vmatrix} 132000 & 6 \\ 0 & 1 \end{vmatrix} - 20000 \begin{vmatrix} 10 & 6 \\ -3 & 1 \end{vmatrix} + 1 \begin{vmatrix} 10 & 132000 \\ -3 & 0 \end{vmatrix}$$

$$= 1(132000) - 20000(28) + 1(396000) = -32000$$

$$N_z = \begin{vmatrix} 1 & 1 & 20000 \\ 10 & 5 & 132000 \\ -3 & 1 & 0 \end{vmatrix} = 1 \begin{vmatrix} 5 & 132000 \\ 1 & 0 \end{vmatrix} - 1 \begin{vmatrix} 10 & 132000 \\ -3 & 0 \end{vmatrix} + 20000 \begin{vmatrix} 10 & 5 \\ -3 & 1 \end{vmatrix}$$

$$= 1(-132000) - 1(396000) + 20000(25) = -28000$$

$$x = \frac{N_x}{D} = \frac{-20000}{-4} = 5000; \quad y = \frac{N_y}{D} = \frac{-32000}{-4} = 8000; \quad z = \frac{N_z}{D} = \frac{-28000}{-4} = 7000$$

He should invest \$5000 in HiTech, \$8000 in SaveTel and \$7000 in HiGas.

61. $\begin{vmatrix} 2 & -3 & 4 \\ -1 & 2 & 4 \\ 3 & -3 & 1 \end{vmatrix} = -23$

63. $\begin{vmatrix} 2 & 1 & -3 \\ -2 & 2 & 4 \\ 1 & -2 & 2 \end{vmatrix} = 26$

65. Answers may vary.

67.

$$\begin{vmatrix} x & y & 1 \\ -2 & 3 & 1 \\ 3 & 5 & 1 \end{vmatrix} = 0$$

$$x \begin{vmatrix} 3 & 1 \\ 5 & 1 \end{vmatrix} - y \begin{vmatrix} -2 & 1 \\ 3 & 1 \end{vmatrix} + 1 \begin{vmatrix} -2 & 3 \\ 3 & 5 \end{vmatrix} = 0$$

$$x(3 - 5) - y(-2 - 3) + 1(-10 - 9) = 0$$

$$-2x + 5y - 19 = 0$$

$$2x - 5y = -19 \qquad \text{Verify that both points satisfy this equation.}$$

69.

$$\begin{vmatrix} 1 & 0 & 2 & 1 \\ 2 & 1 & 1 & 3 \\ 1 & 1 & 1 & 1 \\ 2 & 1 & 1 & 1 \end{vmatrix} = 1 \begin{vmatrix} 1 & 1 & 3 \\ 1 & 1 & 1 \\ 1 & 1 & 1 \end{vmatrix} - 0 \begin{vmatrix} 2 & 1 & 3 \\ 1 & 1 & 1 \\ 2 & 1 & 1 \end{vmatrix} + 2 \begin{vmatrix} 2 & 1 & 3 \\ 1 & 1 & 1 \\ 2 & 1 & 1 \end{vmatrix} - 1 \begin{vmatrix} 2 & 1 & 1 \\ 1 & 1 & 1 \\ 2 & 1 & 1 \end{vmatrix}$$

$$= 1(0) - 0(???) + 2(-2) - 1(0) = -4$$

Exercise 13.5 (page 867)

1. $\sqrt{200x^2} - 3\sqrt{98x^2} = \sqrt{100x^2}\sqrt{2} - 3\sqrt{49x^2}\sqrt{2} = 10x\sqrt{2} - 3(7x)\sqrt{2} = -11x\sqrt{2}$

3. $\dfrac{3t\sqrt{2t} - 2\sqrt{2t^3}}{\sqrt{18t} - \sqrt{2t}} = \dfrac{3t\sqrt{2t} - 2\sqrt{t^2}\sqrt{2t}}{\sqrt{9}\sqrt{2t} - \sqrt{2t}} = \dfrac{3t\sqrt{2t} - 2t\sqrt{2t}}{3\sqrt{2t} - \sqrt{2t}} = \dfrac{t\sqrt{2t}}{2\sqrt{2t}} = \dfrac{t}{2}$

5. graphing; substitution

7. $\begin{cases} 8x^2 + 32y^2 = 256 \\ x = 2y \end{cases}$

9. $\begin{cases} x^2 + y^2 = 10 \\ y = 3x^2 \end{cases}$

11. $\begin{cases} x^2 + y^2 = 25 \\ 12x^2 + 64y^2 = 768 \end{cases}$

$(-4, -2), (4, 2)$

$(-1, 3), (1, 3)$

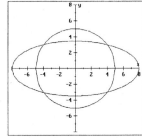

$(-4, 3), (4, 3), (4, -3)$
$(-4, -3)$

13. $\begin{cases} x^2 - 13 = -y^2 \\ y = 2x - 4 \end{cases}$

15. $\begin{cases} x^2 - 6x - y = -5 \\ x^2 - 6x + y = -5 \end{cases}$
$\begin{cases} y = x^2 - 6x + 5 \\ y = -x^2 + 6x - 5 \end{cases}$

$\left(\frac{1}{5}, -\frac{18}{5}\right), (3, 2)$

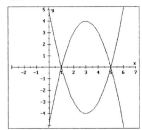

$(1, 0), (5, 0)$

17. $\begin{cases} (1) & 25x^2 + 9y^2 = 225 \\ (2) & 5x + 3y = 15 \end{cases}$

Substitute $x = -\frac{3}{5}y + 3$ from (2) into (1):

$$25x^2 + 9y^2 = 225$$
$$25\left(-\frac{3}{5}y + 3\right)^2 + 9y^2 = 225$$
$$25\left(\frac{9}{25}y^2 - \frac{18}{5}y + 9\right) + 9y^2 = 225$$
$$9y^2 - 90y + 225 + 9y^2 = 225$$
$$18y^2 - 90y = 0$$
$$18y(y - 5) = 0$$

$18y = 0$ **or** $y - 5 = 0$
$\quad y = 0 \qquad\qquad y = 5$

Substitute these and solve for x:

$5x + 3y = 15 \qquad\quad 5x + 3y = 15$
$5x + 3(0) = 15 \qquad 5x + 3(5) = 15$
$\quad 5x = 15 \qquad\qquad\quad 5x = 0$
$\quad\; x = 3 \qquad\qquad\qquad x = 0$

Solutions: $(3, 0), (0, 5)$

19. $\begin{cases} (1) & x^2 + y^2 = 2 \\ (2) & x + y = 2 \end{cases}$

Substitute $x = 2 - y$ from (2) into (1):

$$x^2 + y^2 = 2$$
$$(2 - y)^2 + y^2 = 2$$
$$4 - 4y + y^2 + y^2 = 2$$
$$2y^2 - 4y + 2 = 0$$
$$2(y - 1)(y - 1) = 0$$

$y - 1 = 0$ **or** $y - 1 = 0$
$\quad y = 1 \qquad\qquad y = 1$

Substitute this and solve for x:

$x = 2 - y$
$x = 2 - 1$
$x = 1$

Solution: $(1, 1)$

21. $\begin{cases} (1) & x^2 + y^2 = 5 \\ (2) & x + y = 3 \end{cases}$

Substitute $x = 3 - y$ from (2) into (1):

$$x^2 + y^2 = 5$$
$$(3 - y)^2 + y^2 = 5$$
$$9 - 6y + y^2 + y^2 = 5$$
$$2y^2 - 6y + 4 = 0$$
$$2(y - 2)(y - 1) = 0$$

$y - 2 = 0$ **or** $y - 1 = 0$
$\quad y = 2 \qquad\qquad y = 1$

Substitute these and solve for x:

$x = 3 - y \qquad\quad x = 3 - y$
$x = 3 - 2 \qquad\quad x = 3 - 1$
$x = 1 \qquad\qquad\; x = 2$

Solutions: $(1, 2), (2, 1)$

23. $\begin{cases} (1) & x^2 + y^2 = 13 \\ (2) & y = x^2 - 1 \end{cases}$

Substitute $x^2 = 13 - y^2$ from (1) into (2):

$$y = x^2 - 1$$
$$y = 13 - y^2 - 1$$
$$y^2 + y - 12 = 0$$
$$(y + 4)(y - 3) = 0$$

$y + 4 = 0$ **or** $y - 3 = 0$
$\quad y = -4 \qquad\qquad y = 3$

Substitute these and solve for x:

$x^2 = 13 - y^2 \qquad x^2 = 13 - y^2$
$x^2 = 13 - 16 \qquad x^2 = 13 - 9$
$x^2 = -4 \qquad\qquad x^2 = 4$
$\quad$ complex $\qquad\qquad x = \pm 2$

Solutions: $(2, 3), (-2, 3)$

25. $\begin{cases} (1) & x^2 + y^2 = 30 \\ (2) & y = x^2 \end{cases}$

Substitute $x^2 = y$ from (2) into (1):
$$x^2 + y^2 = 30$$
$$y + y^2 = 30$$
$$y^2 + y - 30 = 0$$
$$(y - 5)(y + 6) = 0$$

$y - 5 = 0$ **or** $y + 6 = 0$
$\qquad y = 5 \qquad\qquad y = -6$

Substitute these and solve for x:

$\begin{aligned} x^2 &= y \\ x^2 &= 5 \\ x &= \pm\sqrt{5} \end{aligned}$ $\qquad\qquad$ $\begin{aligned} x^2 &= y \\ x^2 &= -6 \\ &\text{complex} \end{aligned}$

Solutions: $\left(\sqrt{5}, 5\right), \left(-\sqrt{5}, 5\right)$

27. $\begin{aligned} x^2 + y^2 &= 13 \\ \underline{x^2 - y^2 = 5} \\ 2x^2 &= 18 \\ x^2 &= 9 \\ x &= \pm 3 \end{aligned}$ Substitute and solve for y:

$\begin{aligned} x^2 + y^2 &= 13 \\ 3^2 + y^2 &= 13 \\ y^2 &= 4 \\ y &= \pm 2 \end{aligned}$ $\qquad$ $\begin{aligned} x^2 + y^2 &= 13 \\ (-3)^2 + y^2 &= 13 \\ y^2 &= 4 \\ y &= \pm 2 \end{aligned}$

Solutions: $(3, 2), (3, -2), (-3, 2), (-3, -2)$

29. $\begin{aligned} x^2 + y^2 &= 20 \\ \underline{x^2 - y^2 = -12} \\ 2x^2 &= 8 \\ x^2 &= 4 \\ x &= \pm 2 \end{aligned}$ Substitute and solve for y:

$\begin{aligned} x^2 + y^2 &= 20 \\ 2^2 + y^2 &= 20 \\ y^2 &= 16 \\ y &= \pm 4 \end{aligned}$ $\qquad$ $\begin{aligned} x^2 + y^2 &= 20 \\ (-2)^2 + y^2 &= 20 \\ y^2 &= 16 \\ y &= \pm 4 \end{aligned}$

Solutions: $(2, 4), (2, -4), (-2, 4), (-2, -4)$

31. $\begin{cases} (1) & y^2 = 40 - x^2 \\ (2) & y = x^2 - 10 \end{cases}$

Substitute $x^2 = 40 - y^2$ from (1) into (2):
$$y = x^2 - 10$$
$$y = 40 - y^2 - 10$$
$$y^2 + y - 30 = 0$$
$$(y + 6)(y - 5) = 0$$

$y + 6 = 0$ **or** $y - 5 = 0$
$\qquad y = -6 \qquad\qquad y = 5$

Substitute these and solve for x:

$\begin{aligned} x^2 &= 40 - y^2 \\ x^2 &= 40 - 36 \\ x^2 &= 4 \\ x &= \pm 2 \end{aligned}$ $\qquad$ $\begin{aligned} x^2 &= 40 - y^2 \\ x^2 &= 40 - 25 \\ x^2 &= 15 \\ x &= \pm\sqrt{15} \end{aligned}$

$(2, -6), (-2, -6), \left(\sqrt{15}, 5\right), \left(-\sqrt{15}, 5\right)$

33. $\begin{cases} (1) & y = x^2 - 4 \\ (2) & x^2 - y^2 = -16 \end{cases}$

Substitute $x^2 = y + 4$ from (1) into (2):
$$x^2 - y^2 = -16$$
$$y + 4 - y^2 = -16$$
$$y^2 - y - 20 = 0$$
$$(y + 4)(y - 5) = 0$$

$y + 4 = 0$ **or** $y - 5 = 0$
$\qquad y = -4 \qquad\qquad y = 5$

Substitute these and solve for x:

$\begin{aligned} x^2 &= y + 4 \\ x^2 &= -4 + 4 \\ x^2 &= 0 \\ x &= 0 \end{aligned}$ $\qquad$ $\begin{aligned} x^2 &= y + 4 \\ x^2 &= 5 + 4 \\ x^2 &= 9 \\ x &= \pm 3 \end{aligned}$

Solutions: $(0, -4), (3, 5), (-3, 5)$

35. $x^2 - y^2 = -5 \Rightarrow (\times 2)\ 2x^2 - 2y^2 = -10$ Substitute and solve for y:

$3x^2 + 2y^2 = 30 \Rightarrow$

$$\begin{aligned} 3x^2 + 2y^2 &= 30 \\ 5x^2 &= 20 \\ x^2 &= 4 \\ x &= \pm 2 \end{aligned}$$

$$\begin{aligned} x^2 - y^2 &= -5 \\ 2^2 - y^2 &= -5 \\ -y^2 &= -9 \\ y^2 &= 9 \\ y &= \pm 3 \end{aligned} \qquad \begin{aligned} x^2 - y^2 &= -5 \\ (-2)^2 - y^2 &= -5 \\ -y^2 &= -9 \\ y^2 &= 9 \\ y &= \pm 3 \end{aligned}$$

Solutions: $(2, 3), (2, -3), (-2, 3), (-2, -3)$

37. $\dfrac{1}{x} + \dfrac{2}{y} = 1 \Rightarrow$

$\dfrac{2}{x} - \dfrac{1}{y} = \dfrac{1}{3} \Rightarrow (\times 2)$

$$\begin{aligned} \frac{1}{x} + \frac{2}{y} &= 1 \\ \frac{4}{x} - \frac{2}{y} &= \frac{2}{3} \\ \frac{5}{x} &= \frac{5}{3} \\ 15 &= 5x \\ 3 &= x \end{aligned}$$

Substitute and solve for y:

$$\begin{aligned} \frac{1}{x} + \frac{2}{y} &= 1 \\ \frac{1}{3} + \frac{2}{y} &= 1 \\ \frac{2}{y} &= \frac{2}{3} \\ 6 &= 2y \\ 3 &= y \end{aligned}$$

Solution: $(3, 3)$

39. $\begin{cases} (1) & 3y^2 = xy \\ (2) & 2x^2 + xy - 84 = 0 \end{cases}$

From (1): $3y^2 - xy = 0$

$y(3y - x) = 0$

$y = 0 \quad$ or $\quad y = \dfrac{1}{3}x$

Substitute these into (2):

$$\begin{aligned} 2x^2 + xy - 84 &= 0 \\ 2x^2 + x(0) - 84 &= 0 \\ 2x^2 &= 84 \\ x^2 &= 42 \\ x &= \pm\sqrt{42} \end{aligned} \qquad \begin{aligned} 2x^2 + xy - 84 &= 0 \\ 2x^2 + x\left(\tfrac{1}{3}x\right) - 84 &= 0 \\ 2x^2 + \tfrac{1}{3}x^2 &= 84 \\ 6x^2 + x^2 &= 252 \\ 7x^2 &= 252 \\ x^2 &= 36 \\ x &= \pm 6 \end{aligned}$$

(substitute and solve for y)

Solutions: $\left(\sqrt{42}, 0\right), \left(-\sqrt{42}, 0\right), (6, 2), (-6, -2)$

41. $\begin{cases} (1) & xy = \frac{1}{6} \\ (2) & y + x = 5xy \end{cases}$

Substitute $x = \dfrac{1}{6y}$ from (1) into (2):

$$\begin{aligned} y + \frac{1}{6y} &= \frac{5y}{6y} \\ 6y^2 + 1 &= 5y \\ 6y^2 - 5y + 1 &= 0 \\ (2y - 1)(3y - 1) &= 0 \end{aligned}$$

$2y - 1 = 0 \quad$ **or** $\quad 3y - 1 = 0$

$y = \dfrac{1}{2} \qquad\qquad y = \dfrac{1}{3}$

Substitute these and solve for x:

$$\begin{aligned} x &= \frac{1}{6y} & x &= \frac{1}{6y} \\ x &= \frac{1}{6\left(\frac{1}{2}\right)} & x &= \frac{1}{6\left(\frac{1}{3}\right)} \\ x &= \frac{1}{3} & x &= \frac{1}{2} \end{aligned}$$

Solutions: $\left(\frac{1}{3}, \frac{1}{2}\right), \left(\frac{1}{2}, \frac{1}{3}\right)$

43. Let the integers be x and y. Then the equations are

$$\begin{cases} (1) & xy = 32 \\ (2) & x + y = 12 \end{cases}$$

Substitute $x = \frac{32}{y}$ from (1) into (2):

$$\frac{32}{y} + y = 12$$
$$32 + y^2 = 12y$$
$$y^2 - 12y + 32 = 0$$
$$(y - 4)(y - 8) = 0$$

$$y - 4 = 0 \quad \textbf{or} \quad y - 8 = 0$$
$$y = 4 \qquad\qquad y = 8$$

Substitute these and solve for x:

$$x = \frac{32}{y} = \frac{32}{4} = 8 \qquad x = \frac{32}{y} = \frac{32}{8} = 4$$

The integers are 8 and 4.

45. $\begin{cases} 2x - y > 4 \\ y < -x^2 + 2 \end{cases}$

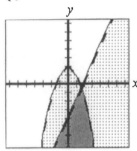

47. $\begin{cases} y > x^2 - 4 \\ y < -x^2 + 4 \end{cases}$

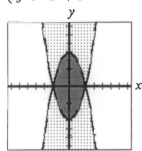

49. Let l = the length of the rectangle, and w = the width of the rectangle. Then the equations are:

$$\begin{cases} (1) & lw = 63 \\ (2) & 2l + 2w = 32 \end{cases}$$

Substitute $l = \frac{63}{w}$ from (1) into (2):

$$2\left(\frac{63}{w}\right) + 2w = 32$$
$$\frac{126}{w} + 2w = 32$$
$$126 + 2w^2 = 32w$$
$$2w^2 - 32y + 126 = 0$$
$$2(w - 7)(w - 9) = 0$$

$$w - 7 = 0 \quad \textbf{or} \quad w - 9 = 0$$
$$w = 7 \qquad\qquad w = 9$$

Substitute these and solve for l:

$$l = \frac{63}{w} = \frac{63}{7} = 9 \qquad l = \frac{63}{w} = \frac{63}{9} = 7$$

The dimensions are 7 cm by 9 cm.

51. Let r = Carol's rate, and let p = the amount Carol invested.
Then Juan invested $p + 150$ at a rate of $r + 0.015$. The equations are

$$\begin{cases} (1)\ pr = 67.50 \Rightarrow p = \frac{67.5}{r} \\ (2)\ (p + 150)(r + 0.015) = 94.5 \end{cases}$$

$$\begin{array}{lll} 100r - 9 = 0 & \text{or} & 120r - 9 = 0 \\ r = 0.09 & & r = 0.075 \end{array}$$

Substitute $p = \frac{67.5}{r}$ from (1) into (2):

Substitute and solve for p:

$$\left(\frac{67.5}{r} + 150\right)(r + 0.015) = 94.5$$

$$p = \frac{67.5}{r} = \frac{67.5}{0.09} = 750 \ \text{ or } \ p = \frac{67.5}{r} = \frac{67.5}{0.075} = 900$$

$$67.5 + \frac{1.0125}{r} + 150r + 2.25 = 94.5$$

Carol invested \$750 at 9% or \$900 at 7.5%.

$$67.5r + 1.0125 + 150r^2 + 2.25r = 94.5r$$

$$150r^2 - 24.75r - 1.0125 = 0$$

$$12{,}000r^2 - 1980r + 81 = 0$$

$$(100r - 9)(120r - 9) = 0$$

53. Let r = Jim's rate and t = Jim's time. Then his brother's rate was $r - 17$ and his time was $t + 1.5$.

$$\begin{cases} (1)\ rt = 306 \Rightarrow t = \frac{306}{r} \\ (2)\ (r - 17)(t + 1.5) = 306 \end{cases}$$

$$\begin{array}{lll} 3r + 153 = 0 & \text{or} & r - 68 = 0 \\ r = -153/3 & & r = 68 \end{array}$$

Substitute $t = \frac{306}{r}$ from (1) into (2):

Substitute and solve for t:

$$(r - 17)\left(\frac{306}{r} + 1.5\right) = 306$$

$$t = \frac{306}{r} = \frac{306}{68} = 4.5$$

$$306 + 1.5r - \frac{5202}{r} - 25.5 = 306$$

Jim drove for 4.5 hours at 68 miles per hour.

$$306r + 1.5r^2 - 5202 - 25.5r = 306r$$

$$1.5r^2 - 25.5r - 5202 = 0$$

$$3r^2 - 51r - 10{,}404 = 0$$

$$(3r + 153)(r - 68) = 0$$

55. Answers may vary.

57. $0, 1, 2, 3, 4$

Chapter 13 Summary   (page 871)

1. $\begin{cases} 2x + y = 11 \\ -x + 2y = 7 \end{cases}$

2. $\begin{cases} 3x + 2y = 0 \\ 2x - 3y = -13 \end{cases}$

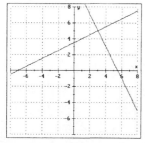

$(3, 5)$ is the solution.

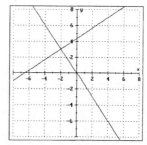

$(-2, 3)$ is the solution.

3. $\begin{cases} \frac{1}{2}x + \frac{1}{3}y = 2 \\ y = 6 - \frac{3}{2}x \end{cases}$

dependent equations

4. $\begin{cases} \frac{1}{3}x - \frac{1}{2}y = 1 \\ 6x - 9y = 2 \end{cases}$

inconsistent system

5. $\begin{cases} (1) \quad y = x + 4 \\ (2) \quad 2x + 3y = 7 \end{cases}$

Substitute $y = x + 4$ from (1) into (2):
$$2x + 3y = 7$$
$$2x + 3(x + 4) = 7$$
$$2x + 3x + 12 = 7$$
$$5x = -5$$
$$x = -1$$

Substitute this and solve for y:
$$y = x + 4$$
$$y = -1 + 4 = 3$$

Solution: $(-1, 3)$

6. $\begin{cases} (1) \quad y = 2x + 5 \\ (2) \quad 3x - 5y = -4 \end{cases}$

Substitute $y = 2x + 5$ from (1) into (2):
$$3x - 5y = -4$$
$$3x - 5(2x + 5) = -4$$
$$3x - 10x - 25 = -4$$
$$-7x = 21$$
$$x = -3$$

Substitute this and solve for y:
$$y = 2x + 5$$
$$y = 2(-3) + 5$$
$$y = -6 + 5 = -1$$

Solution: $(-3, -1)$

7. $\begin{cases} (1) \quad x + 2y = 11 \\ (2) \quad 2x - y = 2 \end{cases}$

Substitute $x = -2y + 11$ from (1) into (2):
$$2x - y = 2$$
$$2(-2y + 11) - y = 2$$
$$-4y + 22 - y = 2$$
$$-5y = -20$$
$$y = 4$$

Substitute this and solve for x:
$$x = -2y + 11$$
$$x = -2(4) + 11$$
$$x = -8 + 11 = 3$$

Solution: $(3, 4)$

8. $\begin{cases} (1) \quad 2x + 3y = -2 \\ (2) \quad 3x + 5y = -2 \end{cases}$

Substitute $x = \frac{-3y-2}{2}$ from (1) into (2):
$$3x + 5y = -2$$
$$3 \cdot \frac{-3y-2}{2} + 5y = -2$$
$$3(-3y - 2) + 10y = -4$$
$$-9y - 6 + 10y = -4$$
$$y = 2$$

Substitute this and solve for y:
$$x = \frac{-3y-2}{2}$$
$$x = \frac{-3(2)-2}{2}$$
$$= \frac{-8}{2} = -4$$

Solution: $(-4, 2)$

9. $x + y = -2 \Rightarrow \times(-2)$ $-2x - 2y = 4$
$2x + 3y = -3 \Rightarrow$ $2x + 3y = -3$
$\underline{}$
$y = 1$

Substitute and solve for x:
$x + y = -2$
$x + 1 = -2$
$x = -3$ Solution: $\boxed{(-3, 1)}$

10. $3x + 2y = 1 \Rightarrow \times 3$ $9x + 6y = 3$
$2x - 3y = 5 \Rightarrow \times 2$ $\underline{4x - 6y = 10}$
$ 13x = 13$
$ x = 1$

Substitute and solve for y:
$3x + 2y = 1$
$3(1) + 2y = 1$
$3 + 2y = 1$
$2y = -2$
$y = -1$ Solution: $\boxed{(1, -1)}$

11. $x + \frac{1}{2}y = 7 \Rightarrow \times 2$ $2x + y = 14 \Rightarrow$ Solve for x:
$\underline{-2x = 3y - 6} \Rightarrow$ $\underline{-2x - 3y = -6} \Rightarrow$ $2x + y = 14$
$ -2y = 8$ $2x + (-4) = 14$
$ y = -4$ $2x = 18$
$ x = 9$ Solution: $\boxed{(9, -4)}$

12. $y = \dfrac{x - 3}{2} \Rightarrow \times 2$ $2y = x - 3 \Rightarrow -x + 2y = -3$ Solve for y:
$x = \dfrac{2y + 7}{2} \Rightarrow \times 2$ $2x = 2y + 7 \Rightarrow 2x - 2y = 7$ $y = \dfrac{x - 3}{2}$
$\underline{}$ $y = \dfrac{4 - 3}{2} = \dfrac{1}{2}$ Solution: $\boxed{\left(4, \frac{1}{2}\right)}$
$ x = 4$

13.
(1) $x + y + z = 6$ (1) $x + y + z = 6$ (1) $x + y + z = 6$
(2) $x - y - z = -4$ (2) $x - y - z = -4$ (3) $-x + y - z = -2$
(3) $-x + y - z = -2$ (4) $\underline{2x = 2}$ (5) $\underline{2y = 4}$
$ x = 1$ $y = 2$

$x + y + z = 6$
$1 + 2 + z = 6$
$3 + z = 6$
$z = 3$ Solution: $\boxed{(1, 2, 3)}$

14.
(1) $2x + 3y + z = -5$ (1) $2x + 3y + z = -5$ (3) $3x + y + 2z = 4$
(2) $-x + 2y - z = -6$ (2) $-x + 2y - z = -6$ $2 \cdot (2) -2x + 4y - 2z = -12$
(3) $3x + y + 2z = 4$ (4) $\underline{x + 5y = -11}$ (5) $\underline{x + 5y = -8}$

$x + 5y = -11 \Rightarrow \times(-1)$ $-x - 5y = 11$
$\underline{x + 5y = -8} \Rightarrow$ $\underline{x + 5y = -8}$
$ 0 = 3$

Since this is an impossible equation, there is no solution. It is an inconsistent system.

15. $\begin{bmatrix} 1 & 2 & | & 4 \\ 2 & -1 & | & 3 \end{bmatrix} \Rightarrow \overset{-2R_1 + R_2 \Rightarrow R_2}{\begin{bmatrix} 1 & 2 & | & 4 \\ 0 & -5 & | & -5 \end{bmatrix}} \Rightarrow \overset{-\frac{1}{5}R_2 \Rightarrow R_2}{\begin{bmatrix} 1 & 2 & | & 4 \\ 0 & 1 & | & 1 \end{bmatrix}}$

From R_2, $y = 1$.　From R_1:　Solution: $\boxed{(2, 1)}$

$$x + 2y = 4$$
$$x + 2(1) = 4$$
$$x = 2$$

16. $\begin{bmatrix} 1 & 1 & 1 & | & 6 \\ 2 & -1 & 1 & | & 1 \\ 4 & 1 & -1 & | & 5 \end{bmatrix}$

$\overset{\substack{-2R_1 + R_2 \Rightarrow R_2 \\ -4R_1 + R_3 \Rightarrow R_3}}{\Rightarrow} \begin{bmatrix} 1 & 1 & 1 & | & 6 \\ 0 & -3 & -1 & | & -11 \\ 0 & -3 & -5 & | & -19 \end{bmatrix} \overset{\substack{-R_2 \Rightarrow R_2 \\ -R_3 + R_2 \Rightarrow R_3}}{\Rightarrow} \begin{bmatrix} 1 & 1 & 1 & | & 6 \\ 0 & 3 & 1 & | & 11 \\ 0 & 0 & 4 & | & 8 \end{bmatrix} \overset{\frac{1}{4}R_3 \Rightarrow R_3}{\Rightarrow} \begin{bmatrix} 1 & 1 & 1 & | & 6 \\ 0 & 3 & 1 & | & 11 \\ 0 & 0 & 1 & | & 2 \end{bmatrix}$

From R_3, $z = 2$.　From R_2:　From R_1:　Solution: $\boxed{(1, 3, 2)}$

$$3y + z = 11 \qquad x + y + z = 6$$
$$3y + 2 = 11 \qquad x + 3 + 2 = 6$$
$$3y = 9 \qquad\qquad x = 1$$
$$y = 3$$

17. $\begin{bmatrix} 1 & 1 & | & 3 \\ 1 & -2 & | & -3 \\ 2 & 1 & | & 4 \end{bmatrix}$

$\overset{\substack{-R_1 + R_2 \Rightarrow R_2 \\ -2R_1 + R_3 \Rightarrow R_3}}{\Rightarrow} \begin{bmatrix} 1 & 1 & | & 3 \\ 0 & -3 & | & -6 \\ 0 & -1 & | & -2 \end{bmatrix} \overset{\substack{-\frac{1}{3}R_2 \Rightarrow R_2 \\ -3R_3 + R_2 \Rightarrow R_3}}{\Rightarrow} \begin{bmatrix} 1 & 1 & | & 3 \\ 0 & 1 & | & 2 \\ 0 & 0 & | & 0 \end{bmatrix}$

From R_2, $y = 2$.　From R_1:　Solution: $\boxed{(1, 2)}$

$$x + y = 3$$
$$x + 2 = 3$$
$$x = 1$$

18. $\begin{bmatrix} 1 & 2 & 1 & | & 2 \\ 2 & 5 & 4 & | & 5 \end{bmatrix} \overset{-2R_1 + R_2 \Rightarrow R_2}{\Rightarrow} \begin{bmatrix} 1 & 2 & 1 & | & 2 \\ 0 & 1 & 2 & | & 1 \end{bmatrix}$

From R_2:　　From R_1:　　Solution: $\boxed{(3z, 1 - 2z, z)}$

$$y + 2z = 1 \qquad\qquad x + 2y + z = 2$$
$$y = 1 - 2z \qquad x + 2(1 - 2z) + z = 2$$
$$\qquad\qquad x + 2 - 4z + z = 2$$
$$\qquad\qquad x - 3z = 0$$
$$\qquad\qquad x = 3z$$

19. $\begin{vmatrix} 2 & 3 \\ -4 & 3 \end{vmatrix} = 2(3) - 3(-4) = 6 + 12 = 18$

20. $\begin{vmatrix} -3 & -4 \\ 5 & -6 \end{vmatrix} = -3(-6) - (-4)(5) = 18 - (-20) = 18 + 20 = 38$

21. $\begin{vmatrix} -1 & 2 & -1 \\ 2 & -1 & 3 \\ 1 & -2 & 2 \end{vmatrix} = -1 \begin{vmatrix} -1 & 3 \\ -2 & 2 \end{vmatrix} - 2 \begin{vmatrix} 2 & 3 \\ 1 & 2 \end{vmatrix} + (-1) \begin{vmatrix} 2 & -1 \\ 1 & -2 \end{vmatrix} = -1(4) - 2(1) - 1(-3) = -3$

22. $\begin{vmatrix} 3 & -2 & 2 \\ 1 & -2 & -2 \\ 2 & 1 & -1 \end{vmatrix} = 3 \begin{vmatrix} -2 & -2 \\ 1 & -1 \end{vmatrix} - (-2) \begin{vmatrix} 1 & -2 \\ 2 & -1 \end{vmatrix} + 2 \begin{vmatrix} 1 & -2 \\ 2 & 1 \end{vmatrix} = 3(4) + 2(3) + 2(5) = 28$

23. $x = \dfrac{\begin{vmatrix} 10 & 4 \\ 1 & -3 \end{vmatrix}}{\begin{vmatrix} 3 & 4 \\ 2 & -3 \end{vmatrix}} = \dfrac{-30 - 4}{-9 - 8} = \dfrac{-34}{-17} = 2; \quad y = \dfrac{\begin{vmatrix} 3 & 10 \\ 2 & 1 \end{vmatrix}}{\begin{vmatrix} 3 & 4 \\ 2 & -3 \end{vmatrix}} = \dfrac{3 - 20}{-17} = \dfrac{-17}{-17} = 1$

Solution: $\boxed{(2,1)}$

24. $x = \dfrac{\begin{vmatrix} -17 & -5 \\ 3 & 2 \end{vmatrix}}{\begin{vmatrix} 2 & -5 \\ 3 & 2 \end{vmatrix}} = \dfrac{-34 + 15}{4 + 15} = \dfrac{-19}{19} = -1; \quad y = \dfrac{\begin{vmatrix} 2 & -17 \\ 3 & 3 \end{vmatrix}}{\begin{vmatrix} 2 & -5 \\ 3 & 2 \end{vmatrix}} = \dfrac{6 + 51}{19} = \dfrac{57}{19} = 3$

Solution: $\boxed{(-1,3)}$

25. $D = \begin{vmatrix} 1 & 2 & 1 \\ 2 & 1 & 1 \\ 1 & 1 & 2 \end{vmatrix} = 1 \begin{vmatrix} 1 & 1 \\ 1 & 2 \end{vmatrix} - 2 \begin{vmatrix} 2 & 1 \\ 1 & 2 \end{vmatrix} + 1 \begin{vmatrix} 2 & 1 \\ 1 & 1 \end{vmatrix} = 1(1) - 2(3) + 1(1) = -4$

$N_x = \begin{vmatrix} 0 & 2 & 1 \\ 3 & 1 & 1 \\ 5 & 1 & 2 \end{vmatrix} = 0 \begin{vmatrix} 1 & 1 \\ 1 & 2 \end{vmatrix} - 2 \begin{vmatrix} 3 & 1 \\ 5 & 2 \end{vmatrix} + 1 \begin{vmatrix} 3 & 1 \\ 5 & 1 \end{vmatrix} = 0(1) - 2(1) + 1(-2) = -4$

$N_y = \begin{vmatrix} 1 & 0 & 1 \\ 2 & 3 & 1 \\ 1 & 5 & 2 \end{vmatrix} = 1 \begin{vmatrix} 3 & 1 \\ 5 & 2 \end{vmatrix} - 0 \begin{vmatrix} 2 & 1 \\ 1 & 2 \end{vmatrix} + 1 \begin{vmatrix} 2 & 3 \\ 1 & 5 \end{vmatrix} = 1(1) - 0(3) + 1(7) = 8$

$N_z = \begin{vmatrix} 1 & 2 & 0 \\ 2 & 1 & 3 \\ 1 & 1 & 5 \end{vmatrix} = 1 \begin{vmatrix} 1 & 3 \\ 1 & 5 \end{vmatrix} - 2 \begin{vmatrix} 2 & 3 \\ 1 & 5 \end{vmatrix} + 0 \begin{vmatrix} 2 & 1 \\ 1 & 1 \end{vmatrix} = 1(2) - 2(7) + 0(1) = -12$

$x = \frac{N_x}{D} = \frac{-4}{-4} = 1; \; y = \frac{N_y}{D} = \frac{8}{-4} = -2; \; z = \frac{N_z}{D} = \frac{-12}{-4} = 3 \Rightarrow$ solution: $(1, -2, 3)$

26. $D = \begin{vmatrix} 2 & 3 & 1 \\ 1 & 3 & 2 \\ 1 & -1 & -1 \end{vmatrix} = 2\begin{vmatrix} 3 & 2 \\ -1 & -1 \end{vmatrix} - 3\begin{vmatrix} 1 & 2 \\ 1 & -1 \end{vmatrix} + 1\begin{vmatrix} 1 & 3 \\ 1 & -1 \end{vmatrix} = 2(-1) - 3(-3) + 1(-4)$

$$= 3$$

$$N_x = \begin{vmatrix} 2 & 3 & 1 \\ 7 & 3 & 2 \\ -7 & -1 & -1 \end{vmatrix} = 2\begin{vmatrix} 3 & 2 \\ -1 & -1 \end{vmatrix} - 3\begin{vmatrix} 7 & 2 \\ -7 & -1 \end{vmatrix} + 1\begin{vmatrix} 7 & 3 \\ -7 & -1 \end{vmatrix}$$

$$= 2(-1) - 3(7) + 1(14) = -9$$

$$N_y = \begin{vmatrix} 2 & 2 & 1 \\ 1 & 7 & 2 \\ 1 & -7 & -1 \end{vmatrix} = 2\begin{vmatrix} 7 & 2 \\ -7 & -1 \end{vmatrix} - 2\begin{vmatrix} 1 & 2 \\ 1 & -1 \end{vmatrix} + 1\begin{vmatrix} 1 & 7 \\ 1 & -7 \end{vmatrix}$$

$$= 2(7) - 2(-3) + 1(-14) = 6$$

$$N_z = \begin{vmatrix} 2 & 3 & 2 \\ 1 & 3 & 7 \\ 1 & -1 & -7 \end{vmatrix} = 2\begin{vmatrix} 3 & 7 \\ -1 & -7 \end{vmatrix} - 3\begin{vmatrix} 1 & 7 \\ 1 & -7 \end{vmatrix} + 2\begin{vmatrix} 1 & 3 \\ 1 & -1 \end{vmatrix}$$

$$= 2(-14) - 3(-14) + 2(-4) = 6$$

$$x = \frac{N_x}{D} = \frac{-9}{3} = -3; \; y = \frac{N_y}{D} = \frac{6}{3} = 2; \; z = \frac{N_z}{D} = \frac{6}{3} = 2 \Rightarrow \text{solution: } (-3, 2, 2)$$

27. $\begin{aligned} 3x^2 + y^2 &= 52 \\ \underline{x^2 - y^2} &= \underline{12} \\ 4x^2 &= 64 \\ x^2 &= 16 \\ x &= \pm 4 \end{aligned}$ Substitute and solve for y:

$\begin{aligned} x^2 - y^2 &= 12 \\ 4^2 - y^2 &= 12 \\ y^2 &= 4 \\ y &= \pm 2 \end{aligned}$ $\qquad \begin{aligned} x^2 - y^2 &= 12 \\ (-4)^2 - y^2 &= 12 \\ y^2 &= 4 \\ y &= \pm 2 \end{aligned}$

Solutions: $(4, 2), (4, -2), (-4, 2), (-4, -2)$.

28. $\begin{aligned} \frac{x^2}{16} + \frac{y^2}{12} = 1 &\Rightarrow \times 48 \\ x^2 - \frac{y^2}{3} = 1 &\Rightarrow \times (-3) \end{aligned}$ $\quad \begin{aligned} 3x^2 + 4y^2 &= 48 \\ \underline{-3x^2 + y^2} &= \underline{-3} \\ 5y^2 &= 45 \\ y^2 &= 9 \\ y &= \pm 3 \end{aligned}$ Substitute and solve for x:

$$3x^2 + 4y^2 = 48$$
$$3x^2 + 4(9) = 48$$
$$3x^2 = 12$$
$$x^2 = 4 \Rightarrow x = \pm 2$$

Solutions: $(2, 3), (2, -3), (-2, 3), (-2, -3)$

29. $\begin{cases} y \geq x^2 - 4 \\ y < x + 3 \end{cases}$

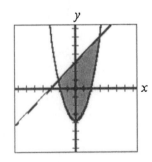

Chapter 13 Test (page 874)

1. $\begin{cases} 2x + y = 5 \\ \quad\quad y = 2x - 3 \end{cases}$

$(2, 1)$ is the solution.

2. $\begin{cases} (1) \quad 2x - 4y = 14 \\ (2) \quad x = -2y + 7 \end{cases}$

Substitute $x = -2y + 7$ from (2) into (1):
$$2x - 4y = 14$$
$$2(\boldsymbol{-2y + 7}) - 4y = 14$$
$$-4y + 14 - 4y = 14$$
$$-8y = 0$$
$$y = 0$$

Substitute this and solve for x:
$$x = -2y + 7$$
$$x = -2(0) + 7 = 7$$

Solution: $(7, 0)$

3.

			Solution:
$2x + 3y = -5 \Rightarrow \times 2$	$4x + 6y = -10$	$2x + 3y = -5$	
$3x - 2y = 12 \Rightarrow \times 3$	$9x - 6y = 36$	$2(2) + 3y = -5$	$\boxed{(2, -3)}$
	$13x = 26$	$4 + 3y = -5$	
	$x = 2$	$3y = -9$	
		$y = -3$	

4.

			Solution:
$\dfrac{x}{2} - \dfrac{y}{4} = -4 \Rightarrow \times 4$	$2x - y = -16$	$x + y = -2$	
$x + y = -2 \Rightarrow$	$x + y = -2$	$-6 + y = -2$	$\boxed{(-6, 4)}$
	$3x = -18$	$y = 4$	
	$x = -6$		

5.

$3(x + y) = x - 3 \Rightarrow \quad 2x + 3y = -3 \quad\quad 2x + 3y = -3 \Rightarrow \quad\quad 2x + 3y = -3$

$-y = \dfrac{2x + 3}{3} \Rightarrow \quad\quad 2x + 3y = -3 \quad\quad 2x + 3y = -3 \Rightarrow \times(-1)\;\underline{-2x - 3y = 3}$

$$0 = 0$$

The equation is an identity, so the system has infinitely many solutions. $\Rightarrow$ dependent equations

6. See #5. Since the system has at least one solution, it is a consistent system.

7. $\begin{bmatrix} 1 & 2 & -1 \\ 2 & -2 & 3 \end{bmatrix} \overset{-3R_1 + R_2 \Rightarrow R_2}{\Rightarrow} \begin{bmatrix} 1 & 2 & -1 \\ -1 & -8 & \mathbf{6} \end{bmatrix}$

8. $\begin{bmatrix} -1 & 3 & 6 \\ 3 & -2 & 4 \end{bmatrix} \overset{-2R_1 + R_2 \Rightarrow R_2}{\Rightarrow} \begin{bmatrix} -1 & 3 & 6 \\ 5 & -8 & -8 \end{bmatrix}$

9. $\left[\begin{array}{ccc|c} 1 & 1 & 1 & 4 \\ 1 & 1 & -1 & 6 \\ 2 & -3 & 1 & -1 \end{array}\right]$

10. $\begin{bmatrix} 1 & 1 & 1 \\ 1 & 1 & -1 \\ 2 & -3 & 1 \end{bmatrix}$

11. $\begin{bmatrix} 1 & 1 & | & 4 \\ 2 & -1 & | & 2 \end{bmatrix} \overset{-2R_1 + R_2 \Rightarrow R_2}{\Rightarrow} \begin{bmatrix} 1 & 1 & | & 4 \\ 0 & -3 & | & -6 \end{bmatrix} \overset{-\frac{1}{3}R_2 \Rightarrow R_2}{\Rightarrow} \begin{bmatrix} 1 & 1 & | & 4 \\ 0 & 1 & | & 2 \end{bmatrix}$

From R_2, $y = 2$. From R_1: Solution: $\boxed{(2, 2)}$
$$x + y = 4$$
$$x + 2 = 4 \Rightarrow x = 2$$

12. $\begin{bmatrix} 1 & 1 & | & 2 \\ 1 & -1 & | & -4 \\ 2 & 1 & | & 1 \end{bmatrix} \overset{\substack{-R_2 + R_1 \Rightarrow R_2 \\ -2R_1 + R_3 \Rightarrow R_3}}{\Rightarrow} \begin{bmatrix} 1 & 1 & | & 2 \\ 0 & 2 & | & 6 \\ 0 & -1 & | & -3 \end{bmatrix} \overset{\substack{\frac{1}{2}R_2 \Rightarrow R_2 \\ 2R_3 + R_2 \Rightarrow R_3}}{\Rightarrow} \begin{bmatrix} 1 & 1 & | & 2 \\ 0 & 1 & | & 3 \\ 0 & 0 & | & 0 \end{bmatrix}$

From R_2, $y = 3$. From R_1: Solution: $\boxed{(-1, 3)}$
$$x + y = 2$$
$$x + 3 = 2 \Rightarrow x = -1$$

13. $\begin{vmatrix} 2 & -3 \\ 4 & 5 \end{vmatrix} = 2(5) - (-3)(4) = 22$

14. $\begin{vmatrix} -3 & -4 \\ -2 & 3 \end{vmatrix} = -3(3) - (-4)(-2) = -17$

15. $\begin{vmatrix} 1 & 2 & 0 \\ 2 & 0 & 3 \\ 1 & -2 & 2 \end{vmatrix} = 1\begin{vmatrix} 0 & 3 \\ -2 & 2 \end{vmatrix} - 2\begin{vmatrix} 2 & 3 \\ 1 & 2 \end{vmatrix} + 0\begin{vmatrix} 2 & 0 \\ 1 & -2 \end{vmatrix} = 1(6) - 2(1) + 0(-4) = 4$

16. $\begin{vmatrix} 2 & -1 & 1 \\ 3 & 1 & 0 \\ 0 & 1 & 2 \end{vmatrix} = 2\begin{vmatrix} 1 & 0 \\ 1 & 2 \end{vmatrix} - (-1)\begin{vmatrix} 3 & 0 \\ 0 & 2 \end{vmatrix} + 1\begin{vmatrix} 3 & 1 \\ 0 & 1 \end{vmatrix} = 2(2) + 1(6) + 1(3) = 13$

17. $\begin{vmatrix} -6 & -1 \\ -6 & 1 \end{vmatrix}$

18. $\begin{vmatrix} 1 & -1 \\ 3 & 1 \end{vmatrix}$

19. $x = \dfrac{\begin{vmatrix} -6 & -1 \\ -6 & 1 \end{vmatrix}}{\begin{vmatrix} 1 & -1 \\ 3 & 1 \end{vmatrix}} = \dfrac{-6 - 6}{1 - (-3)} = \dfrac{-12}{4} = -3$

20. $y = \dfrac{\begin{vmatrix} 1 & -6 \\ 3 & -6 \end{vmatrix}}{\begin{vmatrix} 1 & -1 \\ 3 & 1 \end{vmatrix}} = \dfrac{-6 - (-18)}{1 - (-3)} = \dfrac{12}{4} = 3$

21. $D = \begin{vmatrix} 1 & 1 & 1 \\ 1 & 1 & -1 \\ 2 & -3 & 1 \end{vmatrix} = 1\begin{vmatrix} 1 & -1 \\ -3 & 1 \end{vmatrix} - 1\begin{vmatrix} 1 & -1 \\ 2 & 1 \end{vmatrix} + 1\begin{vmatrix} 1 & 1 \\ 2 & -3 \end{vmatrix} = 1(-2) - 1(3) + 1(-5) = -10$

$N_x = \begin{vmatrix} 4 & 1 & 1 \\ 6 & 1 & -1 \\ -1 & -3 & 1 \end{vmatrix} = 4\begin{vmatrix} 1 & -1 \\ -3 & 1 \end{vmatrix} - 1\begin{vmatrix} 6 & -1 \\ -1 & 1 \end{vmatrix} + 1\begin{vmatrix} 6 & 1 \\ -1 & -3 \end{vmatrix}$

$= 4(-2) - 1(5) + 1(-17) = -30$

$x = \dfrac{N_x}{D} = \dfrac{-30}{-10} = 3$

22. See #21. $D = -10$.

$$N_z = \begin{vmatrix} 1 & 1 & 4 \\ 1 & 1 & 6 \\ 2 & -3 & -1 \end{vmatrix} = 1 \begin{vmatrix} 1 & 6 \\ -3 & -1 \end{vmatrix} - 1 \begin{vmatrix} 1 & 6 \\ 2 & -1 \end{vmatrix} + 4 \begin{vmatrix} 1 & 1 \\ 2 & -3 \end{vmatrix} = 1(17) - 1(-13) + 4(-5) = 10$$

$$z = \frac{N_z}{D} = \frac{10}{-10} = -1$$

23.
$$\begin{aligned} x^2 + y^2 &= 5 \\ x^2 - y^2 &= 3 \\ \hline 2x^2 &= 8 \\ x^2 &= 4 \\ x &= \pm 2 \end{aligned}$$

Substitute and solve for y:

$$\begin{aligned} x^2 + y^2 &= 5 \\ 2^2 + y^2 &= 5 \\ y^2 &= 1 \\ y &= \pm 1 \end{aligned} \qquad \begin{aligned} x^2 + y^2 &= 5 \\ (-2)^2 + y^2 &= 5 \\ y^2 &= 1 \\ y &= \pm 1 \end{aligned}$$

Solutions: $(2, 1), (2, -1), (-2, 1), (-2, -1)$.

24. $\begin{cases} (1) & x^2 + y^2 = 25 \\ (2) & 4x^2 - 9y = 0 \end{cases}$

Substitute $x^2 = 25 - y^2$ from (1) into (2):

$$4x^2 - 9y = 0$$
$$4(25 - y^2) - 9y = 0$$
$$100 - 4y^2 - 9y = 0$$
$$-4y^2 - 9y + 100 = 0$$
$$4y^2 + 9y - 100 = 0$$
$$(y - 4)(4y + 25) = 0$$

$y - 4 = 0$ **or** $4y + 25 = 0$

$\qquad y = 4 \qquad\qquad y = -\frac{25}{4}$

Substitute these and solve for x:

$$\begin{aligned} x^2 &= 25 - y^2 \\ x^2 &= 25 - 4^2 \\ x^2 &= 25 - 16 = 9 \\ x &= \pm 3 \end{aligned} \qquad \begin{aligned} x^2 &= 25 - y^2 \\ x^2 &= 25 - \left(-\frac{25}{4}\right)^2 \\ x^2 &= 25 - \frac{625}{16} = -\frac{225}{16} \\ x &\text{ is nonreal.} \end{aligned}$$

Solutions: $(3, 4), (-3, 4)$

25. $\begin{cases} y \geq x^2 \\ y < x + 3 \end{cases}$

Exercise 14.1 (page 881)

1. $\log_4 16 = x \Rightarrow 4^x = 16 \Rightarrow x = 2$

3. $\log_{25} x = \frac{1}{2} \Rightarrow 25^{1/2} = x \Rightarrow x = 5$

5. one

7. Pascal's

9. 6!

11. 1

13. $3! = 3 \cdot 2 \cdot 1 = 6$

15. $-5! = -1 \cdot 5! = -1 \cdot 5 \cdot 4 \cdot 3 \cdot 2 \cdot 1$
$\qquad = -120$

17. $3! + 4! = 3 \cdot 2 \cdot 1 + 4 \cdot 3 \cdot 2 \cdot 1$
$\qquad = 6 + 24 = 30$

19. $3!(4!) = 3 \cdot 2 \cdot 1 \cdot 4 \cdot 3 \cdot 2 \cdot 1 = 144$

21. $8(7!) = 8 \cdot 7 \cdot 6 \cdot 5 \cdot 4 \cdot 3 \cdot 2 \cdot 1 = 40{,}320$

23. $\dfrac{9!}{11!} = \dfrac{9!}{11 \cdot 10 \cdot 9!} = \dfrac{1}{11 \cdot 10} = \dfrac{1}{110}$

25. $\dfrac{49!}{47!} = \dfrac{49 \cdot 48 \cdot 47!}{47!} = 49 \cdot 48 = 2352$

27. $\dfrac{5!}{3!(5-3)!} = \dfrac{5!}{3!2!} = \dfrac{5 \cdot 4 \cdot 3!}{3! \cdot 2 \cdot 1} = \dfrac{5 \cdot 4}{2 \cdot 1} = 10$

29. $\dfrac{7!}{5!(7-5)!} = \dfrac{7!}{5!2!} = \dfrac{7 \cdot 6 \cdot 5!}{5! \cdot 2 \cdot 1} = \dfrac{7 \cdot 6}{2 \cdot 1} = 21$

31. $\dfrac{5!(8-5)!}{4!7!} = \dfrac{5!3!}{4!7!} = \dfrac{5!}{7!} \cdot \dfrac{3!}{4!} = \dfrac{5!}{7 \cdot 6 \cdot 5!} \cdot \dfrac{3!}{4 \cdot 3!} = \dfrac{1}{7 \cdot 6} \cdot \dfrac{1}{4} = \dfrac{1}{168}$

33. $(x+y)^3 = x^3 + \dfrac{3!}{1!(3-1)!}x^2 y + \dfrac{3!}{2!(3-2)!}xy^2 + y^3 = x^3 + \dfrac{3!}{1!2!}x^2 y + \dfrac{3!}{2!1!}xy^2 + y^3$

$\qquad\qquad = x^3 + \dfrac{3 \cdot 2!}{1!2!}x^2 y + \dfrac{3 \cdot 2!}{2!1!}xy^2 + y^3$

$\qquad\qquad = x^3 + \dfrac{3}{1}x^2 y + \dfrac{3}{1}xy^2 + y^3$

$\qquad\qquad = x^3 + 3x^2 y + 3xy^2 + y^3$

35. $(x-y)^4 = x^4 + \dfrac{4!}{1!(4-1)!}x^3(-y) + \dfrac{4!}{2!(4-2)!}x^2(-y)^2 + \dfrac{4!}{3!(4-3)!}x(-y)^3 + (-y)^4$

$\qquad = x^4 + \dfrac{4!}{1!3!}(-x^3 y) + \dfrac{4!}{2!2!}x^2 y^2 + \dfrac{4!}{3!1!}(-xy^3) + y^4$

$\qquad = x^4 - \dfrac{4 \cdot 3!}{1!3!}x^3 y + \dfrac{4 \cdot 3 \cdot 2!}{2! \cdot 2 \cdot 1}x^2 y^2 - \dfrac{4 \cdot 3!}{3!1!}xy^3 + y^4$

$\qquad = x^4 - \dfrac{4}{1}x^3 y + \dfrac{12}{2}x^2 y^2 - \dfrac{4}{1}xy^3 + y^4$

$\qquad = x^4 - 4x^3 y + 6x^2 y^2 - 4xy^3 + y^4$

37. $(2x+y)^3 = (2x)^3 + \dfrac{3!}{1!(3-1)!}(2x)^2 y + \dfrac{3!}{2!(3-2)!}2xy^2 + y^3$

$= 8x^3 + \dfrac{3!}{1!2!} \cdot 4x^2 y + \dfrac{3!}{2!1!} \cdot 2xy^2 + y^3$

$= 8x^3 + \dfrac{3 \cdot 2!}{1!2!} \cdot 4x^2 y + \dfrac{3 \cdot 2!}{2!1!} \cdot 2xy^2 + y^3$

$= 8x^3 + \dfrac{3}{1} \cdot 4x^2 y + \dfrac{3}{1} \cdot 2xy^2 + y^3$

$= 8x^3 + 12x^2 y + 6xy^2 + y^3$

39. $(x-2y)^3 = x^3 + \dfrac{3!}{1!(3-1)!}x^2(-2y) + \dfrac{3!}{2!(3-2)!}x(-2y)^2 + (-2y)^3$

$= x^3 + \dfrac{3!}{1!2!} \cdot (-2x^2 y) + \dfrac{3!}{2!1!} \cdot 4xy^2 - 8y^3$

$= x^3 - \dfrac{3 \cdot 2!}{1!2!} \cdot 2x^2 y + \dfrac{3 \cdot 2!}{2!1!} \cdot 4xy^2 - 8y^3$

$= x^3 - \dfrac{3}{1} \cdot 2x^2 y + \dfrac{3}{1} \cdot 4xy^2 - 8y^3$

$= x^3 - 6x^2 y + 12xy^2 - 8y^3$

41. $(2x+3y)^3 = (2x)^3 + \dfrac{3!}{1!(3-1)!}(2x)^2(3y) + \dfrac{3!}{2!(3-2)!}2x(3y)^2 + (3y)^3$

$= 8x^3 + \dfrac{3!}{1!2!} \cdot 4x^2(3y) + \dfrac{3!}{2!1!} \cdot 2x(9y^2) + 27y^3$

$= 8x^3 + \dfrac{3 \cdot 2!}{1!2!} \cdot 12x^2 y + \dfrac{3 \cdot 2!}{2!1!} \cdot 18xy^2 + 27y^3$

$= 8x^3 + \dfrac{3}{1} \cdot 12x^2 y + \dfrac{3}{1} \cdot 18xy^2 + 27y^3$

$= 8x^3 + 36x^2 y + 54xy^2 + 27y^3$

43. $\left(\dfrac{x}{2} - \dfrac{y}{3}\right)^3 = \left(\dfrac{x}{2}\right)^3 + \dfrac{3!}{1!(3-1)!}\left(\dfrac{x}{2}\right)^2\left(-\dfrac{y}{3}\right) + \dfrac{3!}{2!(3-2)!}\left(\dfrac{x}{2}\right)\left(-\dfrac{y}{3}\right)^2 + \left(-\dfrac{y}{3}\right)^3$

$= \dfrac{x^3}{8} - \dfrac{3!}{1!2!} \cdot \dfrac{x^2}{4} \cdot \dfrac{y}{3} + \dfrac{3!}{2!1!} \cdot \dfrac{x}{2} \cdot \dfrac{y^2}{9} - \dfrac{y^3}{27}$

$= \dfrac{x^3}{8} - \dfrac{3 \cdot 2!}{1!2!} \cdot \dfrac{x^2 y}{12} + \dfrac{3 \cdot 2!}{2!1!} \cdot \dfrac{xy^2}{18} - \dfrac{y^3}{27}$

$= \dfrac{x^3}{8} - \dfrac{3}{1} \cdot \dfrac{x^2 y}{12} + \dfrac{3}{1} \cdot \dfrac{xy^2}{18} - \dfrac{y^3}{27}$

$= \dfrac{x^3}{8} - \dfrac{x^2 y}{4} + \dfrac{xy^2}{6} - \dfrac{y^3}{27}$

45. $(3 + 2y)^4 = 3^4 + \dfrac{4!}{1!(4-1)!}3^3(2y) + \dfrac{4!}{2!(4-2)!}3^2(2y)^2 + \dfrac{4!}{3!(4-3)!}3(2y)^3 + (2y)^4$

$\qquad = 81 + \dfrac{4!}{1!3!} \cdot 27(2y) + \dfrac{4!}{2!2!} \cdot 9(4y^2) + \dfrac{4!}{3!1!} \cdot 3(8y^3) + 16y^4$

$\qquad = 81 + \dfrac{4 \cdot 3!}{1!3!} \cdot 54y + \dfrac{4 \cdot 3 \cdot 2!}{2! \cdot 2 \cdot 1} \cdot 36y^2 + \dfrac{4 \cdot 3!}{3!1!} \cdot 24y^3 + 16y^4$

$\qquad = 81 + \dfrac{4}{1} \cdot 54y + \dfrac{12}{2} \cdot 36y^2 + \dfrac{4}{1} \cdot 24y^3 + 16y^4$

$\qquad = 81 + 216y + 216y^2 + 96y^3 + 16y^4$

47. $\left(\dfrac{x}{3} - \dfrac{y}{2}\right)^4$

$\qquad = \left(\dfrac{x}{3}\right)^4 + \dfrac{4!}{1!(4-1)!}\left(\dfrac{x}{3}\right)^3\left(-\dfrac{y}{2}\right) + \dfrac{4!}{2!(4-2)!}\left(\dfrac{x}{3}\right)^2\left(-\dfrac{y}{2}\right)^2 + \dfrac{4!}{3!(4-3)!}\left(\dfrac{x}{3}\right)\left(-\dfrac{y}{2}\right)^3$

$\qquad\qquad\qquad + \left(-\dfrac{y}{2}\right)^4$

$\qquad = \dfrac{x^4}{81} - \dfrac{4!}{1!3!}\dfrac{x^3}{27}\cdot\dfrac{y}{2} + \dfrac{4!}{2!2!}\cdot\dfrac{x^2}{9}\cdot\dfrac{y^2}{4} - \dfrac{4!}{3!1!}\cdot\dfrac{x}{3}\cdot\dfrac{y^3}{8} + \dfrac{y^4}{16}$

$\qquad = \dfrac{x^4}{81} - \dfrac{4 \cdot 3!}{1!3!}\dfrac{x^3 y}{54} + \dfrac{4 \cdot 3 \cdot 2!}{2!2!}\cdot\dfrac{x^2 y^2}{36} - \dfrac{4 \cdot 3!}{3!1!}\cdot\dfrac{xy^3}{24} + \dfrac{y^4}{16}$

$\qquad = \dfrac{x^4}{81} - \dfrac{4}{1}\cdot\dfrac{x^3 y}{54} + \dfrac{12}{2}\cdot\dfrac{x^2 y^2}{36} - \dfrac{4}{1}\cdot\dfrac{xy^3}{24} + \dfrac{y^4}{16}$

$\qquad = \dfrac{x^4}{81} - \dfrac{2x^3 y}{27} + \dfrac{x^2 y^2}{6} - \dfrac{xy^3}{6} + \dfrac{y^4}{16}$

49.

```
                              1
                          1       1
                      1       2       1
                  1       3       3       1
              1       4       6       4       1
          1       5      10      10       5       1
      1       6      15      20      15       6       1
  1       7      21      35      35      21       7       1
1     8     28      56      70      56      28       8      1
1   9    36     84     126     126      84     36      9      1
```

51. $1, 1, 2, 3, 5, 8, 13, \ldots$; Beginning with 2, each number is the sum of the previous two numbers.

53. **Answers may vary.**

55. $\dfrac{n!}{n!(n-n)!} = \dfrac{n!}{n!(0)!} = \dfrac{n!}{n! \cdot 1} = 1$

Exercise 14.2 (page 884)

1. $3x + 2y = 12 \Rightarrow \qquad 3x + 2y = 12$ Substitute and solve for y:

$\underline{2x - \; y = \; 1} \Rightarrow \times 2 \; \underline{4x - 2y = \; 2}$

$\qquad\qquad\qquad\qquad 7x \qquad = 14 \qquad 2x - y = 1$

$\qquad\qquad\qquad\qquad\qquad x = \; 2 \qquad 2(2) - y = 1$

$\qquad\qquad\qquad\qquad\qquad\qquad\qquad 4 - y = 1$

$\qquad\qquad\qquad\qquad\qquad\qquad\qquad -y = -3$

$\qquad\qquad\qquad\qquad\qquad\qquad\qquad y = 3 \qquad$ Solution: $(2, 3)$

3. $\begin{vmatrix} 2 & -3 \\ 4 & -2 \end{vmatrix} = 2(-2) - (-3)(4) = -4 - (-12) = -4 + 12 = 8$

5. 3

7. 7

9. In the 2nd term, the exponent on b is 1.

Variables: $a^2 b^1 = a^2 b$

Coef. $= \dfrac{n!}{r!(n - r)!} = \dfrac{3!}{1!2!} = 3$

Term $= 3a^2 b$

11. In the 4th term, the exponent on $-y$ is 3.

Variables: $x^1(-y)^3 = -xy^3$

Coef. $= \dfrac{n!}{r!(n - r)!} = \dfrac{4!}{3!1!} = 4$

Term $= 4(-xy^3) = -4xy^3$

13. In the 5th term, the exponent on y is 4.

Variables: $x^2 y^4$

Coef. $= \dfrac{n!}{r!(n - r)!} = \dfrac{6!}{4!2!} = 15$

Term $= 15x^2 y^4$

15. In the 3rd term, the exponent on $-y$ is 2.

Variables: $x^6(-y)^2 = x^6 y^2$

Coef. $= \dfrac{n!}{r!(n - r)!} = \dfrac{8!}{2!6!} = 28$

Term $= 28x^6 y^2$

17. In the 3rd term, the exponent on 3 is 2.

Variables: $x^3(3)^2 = 9x^3$

Coef. $= \dfrac{n!}{r!(n - r)!} = \dfrac{5!}{2!3!} = 10$

Term $= 10(9x^3) = 90x^3$

19. In the 3rd term, the exponent on y is 2.

Variables: $(4x)^3 y^2 = 64x^3 y^2$

Coef. $= \dfrac{n!}{r!(n - r)!} = \dfrac{5!}{2!3!} = 10$

Term $= 10(64x^3 y^2) = 640x^3 y^2$

21. In the 2nd term, the exponent on $-3y$ is 1. Variables: $x^3(-3y)^1 = -3x^3 y$

Coef. $= \dfrac{n!}{r!(n - r)!} = \dfrac{4!}{1!3!} = 4$; Term $= 4(-3x^3 y) = -12x^3 y$

23. In the 4th term, the exponent on -5 is 3. Variables: $(2x)^4(-5)^3 = (16x^4)(-125) = -2000x^4$

Coef. $= \dfrac{n!}{r!(n - r)!} = \dfrac{7!}{3!4!} = 35$; Term $= 35(-2000x^4) = -70{,}000x^4$

25. In the 5th term, the exponent on $-3y$ is 4. Variables: $(2x)^1(-3y)^4 = 2x(81y^4) = 162xy^4$

Coef. $= \dfrac{n!}{r!(n - r)!} = \dfrac{5!}{4!1!} = 5$; Term $= 5(162xy^4) = 810xy^4$

27. In the 3rd term, the exponent on $\sqrt{3}y$ is 2. Variables: $(\sqrt{2}x)^4(\sqrt{3}y)^2 = (4x^4)(3y^2) = 12x^4y^2$

Coef. $= \dfrac{n!}{r!(n-r)!} = \dfrac{6!}{2!4!} = 15$; Term $= 15(12x^4y^2) = 180x^4y^2$

29. In the 2nd term, the exponent on $-\dfrac{y}{3}$ is 1. Variables: $\left(\dfrac{x}{2}\right)^3\left(-\dfrac{y}{3}\right)^1 = \left(\dfrac{x^3}{8}\right)\left(-\dfrac{y}{3}\right) = -\dfrac{x^3y}{24}$

Coef. $= \dfrac{n!}{r!(n-r)!} = \dfrac{4!}{1!3!} = 4$; Term $= 4\left(-\dfrac{x^3y}{24}\right) = -\dfrac{x^3y}{6} = -\dfrac{1}{6}x^3y$

31. In the 4th term, the exponent on b is 3.

Variables: $a^{n-3}b^3$

Coef. $= \dfrac{n!}{r!(n-r)!} = \dfrac{n!}{3!(n-3)!}$

Term $= \dfrac{n!}{3!(n-3)!}a^{n-3}b^3$

33. In the 5th term, the exponent on $-b$ is 4.

Variables: $a^{n-4}(-b)^4 = a^{n-4}b^4$

Coef. $= \dfrac{n!}{r!(n-r)!} = \dfrac{n!}{4!(n-4)!}$

Term $= \dfrac{n!}{4!(n-4)!}a^{n-4}b^4$

35. In the rth term, the coefficient on b is $r-1$. Variables: $a^{n-(r-1)}b^{r-1} = a^{n-r+1}b^{r-1}$

Coef. $= \dfrac{n!}{r!(n-r)!} = \dfrac{n!}{(r-1)![n-(r-1)]!} = \dfrac{n!}{(r-1)!(n-r+1)!}$

Term $= \dfrac{n!}{(r-1)!(n-r+1)!}a^{n-r+1}b^{r-1}$

37. Answers may vary.

39. $\left(x+\dfrac{1}{x}\right)^{10} = \left(x+x^{-1}\right)^{10}$. The constant term occurs when the exponent is 0.

The $(r+1)$th term of $(x+x^{-1})^{10}$ is $\dfrac{10!}{r!(10-r)!}x^{10-r}\left(x^{-1}\right)^r = \dfrac{10!}{r!(10-r)!}x^{10-r}x^{-r}$.

But $\dfrac{10!}{r!(10-r)!}x^{10-r}x^{-r} = \dfrac{10!}{r!(10-r)!}x^{10-2r}$. If $10-2r = 0$, then $r = 5$.

The term is $\dfrac{10!}{5!(10-5)!}x^{10-5}x^{-5} = \dfrac{10!}{5!5!} = \dfrac{10\cdot 9\cdot 8\cdot 7\cdot 6\cdot 5!}{5!\cdot 5\cdot 4\cdot 3\cdot 2\cdot 1} = 252$

Exercise 14.3 (page 890)

1. $3(2x^2-4x+7)+4(3x^2+5x-6) = 6x^2-12x+21+12x^2+20x-24 = 18x^2+8x-3$

3. $\dfrac{3a+4}{a-2}+\dfrac{3a-4}{a+2} = \dfrac{(3a+4)(a+2)}{(a-2)(a+2)}+\dfrac{(3a-4)(a-2)}{(a+2)(a-2)}$

$= \dfrac{3a^2+10a+8+3a^2-10a+8}{(a+2)(a-2)} = \dfrac{6a^2+16}{(a+2)(a-2)}$

5. sequence

7. arithmetic; difference

9. arithmetic mean

11. $1+2+3+4+5$

13. $3, 5, 7, 9, 11$

15. $-5, -8, -11, -14, -17$

17. nth term $= a + (n-1)d$
$29 = 5 + (5-1)d$
$29 = 5 + 4d$
$24 = 4d$
$6 = d$
$5, 11, 17, 23, 29$

19. nth term $= a + (n-1)d$
$-39 = -4 + (6-1)d$
$-39 = -4 + 5d$
$-35 = 5d$
$-7 = d$
$-4, -11, -18, -25, -32$

21. nth term $= a + (n-1)d$
$-83 = a + (6-1)7$
$-83 = a + 5(7)$
$-83 = a + 35$
$-118 = a$
$-118, -111, -104, -97, -90$

23. nth term $= a + (n-1)d$
$16 = a + (7-1)(-3)$
$16 = a + 6(-3)$
$16 = a - 18$
$34 = a$
$34, 31, 28, 25, 22$

25. nth term $= a + (n-1)d$
$131 = a + (19-1)d$
$131 = a + 18d$
nth term $= a + (n-1)d$
$138 = a + (20-1)d$
$138 = a + 19d$

$a + 18d = 131 \Rightarrow \times(-1) \quad -a - 18d = -131$
$\underline{a + 19d = 138} \Rightarrow \qquad \underline{a + 19d = \quad 138}$
$\qquad\qquad\qquad\qquad\qquad\qquad\quad d = \qquad 7$

Substitute and solve for a:
$a + 18d = 131$
$a + 18(7) = 131$
$a + 126 = 131$
$a = 5 \Rightarrow 5, 12, 19, 26, 33$

27. nth term $= a + (n-1)d$
$= 7 + (30-1)12$
$= 7 + 29(12)$
$= 7 + 348 = 355$

29. nth term $= a + (n-1)d$
$-4 = a + (2-1)d$
$-4 = a + d$
nth term $= a + (n-1)d$
$-9 = a + (3-1)d$
$-9 = a + 2d$

$a + \quad d = -4 \Rightarrow \times(-1) \quad -a - \quad d = \quad 4$
$\underline{a + 2d = -9} \Rightarrow \qquad \underline{a + 2d = -9}$
$\qquad\qquad\qquad\qquad\qquad\qquad\quad d = -5$

Substitute and solve for a:
$a + d = -4$
$a + (-5) = -4$
$a = 1$
nth term $= a + (n-1)d$
$= 1 + (37-1)(-5)$
$= 1 + 36(-5)$
$= 1 - 180 = -179$

31. nth term $= a + (n-1)d$
$$263 = a + (27-1)(11)$$
$$263 = a + 26(11)$$
$$263 = a + 286$$
$$-23 = a$$

33. nth term $= a + (n-1)d$
$$556 = 40 + (44-1)d$$
$$556 = 40 + 43d$$
$$516 = 43d$$
$$12 = d$$

35. Form an arithmetic sequence with a 1st term of 2 and a 5th term of 11:
nth term $= a + (n-1)d$
$$11 = 2 + (5-1)d$$
$$11 = 2 + 4d$$
$$9 = 4d$$
$$\frac{9}{4} = d$$
$2, \boxed{\frac{17}{4}, \frac{13}{2}, \frac{35}{4}}, 11$

37. Form an arithmetic sequence with a 1st term of 10 and a 6th term of 20:
nth term $= a + (n-1)d$
$$20 = 10 + (6-1)d$$
$$20 = 10 + 5d$$
$$10 = 5d$$
$$2 = d$$
$10, \boxed{12, 14, 16, 18}, 20$

39. Form an arithmetic sequence with a 1st term of 10 and a 3rd term of 19:
nth term $= a + (n-1)d$
$$19 = 10 + (3-1)d$$
$$19 = 10 + 2d$$
$$9 = 2d$$
$$\frac{9}{2} = d$$
$10, \boxed{\frac{29}{2}}, 19$

41. Form an arithmetic sequence with a 1st term of -4.5 and a 3rd term of 7:
nth term $= a + (n-1)d$
$$7 = -4.5 + (3-1)d$$
$$7 = -4.5 + 2d$$
$$11.5 = 2d$$
$$5.75 = d$$
$-4.5, \boxed{1.25}, 7$

43. $a = 1, d = 3, n = 30$
$l = a + (n-1)d = 1 + 29(3) = 88$
$S_n = \dfrac{n(a+l)}{2} = \dfrac{30(1+88)}{2} = 1335$

45. $a = -5, d = 4, n = 17$
$l = a + (n-1)d = -5 + 16(4) = 59$
$S_n = \dfrac{n(a+l)}{2} = \dfrac{17(-5+59)}{2} = 459$

47. nth term $= a + (n-1)d$
$$7 = a + (2-1)d$$
$$7 = a + d$$
nth term $= a + (n-1)d$
$$12 = a + (3-1)d$$
$$12 = a + 2d$$

$\begin{aligned} a + d &= 7 \Rightarrow \times(-1) \\ a + 2d &= 12 \Rightarrow \end{aligned}$ $\quad \begin{aligned} -a - d &= -7 \\ \underline{a + 2d} &= \underline{12} \\ d &= 5 \end{aligned}$

Substitute and solve for a:
$$a + d = 7$$
$$a + 5 = 7$$
$$a = 2, d = 5, n = 12$$
$l = a + (n-1)d = 2 + 11(5) = 57$
$S_n = \dfrac{n(a+l)}{2} = \dfrac{12(2+57)}{2} = 354$

49. $f(n) = 2n + 1 \Rightarrow f(1) = 3$
$f(n) = 2n + 1 = 31$
$2n = 30$
$n = 15$
$S_n = \dfrac{n(a + l)}{2} = \dfrac{15(3 + 31)}{2} = 255$

51. $a = 1, d = 1, n = 50$
$l = a + (n - 1)d = 1 + 49(1) = 50$
$S_n = \dfrac{n(a + l)}{2} = \dfrac{50(1 + 50)}{2} = 1275$

53. $a = 1, d = 2, n = 50$
$l = a + (n - 1)d = 1 + 49(2) = 99$
$S_n = \dfrac{n(a + l)}{2} = \dfrac{50(1 + 99)}{2} = 2500$

55. $\displaystyle\sum_{k=1}^{4} 6k = 6(1) + 6(2) + 6(3) + 6(4)$
$= 6 + 12 + 18 + 24 = 60$

57. $\displaystyle\sum_{k=3}^{4} (k^2 + 3) = (3^2 + 3) + (4^2 + 3)$
$= 9 + 3 + 16 + 3 = 31$

59. $\displaystyle\sum_{k=4}^{4} (2k + 4) = 2(4) + 4 = 8 + 4 = 12$

61. $a = 60, d = 50 \Rightarrow 60, 110, 160, 210, 260, 310; n = 121$
nth term $= a + (n - 1)d = 60 + (121 - 1)(50) = 60 + 120(50) = \$6060$

63. $a = 1, d = 1, n = 150, l = 150 \Rightarrow S_n = \dfrac{n(a + l)}{2} = \dfrac{150(1 + 150)}{2} = 11{,}325$ bricks

65. After 1 sec.: $s = 16(1)^2 = 16$; After 2 sec.: $s = 16(2)^2 = 64$; After 3 sec.: $s = 16(3)^2 = 144$
During 2nd second $\Rightarrow$ falls $64 - 16 = 48$ ft; During 3rd second $\Rightarrow$ falls $144 - 64 = 80$ ft
The sequence of the amounts fallen during each second is $16, 48, 80 \Rightarrow a = 16, d = 32$
nth term $= a + (n - 1)d = 16 + (12 - 1)(32) = 16 + 11(32) = 368$ ft

67. Answers may vary.

69. $\displaystyle\sum_{n=1}^{6} \left(\tfrac{1}{2}n + 1\right)$: $\tfrac{3}{2}, 2, \tfrac{5}{2}, 3, \tfrac{7}{2}, 4$

71. Form an arithmetic sequence with a 1st term of a and a 3rd term of b:
nth term $= a + (n - 1)d$
$b = a + (3 - 1)d$
$b = a + 2d$
$b - a = 2d$
$\dfrac{b - a}{2} = d \Rightarrow$ mean $= a + \dfrac{b - a}{2} = \dfrac{2a}{2} + \dfrac{b - a}{2} = \dfrac{a + b}{2}$

73. $\displaystyle\sum_{k=1}^{5} 5k = 5(1) + 5(2) + 5(3) + 5(4) + 5(5) = 5(1 + 2 + 3 + 4 + 5) = 5\sum_{k=1}^{5} k.$

75. $\displaystyle\sum_{k=1}^{n} 3 = \sum_{k=1}^{n} 3k^0 = 3(1)^0 + 3(2)^0 + \cdots + 3(n)^0 = 3 + 3 + \cdots + 3 = 3n$

Exercise 14.4 (page 898)

1. $x^2 - 5x - 6 \leq 0$
$(x - 6)(x + 1) \leq 0$
$x - 6$ `- - - - - - - - 0++++`
$x + 1$ `- - -  0++++++++++`

solution set: $[-1, 6]$

3. $\dfrac{x - 4}{x + 3} > 0$

$x - 4$ `- - - - - - - - 0++++`
$x + 3$ `- - -0+++++++++++++`

solution set: $(-\infty, -3) \cup (4, \infty)$

5. geometric

7. common ratio

9. $S_n = \dfrac{a - ar^n}{1 - r}$

11. $3, 6, 12, 24, 48$

13. $-5, -1, -\frac{1}{5}, -\frac{1}{25}, -\frac{1}{125}$

15. nth term $= ar^{n-1}$
$32 = 2r^{3-1}$
$32 = 2r^2$
$16 = r^2$
$\pm 4 = r$, so $r = 4$ $(r > 0)$
$2, 8, 32, 128, 512$

17. nth term $= ar^{n-1}$
$-192 = -3r^{4-1}$
$-192 = -3r^3$
$64 = r^3$
$4 = r$
$-3, -12, -48, -192, -768$

19. nth term $= ar^{n-1}$
$-4 = -64r^{5-1}$
$-4 = -64r^4$
$\frac{1}{16} = r^4$
$\pm \frac{1}{2} = r$, so $r = -\frac{1}{2}$ $(r < 0)$
$-64, 32, -16, 8, -4$

21. nth term $= ar^{n-1}$
$-2 = -64r^{6-1}$
$-2 = -64r^5$
$\frac{1}{32} = r^5$
$\frac{1}{2} = r$
$-64, -32, -16, -8, -4$

23. If the 3rd term is 50 and the 2nd term is 10, then the common ratio $r = 50 \div 10 = 5$.
1st term = 2nd term $\div r = 10 \div 5 = 2$
$2, 10, 50, 250, 1250$

25. nth term $= ar^{n-1}$
$= 7 \cdot 2^{10-1} = 7 \cdot 2^9 = 7 \cdot 512 = 3584$

27. nth term $= ar^{n-1}$
$-81 = a(-3)^{8-1}$
$-81 = a(-3)^7$
$-81 = -2187a$
$\frac{1}{27} = a$

29. nth term $= ar^{n-1}$
$-1944 = -8r^{6-1}$
$-1944 = -8r^5$
$243 = r^5$
$3 = r$

31. 1st term $= 2$, 5th term $= 162$

nth term $= ar^{n-1}$

$162 = 2r^{5-1}$

$81 = r^4$

$3 = r$

$2, \boxed{6, 18, 54}, 162$

33. 1st term $= -4$, 6th term $= -12500$

nth term $= ar^{n-1}$

$-12500 = -4r^{6-1}$

$3125 = r^5$

$5 = r$

$-4, \boxed{-20, -100, -500, -2500}, -12500$

35. 1st term $= 2$, 3rd term $= 128$

nth term $= ar^{n-1}$

$128 = 2r^{3-1}$

$64 = r^2$

$-8 = r$

$2, \boxed{-16}, 128$

37. 1st term $= 10$, 3rd term $= 20$

nth term $= ar^{n-1}$

$20 = 10r^{3-1}$

$2 = r^2$

$\sqrt{2} = r$

$10, \boxed{10\sqrt{2}}, 20$

39. 1st term $= -50$, 3rd term $= 10$

nth term $= ar^{n-1}$

$10 = -50r^{3-1}$

$-\frac{1}{5} = r^2 \Rightarrow$ No such mean exists.

41. $a = 2, r = 3, n = 6$; $\quad S_n = \dfrac{a - ar^n}{1 - r} = \dfrac{2 - 2(3)^6}{1 - 3} = \dfrac{2 - 2(729)}{-2} = \dfrac{-1456}{-2} = 728$

43. $a = 2, r = -3, n = 5$; $\quad S_n = \dfrac{a - ar^n}{1 - r} = \dfrac{2 - 2(-3)^5}{1 - (-3)} = \dfrac{2 - 2(-243)}{4} = \dfrac{488}{4} = 122$

45. $a = 3, r = -2, n = 8$; $\quad S_n = \dfrac{a - ar^n}{1 - r} = \dfrac{3 - 3(-2)^8}{1 - (-2)} = \dfrac{3 - 3(256)}{3} = \dfrac{-765}{3} = -255$

47. $a = 3, r = 2, n = 7$; $\quad S_n = \dfrac{a - ar^n}{1 - r} = \dfrac{3 - 3(2)^7}{1 - 2} = \dfrac{3 - 3(128)}{-1} = \dfrac{-381}{-1} = 381$

49. If the 3rd term is $\frac{1}{5}$ and the 2nd term is 1, then the common ratio $r = \frac{1}{5} \div 1 = \frac{1}{5}$.

$a = 1 \div \frac{1}{5} = 5, r = \frac{1}{5}, n = 4$; $\quad S_n = \dfrac{a - ar^n}{1 - r} = \dfrac{5 - 5\left(\frac{1}{5}\right)^4}{1 - \frac{1}{5}} = \dfrac{5 - \frac{1}{125}}{\frac{4}{5}} = \dfrac{\frac{624}{125}}{\frac{4}{5}} = \dfrac{156}{25}$

51. If the 4th term is 1 and the 3rd term is -2, then the common ratio $r = 1 \div (-2) = -\frac{1}{2}$.

2nd term $= -2 \div \left(-\frac{1}{2}\right) = 4$; 1st term $= 4 \div \left(-\frac{1}{2}\right) = -8$; $a = -8, r = -\frac{1}{2}, n = 6$

$S_n = \dfrac{a - ar^n}{1 - r} = \dfrac{-8 - (-8)\left(-\frac{1}{2}\right)^6}{1 - \left(-\frac{1}{2}\right)} = \dfrac{-8 + \frac{1}{8}}{\frac{3}{2}} = \dfrac{-\frac{63}{8}}{\frac{3}{2}} = -\dfrac{21}{4}$

53. Sequence of population: $500, 500(1.06), 500(1.06)^2, \dots$

$a = 500, r = 1.06, n = 6 \Rightarrow n$th term $= ar^{n-1} = 500(1.06)^5 \approx 669$

55. Sequence of amounts: $10000, 10000(0.88), 10000(0.88)^2, \dots$

$a = 10000, r = 0.88, n = 16 \Rightarrow n$th term $= ar^{n-1} = 10000(0.88)^{15} \approx \$1{,}469.74$

57. Sequence of values: $70000, 70000(1.06), 70000(1.06)^2, \dots$

$a = 70000, r = 1.06, n = 13 \Rightarrow n$th term $= ar^{n-1} = 70000(1.06)^{12} \approx \$140{,}853.75$

59. Sequence of areas: $1, \frac{1}{2}, \frac{1}{4}, \dots$

$a = 1, r = \frac{1}{2}, n = 12 \Rightarrow n$th term $= ar^{n-1}1 = 1\left(\frac{1}{2}\right)^{11} = \left(\frac{1}{2}\right)^{11} \approx 0.0005$

61. Sequence of amounts: $1000(1.03), 1000(1.03)^2, 1000(1.03)^3, \dots$

$a = 1030, r = 1.03, n = 4 \Rightarrow S_n = \dfrac{a - ar^n}{1 - r} = \dfrac{1030 - 1030(1.03)^4}{1 - 1.03} = \dfrac{-129.2740743}{-0.03}$
$\approx \$4{,}309.14$

63. Answers may vary.   **65.** Answers may vary.

67. arithmetic mean   **69.** Answers may vary.

Exercise 14.5 (page 903)

1. $y = 3x^3 - 4$
function

3. $3x = y^2 + 4$
not a function

5. infinite

7. $S = \dfrac{a}{1 - r}$

9. $a = 8, r = \dfrac{1}{2}$

$S = \dfrac{a}{1 - r} = \dfrac{8}{1 - \frac{1}{2}} = \dfrac{8}{\frac{1}{2}} = 16$

11. $a = 54, r = \dfrac{1}{3}$

$S = \dfrac{a}{1 - r} = \dfrac{54}{1 - \frac{1}{3}} = \dfrac{54}{\frac{2}{3}} = 81$

13. $a = 12, r = -\dfrac{1}{2}$

$S = \dfrac{a}{1 - r} = \dfrac{12}{1 - \left(-\frac{1}{2}\right)} = \dfrac{12}{\frac{3}{2}} = 8$

15. $a = -45, r = -\dfrac{1}{3}$

$S = \dfrac{a}{1 - r} = \dfrac{-45}{1 - \left(-\frac{1}{3}\right)} = \dfrac{-45}{\frac{4}{3}} = -\dfrac{135}{4}$

17. $a = \dfrac{9}{2}, r = \dfrac{4}{3} \Rightarrow$ no sum $(|r| > 1)$

19. $a = -\dfrac{27}{2}, r = \dfrac{2}{3}$

$S = \dfrac{a}{1 - r} = \dfrac{-\frac{27}{2}}{1 - \frac{2}{3}} = \dfrac{-\frac{27}{2}}{\frac{1}{3}} = -\dfrac{81}{2}$

21. $0.\overline{1} = \dfrac{1}{10} + \dfrac{1}{100} + \dfrac{1}{1000} + \cdots \Rightarrow a = \dfrac{1}{10}, r = \dfrac{1}{10} \Rightarrow S = \dfrac{a}{1 - r} = \dfrac{\frac{1}{10}}{1 - \frac{1}{10}} = \dfrac{\frac{1}{10}}{\frac{9}{10}} = \dfrac{1}{9}$

23. $-0.\overline{3} = -\frac{3}{10} - \frac{3}{100} - \frac{3}{1000} + \cdots \Rightarrow a = -\frac{3}{10}, r = \frac{1}{10} \Rightarrow S = \frac{a}{1-r} = \frac{-\frac{3}{10}}{1-\frac{1}{10}} = \frac{-\frac{3}{10}}{\frac{9}{10}} = -\frac{1}{3}$

25. $0.\overline{12} = \frac{12}{100} + \frac{12}{10,000} + \frac{12}{1,000,000} + \cdots \Rightarrow a = \frac{12}{100}, r = \frac{1}{100}$

$S = \frac{a}{1-r} = \frac{\frac{12}{100}}{1-\frac{1}{100}} = \frac{\frac{12}{100}}{\frac{99}{100}} = \frac{12}{99} = \frac{4}{33}$

27. $0.\overline{75} = \frac{75}{100} + \frac{75}{10,000} + \frac{75}{1,000,000} + \cdots \Rightarrow a = \frac{75}{100}, r = \frac{1}{100}$

$S = \frac{a}{1-r} = \frac{\frac{75}{100}}{1-\frac{1}{100}} = \frac{\frac{75}{100}}{\frac{99}{100}} = \frac{75}{99} = \frac{25}{33}$

29. Distance ball travels down $= 10 + 5 + 2.5 + \cdots = \frac{a}{1-r} = \frac{10}{1-\frac{1}{2}} = \frac{10}{\frac{1}{2}} = 20$

Distance ball travels up $= 5 + 2.5 + 1.25 + \cdots = \frac{a}{1-r} = \frac{5}{1-\frac{1}{2}} = \frac{5}{\frac{1}{2}} = 10$

Total distance $= 20 + 10 = 30$ m

31. $S = \frac{a}{1-r} = \frac{1000}{1-0.8} = \frac{1000}{0.2}$

$= 5,000$ moths

33. **Answers may vary.**

35. $S = \frac{a}{1-r}$

$5 = \frac{1}{1-r}$

$5(1-r) = 1$

$5 - 5r = 1$

$4 = 5r$

$\frac{4}{5} = r$

37. $0.\overline{9} = \frac{9}{10} + \frac{9}{100} + \frac{9}{1000} + \cdots$

$a = \frac{9}{10}, r = \frac{1}{10}$

$S = \frac{a}{1-r} = \frac{\frac{9}{10}}{1-\frac{1}{10}} = \frac{\frac{9}{10}}{\frac{9}{10}} = \frac{9}{9} = 1$

39. No. $0.999999 = \frac{999,999}{1,000,000} < 1$

Exercise 14.6 (page 912)

1. $|2x - 3| = 9$

$2x - 3 = 9$ or $2x - 3 = -9$

$2x = 12$ $2x = -6$

$x = 6$ $x = -3$

3. $\frac{3}{x-5} = \frac{8}{x}$

$3x = 8(x - 5)$

$3x = 8x - 40$

$-5x = -40$

$x = 8$

5. $p \cdot q$

7. $P(n, r)$

9. $n!$

11. $\binom{n}{r}$; combinations

13. 1

15. $P(3,3) = \dfrac{3!}{(3-3)!} = \dfrac{3!}{0!} = \dfrac{6}{1} = 6$

17. $P(5,3) = \dfrac{5!}{(5-3)!} = \dfrac{5!}{2!} = \dfrac{120}{2} = 60$

19. $P(2,2) \cdot P(3,3) = \dfrac{2!}{(2-2)!} \cdot \dfrac{3!}{(3-3)!} = \dfrac{2!}{0!} \cdot \dfrac{3!}{0!} = \dfrac{2}{1} \cdot \dfrac{6}{1} = 12$

21. $\dfrac{P(5,3)}{P(4,2)} = \dfrac{\frac{5!}{(5-3)!}}{\frac{4!}{(4-2)!}} = \dfrac{\frac{5!}{2!}}{\frac{4!}{2!}} = \dfrac{\frac{120}{2}}{\frac{24}{2}} = \dfrac{60}{12} = 5$

23. $\dfrac{P(6,2) \cdot P(7,3)}{P(5,1)} = \dfrac{\frac{6!}{(6-2)!} \cdot \frac{7!}{(7-3)!}}{\frac{5!}{(5-1)!}} = \dfrac{\frac{6!}{4!} \cdot \frac{7!}{4!}}{\frac{5!}{4!}} = \dfrac{\frac{720}{24} \cdot \frac{5040}{24}}{\frac{120}{24}} = \dfrac{30 \cdot 210}{5} = 1{,}260$

25. $C(5,3) = \dfrac{5!}{3!(5-3)!} = \dfrac{5!}{3!2!} = \dfrac{120}{6 \cdot 2} = \dfrac{120}{12} = 10$

27. $\binom{6}{3} = \dfrac{6!}{3!(6-3)!} = \dfrac{6!}{3!3!} = \dfrac{720}{6 \cdot 6} = \dfrac{720}{36} = 20$

29. $\binom{5}{4}\binom{5}{3} = \dfrac{5!}{4!(5-4)!} \cdot \dfrac{5!}{3!(5-3)!} = \dfrac{5!}{4!1!} \cdot \dfrac{5!}{3!2!} = \dfrac{120}{24 \cdot 1} \cdot \dfrac{120}{6 \cdot 2} = \dfrac{120}{24} \cdot \dfrac{120}{12} = 5 \cdot 10 = 50$

31. $\dfrac{C(38,37)}{C(19,18)} = \dfrac{\frac{38!}{37!(38-37)!}}{\frac{19!}{18!(19-18)!}} = \dfrac{\frac{38 \cdot 37!}{37!1!}}{\frac{19 \cdot 18!}{18!1!}} = \dfrac{\frac{38}{1}}{\frac{19}{1}} = \dfrac{38}{19} = 2$

33. $C(12,0)C(12,12) = \dfrac{12!}{0!(12-0)!} \cdot \dfrac{12!}{12!(12-12)!} = \dfrac{12!}{0!12!} \cdot \dfrac{12!}{12!0!} = \dfrac{12!}{12!} \cdot \dfrac{12!}{12!} = 1 \cdot 1 = 1$

35. $C(n,2) = \dfrac{n!}{2!(n-2)!}$

37. $(x+y)^4 = \binom{4}{0}x^4y^0 + \binom{4}{1}x^3y^1 + \binom{4}{2}x^2y^2 + \binom{4}{3}x^1y^3 + \binom{4}{4}x^0y^4$

$\qquad = \dfrac{4!}{0!4!}x^4 + \dfrac{4!}{1!3!}x^3y + \dfrac{4!}{2!2!}x^2y^2 + \dfrac{4!}{3!1!}xy^3 + \dfrac{4!}{4!0!}y^4 = x^4 + 4x^3y + 6x^2y^2 + 4xy^3 + y^4$

39. $(2x+y)^3 = \binom{3}{0}(2x)^3y^0 + \binom{3}{1}(2x)^2y^1 + \binom{3}{2}(2x)^1y^2 + \binom{3}{3}(2x)^0y^3$

$\qquad = \dfrac{3!}{0!3!} \cdot 8x^3 + \dfrac{3!}{1!2!} \cdot 4x^2y + \dfrac{3!}{2!1!} \cdot 2xy^2 + \dfrac{3!}{3!0!}y^3$

$\qquad = 8x^3 + 12x^2y + 6xy^2 + y^3$

41. $(3x - 2)^4$

$$= \binom{4}{0}(3x)^4(-2)^0 + \binom{4}{1}(3x)^3(-2)^1 + \binom{4}{2}(3x)^2(-2)^2 + \binom{4}{3}(3x)^1(-2)^3$$

$$+ \binom{4}{4}(3x)^0(-2)^4$$

$$= \frac{4!}{0!4!} \cdot 81x^4(1) + \frac{4!}{1!3!} \cdot 27x^3(-2) + \frac{4!}{2!2!} \cdot 9x^2(4) + \frac{4!}{3!1!} \cdot 3x(-8) + \frac{4!}{4!0!} \cdot 1(16)$$

$$= 81x^4 - 216x^3 + 216x^2 - 96x + 16$$

43. $\binom{5}{3}x^2(-5y)^3 = \frac{5!}{3!2!}x^2(-125y^3) = 10x^2(-125y^3) = -1,250x^2y^3$

45. $\binom{4}{1}(x^2)^3(-y^3)^1 = \frac{4!}{1!3!}x^6(-y^3) = -4x^6y^3$

47. $7 \cdot 5 = 35$

49. $10 \cdot 10 \cdot 10 \cdot 10 \cdot 10 \cdot 10 = 1,000,000$

51. $9 \cdot 9 \cdot 8 \cdot 7 \cdot 6 \cdot 5 = 136,080$

53. $8 \cdot 10 \cdot 10 \cdot 10 \cdot 10 \cdot 10 \cdot 10 = 8,000,000$

55. $6! = 720$

57. $4! \cdot 5! = 24 \cdot 120 = 2,880$

59. $25 \cdot 24 \cdot 23 = 13,800$

61. $P(10, 3) = \dfrac{10!}{(10 - 3)!} = \dfrac{10!}{7!} = 720$

63. $9 \cdot 10 \cdot 10 \cdot 1 \cdot 1 = 900$

65. $C(14, 3) = \dfrac{14!}{3!(14 - 3)!} = \dfrac{14!}{3!11!} = 364$

67. $C(5, 3) = 10 \Rightarrow 5$ persons

69. $C(100, 6) = \dfrac{100!}{6!(100 - 6)!} = \dfrac{100!}{6!94!} = \dfrac{100 \cdot 99 \cdot 98 \cdot 97 \cdot 96 \cdot 95 \cdot 94!}{6!94!} = 1,192,052,400$

71. $C(3, 2) \cdot C(4, 2) = \dfrac{3!}{2!1!} \cdot \dfrac{4!}{2!2!} = 3 \cdot 6 = 18$

73. $C(12, 2) \cdot C(10, 3) = \dfrac{12!}{2!10!} \cdot \dfrac{10!}{3!7!}$
$$= 66 \cdot 120 = 7,920$$

75. Answers may vary.

77. Consider the two people who insist on standing together as one person. Then there are a total of 4 "persons" to be arranged. This can be done in $4! = 24$ ways. However, the two people who are standing together can be arranged in 2 different ways, so there are $24 \cdot 2 = 48$ arrangements.

Chapter 14 Summary (page 916)

1. $(4!)(3!) = 4 \cdot 3 \cdot 2 \cdot 1 \cdot 3 \cdot 2 \cdot 1 = 144$

2. $\dfrac{5!}{3!} = \dfrac{5 \cdot 4 \cdot 3!}{3!} = 5 \cdot 4 = 20$

3. $\dfrac{6!}{2!(6-2)!} = \dfrac{6!}{2!4!} = \dfrac{6 \cdot 5 \cdot 4!}{2 \cdot 1 \cdot 4!} = \dfrac{30}{2} = 15$

4. $\dfrac{12!}{3!(12-3)!} = \dfrac{12!}{3!9!} = \dfrac{12 \cdot 11 \cdot 10 \cdot 9!}{3 \cdot 2 \cdot 1 \cdot 9!}$

$$= \dfrac{1320}{6} = 220$$

5. $(n-n)! = 0! = 1$

6. $\dfrac{8!}{7!} = \dfrac{8 \cdot 7!}{7!} = 8$

7. $(x+y)^5 = x^5 + \dfrac{5!}{1!(5-1)!}x^4y + \dfrac{5!}{2!(5-2)!}x^3y^2 + \dfrac{5!}{3!(5-3)!}x^2y^3 + \dfrac{5!}{4!(5-4)!}xy^4 + y^5$

$$= x^5 + \dfrac{5!}{1!4!}x^4y + \dfrac{5!}{2!3!}x^3y^2 + \dfrac{5!}{3!2!}x^2y^3 + \dfrac{5!}{4!1!}xy^4 + y^5$$

$$= x^5 + \dfrac{5 \cdot 4!}{1!4!}x^4y + \dfrac{5 \cdot 4 \cdot 3!}{2 \cdot 1 \cdot 3!}x^3y^2 + \dfrac{5 \cdot 4 \cdot 3!}{3! \cdot 2 \cdot 1}x^2y^3 + \dfrac{5 \cdot 4!}{4! \cdot 1}xy^4 + y^5$$

$$= x^5 + \dfrac{5}{1}x^4y + \dfrac{20}{2}x^3y^2 + \dfrac{20}{2}x^2y^3 + \dfrac{5}{1}xy^4 + y^5$$

$$= x^5 + 5x^4y + 10x^3y^2 + 10x^2y^3 + 5xy^4 + y^5$$

8. $(x-y)^4 = x^4 + \dfrac{4!}{1!(4-1)!}x^3(-y) + \dfrac{4!}{2!(4-2)!}x^2(-y)^2 + \dfrac{4!}{3!(4-3)!}x(-y)^3 + (-y)^4$

$$= x^4 + \dfrac{4!}{1!3!}(-x^3y) + \dfrac{4!}{2!2!}x^2y^2 + \dfrac{4!}{3!1!}(-xy^3) + y^4$$

$$= x^4 - \dfrac{4 \cdot 3!}{1!3!}x^3y + \dfrac{4 \cdot 3 \cdot 2!}{2! \cdot 2 \cdot 1}x^2y^2 - \dfrac{4 \cdot 3!}{3!1!}xy^3 + y^4$$

$$= x^4 - \dfrac{4}{1}x^3y + \dfrac{12}{2}x^2y^2 - \dfrac{4}{1}xy^3 + y^4 = x^4 - 4x^3y + 6x^2y^2 - 4xy^3 + y^4$$

9. $(4x-y)^3 = (4x)^3 + \dfrac{3!}{1!(3-1)!}(4x)^2(-y) + \dfrac{3!}{2!(3-2)!}(4x)(-y)^2 + (-y)^3$

$$= 64x^3 + \dfrac{3!}{1!2!} \cdot (-16x^2y) + \dfrac{3!}{2!1!} \cdot 4xy^2 - y^3$$

$$= 64x^3 - \dfrac{3 \cdot 2!}{1!2!} \cdot 16x^2y + \dfrac{3 \cdot 2!}{2!1!} \cdot 4xy^2 - y^3$$

$$= 64x^3 - \dfrac{3}{1} \cdot 16x^2y + \dfrac{3}{1} \cdot 4xy^2 - y^3 = 64x^3 - 48x^2y + 12xy^2 - y^3$$

10. $(x + 4y)^3 = x^3 + \dfrac{3!}{1!(3-1)!}x^2(4y) + \dfrac{3!}{2!(3-2)!}x(4y)^2 + (4y)^3$

$\qquad = x^3 + \dfrac{3!}{1!2!} \cdot 4x^2y + \dfrac{3!}{2!1!} \cdot 16xy^2 + 64y^3$

$\qquad = x^3 + \dfrac{3 \cdot 2!}{1!2!} \cdot 4x^2y + \dfrac{3 \cdot 2!}{2!1!} \cdot 16xy^2 + 64y^3$

$\qquad = x^3 + \dfrac{3}{1} \cdot 4x^2y + \dfrac{3}{1} \cdot 16xy^2 + 64y^3 = x^3 + 12x^2y + 48xy^2 + 64y^3$

11. In the 3rd term, the exponent on 3 is 2.
    Variables: x^2y^2
    Coef. $= \dfrac{n!}{r!(n-r)!} = \dfrac{4!}{2!2!} = 6$
    Term $= 6x^2y^2$

12. In the 4th term, the exponent on $-y$ is 3.
    Variables: $x^2(-y)^3 = -x^2y^3$
    Coef. $= \dfrac{n!}{r!(n-r)!} = \dfrac{5!}{3!2!} = 10$
    Term $= 10(-x^2y^3) = -10x^2y^3$

13. 2nd term: The exponent on $-4y$ is 1. Variables: $(3x)^2(-4y)^1 = (9x^2)(-4y) = -36x^2y$
    Coef. $= \dfrac{n!}{r!(n-r)!} = \dfrac{3!}{2!1!} = 3$; Term $= 3(-36x^2y) = -108x^2y$

14. 3rd term: The exponent on $3y$ is 2. Variables: $(4x)^2(3y)^2 = (16x^2)(9y^2) = 144x^2y^2$
    Coef. $= \dfrac{n!}{r!(n-r)!} = \dfrac{4!}{2!2!} = 6$; Term $= 6(144x^2y^2) = 864x^2y^2$

15. nth term $= a + (n-1)d = 7 + (8-1)5 = 7 + 7 \cdot 5 = 42$

16. nth term $= a + (n-1)d$ $a + 8d = 242 \Rightarrow \times(-1)$ $-a - 8d = -242$
    $242 = a + (9-1)d$ $\underline{a + 6d = 212} \Rightarrow$ $\underline{a + 6d = 212}$
    $242 = a + 8d$ $-2d = -30$
    nth term $= a + (n-1)d$ $d = 15$
    $212 = a + (7-1)d$ Substitute and solve for a:
    $212 = a + 6d$ $a + 6d = 212$
                              $a + 6(15) = 212$
                                 $a + 90 = 212$
                                   $a = 122 \Rightarrow 122, 137, 152, 167, 182$

17. Form an arithmetic sequence with a 1st term of 8 and a 4th term of 25:
    nth term $= a + (n-1)d$
    $\qquad 25 = 8 + (4-1)d$
    $\qquad 17 = 3d$
    $\qquad \frac{17}{3} = d \Rightarrow 8, \boxed{\frac{41}{3}, \frac{58}{3}}, 25$

18. $a = 11, d = 7, n = 20$
    $l = a + (n-1)d = 11 + 19(7) = 144$
    $S_n = \dfrac{n(a+l)}{2} = \dfrac{20(11+144)}{2} = 1{,}550$

19. $a = 9, d = -\frac{5}{2}, n = 10$
    $l = a + (n-1)d = 9 + 9\left(-\frac{5}{2}\right) = -\frac{27}{2}$
    $S_n = \dfrac{n(a+l)}{2} = \dfrac{10\left(9 - \frac{27}{2}\right)}{2} = -\dfrac{45}{2}$

20. $\sum\limits_{k=4}^{6} \frac{1}{2}k = \frac{1}{2}(4) + \frac{1}{2}(5) + \frac{1}{2}(6) = 2 + \frac{5}{2} + 3 = \frac{15}{2}$

21. $\sum\limits_{k=2}^{5} 7k^2 = 7(2)^2 + 7(3)^2 + 7(4)^2 + 7(5)^2 = 28 + 63 + 112 + 175 = 378$

22. $\sum\limits_{k=1}^{4} (3k - 4) = (3(1) - 4) + (3(2) - 4) + (3(3) - 4) + (3(4) - 4) = -1 + 2 + 5 + 8 = 14$

23. $\sum\limits_{k=10}^{10} 36k = 36(10) = 360$

24. If the 5th term is $\frac{3}{2}$ and the 4th term is 3, then the common ratio $r = \frac{3}{2} \div 3 = \frac{1}{2}$.

3rd term = 4th term $\div r = 3 \div \frac{1}{2} = 6$; 2nd term = 3rd term $\div r = 6 \div \frac{1}{2} = 12$

1st term = 2nd term $\div r = 12 \div \frac{1}{2} = 24 \Rightarrow \boxed{24, 12, 6, 3, \frac{3}{2}}$

25. nth term $= ar^{n-1} = \frac{1}{8}(2)^{6-1} = \frac{1}{8}(32) = 4$

26. 1st term $= -6$, 4th term $= 384$ nth term $= ar^{n-1}$

$$384 = -6r^{4-1}$$
$$-64 = r^3$$
$$-4 = r \Rightarrow -6, \boxed{24, -96}, 384$$

27. $a = 162, r = \frac{1}{3}, n = 7;\ S_n = \dfrac{a - ar^n}{1 - r} = \dfrac{162 - 162\left(\frac{1}{3}\right)^7}{1 - \frac{1}{3}} = \dfrac{162 - 162\left(\frac{1}{2187}\right)}{-\frac{2}{3}} = \dfrac{\frac{4372}{27}}{-\frac{2}{3}} = \dfrac{2186}{9}$

28. $a = \frac{1}{8}, r = -2, n = 8;\ S_n = \dfrac{a - ar^n}{1 - r} = \dfrac{\frac{1}{8} - \frac{1}{8}(-2)^8}{1 - (-2)} = \dfrac{\frac{1}{8} - \frac{1}{8}(256)}{3} = \dfrac{-\frac{255}{8}}{3} = -\dfrac{85}{8}$

29. $a = 25, r = \frac{4}{5};\ S = \dfrac{a}{1 - r} = \dfrac{25}{1 - \frac{4}{5}} = \dfrac{25}{\frac{1}{5}} = 125$

30. $0.0\overline{5} = \frac{5}{100} + \frac{5}{10,000} + \frac{5}{1,000,000} + \cdots \Rightarrow a = \frac{5}{100}, r = \frac{1}{100};\ S = \frac{a}{1-r} = \frac{\frac{5}{100}}{1 - \frac{1}{100}} = \frac{\frac{5}{100}}{\frac{99}{100}} = \frac{5}{99}$

31. $17 \cdot 8 = 136$

32. $P(7, 7) = \dfrac{7!}{(7 - 7)!} = \dfrac{7!}{0!} = 7! = 5,040$

33. $P(7, 0) = \dfrac{7!}{(7 - 0)!} = \dfrac{7!}{7!} = 1$

34. $P(8, 6) = \dfrac{8!}{(8 - 6)!} = \dfrac{8!}{2!} = \dfrac{40,320}{2} = 20,160$

35. $\dfrac{P(9, 6)}{P(10, 7)} = \dfrac{\frac{9!}{(9-6)!}}{\frac{10!}{(10-7)!}} = \dfrac{\frac{9!}{3!}}{\frac{10!}{3!}} = \dfrac{9!}{3!} \cdot \dfrac{3!}{10!} = \dfrac{9!}{10!} = \dfrac{9!}{10 \cdot 9!} = \dfrac{1}{10}$

36. $C(7,7) = \dfrac{7!}{7!(7-7)!} = \dfrac{7!}{7!0!} = \dfrac{7!}{7!} = 1$

37. $C(7,0) = \dfrac{7!}{0!(7-0)!} = \dfrac{7!}{0!7!} = \dfrac{7!}{7!} = 1$

38. $\dbinom{8}{6} = \dfrac{8!}{6!(8-6)!} = \dfrac{8 \cdot 7 \cdot 6!}{6!2!} = \dfrac{56}{2} = 28$

39. $\dbinom{9}{6} = \dfrac{9!}{6!(9-6)!} = \dfrac{9 \cdot 8 \cdot 7 \cdot 6!}{6!3!} = \dfrac{504}{6} = 84$

40. $C(6,3) \cdot C(7,3) = \dfrac{6!}{3!(6-3)!} \cdot \dfrac{7!}{3!(7-3)!} = \dfrac{6!}{3!3!} \cdot \dfrac{7!}{3!4!} = 20 \cdot 35 = 700$

41. $\dfrac{C(7,3)}{C(6,3)} = \dfrac{\frac{7!}{3!(7-3)!}}{\frac{6!}{3!(6-3)!}} = \dfrac{\frac{7!}{3!4!}}{\frac{6!}{3!3!}} = \dfrac{35}{20} = \dfrac{7}{4}$

42. Sequence of amounts: $5000, 5000(0.80), 5000(0.80)^2, \ldots$

$a = 5000, r = 0.80, n = 6 \Rightarrow n$th term $= ar^{n-1} = 5000(0.80)^5 \approx \$1,638.40$

43. Sequence of amounts: $25700, 25700(1.18), 25700(1.18)^2, \ldots$

$a = 25700, r = 1.18, n = 11 \Rightarrow n$th term $= ar^{n-1} = 25700(1.18)^{10} \approx \$134,509.57$

44. $a = 300, d = 75;$ nth term $= a + (n-1)d$

$1200 = 300 + (n-1)(75)$

$900 = 75n - 75$

$975 = 75n \Rightarrow n = 13 \Rightarrow$ in 12 years

45. $16, d = 32;$ nth term $= a + (n-1)d = 16 + (10-1)(32) = 16 + 9(32) = 304$

$S_n = \dfrac{n(a+l)}{2} = \dfrac{10(16+304)}{2} = 1600$ ft

46. $5! = 120$

47. $5! \cdot 3! = 720$

48. $\dbinom{10}{3} = \dfrac{10!}{3!7!} = 120$

49. $\dbinom{5}{2}\dbinom{6}{2} = 10 \cdot 15 = 150$

Chapter 14 Test (page 919)

1. $\dfrac{7!}{4!} = \dfrac{7 \cdot 6 \cdot 5 \cdot 4!}{4!} = 7 \cdot 6 \cdot 5 = 210$

2. $0! = 1$

3. In the 2nd term, the exponent on $-y$ is 1. Variables: $x^4(-y)^1 = -x^4y$

Coef. $= \dfrac{n!}{r!(n-r)!} = \dfrac{5!}{1!4!} = 5;$ Term $= 5(-x^4y) = -5x^4y$

4. In the 3rd term, the exponent on $2y$ is 2. Variables: $(x)^2(2y)^2 = 4x^2y^2$

Coef. $= \dfrac{n!}{r!(n-r)!} = \dfrac{4!}{2!2!} = 6;$ Term $= 6(4x^2y^2) = 24x^2y^2$

5. $a = 3, d = 7, n = 10$; nth term $= a + (n-1)d = 3 + (10-1)(7) = 3 + 9(7) = 66$

6. $a = -2, d = 5, n = 12$; nth term $= a + (n-1)d = -2 + (12-1)(5) = -2 + 11(5) = 53$

$S_n = \dfrac{n(a+l)}{2} = \dfrac{12(-2+53)}{2} = \dfrac{12(51)}{2} = 306$

7. Form an arithmetic sequence with a 1st term of 2 and a 4th term of 98:

nth term $= a + (n-1)d$

$98 = 2 + (4-1)d$

$96 = 3d$

$32 = d \Rightarrow 2, \boxed{34, 66}, 98$

8. $\displaystyle\sum_{k=1}^{3}(2k-3) = (2(1)-3) + (2(2)-3) + (2(3)-3) = -1 + 1 + 3 = 3$

9. $a = -\frac{1}{9}, r = 3, n = 7$; nth term $= ar^{n-1} = -\frac{1}{9}(3)^{7-1} = -\frac{1}{9}(3)^6 = -\frac{1}{9}(729) = -81$

10. $a = \frac{1}{27}, r = 3, n = 6$; $S_n = \dfrac{a - ar^n}{1-r} = \dfrac{\frac{1}{27} - \frac{1}{27}(3)^6}{1-3} = \dfrac{\frac{1}{27} - \frac{1}{27}(729)}{-2} = \dfrac{-\frac{728}{27}}{-2} = \dfrac{364}{27}$

11. 1st term $= 3$, 4th term $= 648$

nth term $= ar^{n-1}$

$648 = 3r^{4-1}$

$216 = r^3$

$6 = r \Rightarrow 3, \boxed{18, 108}, 648$

12. $a = 9, r = \dfrac{1}{3}$

$S = \dfrac{a}{1-r} = \dfrac{9}{1-\frac{1}{3}} = \dfrac{9}{\frac{2}{3}} = \dfrac{27}{2}$

13. $P(5,4) = \dfrac{5!}{(5-4)!} = \dfrac{5!}{1!} = 5! = 120$

14. $P(8,8) = \dfrac{8!}{(8-8)!} = \dfrac{8!}{0!} = 8! = 40{,}320$

15. $C(6,4) = \dfrac{6!}{4!2!} = \dfrac{6 \cdot 5 \cdot 4!}{4! \cdot 2 \cdot 1} = \dfrac{30}{2} = 15$

16. $C(8,3) = \dfrac{8!}{3!5!} = \dfrac{8 \cdot 7 \cdot 6 \cdot 5!}{3 \cdot 2 \cdot 1 \cdot 5!} = \dfrac{56 \cdot 6}{6} = 56$

17. $C(6,0) \cdot P(6,5) = \dfrac{6!}{0!6!} \cdot \dfrac{6!}{1!} = 1 \cdot 6!$
$= 720$

18. $P(8,7) \cdot C(8,7) = \dfrac{8!}{1!} \cdot \dfrac{8!}{7!1!} = 8! \cdot 8$
$= 322{,}560$

19. $\dfrac{P(6,4)}{C(6,4)} = \dfrac{\frac{6!}{2!}}{\frac{6!}{4!2!}} = \dfrac{6!}{2!} \cdot \dfrac{4!2!}{6!} = 4! = 24$

20. $\dfrac{C(9,6)}{P(6,4)} = \dfrac{\frac{9!}{6!3!}}{\frac{6!}{2!}} = \dfrac{84}{360} = \dfrac{7}{30}$

21. $\dbinom{7}{3} = \dfrac{7!}{3!4!} = 35$

22. $\dbinom{5}{1}\dbinom{4}{2} = 5 \cdot 6 = 30$

Cumulative Review Exercises (page 920)

1. $\begin{cases} 2x + y = 5 \\ x - 2y = 0 \end{cases}$

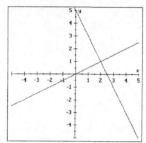

Solution: $(2, 1)$

2. $\begin{cases} (1) \quad 3x + y = 4 \\ (2) \quad 2x - 3y = -1 \end{cases}$

Substitute $y = 4 - 3x$ from (1) into (2):
$$2x - 3y = -1$$
$$2x - 3(4 - 3x) = -1$$
$$2x - 12 + 9x = -1$$
$$11x = 11$$
$$x = 1$$

Substitute this and solve for y:
$$y = 4 - 3x = 4 - 3(1) = 1$$
Solution: $(1, 1)$

3.

$x + 2y = -2$

$2x - y = 6 \Rightarrow \times (2)$

$\begin{array}{l} x + 2y = -2 \\ 4x - 2y = 12 \\ \hline 5x \quad = 10 \\ x \quad = 2 \end{array}$

$\begin{array}{l} x + 2y = -2 \\ 2 + 2y = -2 \\ 2y = -4 \\ y = -2 \end{array}$

Solution:

$\boxed{(2, -2)}$

4.

$\dfrac{x}{10} + \dfrac{y}{5} = \dfrac{1}{2} \Rightarrow \times 10 \quad x + 2y = 5$

$\dfrac{x}{2} - \dfrac{y}{5} = \dfrac{13}{10} \Rightarrow \times 10 \quad 5x - 2y = 13$

$\begin{array}{l} \hline 6x \quad = 18 \\ x \quad = 3 \end{array}$

$\begin{array}{l} x + 2y = 5 \\ 3 + 2y = 5 \\ 2y = 2 \\ y = 1 \Rightarrow \text{Solution: } \boxed{(3, 1)} \end{array}$

5. $\begin{vmatrix} 3 & -2 \\ 1 & -1 \end{vmatrix} = 3(-1) - (-2)(1)$

$= -3 + 2 = -1$

6. $y = \dfrac{\begin{vmatrix} 4 & -1 \\ 3 & -7 \end{vmatrix}}{\begin{vmatrix} 4 & -3 \\ 3 & 4 \end{vmatrix}} = \dfrac{4(-7) - (-1)(3)}{4(4) - (-3)(3)}$

$= \dfrac{-25}{25} = -1$

7.

(1) $x + y + z = 1$
(2) $2x - y - z = -4$
(3) $x - 2y + z = 4$

(1) $x + y + z = 1$
(2) $2x - y - z = -4$
(4) $3x \qquad = -3$
$x \qquad = -1$

(2) $2x - y - z = -4$
(3) $x - 2y + z = 4$
(5) $3x - 3y \quad = 0$

$\begin{array}{l} 3x - 3y = 0 \\ 3(-1) - 3y = 0 \\ -3 - 3y = 0 \\ -3y = 3 \\ y = -1 \end{array}$

$\begin{array}{l} x + y + z = 1 \\ -1 + (-1) + z = 1 \\ -2 + z = 1 \\ z = 3 \end{array}$ $\boxed{\text{The solution is } (-1, -1, 3).}$

8.
$$z = \frac{\begin{vmatrix} 1 & 2 & 6 \\ 3 & 2 & 6 \\ 2 & 3 & 6 \end{vmatrix}}{\begin{vmatrix} 1 & 2 & 3 \\ 3 & 2 & 1 \\ 2 & 3 & 1 \end{vmatrix}} = \frac{1\begin{vmatrix} 2 & 6 \\ 3 & 6 \end{vmatrix} - 2\begin{vmatrix} 3 & 6 \\ 2 & 6 \end{vmatrix} + 6\begin{vmatrix} 3 & 2 \\ 2 & 3 \end{vmatrix}}{1\begin{vmatrix} 2 & 1 \\ 3 & 1 \end{vmatrix} - 2\begin{vmatrix} 3 & 1 \\ 2 & 1 \end{vmatrix} + 3\begin{vmatrix} 3 & 2 \\ 2 & 3 \end{vmatrix}} = \frac{1(-6) - 2(6) + 6(5)}{1(-1) - 2(1) + 3(5)} = \frac{12}{12} = 1$$

9. $\begin{cases} 3x - 2y < 6 \\ y < -x + 2 \end{cases}$

10. $\begin{cases} y < x + 2 \\ 3x + y \le 6 \end{cases}$

11. $y = \left(\frac{1}{2}\right)^x$

12. $y = \log_2 x \Rightarrow 2^y = x$

13. $\log_x 25 = 2 \Rightarrow x^2 = 25 \Rightarrow x = 5$

14. $\log_5 125 = x \Rightarrow 5^x = 125 \Rightarrow x = 3$

15. $\log_3 x = -3 \Rightarrow 3^{-3} = x \Rightarrow x = \frac{1}{27}$

16. $\log_5 x = 0 \Rightarrow 5^0 = x \Rightarrow x = 1$

17. $y = 2^x$

18. x

19. $\log 98 = \log(14 \cdot 7) = \log 14 + \log 7 = 1.1461 + 0.8451 = 1.9912$

20. $\log 2 = \log \frac{14}{7} = \log 14 - \log 7 = 1.1461 - 0.8451 = 0.3010$

21. $\log 49 = \log 7^2 = 2 \log 7 = 2(0.8451) = 1.6902$

22. $\log \frac{7}{5} = \log \frac{14}{10} = \log 14 - \log 10 = 1.1461 - 1 = 0.1461$

23.
$$2^{x+5} = 3^x$$
$$\log 2^{x+2} = \log 3^x$$
$$(x+5)\log 2 = x\log 3$$
$$x\log 2 + 5\log 2 = x\log 3$$
$$5\log 2 = x\log 3 - x\log 2$$
$$5\log 2 = x(\log 3 - \log 2)$$
$$\frac{5\log 2}{\log 3 - \log 2} = x$$

24. $\log 5 + \log x - \log 4 = 1$
$$\log \frac{5x}{4} = 1$$
$$10^1 = \frac{5x}{4}$$
$$40 = 5x$$
$$8 = x$$

25. $A = P\left(1 + \dfrac{r}{k}\right)^{kt} = 9000\left(1 + \dfrac{-0.12}{1}\right)^{1(9)} \approx \$2,848.31$

26. $\log_6 8 = \dfrac{\log 8}{\log 6} \approx 1.16056$

27. $\dfrac{6!7!}{5!} = \dfrac{6 \cdot 5! \cdot 7!}{5!} = 6 \cdot 7! = 30{,}240$

28. $(3a - b)^4$
$$= (3a)^4 + \frac{4!}{1!(4-1)!}(3a)^3(-b) + \frac{4!}{2!(4-2)!}(3a)^2(-b)^2 + \frac{4!}{3!(4-3)!}(3a)(-b)^3 + (-b)^4$$
$$= 81a^4 + \frac{4!}{1!3!}(-27a^3b) + \frac{4!}{2!2!}\left(9a^2b^2\right) + \frac{4!}{3!1!}(-3ab^3) + b^4$$
$$= 81a^4 - \frac{4 \cdot 3!}{1!3!}\left(27a^3b\right) + \frac{4 \cdot 3 \cdot 2!}{2! \cdot 2 \cdot 1}\left(9a^2b^2\right) - \frac{4 \cdot 3!}{3!1!}\left(3ab^3\right) + b^4$$
$$= 81a^4 - \frac{4}{1}\left(27a^3b\right) + \frac{12}{2}\left(9a^2b^2\right) - \frac{4}{1}\left(3ab^3\right) + b^4$$
$$= 81a^4 - 108a^3b + 54a^2b^2 - 12ab^3 + b^4$$

29. In the 7th term, the exponent on $-y$ is 6.
Variables: $(2x)^2(-y)^6 = 4x^2y^6$
Coef. $= \dfrac{n!}{r!(n-r)!} = \dfrac{8!}{6!2!} = 28$
Term $= 28(4x^2y^6) = 112x^2y^6$

30. $a = -11, d = 6, n = 20$
nth term $= a + (n-1)d$
$$= -11 + (19)(6)$$
$$= -11 + 114 = 103$$

31. $a = 6, d = 3, n = 20$; nth term $= a + (n-1)d = 6 + (20-1)(3) = 6 + 19(3) = 63$
$$S_n = \frac{n(a+l)}{2} = \frac{20(6+63)}{2} = \frac{20(69)}{2} = 690$$

32. 1st term $= -3$; 4th term $= 30$:
nth term $= a + (n-1)d$
$$30 = -3 + (4-1)d$$
$$33 = 3d$$
$$11 = d \Rightarrow -3, \boxed{8, 19}, 30$$

33. $\displaystyle\sum_{k=1}^{3} 3k^2 = 3(1)^2 + 3(2)^2 + 3(3)^2$
$$= 3 + 12 + 27 = 42$$

34. $\sum\limits_{k=3}^{5}(2k+1)=(2(3)+1)+(2(4)+1)+(2(5)+1)=7+9+11=27$

35. $a=\frac{1}{27},r=3,n=7$; nth term $=ar^{n-1}=\frac{1}{27}(3)^{7-1}=\frac{1}{27}(3)^6=\frac{1}{27}(729)=27$

36. $a=\frac{1}{64},r=2,n=10;\ S_n=\dfrac{a-ar^n}{1-r}=\dfrac{\frac{1}{64}-\frac{1}{64}(2)^{10}}{1-2}=\dfrac{\frac{1}{64}-\frac{1}{64}(1024)}{-1}=\dfrac{-\frac{1023}{64}}{-1}=\dfrac{1023}{64}$

37. 1st term $=-3$, 4th term $=192$

nth term $=ar^{n-1}$

$192=-3r^{4-1}$

$-64=r^3$

$-4=r\Rightarrow-3,\boxed{12,-48},192$

38. $a=9,r=\dfrac{1}{3}$

$S=\dfrac{a}{1-r}=\dfrac{9}{1-\frac{1}{3}}=\dfrac{9}{\frac{2}{3}}=\dfrac{27}{2}$

39. $P(9,3)=\dfrac{9!}{(9-3)!}=\dfrac{9!}{6!}=\dfrac{9\cdot8\cdot7\cdot6!}{6!}=9\cdot8\cdot7=504$

40. $C(7,4)=\dfrac{7!}{4!(7-4)!}=\dfrac{7!}{4!3!}=\dfrac{7\cdot6\cdot5\cdot4!}{4!\cdot3\cdot2\cdot1}=\dfrac{210}{6}=35$

41. $\dfrac{C(8,4)C(8,0)}{P(6,2)}=\dfrac{\frac{8!}{4!4!}\cdot\frac{8!}{0!8!}}{\frac{6!}{4!}}=\dfrac{70\cdot1}{30}=\dfrac{7}{3}$

42. $C(n,n)=1$ is smaller than $P(n,n)=n!$.

43. $7!=5{,}040$

44. $\dbinom{9}{3}=\dfrac{9!}{3!6!}=84$

Appendix 1 (page A-4)

1.

$$y=x^2-1$$

x-axis	y-axis	origin
$-y=x^2-1$	$y=(-x)^2-1$	$-y=(-x)^2-1$
not equivalent: no symmetry	$y=x^2-1$	$-y=x^2-1$
	equivalent: $\boxed{\text{symmetry}}$	not equivalent: no symmetry

2.

$$y=x^3$$

x-axis	y-axis	origin
$-y=x^3$	$y=(-x)^3$	$-y=(-x)^3$
not equivalent: no symmetry	$y=-x^3$	$-y=-x^3$
	not equivalent: no symmetry	$y=x^3$
		equivalent: $\boxed{\text{symmetry}}$

3.

$$y = x^5$$

x-axis	y-axis	origin
$-y = x^5$	$y = (-x)^3$	$-y = (-x)^3$
not equivalent: no symmetry	$y = -x^5$	$-y = -x^5$
	not equivalent: no symmetry	$y = x^5$
		equivalent: $\boxed{\text{symmetry}}$

4.

$$y = x^4$$

x-axis	y-axis	origin
$-y = x^4$	$y = (-x)^2$	$-y = (-x)^2$
not equivalent: no symmetry	$y = x^4$	$-y = x^4$
	equivalent: $\boxed{\text{symmetry}}$	not equivalent: no symmetry

5.

$$y = -x^2 + 2$$

x-axis	y-axis	origin
$-y = -x^2 + 2$	$y = (-x)^2 + 2$	$-y = (-x)^2 + 2$
not equivalent: no symmetry	$y = -x^2 + 2$	$-y = x^2 + 2$
	equivalent: $\boxed{\text{symmetry}}$	not equivalent: no symmetry

6.

$$y = x^3 + 1$$

x-axis	y-axis	origin
$-y = x^3 + 1$	$y = (-x)^3 + 1$	$-y = (-x)^3 + 1$
not equivalent: no symmetry	$y = -x^3 + 1$	$-y = -x^3 + 1$
	not equivalent: no symmetry	$y = x^3 - 1$
		not equivalent: no symmetry

7.

$$y = x^2 - x$$

x-axis	y-axis	origin
$-y = x^2 - x$	$y = (-x)^2 - (-x)$	$-y = (-x)^2 - (-x)$
not equivalent: no symmetry	$y = x^2 + x$	$-y = x^2 + x$
	not equivalent: no symmetry	not equivalent: no symmetry

8.

$$y^2 = x + 7$$

x-axis	y-axis	origin
$(-y)^2 = x + 7$	$y^2 = -x + 7$	$(-y)^2 = -x + 7$
$y^2 = x + 7$	not equivalent: no symmetry	$y^2 = -x + 7$
equivalent: $\boxed{\text{symmetry}}$		not equivalent: no symmetry

9.

$$y = -|x + 2|$$

x-axis	y-axis	origin
$-y = -|x + 2|$	$y = -|-x + 2|$	$-y = -|-x + 2|$
not equivalent: no symmetry	not equivalent: no symmetry	not equivalent: no symmetry

10.

$$y = |x| - 3$$

x-axis	y-axis	origin
$-y = |x| - 3$	$y = |-x| - 3$	$-y = |-x| - 3$
not equivalent: no symmetry	$y = |-1||x| - 3$	$-y = |-1||x| - 3$
	$y = |x| - 3$	$-y = |x| - 3$
	equivalent: $\boxed{\text{symmetry}}$	not equivalent: no symmetry

11.

$$|y| = x$$

x-axis	y-axis	origin
$|-y| = x$	$|y| = -x$	$|-y| = -x$
$|-1||y| = x$	not equivalent: no symmetry	$|-1||y| = -x$
$|y| = x$		$|y| = -x$
equivalent: $\boxed{\text{symmetry}}$		not equivalent: no symmetry

12.

$$y = 2\sqrt{x}$$

x-axis	y-axis	origin
$-y = 2\sqrt{x}$	$y = 2\sqrt{-x}$	$-y = 2\sqrt{-x}$
not equivalent: no symmetry	not equivalent: no symmetry	not equivalent: no symmetry

13. $y = x^4 - 4$

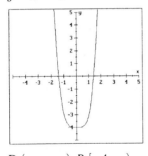

D $(-\infty, \infty)$; R $[-4, \infty)$

14. $y = \dfrac{1}{2}x^4 - 1$

D $(-\infty, \infty)$; R $[-1, \infty)$

15. $y = -x^3$

D $(-\infty, \infty)$; R $(-\infty, \infty)$

16. $y = x^3 + 2$

D $(-\infty, \infty)$; R $(-\infty, \infty)$

17. $y = x^4 + x^2$

D $(-\infty, \infty)$; R $[0, \infty)$

18. $y = 3 - x^4$

D $(-\infty, \infty)$; R $(-\infty, 3]$

19. $y = x^3 - x$

D $(-\infty, \infty)$; R $(-\infty, \infty)$

20. $y = x^3 + x$

D $(-\infty, \infty)$; R $(-\infty, \infty)$

21. $y = \dfrac{1}{2}|x| - 1$

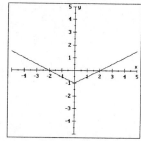

D $(-\infty, \infty)$; R $[-1, \infty)$

22. $y = -|x| + 1$

D $(-\infty, \infty)$; R $(-\infty, 1]$

23. $y = -|x + 2|$

D $(-\infty, \infty)$; R $(-\infty, 0]$

24. $y = |x - 2|$

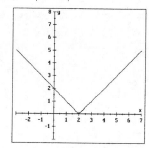

D $(-\infty, \infty)$; R $[0, \infty)$